Computer Aided Design with

Unigraphics® NX5

Engineering Design
in
Computer Integrated Design and Manufacturing

Seventh Edition

H. Felix Lee
David W. Fulton

Department of Mechanical and Industrial Engineering
Southern Illinois University Edwardsville

KENDALL/HUNT PUBLISHING COMPANY
4050 Westmark Drive Dubuque, Iowa 52002

Book Team

Chairman and Chief Executive Officer Mark C. Falb
President and Chief Operating Office Chad M. Chandlee
Vice President, Higher Education David L. Tart
Director of National Book Program Paul B. Carty
Editorial Development Manager Georgia Botsford
Assistant Developmental Editor Melissa M. Tittle
Assistant Production Editor Abby Davis
Permissions Editor Renae Horstman
Cover Designer Janell Edwards

ISBN 978-0-7575-6031-6

Printed in the United States of America
10 9 8 7 6 5 4 3 2 1

Dedication

To my parents and wife, Guim, and children, Jake and Joshua.

- HFL

For my wife, Barbara, our grown up, married children and children-in-law, Keith and Emmanuelle, Kyle and Ginger, and Kara and Cris, and our precious grandchildren, Matthew, Elizabeth, Natasha, Nicholas, and Stella.

- DWF

TABLE OF CONTENTS

Chapter 10. Introduction to Sketch

Chapter 11. Instance, Trim Body, and Thread Features

Appendix A. Guided Student Assembly Project: Geneva CAM Assembly Model

Appendix B. Additional Design and Assembly Projects

Appendix C. Glossary

Preface

The twenty-first century manufacturing environment can be characterized by the paradigm of delivering products of increasing variety, lower demand and higher quality in the context of expanding global competition. In order to realize this manufacturing paradigm and facilitate coordinated efforts across different departments, more and more companies are moving into computer-integrated production systems where computer technology is widely used to support design, manufacturing, and business operation in an integrated fashion. The core of this computer technology includes computer-aided design (CAD), computer-aided manufacturing (CAM), and computer-aided analysis (CAE).

This book is intended to introduce CAD using Unigraphics in order to support the design process. Unigraphics is a software product of EDS (Electronic Data Systems). Unigraphics is 3D solid modeling CAD/CAM/CAE software used by thousands of companies worldwide in diverse industries. They include automobiles, aircraft and spacecraft, and various consumer products including shavers, phones, appliances, motorbikes etc. Unigraphics is considered one of the top 3 mechanical design systems in the world. According to an independent study (Business Week Magazine) conducted in 1999, Unigraphics was selected as one of the top-10 software companies in the world.

While the intent of this book is to introduce CAD with usage of Unigraphics, the focus of the book is on 3-D solid and assembly modeling to support various engineering design projects. We also introduce parametric and associative modeling to facilitate capturing the design intent and easy changes of designs.

This book can serve as a stand-alone text for Engineering Design or entry-level CAD/CAM courses in colleges and universities. Also this book can be used for Engineering Drawing courses that cover 3D modeling. No prerequisite is required. Instead of presenting all available functionalities and options of Unigraphics CAD like a look-up dictionary format, this book coherently organizes the key features and functionalities of Unigraphics CAD that are most commonly used in engineering design modeling. Thus, this book is suitable for one or two semester usage for a 3-credit course.

This book is application-oriented and design-project based rather than involved in theories and mathematical derivations. It provides many projects to which students can apply the design and modeling concepts. This book can also serve as a supplement to upper-level CAD/CAM theory-oriented books by supplying applications to real design problems in industries. From our past teaching experience, we find these applications to motivate students to learn more about the subject.

The book consists of 14 chapters and 3 appendixes. Most chapters follow a common format. First, concepts of a modeling feature are presented, which is followed by activities, which demonstrates step-by-step how to carry out the concepts in Unigraphics. And then design projects are presented with brief guidelines so that students can further practice and solidify the concepts in the context of engineering design. These design projects can be used as classroom homework or test. There is usually more than one modeling approach to design projects. We find it very educational to compare different modeling approaches. Each chapter concludes with exercise problems to help students review what they learn in each chapter.

In Chapter 1 we give an overview of computer integrated design and manufacturing. We describe a product-life cycle, a product design process and CAD, and three geometric modeling methods for CAD to represent objects with a focus on solid modeling. We also present steps necessary to get started with Unigraphics.

In Chapter 2 we present a case study on CAD/CAM/CAE using Unigraphics. A simple part is used to take you though step-by-step from design, analysis, then to manufacturing of the part. The objective is to demonstrate how Unigraphics can support computer integrated design and manufacturing. Chapter 2 can be skipped if desired without losing any continuity. Alternatively, Chapter 2 can be covered after Chapter 13, since the case study of Chapter 2 uses the Master Modeling concept addressed in Chapter 13.

The rest of chapters deal with Unigraphics CAD. In Chapter 3 we present fundamental UG essentials that will be frequently used during a session with

Unigraphics. In Chapters 4 to 11 we discuss different features used to build a 3D solid part. We suggest covering Chapters 3 to 11 in sequence since later chapters build on the previous chapter materials. In Chapter 12 we introduce different approaches to assembly modeling, putting together more than one part into an assembly. In Chapter 13 we discuss the concept of Master Model, which enables different engineering departments to concurrently work together with use of 3D solid models. In Chapter 14, we introduce Drafting, which is one application of the Master Model concept, showing how to create 2D drawings and dimensions from 3D solid models.

In Appendixes A and B, we provide extra design and assembly projects. Students use 2D drawings to build 3D solid models for individual parts from a scratch and then assemble them. In Appendix A, we provide detail steps for students to easily follow to build a Geneva Cam assembly model. These projects will serve as good opportunities to practice textbook materials in total and real setting. In Appendix C, we provide glossary of terms used in this book.

In order to assist instructors who adopt this book for their class teaching, we have prepared the following items: (a) the solution WORD file for chapter review questions, (b) the Unigraphics part files for completed chapter activities and projects, (c) the Unigraphics part files for completed Appendix A and B projects. These items are available on a separate instructor CD upon request to the book publisher by the instructors.

We use Unigraphics NX5 version throughout this book (7th edition). The first edition of this book was written in V.16, while the second, third, fourth, fifth and sixth editions written in V.18, NX, NX2, NX3, and NX4 respectively. Unigraphics V16 or later version is fully Windows compliant and uses native Windows user interface.

We are thankful to our colleagues who used the earlier edition of this book and provided valuable feedback. In this new edition, we have made efforts to incorporate their feedback to further improve the presentation and contents of the book. We are grateful to Hulas King of UGS/Siemens for his constant support and encouragement, which made it possible to complete this book, and to

UGS/Siemens for providing an excellent source of reference material through various training documentations. Sean Kenney, who is a Boeing engineer, deserves a special mention for his significant contribution to Chapters 3 and 6. Thanks also go to several students at our program, Kyle Knowlson, Matthew Grainger, Bryon Belter, Sam Hall, and Steve Arana for testing out design projects in Appendixes A and B, to Srinivas Sambaraju, Shujath Mohammed, Terry Goble, and Chris Anderson for assistance with this revision work, and to many Boeing Engineers who took classes on this subject and provided useful feedback on course material. Lastly we always remember our spouses and children for generously giving in their special time with us for our work toward this book.

H. Felix Lee
David W. Fulton

About the Authors

H. Felix Lee

H. Felix Lee is Professor of Department of Mechanical and Industrial Engineering at Southern Illinois University Edwardsville. He holds a Ph.D. in Industrial and Operations Engineering from the University of Michigan, an M.S. in Industrial Engineering and Management from Oklahoma State University, and a B.S. in Industrial Engineering from Hanyang University in S. Korea. His area of interest is in computer integrated design and manufacturing. He renovated Manufacturing Engineering curriculum and courses with introduction of Unigraphics CAD/CAM/CAE, and has taught students and Boeing engineers with these courses. He was a recipient of a NSF research grant on developing an integrated design-aid tool for flexible manufacturing systems. Dr. Lee is a member of SME, IIE, Tau Beta Pi, and Phi Kappa Phi. His papers appear in numerous refereed journals, books, and proceedings. In 2004, he organized the CAD/CAM/CAE student design contest in the PLM (Product Lifecycle Management) World Conference and chaired the first five contests in the 2004 through 2008 conferences that were held in Anaheim, CA, Dallas, TX, Long Beach, CA, and Orlando, FL.

David W. Fulton

David W. Fulton holds a B.S. in Mechanical Engineering from University of Missouri-Columbia. He has many years of industry experience as a design engineer and has taken various roles in computer engineering application customer support. The application support includes 15 years experience in computer aided design engineering in McDonnell Douglas and 12 years with Unigraphics Solutions (UGS) as a Unigraphics trainer, courseware development, curriculum development, and training manager.

Chapter 1. Introduction

This chapter consists of three sections. In section 1.1, we give an overview of computer integrated design and manufacturing. We describe a product-life cycle, a product design process, and computer-aided design (CAD) and three geometric modeling methods for CAD to represent objects with a focus on solid modeling. We also give a brief introduction to computer-aided engineering analysis, and computer-aided manufacturing. In section 1.2 we give an introduction to Unigraphics to support computer integrated design and manufacturing. In section 1.3, we give a description and steps necessary to get started with Unigraphics.

1.1 Overview of Computer Integrated Design and Manufacturing

Traditional view of manufacturing is just the transformation of raw materials into value-added products meeting specifications. Present-day's view of manufacturing takes a much broader meaning. According to the CAM-I definition, manufacturing is defined as a series of interrelated activities and operations involving design, material selection, planning, production, quality assurance, management, and marketing of discrete consumer and durable goods. This broader view puts more emphasis on coordinated efforts across different departments within manufacturing companies.

The twenty-first century manufacturing environment can be characterized by the paradigm of delivering products of increasing variety, lower demand and higher quality in the context of expanding global competition. This implies manufacturing products for mass market in such a way that products are customized for each individual in that market. In order to realize this manufacturing paradigm and facilitate coordinated efforts across different departments, more and more companies are moving into computer-integrated production systems where computer technology is widely used to support design, manufacturing, and business operation in an integrated fashion. The core of this computer technology includes computer-aided design (CAD) and computer-aided manufacturing (CAM), and computer-aided analysis (CAE).

In order for a manufacturing company to succeed, it must deliver products to customers at the minimum possible cost, the best possible quality, and the minimum lead time starting from the product conception stage to final delivery, service, and disposal. As a result, a product has to be understood in a much broader context and the overall performance of the product during its life cycle becomes important. The product life cycle includes the following four phases: design phase, manufacturing phase, product usage phase, and disposal phase. Opportunities for reduction in cost and leadtime and improvements in product quality must be sought from all the areas of a product-life cycle. However, the design phase is most critical, determining the product success or failure in the market place. This is because it has a direct impact on product cost, quality, and manufacturing process yields, future maintenance costs, and cost to dispose of the product after its usage. According to Dransfield (1994), about 80% of the product cost is fixed at the design stage, and it is easier and less costly to change the design at the early design phase.

Product Design Process

Product design is a complex process. It requires a cross-functional team consisting of not only design engineers but also various individuals from manufacturing, financing, marketing, and etc. There are different approaches to product design but we give one adopted by Earle (1992). This approach consists of the following six steps.

1. Problem Identification. This step is to identify the attributes of the need for which the product is being designed (what customers want). This involves collection of field data via field surveys, personal observations, and physical measurements.
2. Preliminary Ideas. With the product attributes of specific need identified from the previous step, this step is to generate preliminary ideas with

respect to technical choices, materials, design complexities, and so on. Usually brain storming sessions among the cross-functional team members are needed in this step.

3. Refinement Process. This step is to refine product ideas using geometric models. Among the preliminary ideas from step 2, several good ideas are pursued in this step in terms of space requirements and critical measurement and dimensions. The use of geometric modeling is useful in visualizing and determining the identity of the product.
4. Analysis Process. This step is to analyze product designs from the point of view of many criteria such as cost, weight, functional requirements, and marketability. Engineering analysis tools such as finite-element methods (for strength, stress, deformation analyses) and assembly analysis tools (for tolerance and weight analyses) can be used to evaluate alternative designs.
5. Decision Process. This step is to select a design that has all the desirable characteristics including ease of assembly, manufacturability, maintainability, etc. A decision matrix approach can be adopted to make a best trade-off solution by using proper weighting schemes for various attributes among design alternatives.
6. Implementation Process. This step is to provide a detailed design providing detailed specifications with respect to materials, dimensions, tolerances, surface roughness, and so on. This involves creating the drawings to be used directly for developing process plans so that the product can be manufactured.

The above design process is an iterative process, usually engaging these six steps several times before the final design is selected that best meets design criteria.

Computer-Aided Design (CAD)

CAD refers to the design process with the aid of computers. Thus, computers do not change the nature of the design process. Design engineers provide knowledge and creativity and control the entire design process from the problem identification step to the implementation step. Computers contribute to improving the efficiency and productivity of the design process. This is because computers can accurately generate easily modifiable geometry models of products, perform complex design analysis at fast speed, and manage information and design knowledge with consistency and speed.

CAD technology has evolved over the last three decades. The development of the SKETCHPAD system by Ivan Sutherland at MIT in 1962 could be considered the beginning of CAD. This system allowed a user to interact graphically with the computer, via a visual display and light pen. CAD in the early 1960s was mainly a two-dimensional (2D) drafting system. Then it was extended to three-dimensional (3D) models with development of wireframe-based modeling systems. It was not possible, however, to represent a higher-order geometry data such as surface data. As a result, surface-based models were developed in the early 1970s. These surface-based models had also limitations. They could not represent solid or volume enclosure information. The need for solid modeling grew with increasing demand of numerical control (NC) code generation for manufacturing and mesh generation for engineering analysis. A volume representation of parts was also necessary to perform topological validity checks. The solid modeling technology has evolved since the mid-1970s.

In the 1980s, assembly modeling was introduced in response to the need of automobile and aircraft industries. Assembly modeling allows putting together various parts of 3-D solid models into an assembly, and managing and performing analyses for an assembly. Soon after, parametric modeling was introduced to make it easy to modify designs and geometric models. At the early design phase, product requirements and product designs are prone to change, which leads to frequent changes in product geometries. Parametric modeling allows dimension-driven capability. This means that an object is defined by a set of dimensions (parameters) for which designers can change their values at any time during the design process. The latest development has been knowledge-based engineering systems that can capture both geometric and non-geometric product information such as engineering rules, part dependences, and manufacturing constraints, which result in more complete product definitions. CAD is becoming more popular as computer technology with higher speed, larger memory, and smaller size is more available and affordable.

Geometric Modeling

We discussed earlier that the third step of the product design process is the refinement process, which involves developing geometric models of a product from the conceptual ideas. Geometric modeling refers to a set of techniques concerned mainly with developing efficient representations of geometric aspects of a design. The geometric information about a part includes types of surfaces and edges and their dimensions and tolerances. Thus, geometry modeling is a heart of CAD. Traditionally, the geometric information about a part was provided on blueprints by a drafter. Today, such information is directly transferred from the CAD files to the CAM files to perform subsequent manufacturing of the part or to the CAE files to conduct subsequent analyses of the part. Clearly this helps to significantly reduce product development and manufacturing leadtime. In Chapter 2, we will present a case study to demonstrate this notion.

There are three basic geometric modeling approaches available in CAD. They are wireframe, surface, and solid modeling. Wireframe modeling is one of the most basic methods among these. It uses points and curves (lines, circles, arcs, conics, and spline curves) to define objects. For example, a user may enter 3D vertices, say (x, y, z)'s, and then join the vertices to form a 3D object. Figure 1.1 shows a wireframe model of a part.

Part A

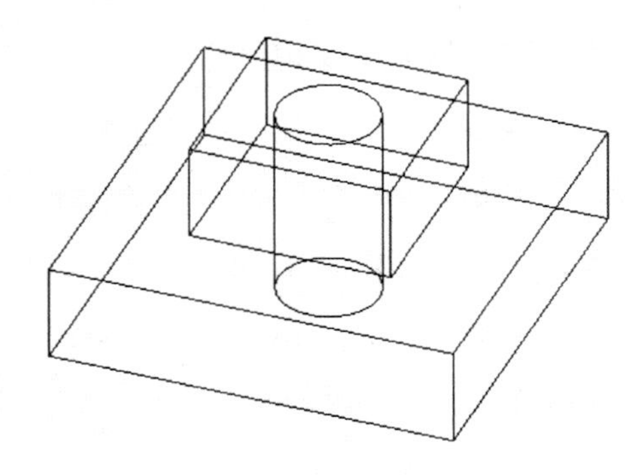

Figure 1.1 Wireframe model of Part A

This is simple and straightforward in concept but has limitations such as:

a. With wireframe modeling, it is not possible to calculate volume and mass properties of a design. It cannot support numeric control (NC) path generation, cross-sections, and interference detection since wireframe modeling contains only low-level information such as points and lines.
b. It can create ambiguous representations of real objects. For example, it is not clear if the wireframe model of Figure 1.2 represents a block or an inward corner. A similar problem occurs between a boss and a hole.

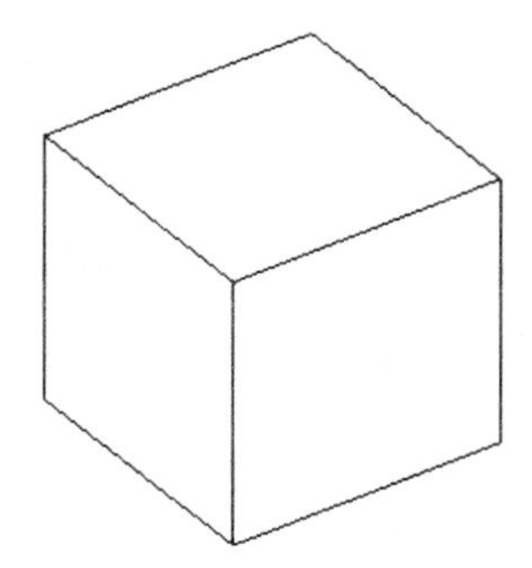

Figure 1.2 Ambiguous Representation: block or inward corner

c. Wireframe modeling usually requires more user input to create models than solid modeling. For example, a 3D wireframe model of a simple cube requires12 lines while a 3D solid model of the same cube requires positions of only 3 corner points.

In wireframe modeling, we take advantage of the simplicity of certain surfaces. For example, a plane is represented by its boundaries. Nothing is said about the middle of the plane. Shapes of cars, airplanes, ships and so forth do not simply consist of standard geometry such as planes and cylinders but curved surfaces. Surface modeling contains definitions of surfaces, edges and vertices. Parametric techniques popularized by Coons were adopted as a way of precisely describing the curvature of a surface in all three dimensions. The Bezier, B-spline, and NURBS surfaces are three types of surfaces based on third-degree parametric curves. Precise, mathematically defined curved surfaces are a must in the aerospace, automotive, and shipbuilding industries. In many cases, information about the surface of the object is the most critical element in manufacturing the product, and can be used directly by analysis and manufacturing software tools.

One general weakness of surface modelers is their inability to handle topology. For example, if you have a series of patches that form a closed-volume model and you drill a hole in the model, you will pass through an infinitely thin surface into a void, for which no information is available. To handle this situation, the user or the system must cover over the inside of the hole. Some surface models simply do not have the flexibility to make this sort of topological modification easily. Surface modeling does not provide information on the component's inside and outside. As a result, a surface modeling system may not guarantee that the user has designed a realizable object. That is, the collection of surfaces may not define a physical part.

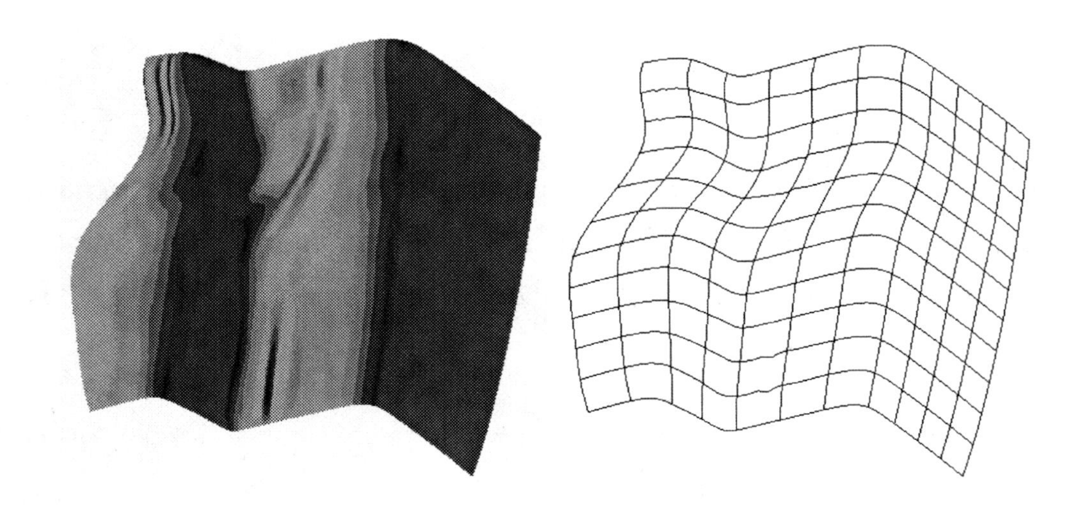

Figure 1.3 Example of a curved surface: shaded and wireframe views

Solid modeling becomes more popular because it has many advantages over wireframe modeling and surface modeling. Solid modeling can support applications to manufacturing and engineering analysis such as finite-element analysis, NC part programming, and generation of CAPP (computer aided process planning). It can be used to evaluate size, shape, and weight of products early during the conceptual design phase. It can give complete and unambiguous representation of objects. There are several solid modeling techniques available, but here we will introduce two common ones. These two are boundary representation (BREP) and constructive solid geometry (CSG).

BREP describes the geometry of an object in terms of its boundaries, namely vertices, edges and surfaces, and direction of the face normal to indicate inside or outside the material. A real object is formed and bounded. For example, three planes do not form a solid object. Euler's equations are used to ensure validity of BREP models.

CSG constructs a model by simple solid objects, called primitives or entities such as blocks, cylinder, cone, and sphere. These solid entities are not simply objects by lines and vertices but objects that occupy space. Thus, intersection of two solid entities creates a more complex object. These primitives are arranged using Boolean operators. There are three types of Boolean operations: Union (U), difference (—), and intersection (∩). For example, CSG models the part of Figure 1.1 consisting of three primitives: a large block A, a small block B, and a cylinder C. CSG constructs the part as {A U B} — C.

The structure is concise and relatively less storage but slow in displaying the objects. It is usually converted internally into BREP to display the model or generate the wireframe drawing. CSG provides easier input and BREP provides faster display and line drawing. CSG serves as an underlying modeling method that will be used to create various models throughout this book.

Computer-Aided Engineering (CAE) Analysis

Engineering analysis is concerned with analysis and evaluation of engineering product designs. There are a number of computer-aided engineering analysis techniques available for this purpose. They include finite-element analysis (FEA), tolerance analysis, design optimization, mechanism analysis, and mass property analysis. Mass information of 3-D solid models helps calculate the forces acting on a part. A real-world object is a continuous mass that responds in a very complex manner to forces acting on it. FEA is one of the most frequently used analysis techniques in engineering applications. FEA is a powerful numerical analysis. It divides an object into a number of small building blocks, called finite elements in order to evaluate the functional performances of the object such as stresses and deflections. This division process is called meshing. Each element is a simple shape such as a triangle, a square, or cube, etc. These elements are connected at node points. The unknowns for each element are the displacements at the node points. The finite-element program develops governing equations in the form of a matrix such that it assembles the matrices for these simple elements to form the global matrix for the entire model. This matrix is solved for the unknown displacements, given the known forces and boundary conditions. From the displacement at the nodes, the functional performances in each element are then calculated.

FEA usually employs the following steps.

1. Meshing
2. Selection of the solution approximation
3. Development of element matrices and equations
4. Assembly of the element equations
5. Solutions for the unknown at the nodes
6. Interpretation of the result

Instead of having real forces exerted on real models, hypothetical forces can be applied to a solid model in FEA. This helps to avoid the often-destructive results for physical prototypes and to reduce costs and lead times for design and manufacturing. We give a case study in Chapter 2 for demonstration of CAE and FEA.

Computer-Aided Manufacturing (CAM)

Solid modeling techniques can be combined with CAM capabilities to ensure that a product design satisfies the desired manufacturability requirement as closely as possible. In addition, these techniques speed the transition from design to manufacturing by reducing or eliminating the need for traditional working or production drawings. Solid models can be made to what each part will look like at each stage of the manufacturing process. These models provide information used to determine how much time, material, and labor would be required to manufacture the product as modeled. If special types of tooling (e.g., cutters, jigs, etc.) are required, solid models can also be made for them.

More and more, the machinery used for fabrication is programmed using the computer models of each part. The information related to the model is translated into manufacturing operations by specialized programs. These programs control the machine tools through a process called numeric control (NC). Originally, information was provided NC machines by punched tapes. Improvements in technology have led to the full-scale integration of CAD and CAM. This means less translation and less chance for error. Simulations of the tool cutting action is created and tested on solid models before actual materials and equipment are used. This cuts down on material waste and reduces troubleshooting time, freeing up the equipment for greater use in production.

1. 2 Introduction to Unigraphics

Unigraphics is an interactive CAD/CAM/CAE system. The CAD functions automate the normal engineering, design, and drafting capabilities found in today's manufacturing companies. The CAM functions provide NC programming for modern machine tools using the Unigraphics design model to describe the finished part. The CAE functions provide the analyses of the design model to check if the part can satisfy design requirements in mechanical, thermal, or dynamic aspects.

Unigraphics functions are divided into "applications" of common capabilities. These applications are supported by a prerequisite application called Unigraphics Gateway. Every Unigraphics user must have Unigraphics Gateway; however, the other applications are optional and may be configured to meet the needs of each individual user.

Unigraphics is a fully three-dimensional, double precision system that can accurately describe almost any geometric shape. By combining these shapes, you can design, analyze, and create drawings of your products.

Once the design is complete and analyzed, the Manufacturing application allows you to select the geometry describing the part, enter manufacturing information such as cutter diameter, and automatically generate a *cutter location source file* (CLSF), which can be used to drive most NC machines.

1.3 Getting Started with Unigraphics (UG)

On starting Unigraphics, you first find yourself in the **Gateway** environment. This is like standing in the entrance lobby of an office building; you are within the realm of Unigraphics, but no part file is loaded and no application has started. You are only just within the front door.

The menu bar is the horizontal menu of options displayed at the top of the main window directly below the title bar. Menu bar options are called menu titles and each corresponds to a Unigraphics functional category. Each menu title provides access to a pull-down menu of menu choices. When a command on a pulldown menu or a series of submenus is selected, the path is often expressed by arrows. For example: **Insert → Design Feature→ Extrude**. To begin work you must first load a part file. To load a part file, you can either create a new part file or retrieve one that already exists. Both options are available from **File** on the menu bar. The most frequently used **File** options are:

- Create a new part (**File → New**)
- Retrieve an existing part (**File → Open**)
- Saves the file (**File → Save**)
- Exit from Unigraphics without saving parts (**File → Exit**)

After you create or retrieve a part and it has been loaded, choose **Start** from the menu bar. Then you can start the application you wish to use, such as **Modeling**, **Manufacturing**, **Motion Simulation**, or **Shape Studio** by picking the respective application.

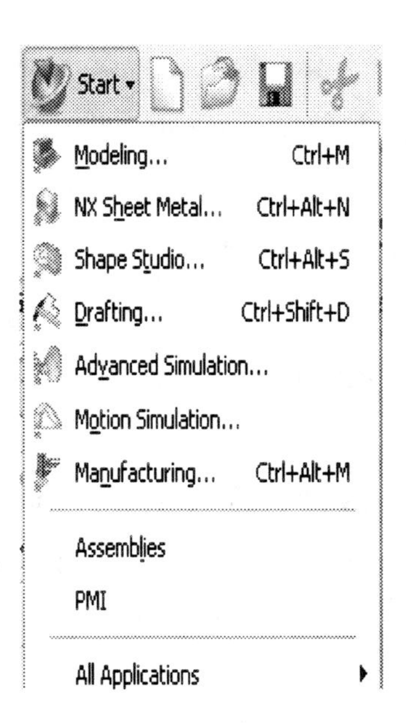

Certain applications are not available on all platforms that run Unigraphics. If an application is not supported on the platform, it will not display on the Application pull-down menu. The application you choose determines which toolbars are available. For example, if you choose Modeling, some available toolbars are: Curve, Edit Curve, Form Feature, Feature Operation, Edit Feature, Surface, Edit Surface, and Sheet Metal Feature. Choosing an icon from the Feature toolbar lets you generate 3D solid models.

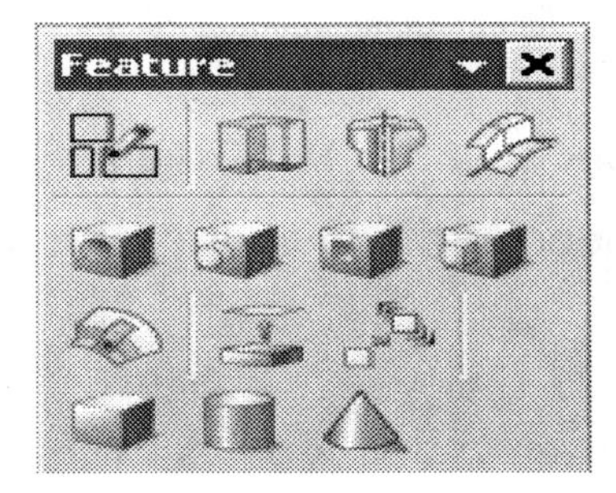

The Feature Toolbar

Note that the minimum screen resolution to display Unigraphics on PC and MAC are 1024 x 768. (For the PC you must also be running in Super VGA mode.) If this minimum resolution is not met, dialogs may not fit entirely on the screen.

NX Templates

A new feature starting with the NX5 version of Unigraphics is design templates. The three template types are mentioned as Model, Drawing, or Simulation. Templates are accessed by choosing **File →New** and instead of a blank modeling screen; the Template dialog pops up to allow the user to choose the appropriate application for what he or she wants to do. Also, a master model part can be determined at this time to reference a new drawing too.

A master model is used in the Drawing section of the templates. Basically, the user uses a regular 3D model of an object to create a top, side, isometric, front, or even no views at all of the specified master model.

NX will create a default name and location for each application, which can be altered by the user during the same process; plus a master model part can be determined at this time also. But if users want to use the same setup as NX4 and previous versions, they can simply choose "**Blank**" as a template.

- Templates help companies follow their set of established standards.
- NX starts the proper application based on selected template, so the user doesn't have to set up the application manually, thus speeding up production.
- By defining a reference master part when creating a new file, one can follow the master model drafting approach more easily.

Pull-down Menu

Pull-down menus appear below the menu bar when you select their title. They remain open without further action until you either choose a menu option or click somewhere outside of the menu. You can open a pull-down menu by pointing at its menu title and clicking mouse button 1 (MB1). Once a pull-down menu is displayed, you can move the focus to individual options by dragging the mouse pointer down the menu (if the menu was displayed with a mouse press), or by using the keyboard arrow keys (if the menu was displayed with a mouse click).

Keyboard mnemonics can also be used as shortcuts for menu bar navigation and selection. You can use a menu's mnemonic to display the menu by holding down the ALT key and pressing the mnemonic letter key for the menu. The mnemonic letter for each menu is underlined in the menu title. Once the menu is displayed, pressing another mnemonic key selects the menu item associated with that key.

Keyboard accelerators (**shortcut keys**) can also be used to make menu selections. Accelerator keys for menu selections are displayed on the menu. Pressing the accelerator keys together, such as <**Ctrl** + **N**> for **File** → **New**, will cause the selected command to be executed directly. Pull-down menu options fall into three categories:

When a menu option name is followed by three periods (**...**), it indicates that the option will display a dialog. A small triangle to the right of an option name indicates that a cascade menu will be displayed.

Toolbars

Toolbars are a row of icons you can use to launch standard Unigraphics menu items. Unigraphics comes with a large selection of toolbars, several of which are displayed when you start Unigraphics. Most toolbars are initially docked in the container area along the border of main Unigraphics window. To undock a toolbar, position the cursor over the grip handle located on the left side of the docked toolbar and use MB1 to drag it. An undocked toolbar has its name displayed in its title area. When you remove your finger from MB1, the toolbar becomes a free-floating icon palette that you can move anywhere on the desktop. When a toolbar is undocked on the graphics display, you can also turn it OFF by choosing the "x" in the upper right corner.

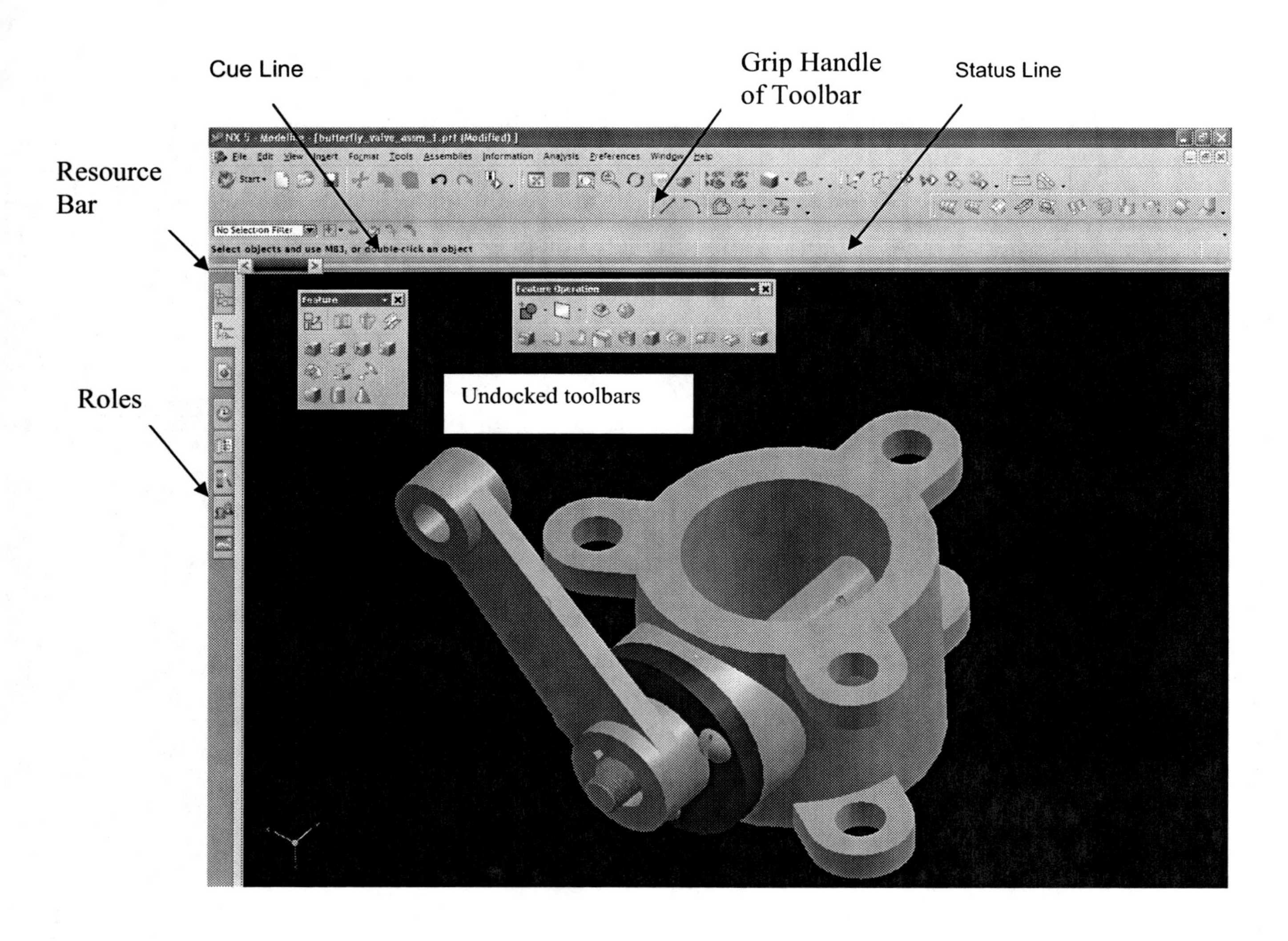

If the toolbar extends beyond the window frame, the toolbar will be clipped. Any adjustments that you make to the toolbars' location, visibility, and docked state will be remembered and restored the next time Unigraphics is invoked. You can customize a toolbar's visibility and content. Please refer to **Tools** → **Customize** for further details.

Saving toolbar layout between sessions

If you want to save the current layout of toolbars and icons when you exit an NX session, make the following setting:

1. Choose **Preferences** → **User Interface**.
2. On the General page, select **Save layout at exit**.

Next time when you start a new NX session, you will have the saved toolbar layout.

Cue/Status Line

The Cue/Status line appears above the graphic window by default. This line can move near bottom by **Tools** → **Customize** → **Layout**. The Cue area displays prompt messages on input expected by the current option. These messages indicate the next action you need to take. The Status area, located to the right of the Cue area, displays information messages about the current option or the most recently completed function. These messages do not require a response.

Roles

Roles let you control the appearance of the user interface in a number of ways. Some examples are:

- What items are displayed on the menu bar
- What icons are displayed on the toolbars
- Whether or not icon names are displayed below the icons

Choosing a Role

NX-5 comes with a number of built-in roles. There are five **System Defaults Roles**: Advanced, Advanced with full menus, Essentials, Essentials with full menus, I-deas. Their icons look like in order.

To activate a role:

- Click the Roles tab to open the palette on the resource bar (see the figure in the previous page).

- Click the role you want. Throughout this book, we use role Advanced with full menus, which provides the wider set of tools to support all UG tasks. If it contains too many options you do not need, you could try Essentials roles.
- In the warning dialog, choose **OK** to accept the new role or choose **Cancel** to stop the change from occurring

View Popup Menu

Certain **View** options are also available from a special **View Popup** menu; which can be opened from the graphics window by pressing mouse button 3 (MB3). The location of the cursor at the time you initiate the popup menu determines the view in which certain options are performed. Other options are either automatically applied to all views, or can be made active in any view by moving the cursor. To activate an option, drag the cursor down the menu with MB3 held down, releasing it when the desired option is covered.

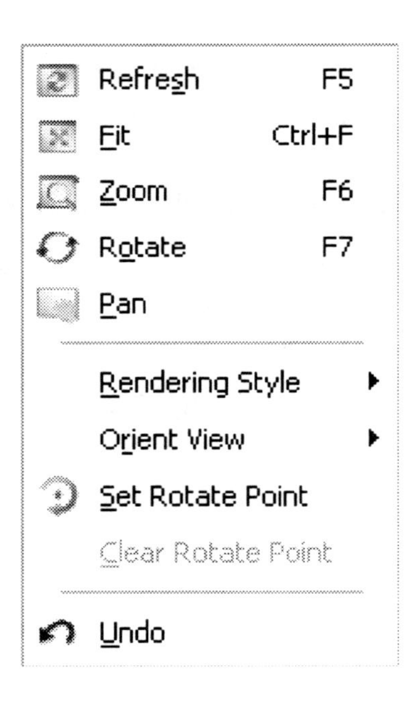

View Popup Options	
Apply	Apply changes made in the current dialog. **Apply** on the View Popup menu is synchronized with the **Apply** option on the current dialog, so that it has the same status on both dialogs and menus (e.g., either active on both or grayed out on both).
Back	Goes back to a previous dialog. **Back** option on the View Popup menu is synchronized with the **Back** option on the current dialog.
Cancel	Cancels the current operation and dismisses the dialog. This option is synchronized with its counterpart on the current dialog.
Refresh	Cleans up the entire graphics screen by eliminating any holes in the graphics display left by blanked or deleted objects, and to display the results of some modification functions. **Refresh** removes temporarily displayed items such as asterisks, conehead vectors.
Fit and Fit All	Fit is available for single view, Fit All is available for multiple views. Fit fits the model in the view where the cursor is positioned by the fit percentage you have set **Preferences → Visualization → View/Screen**. Fit All does same for all views.
Zoom	When this toggle is ON, Zoom mode is active, and you can zoom in any view by positioning the *Zoom Cursor* (a magnifying glass) and dragging a rectangle while pressing MB1. To cancel Zoom mode, choose Zoom again or click MB2.
Zoom In/Out	Position the magnifying glass on the current display (if there are multiple displays, the view the magnifying glass is positioned over will be affected). Hold down MB1, and drag the cursor towards the top of the screen to reduce the view (Zoom Out) or, drag the cursor towards the bottom of the screen to magnify (Zoom In) the view.
Rotate	When this toggle is ON, the **Mouse-Driven XY Axes** option from the **View → Operation → Rotate** dialog becomes active. In this mode the cursor in the graphics window changes to the *Mouse-Driven*

	Rotation Cursor. Simply position this cursor in a view and press and hold MB1 while dragging. The view will rotate around the XY axes. How to cancel Rotate mode works in the same as canceling Zoom.
Pan	Turn this option toggle ON to activate the *Mouse-Driven Pan Cursor*. Once the cursor is visible and you are in Pan mode, position it over a desired starting location for a pan in a view. Next, while pressing MB1, drag the cursor to dynamically pan the view.
Update Display	Performs a display cleanup. This includes updating curves and edges displayed by line segment approximations, if the scale has increased such that the approximations are no longer smooth. Rotations cause Update Display to redisplay silhouette curves of faces and hidden edges of solids. Update Display also performs the tasks of the Refresh option, the erasure of temporary displays and the redrawing of the entire screen.
Restore	Restores the original view immediately after these operations: **Format → Layout → New; Format → Layout → Open; Drawing → New; Drawing → Open; View → Operation → Fit; View → Fit All; View → Operation → Regenerate Work; Format → Layout → Regenerate; Format → Layout → Fit All Views**, or **Fit** from the View Popup menu.
Rendering Style	Specifies display options for the selected view. You can set the mode to one of eight options--3 wireframe options, 3 shaded options, Face Analysis and Studio. NOTE: The setting for the curve tolerance (**Preferences → Visualization → Line → Curve Tolerance**) may affect both the image quality and the performances of the display modes. For the Partially Shaded option, you must have the Partially Shaded attribute checked ON, in **Edit → Object Display**, for selected geometry to function.
Expand	This is available when multiple views are present. When the cursor is positioned over a specific view, the view under the cursor position is

	expanded to a full display area view and now becomes the work view.
Orient View	Lets you change the alignment (orientation) of the view in which the pointer currently resides to one of canned views such as Top, Right, Trimetric, etc. Only the alignment of the view changes, and not the view name. The view name is displayed at lower left corner of graphics window.
Replace View	Lets you switch the view in which the pointer currently resides with a canned view. The view name changes.
Set Rotate Point	Sets (Clears) a point around which a part on the graphics window rotates. Once you click this Set Rotate Point option, you need to pick a point in the graphics window.
Undo	Cancels the last operation performed.

Terminologies and Unigraphics Capabilities

The Unigraphics Modeling application provides the capabilities to help design engineers quickly perform conceptual and detailed designs. It is a feature and constraint based solid modeler that allows the user to create and edit complex solid models interactively. With this feature, the design engineer can create and edit more complex models with far less effort than traditional wireframe and solid based systems.

Early CAD systems built models using geometry construction and editing techniques. These techniques created models made up of wireframe and surfaces, which could be edited by geometry editing. However, the system had no understanding of relationships between the geometry.

Later parametric systems introduced conceptual modeling techniques, which could be used to build relationships between geometric entities. These techniques relied on constraints and parametric expressions defined by the user to build the model. This approach built a more intelligent model, but could become restrictive. The constraints would sometimes stop a user from carrying out a modification, which had not been originally foreseen. Parametric systems also relied heavily on bodies created from swept sketches, and did not have a wide variety of construction and editing techniques.

The Modeling application is a new generation of modeler, combining traditional and parametric approaches to modeling. This gives you the freedom to choose the design approach that best suits your needs. There are times when a simple wireframe model is adequate for the job, and there is no need for the complexity that a constrained solid model introduces. However, it also provides a rich set of parametric and traditional solid modeling capabilities, which enable realistic solid models to be developed and manipulated easily.

Once a model has been built, dimension driven editing techniques can be used, even if it was not built using parametric functions - constraints are automatically inferred from the model. These constraints are not imposed, however, if you decide to modify the model so that they are no longer valid. This means that the tendency to "model yourself into a corner," which can happen with old parametric systems, is avoided using Unigraphics Modeling.

There are many other Unigraphics capabilities. We give brief descriptions below for some of them that are going to be covered in the later chapters in detail.

Start with a Sketch

You can use the Sketcher to freehand sketch and dimension an "outline" of curves. The sketch can then be swept (extruded or revolved) to create a solid or sheet body. The sketch can later be refined to precisely represent the object of interest by editing the dimensions and by creating relationships between geometric objects. Editing a dimension of the sketch not only modifies the geometry of the sketch, but also the body created from the sketch. This is discussed in detail in Chapter 10.

Creating and Editing Features

Feature based solid modeling lets you create features, such as holes, slots, and grooves, on a model. You can then directly edit the dimensions of the feature and locate this feature by dimensions. For example, a hole is defined by its diameter and length. You can directly edit all of these parameters by entering new values. This is addressed in detail in Chapters 5 and 6.

Associativity

Associativity is the term used to indicate geometric relationships between individual portions of a model. These relationships are established as the designer uses various functions for model creation. Constraints and relationships are captured automatically as the model is developed. For example, a through hole is associated with the faces in a model that the hole penetrates. If the model is later changed so that one or both of those faces moves, the hole updates automatically due to its association to the faces. This concept is addressed throughout this book.

Positioning a Feature

Within Modeling, you can position a feature relative to the geometry on your model using positioning dimensions. The feature is then associated with that geometry and will maintain those associations whenever you edit the model. You can also edit the position of the feature by changing the values of the positioning dimensions. This is discussed in Chapter 5

Reference Features

You can create reference features, such as a datum plane or datum axis, that can be used as reference geometry when needed. These datum features can be used as construction devices for other features. Any feature created using a reference feature is associated to that reference feature and retains that association during edits to the model. Datum planes can be used as a reference plane in constructing sketches, creating features, and positioning features. Datum axes can be used to create datum planes, for placing items concentrically, or for creating radial patterns. Chapter 8 is devoted to this subject.

Expressions

You can also incorporate your requirements and design restrictions by defining mathematical relationships between different parts of the design. For example, you can define the height of a boss as three times its diameter, so that when the diameter changes, the height changes also. This is addressed in Chapter 6.

Boolean Operations

Modeling provides the following Boolean operations: *unite*, *subtract*, and *intersect*. A unite operation combines bodies; for example, uniting two rectangular blocks to form a T-shaped solid body. A subtract operation removes one body from another. An intersect operation creates a solid body from material shared by two solid bodies. These operations can also be used with free form features called sheets. This is studied in Chapter 9.

Undo and Redo

A design can be returned to its previous state any number of times using the **Undo** function. You do not have to take a great deal of time making sure each operation is absolutely correct, because a mistake can be easily undone. You can also use the **Redo** function to reverse what you undid. This freedom to change the model liberates you from worrying about getting it wrong and lets you explore many more iterations in order to get it right. You could launch this function by choosing **Edit→Undo List** or **Edit→Redo** or selecting the corresponding icon as shown below.

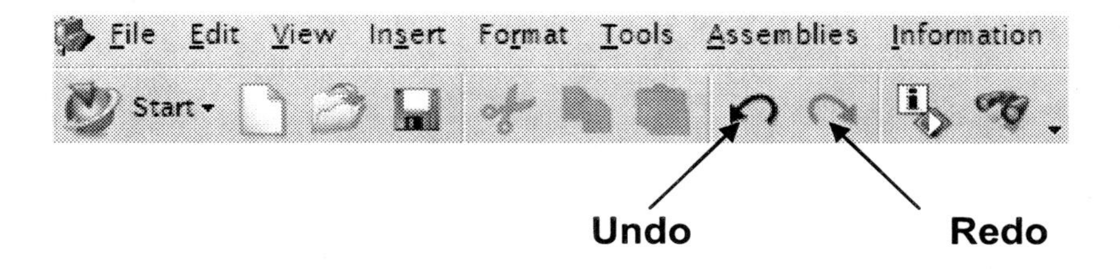

Other Capabilities

Other Unigraphics applications can operate directly on solid objects created within Modeling without any translation of the solid body. For example, you can perform drafting, engineering analysis, and NC machining functions by accessing the appropriate application. Using Modeling, you can design a complete, unambiguous, three dimensional model to describe an object. You can extract a wide range of physical properties from the solid bodies, including mass properties. Shading and hidden line capabilities help you visualize complex assemblies. You can identify interferences automatically, eliminating the need to attempt to do so manually. Hidden edge views can later be generated and placed on drawings. Fully associative dimensioned drawings can be created from solid models using the appropriate options of the Drafting application. If the solid model is edited later, the drawing and dimensions are updated automatically. This concept is called the Master Model and is discussed in Chapter 13.

Exercise Problems

1.1. What is the paradigm of the twenty-first century manufacturing companies?

1.2. What are the four phases of the product-life cycle?

1.3. Give typical steps taken in the product design process.

1.4. CAD significantly changes the design process steps listed in the above question. (True/False)

1.5. Which design process step uses geometric modeling most?

1.6. What are three common geometric modeling approaches?

1.7. Name two common methods for solid modeling discussed in this chapter.

1.8. Between the two methods for solid modeling, which one will be used in this book to model objects?

1.9. What are the limitations of wireframe modeling?

1.10. Describe CAE briefly.

1.11. Describe CAM briefly.

1.12. What is the relation between the paradigm of the twenty-first century manufacturing companies and CAD/CAM/CAE?

1.13. What is Unigraphics?

References

Dransfield, J. (1994), Design for manufacturability at northern telecom., In Successful Implementation of Concurrent Engineering Products and Processes (S.G. Shina, ed.), Van Nostrand Reinhold, New York.

Earle, J.H. (1992), Graphics for Engineers: AutoCAD Release 11, Addison – Wesley, Reading, Massachusetts.

Singh, N. (1996), Systems Approach to Computer-Integrated Design and Manufacturing, John Wiley & Sons, Inc., New York.

Chapter 2. A Case Study for CAD/CAM/CAE

The objective of this chapter is to give a quick overview on how UG can support CAD/CAM/CAE. A simple part is used to take you through step-by-step from design, analysis, then to manufacturing of the part. The figure of the part appears in Figure 2.1. The part is made of stainless steel. One functional performance requirement of the part is that the maximum displacement of the part (i.e., deformation) is less than a threshold value, .01 inch, under the following two conditions. The first condition is that 10,000 lb of force is applied to the hole downward. The second condition is that the fixed constraint is applied to the back long rectangular face such that this face is not allowed to move.

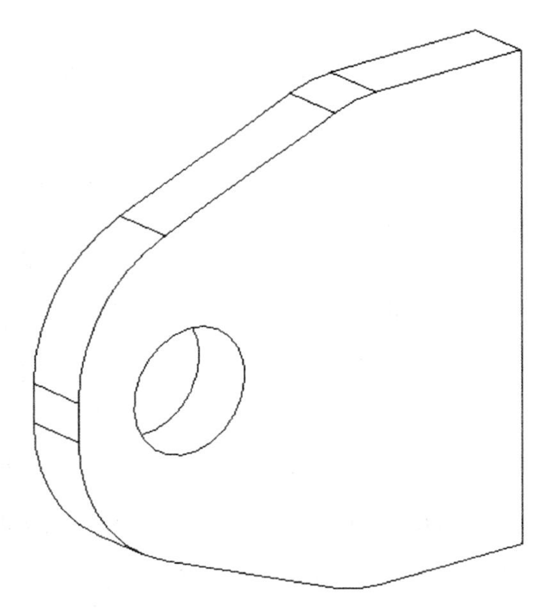

Figure 2.1 A case study part

2.1 CAD for 3-D solid modeling

You will first create a 3-D solid model for this part.

Step 1. Create a block.

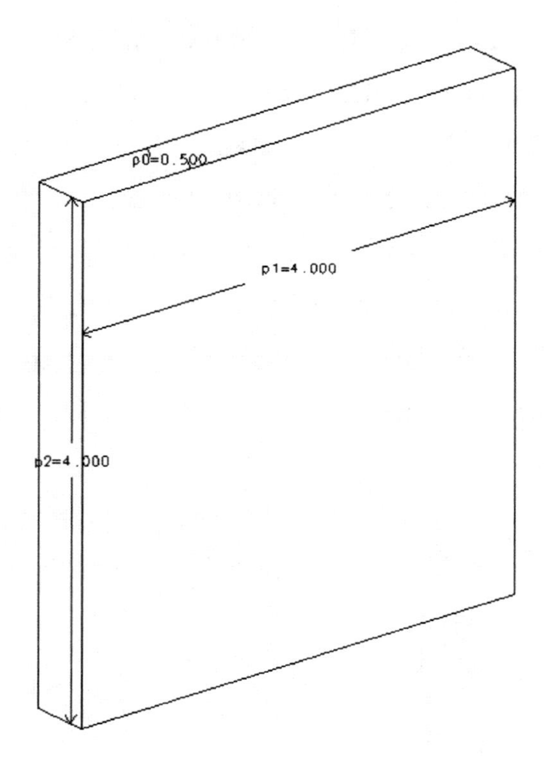

Figure 2.2 Creating a block

1.1 Create a new part (Units: Inches) with filename ***_casestudy.prt, where *** is your three letter name initials. (File → New→Model Template). The file unit choice appears in the lower left corner of dialog “New Part File”. Change your role setting to Advanced Role (see page. 1-16 for the detail).

1.2 Enter Modeling under Start (Start → Modeling).

1.3 Replace the current view with TFR-TRI view if necessary (Move your cursor into the graphics window, click the right mouse button (MB3) → Replace view → TFR-TRI).

1.4 Click the block icon from the tool bar (alternatively, Insert → Design Feature → Block). A block dialog appears.

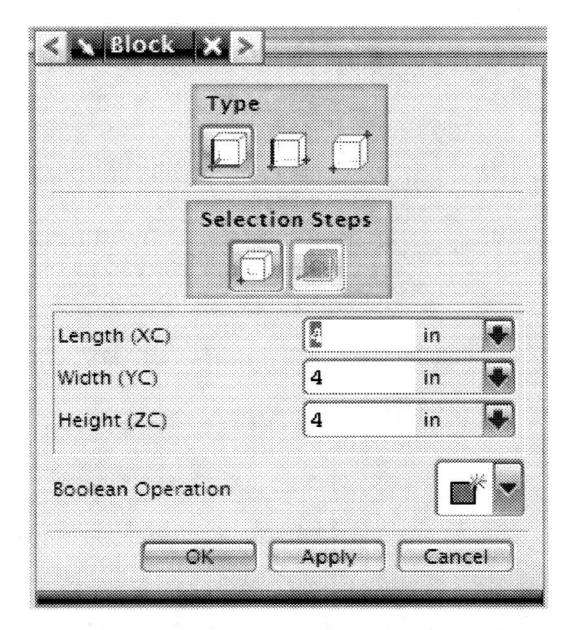

1.5 Click the left icon on the first row if necessary, named as "Origin, Edge Lengths"

1.6 Enter

Length (XC) = 0.5 (Use the Tab key to move to the next field)

Width (YC) = 4

Height (ZC) = 4

1.7 Click "OK" or the middle mouse button (MB2). You will see the block created and located at the origin of Work Coordinate System (WCS).

Step 2. Create a thru hole in the block.

2.1 Choose the pre-NX5 Hole icon from a toolbar, typically either the feature or feature operation toolbar as shown below. If you do not see this icon from any toolbar, refer to page 5-10 for the instruction of adding the icon to the toolbar. Dialog "Hole" opens.

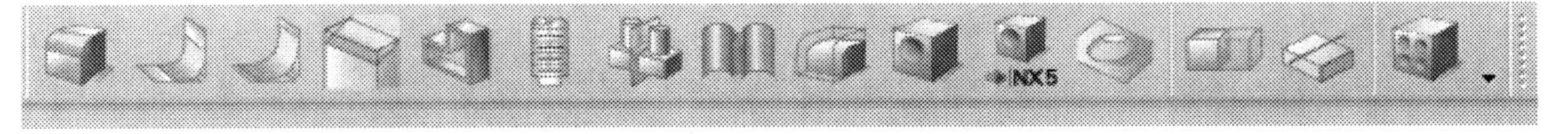

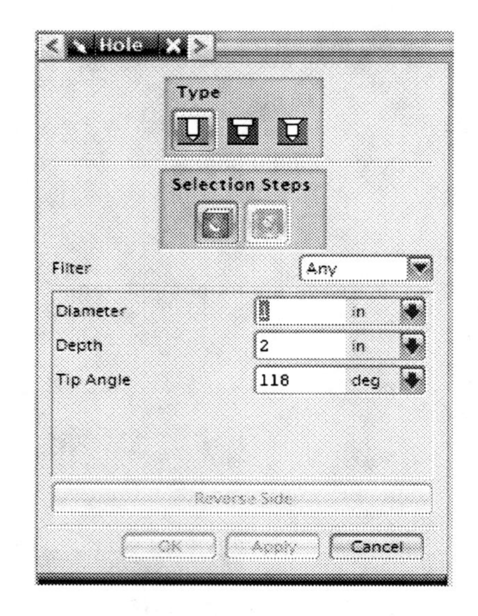

The default hole type (simple hole) and parameter values are already in place.

2.2 The Cue Line is prompting to select the planar placement face, so select the large square face of the block to place a hole on the face. As you click the face, you may see a small dialog pop out to show that there are two faces that can be selectable. If this is the case, click "1." Otherwise, move on to next.

2.3 The Cue Line is prompting to select the thru face so select the other large square face to specify the thru face of the hole.

2.4 The hole parameters change to only requiring the diameter so enter Diameter = 1 and click "OK." A hole with diameter =1 is created.

2.5 Position the hole on the face. Dialog "Positioning" appears to guide where to place the hole on the face.

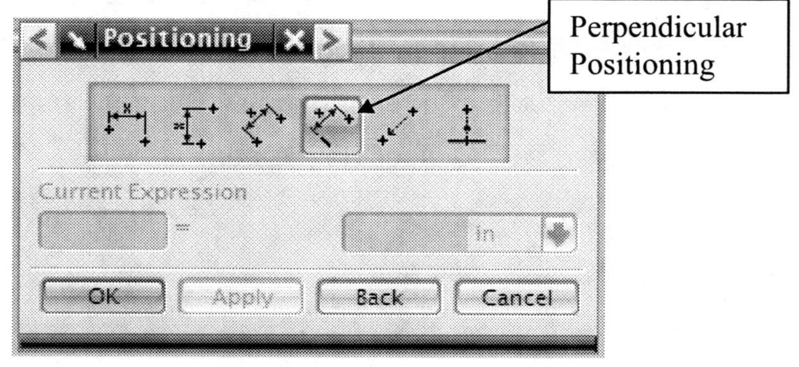

The hole center will be located 2 inch from the top edge and 1 inch from the left edge as shown below.

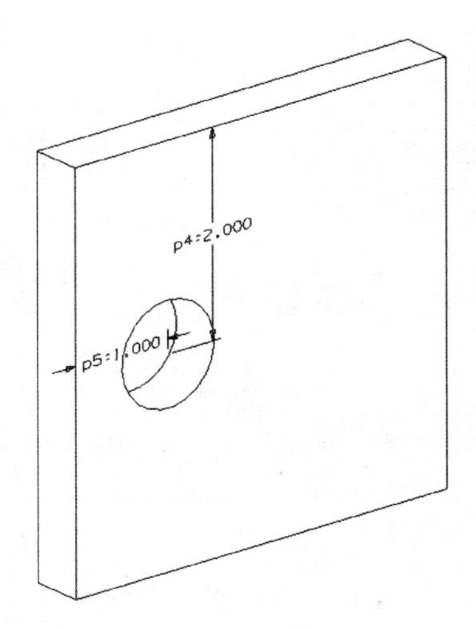

Figure 2.3 Positioning a hole on the block

You will use two perpendicular positions to locate the hole. Use the first perpendicular position by picking the fourth icon from the 1st row. Pick the upper edge of the face and enter 2 as shown in figure and click "Apply." Use the second perpendicular position by picking the same icon. This time pick the left edge of the face and enter 1 and click "OK."

Step 3. Apply a chamfer operation to two edges.

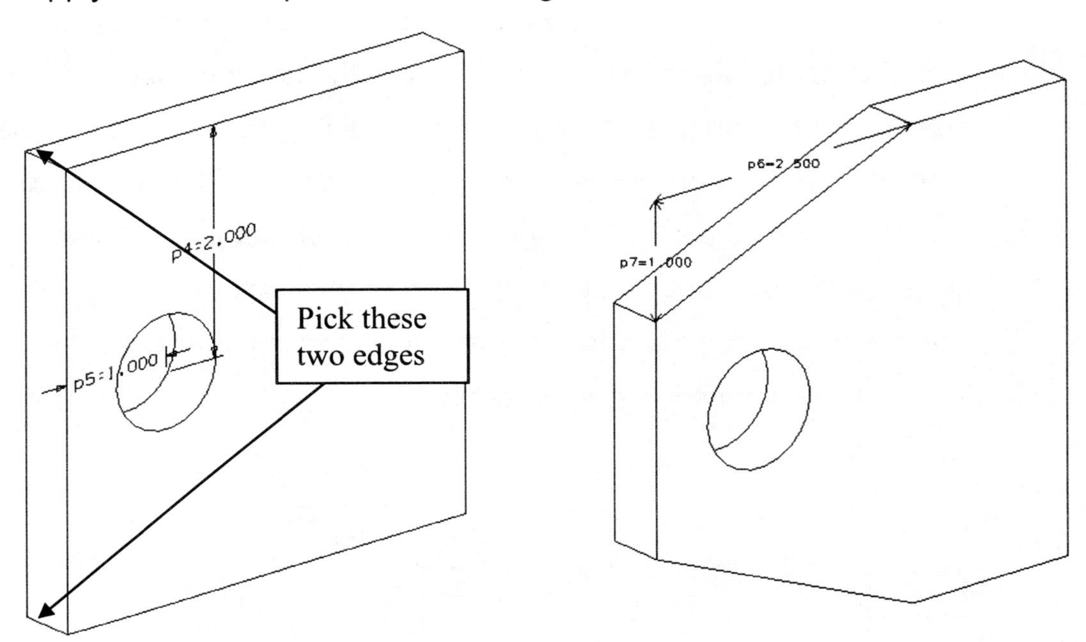

Figure 2.4 Before and after chamfering two left corner edges

3.1 Insert → Detail Feature → Chamfer or choose the chamfer icon .

3.2 Chamfer dialog appears. Click the input option icon "Asymmetric Offsets."

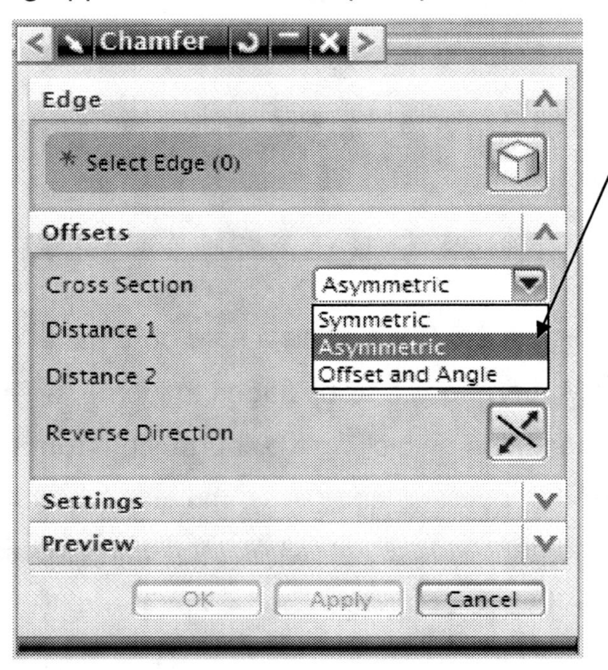

3.3 Pick two small edges that you want to chamfer, one upper left and the other bottom left as shown in Figure 2.4 on the previous page. As you click the edge, you may see another small dialog pop out to show that there is more than one edge that can be selectable with your last click. If so, click among the selectable numbers one number that indicates the edge. Otherwise, move on to next. Also as you click the edge, you will see the previewed chamfer image appear with the current offset values. The preview feature is a default setting.

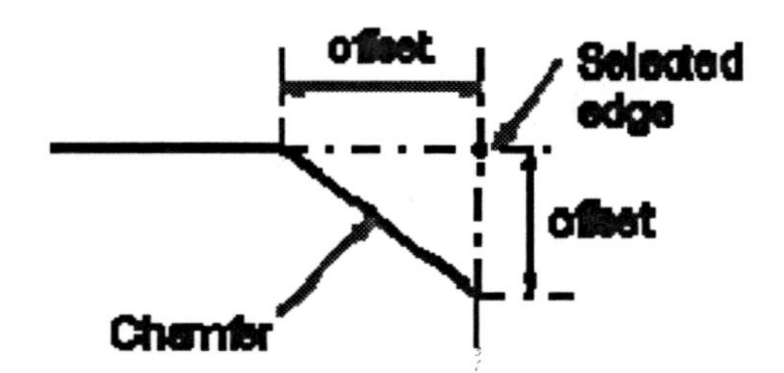

3.4 Enter the following chamfer values and click “Apply.”

First Offset = 2.5

Second Offset = 1

If the chamfer offset direction is reversed, click the reverse offsets icon in the dialog and click “OK.” Otherwise, choose “Cancel.”

Step 4. Apply two edge blend operations to four small edges created from the chamfer operation.

4.1 Insert → Detail Feature → Edge Blend or choose the edge blend icon . This dialog allows the type of blend to be defined with a constant radius blend or a variable radius, as well as other options.

This dialog is an interactive dialog to input the radius of the blend. The down arrow allows other options to define the blend.

4.2 Pick two small edges to blend, one leftmost upper and the other leftmost bottom as shown in Figure 2.5.

4.3 Enter Radius = 1.25 and click the complete set icon on the above dialog. These two edges are blended and the Edge Blend dialog is still open.

4.4 Do the edge blend operation one more time. Pick another two small edges resulting from the chamfer operation: one small edge on the top face and the other on the bottom face.

4.5 Enter Radius= 1 and click OK in the dialog. Now the 3-D CAD model for this part is created.

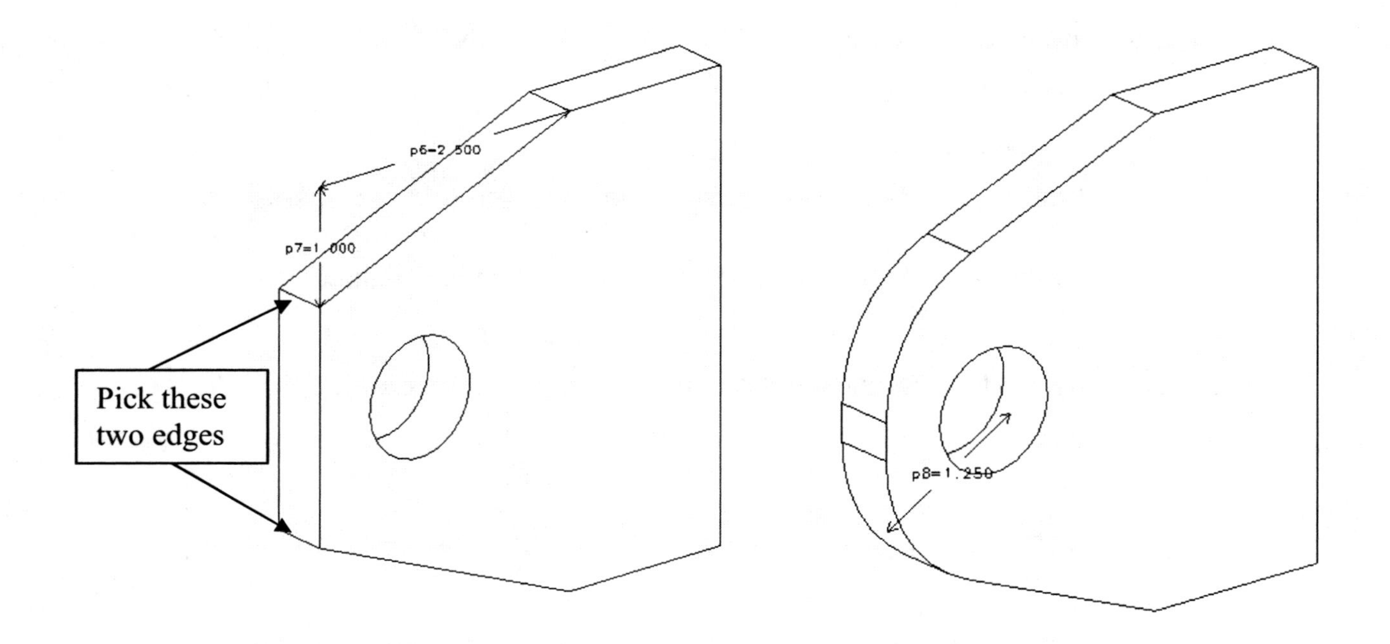

Figure 2.5 Before and after blending two left corner edges

Step 5. Assign material type, stainless steel, to this part.

5.1 Tools → Material Properties

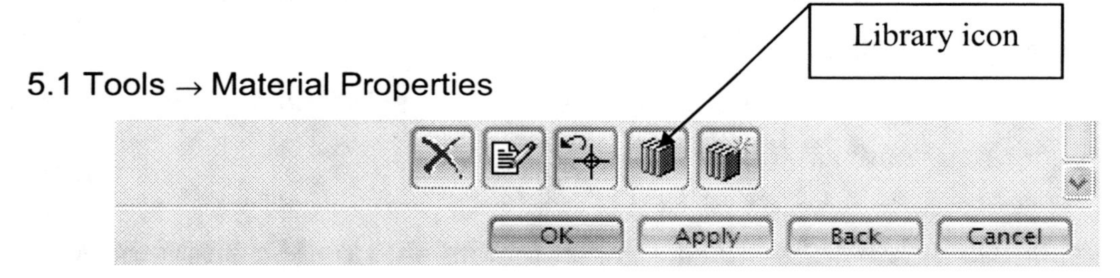

5.2 Materials dialog appears. Click "Library" on the near bottom of the dialog.

5.3 Another dialog "Search Criteria" appears. Accept the default setting by just clicking "OK." This will give all the list of isotropic metal material types.

5.4 Pick Ref 16 "S/Steel PH15-5" among the list of metals found in the library. If you do not find this material in your library, pick any "Steel" item instead. Click "OK".

5.5 Pick the part in the graphics window and click "OK." The selected material, S/Steel, has been assigned to the part.

2.2 CAE for Design Analysis

You will find the part weight and conduct the structure analysis of the part.

Step 1. Save your model by choosing File→Save.

Step 2. Find the Volume/Area/Weight of the block.

2.1. Choose Analysis → Measure Bodies.

2.2. Select the Work Part in the graphic window. A small window will pop up with its volume info. Use the pull down menu to select the weight.

2.3. If you want more detail info, click the information icon . The information window pops out, providing the surface area, volume and mass about the part. Note that this mass info is based on the material property (S/Steel) assigned to the part as done in Step 5 of the previous page. Other useful information available includes the mass center, and moments, and inertia. Close the information window.

Step 3. Select Start → All Applications → Design Simulation to enter the analysis application. Dialog "New FEM and Simulation" appears. Click OK by accepting defaults.

Step 4. Dialog "Create Solution" appears. Keep the solver type to NX Nastran Design as default.

Choose OK. You will have the simulation navigator window that looks like

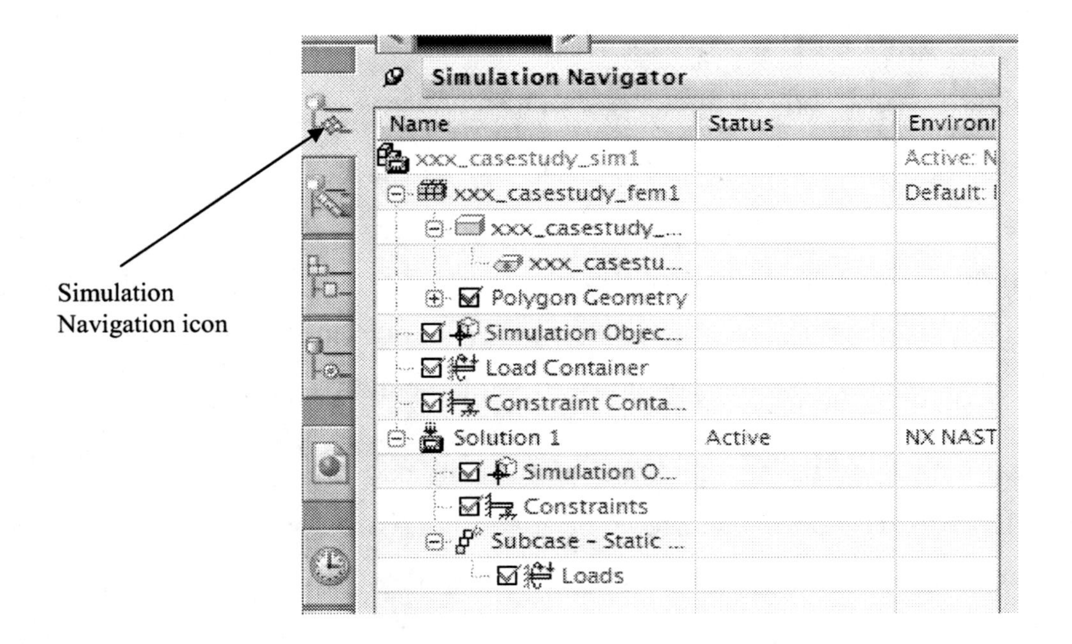

If not, click the Simulation Navigation icon on the resource bar (see the above figure) to open the window. Note the icons that are dedicated to structure analysis are added to the left side of the graphic window and you are ready to enter input for the structure analysis.

Step 5. Apply a vertical load.

5.1 Under the Load Type Icon select the Force icon which will bring up the following dialog.

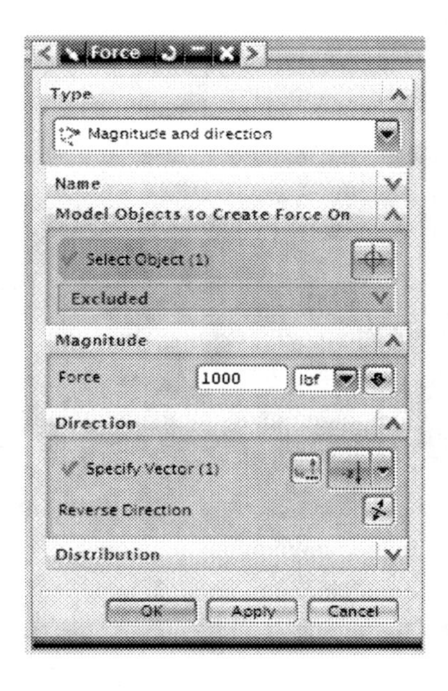

5.2 Select the cylindrical face where the bearing load will be applied as shown below.

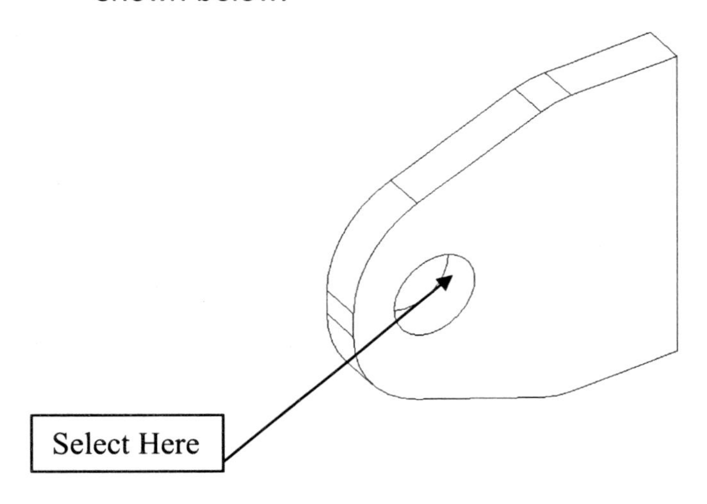

5.3 Enter 10000 pounds in the Force field. Specify the force direction in a negative Z direction as follows: in the selection step at the top of the dialog select –ZC axis in the pull down submenu right next to the inferred vector icon. Your dialog now should look like one shown on the previous page. Click OK. You will see the force direction arrows displayed on the cylindrical face.

Step 6. Apply the Boundary Condition.

6.1. Under the Constraints icon select the fixed constraint icon to bring up the following dialog.

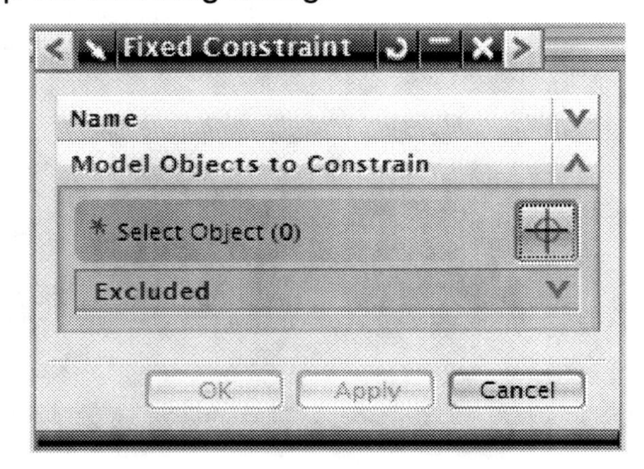

6.2. Select the back face as shown below.

6.3. Click OK.

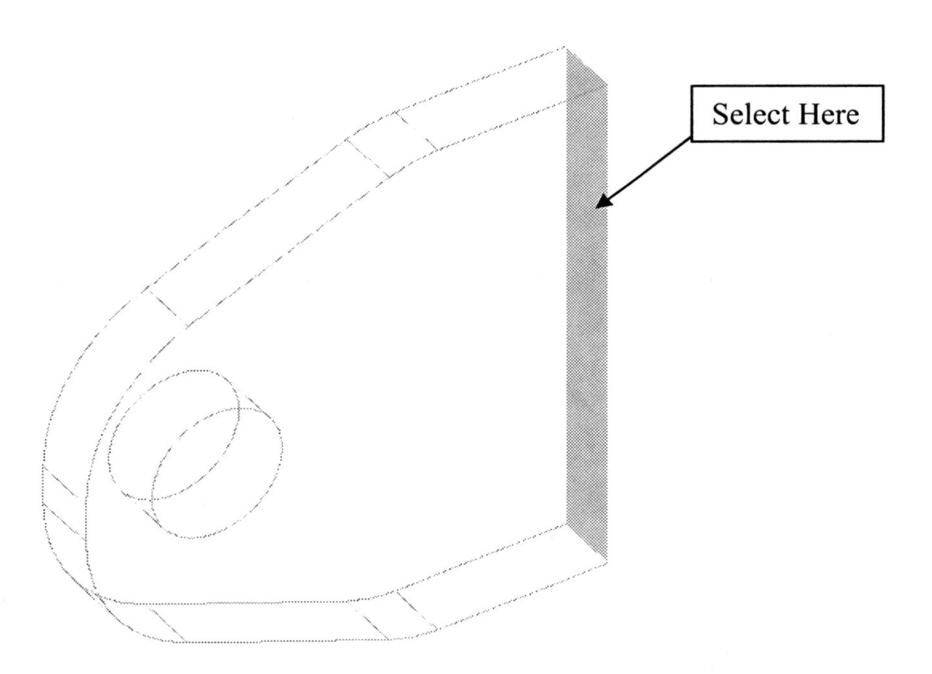

At this point, your graphics screen should look as shown below. Next, we will apply the solid mesh that will carry the load.

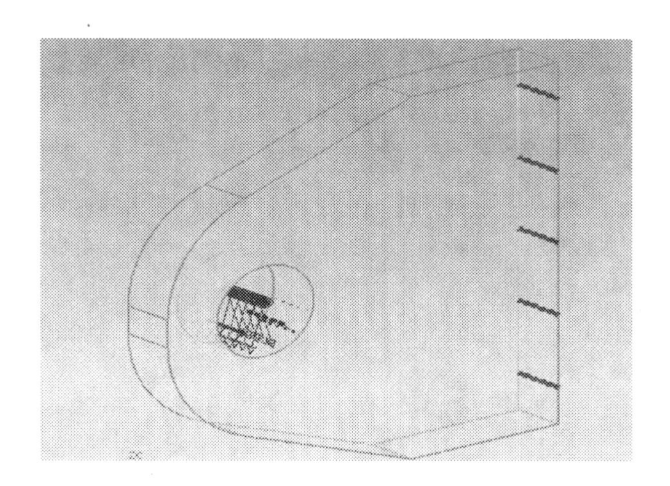

Step 7. Create the Mesh

7.1. Select the 3D Tetrahedral Mesh Icon 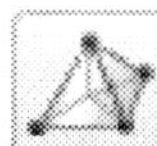 to bring up the following dialog.

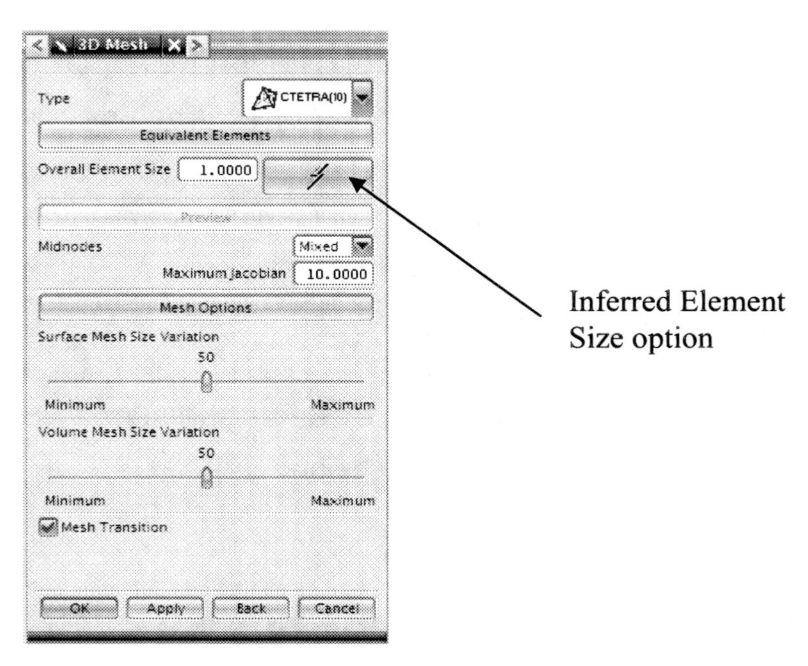

7.2. Select the solid body and select the inferred element size icon (see above) which will let the system determine the mesh element size for the solid body. Click OK to create the mesh. When complete, your model looks similar to one as shown below:

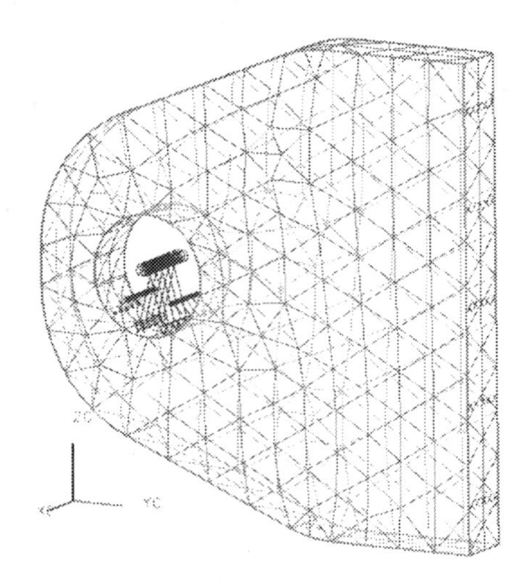

At this point, we are ready to run the analysis.

Step 8. Run the Analysis

8.1 Select the Solve Icon followed by OK to submit the analysis. The information window will open. The Analysis Job Monitor window will come up so wait for it to notify the completion. Once completed, click "Cancel". Close the information window.

Step 9. View the Results

9.1. Select the Enter Post Processing icon to see the deflected shape as shown below.

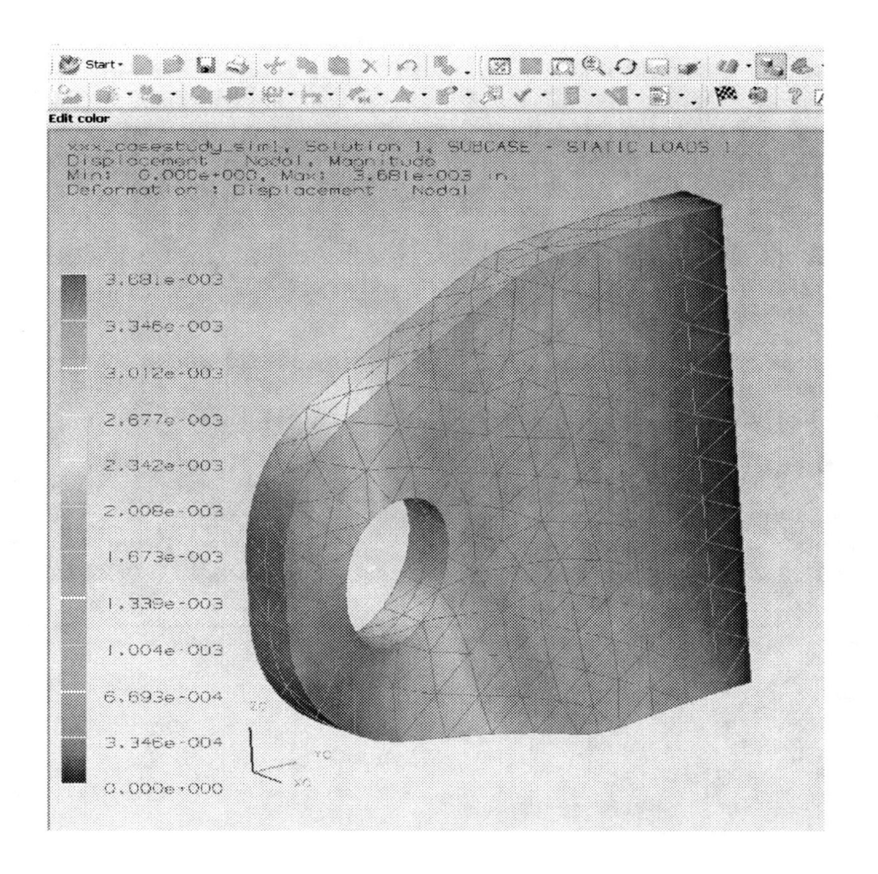

As you can see the graphical results above from the analysis, you can observe a couple of points. Relatively large displacement, colored in red, occurs around the hole where the force of 10,000 pound is applied as specified in Step 4. The maximum displacement estimated by this analysis is 3.68 x 10^{-3} inch (your number may slightly be different). This is less than the threshold value, .01 inch, thus meeting the functional performance requirements. Very little displacement, colored in blue, occurs in the back rectangular face, where the fixed boundary condition is applied in Step 5.

9.1. Animate the deflections by selecting the animation icon followed by selecting the play button at the bottom. When you are through viewing the animation, select the stop button and choose Cancel.

9.2. Quit the result displaying by selecting the Finish Post Processing icon .

Step 10. Return to Modeling.

10.1 Double-click xxx_casestudy.prt in the Simulation Navigator window as shown below to display the original solid model.

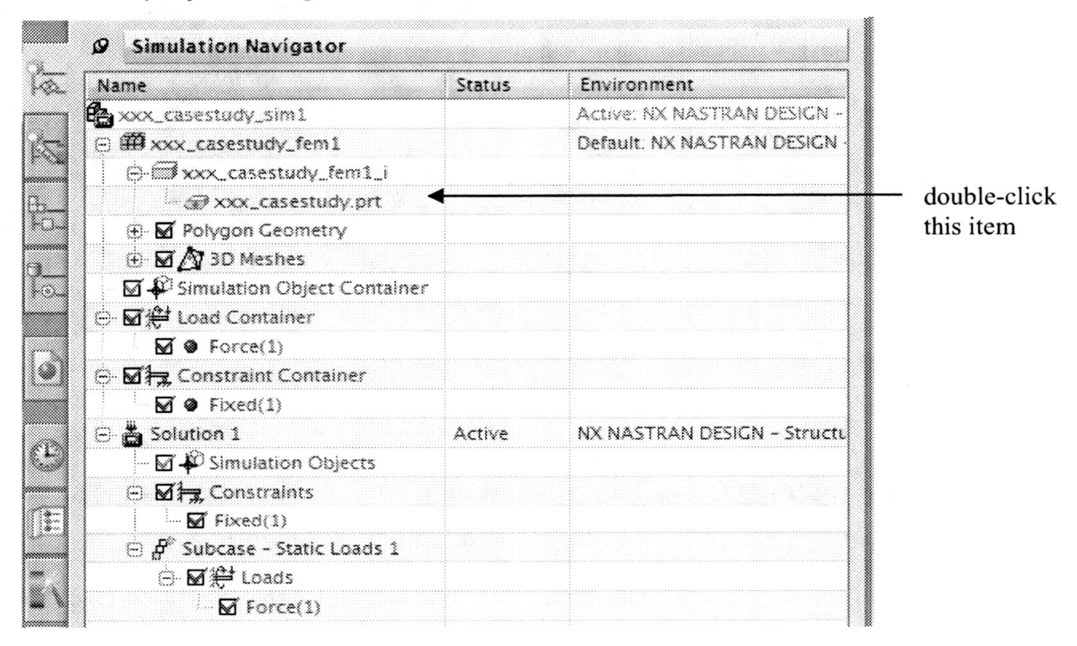

10.2 To return to Application Modeling, choose Start → Modeling.

2.3 CAM for Cutting Tool Path Generation

Step 1. Start → Modeling if necessary. Create a new part file (Units: Inches) and name it as ***_casestudy_block.prt, where *** is your three letter name initials. Create a block and assign the same material type (i.e., stainless steel) to it by following Steps 1 and 5 (pages 2-2 and 2-8) of Section 2.1. This part serves as a blank part. Save this file. You will be guided to create a cutting tool path to cut this blank part into the final part shown in Figure 2.1.

Step 2. Create another new part in inches and name it as ***_casestudy_mfg.prt. This part will contain the information required to generate the tool path.

Step 3. Add the two parts, the blank part (***_casestudy_block.prt) and the final part (***_casestudy.prt), into the manufacturing part (***_casestudy_mfg.prt).

3.1 Start →Assemblies if necessary to turn on the assembly functions.

3.2 Assemblies → Components

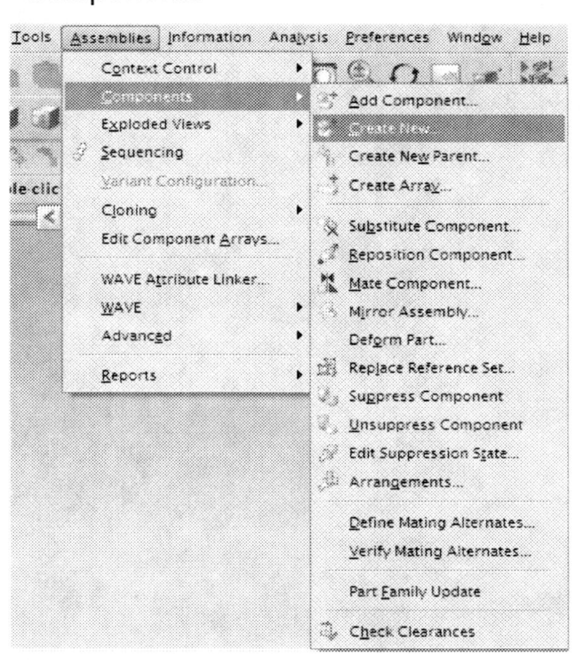

3.3 Slide out and click "Add Component" in the submenu.

3.4 "Select Part" dialog appears. If you see ***_casestudy_block in the list of loaded parts, pick it and then click "OK". If not, click "Choose Part File" button and find the part file by navigating the file directories in the same manner as you do in "File → Open".

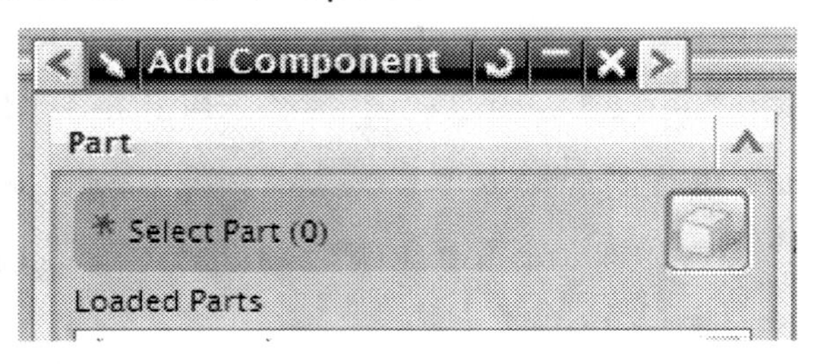

3.5 Accept the default setting on "Add Existing Part" dialog by clicking "OK."

3.6 "Point Constructor" dialog appears. Click "Reset" button and then click "OK." Now the blank part has been added.

3.7 Replace the current view with view TFR-TRI if necessary by clicking Right Mouse Button (MB3) on the graphics window → Replace View → TFR-TRI.

3.8 Repeat 3.4 to 3.6 above except adding ***_casestudy.prt in place of ***_casestudy_block.prt. When you repeat 3.5, make it sure to have Positioning set to Absolute as follows:

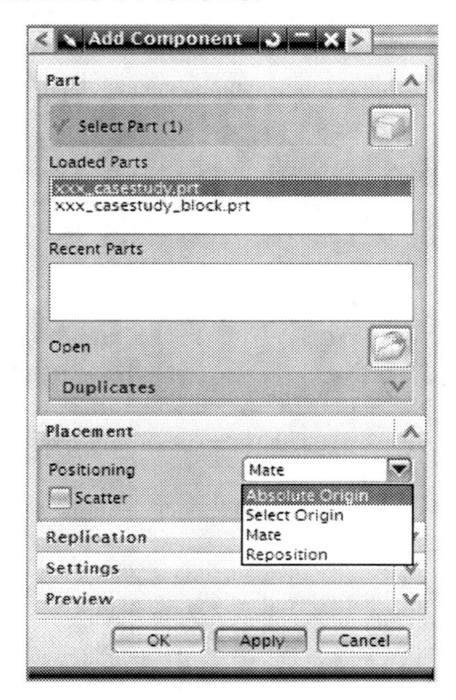

3.9 Click “Cancel” to stop adding any more parts and save this file.

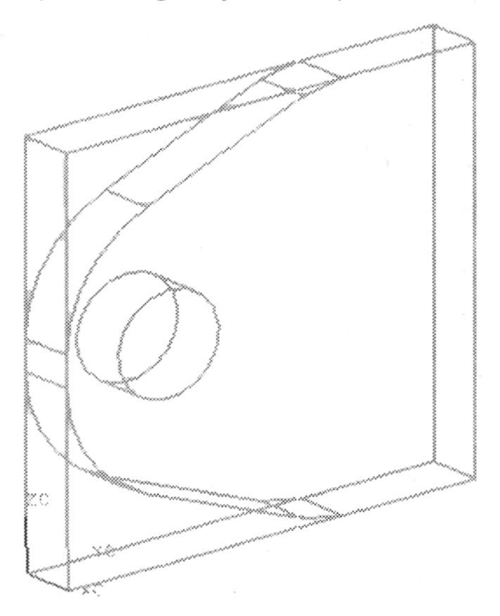

Figure 2.6 ***_casestudy_mfg.prt after adding the two parts in Step 3

Step 4. Start → Manufacturing to enter the Manufacturing Application. “Machining Environment” dialog appears.

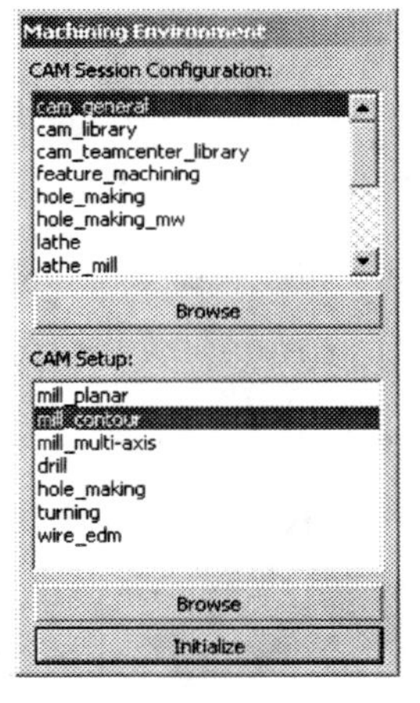

Step 5. Initialize the machining environment.

5.1 Pick “mill_contour” from a list of options available in the second menu “CAM Setup”. Click “Initialize” from the bottom of the dialog.

The system displays manufacturing-related tool bars including the Operation Navigator tool bar as shown in the below figure.

5.2 Click Operation Navigator icon from the resource bar on the left side of the screen. The Operation Navigator window opens.

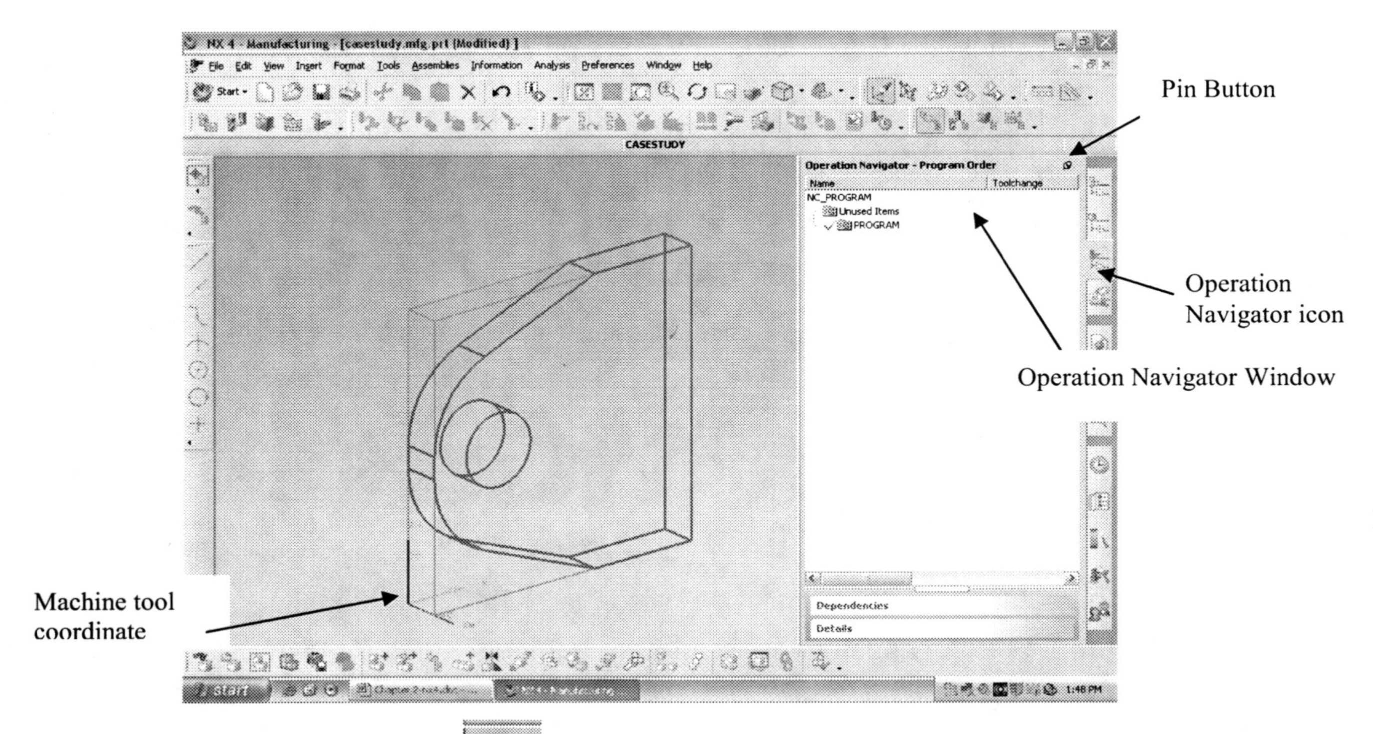

Click the pin button in the right upper corner of the Operation Navigator window to keep this window remaining open.

Step 6. Specify the correct machine tool coordinate system (MCS). The part display window shows that the current ZM direction, which is a cutting tool access direction to the part, points upward. It needs to rotate by 90 degrees such that it points right.

6.1 Select Tools → Operation Navigator → View → Geometry View or click the Geometry View icon located on the Operation Navigator toolbar, which changes window "Operation Navigator" into the Geometry view.

6.2 In window "Operation Navigator – Geometry," double click "MCS_MILL" to edit MCS.

6.3 "Mill_Orient" dialog appears. In the pull down menu for "Specify MCS" select "X-Axis, Y-Axis, Origin."

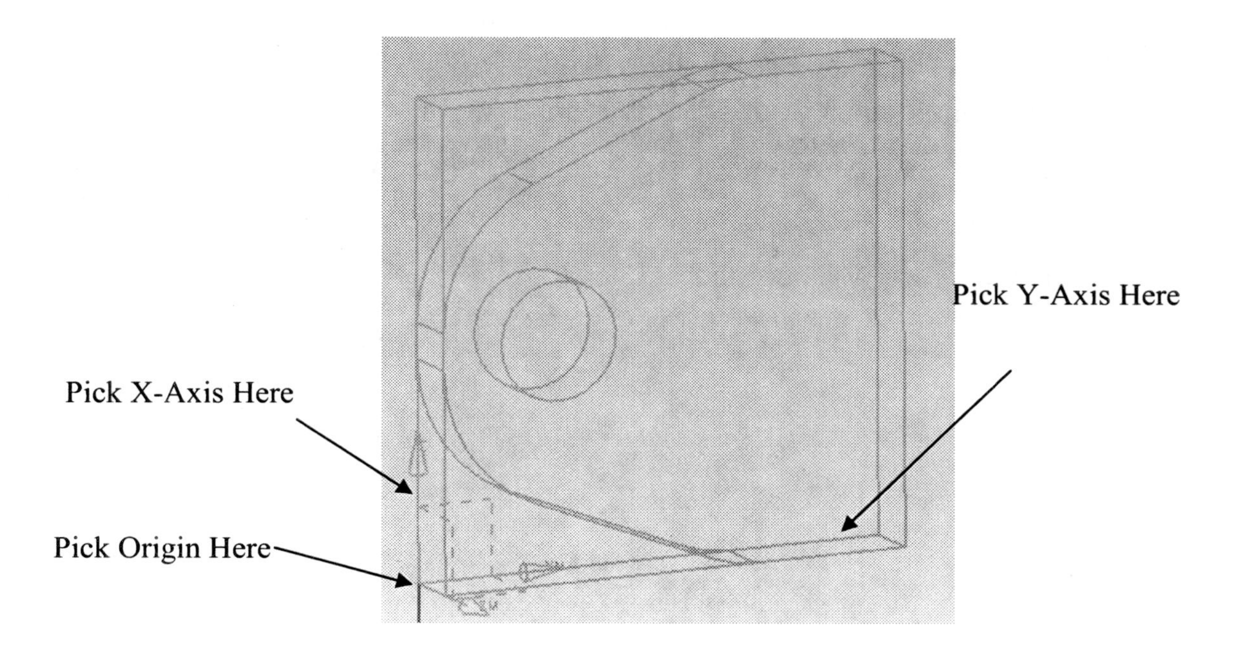

6.4 When the queue line prompts you, pick the X and Y axis and the origin as indicated above. Click OK to accept.

Step 7. Specify a 0.5-inch end mill tool for the cutting tool.

7.1 Select Insert → Tool.

7.2 Dialog "Create Tool" appears. The tool icon MILL is highlighted (a default tool) and click OK. Dialog Milling Tool-5 Parameters appears.

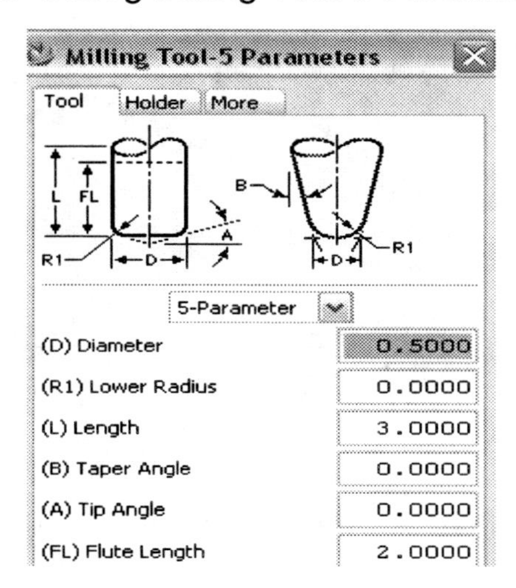

7.3 Enter (D) Diameter = .5 to cut the 1” diameter hole, and click “OK” by accepting the default values for other parameters.

7.4 Select Insert → Operation. Dialog “Create Operation” appears.

Step 8. Setup the Cavity Mill operation.

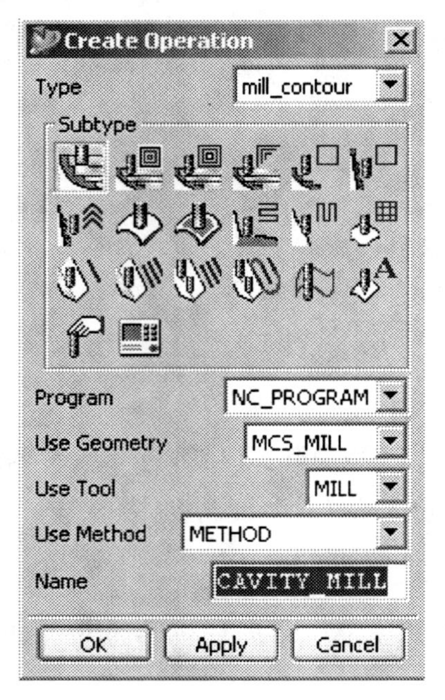

8.1 The default operation selected among a set of operations listed under “Subtype” is CAVITY_MILL (the upper left corner icon highlighted).

8.2 Choose OK. Cavity_Mill dialog appears.

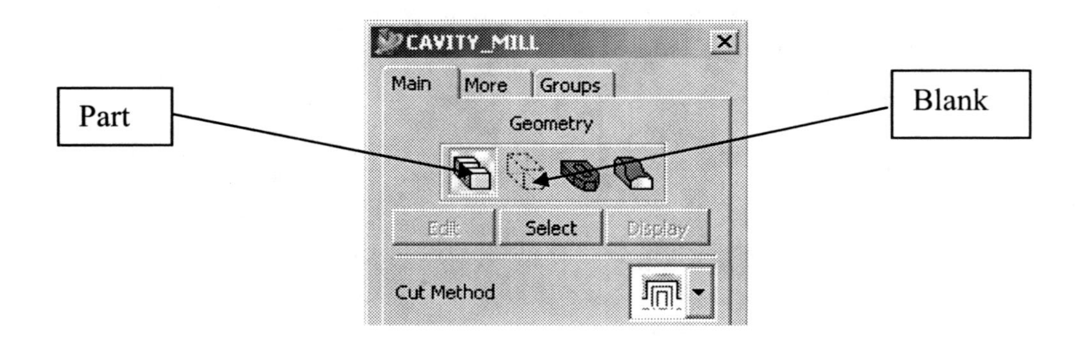

Next, you will assign which part among the two in the graphics window is the Blank and the final Part, respectively.

8.3 Click “Part” icon under title “Geometry.” Click “Select” button on the next line.

8.4 “Part Geometry” dialog appears. Select the final part in the graphics window(***_casestudy.prt.) If more than one is selectable, you will see the selection dialogue looking like appear on the graphics window. Click the correct number between the two to pick the final part, but not the blank part. Click “OK” to close the Part Geometry dialog.

8.5 Click “Blank” icon right next to “Part” icon as shown above. Click “Select” button.

8.6 Select the blank part in the graphics window. Click “OK” to close the Part Geometry dialog.

8.7 For the tool to be displayed, click on the Edit Display icon under the Options near the bottom of the Cavity_Mill dialog.

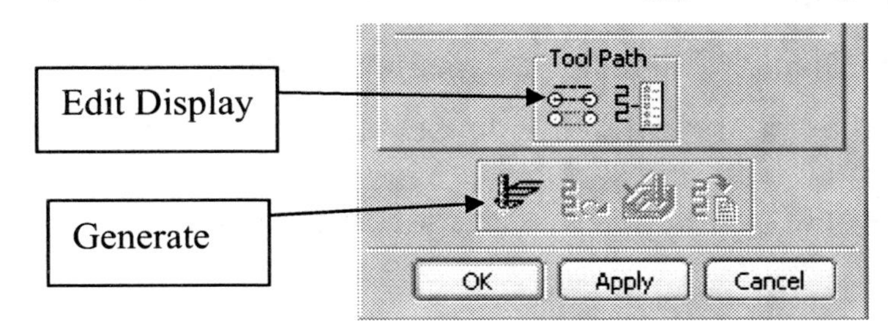

This will bring up the Display Options window and under the Tool Display change the option to 3D. Click OK.

Step 9. Generate the tool path.

9.1 On the Cavity_Mill dialog, click the “Generate” icon near the bottom of the dialog.

9.2 Dialog “Display Parameters” appears. Click “OK”. The tool path is shown on the part. Continue to click “OK” until it finishes generating the tool path.

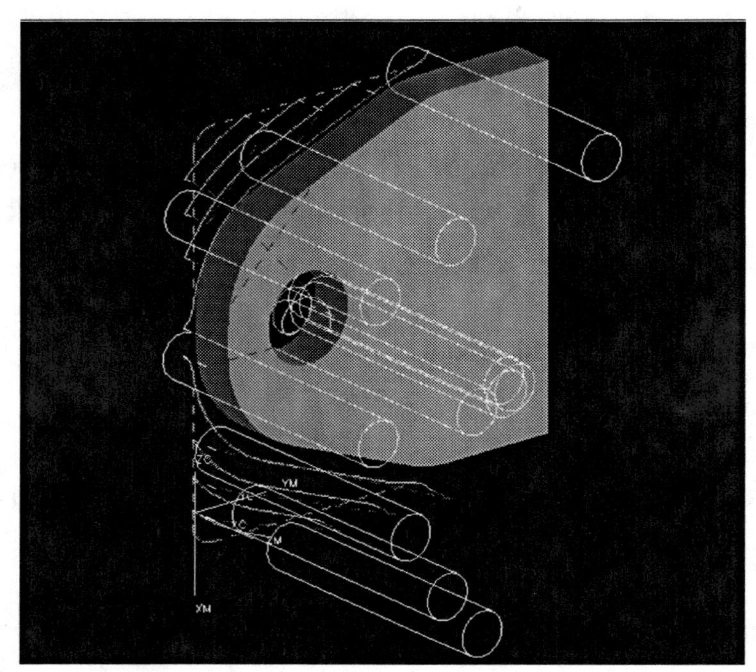

9.3 Animate the tool path for verification. Click “Verify” icon right to the “Generate” icon. Dialog “Toolpath Visualization” appears. Click tab “2D Dynamic” located in the middle of the dialog. Click the play forward button at the bottom and observe animation of the cutting process as shown below.

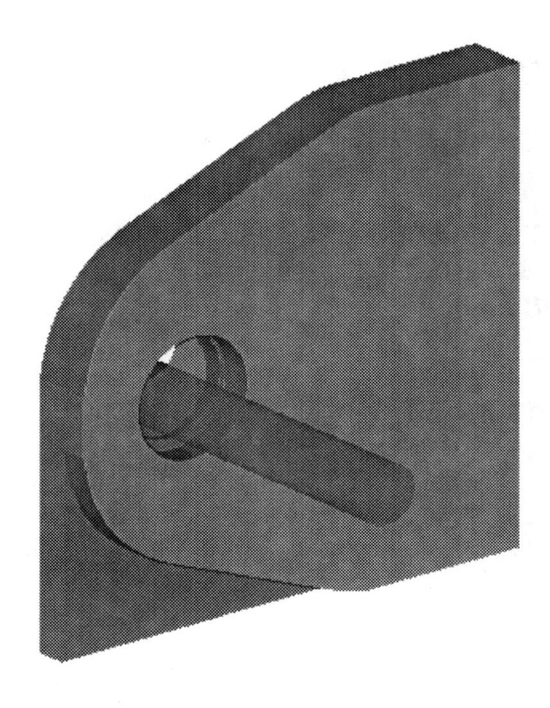

Click “OK” to take it back to the Cavity_Mill dialog.

9.4 Click “List” icon right next to “Verify” icon. This lists the generic NC program list called the cutter location source file (CLSF). This can be post-processed to a specific NC machine tool. Close the list window.

A top portion of CLSF

```
TOOL PATH/CAVITY_MILL_1,TOOL,MILL
TLDATA/MILL,0.5000,0.0000,75.0000,0.0000,0.0000
MSYS/0.0000,0.0000,0.0000,0.0000000,0.0000000,1.0000000,0.0000000,1.000
0000,0.0000000
$$ centerline data
PAINT/PATH
PAINT/SPEED,10
PAINT/COLOR,186
RAPID
GOTO/-100.4027,-2.9481,15.7000,0.0000000,0.0000000,1.0000000
PAINT/COLOR,211
RAPID
GOTO/-100.4027,-2.9481,11.4667
PAINT/COLOR,42
```

```
FEDRAT/MMPM,250.0000
GOTO/-100.4027,-2.9481,8.4667
GOTO/-101.4855,0.0563,8.4667
PAINT/COLOR,31
CIRCLE/-
55.9303,31.0600,8.4667,0.0000000,0.0000000,1.0000000,55.1045,0.0600,0.500
0,0.5000,0.0000
GOTO/-101.5481,0.1485,8.4667
PAINT/COLOR,37
RAPID
GOTO/-104.5481,1.9513,8.4667
```

9.5 In dialog “Cavity Mill”, click “OK” to accept this tool path. Save and close all the files.

Exercise Problems

2.1. Suppose the block size of Section 2.1 changes from .5 x 4 x 4 to .5 x 4 x 5. Redo all the steps of Sections 2.1, 2.2, and 2.3 for CAD/CAE/CAM.

2.2. In Problem 2.1, you are asked to redo all the steps from scratch in order to address the design change in the block size. Is there an easier way?

Chapter 3. Unigraphics Essentials

In this chapter, we will present some of the fundamental UG essentials that will be frequently used during a session with Unigraphics. These include using and manipulating coordinate systems as references, how to choose points in space, layer standards, a selection method to select a desired object among many, and other user interfaces of Unigraphics. We will also discuss default files that determine how Unigraphics is set up, how to obtain help when you need it, and how to set or change the unit system.

3.1 Different Approaches to Modeling Objects in Space

It is important to be familiar with the fundamental aspects of Unigraphics in order to take full advantage of the functionality that lies within the software. First we discuss different ways objects are modeled in industries and their implications. Although companies use the same software, they can differ in the way they manage design information and models. These different aspects and edicts are generally referred to as company standards. These standards include layering conventions, part file storage, views in a model, colors of objects, and how objects are modeled in space. We will discuss the last subject more in detail in this section.

There are three ways that companies model their parts in space: in the exact location they are assembled in, without regard to the location where they are assembled in, or a mix of the two approaches. In aerospace industries, parts are typically modeled in their final assembled location. For detail drawings of parts, the parts themselves are generally defined by using datums as references for dimensions. As you begin to put the details together in an assembly fashion, the assembly drawings will also have datums. In fact, in most aircraft programs today, these same datums are also used for purposes of generating tooling to support the synthesis of the assembly. This concept is referred to as **coordinated datums** whereby the tooling uses the same references as the assembly it is designed to build.

This approach generally insures that items used in defining the aircraft are in their correct place (and are actually referenced by their location) and that all tooling usages for the synthesis of the design are in their correct place as well. So does this mean that when parts are modeled without regard to their correct assembled location, they do not end up in their correct location at the assembly level? No, it does not. The result of modeling parts of an assembly without regard to location is generally easier than modeling to true, assembled space but the final result can be the same. This approach is typically employed on simpler assemblies, but is still scalable and viable to be used on larger assemblies as well. For example, take the assembly of a door slab and three hinges. In this situation, you would have to provide a detail model of the door itself (the slab), a model of the hinge plate, and a final model of the hinge pin. While we do need three hinges, they are all the exact same hinge. Therefore, we will model them only once and then add three complete hinge assemblies to our door at the assembly level. Below is a sample drawing tree of this approach to further explain this concept.

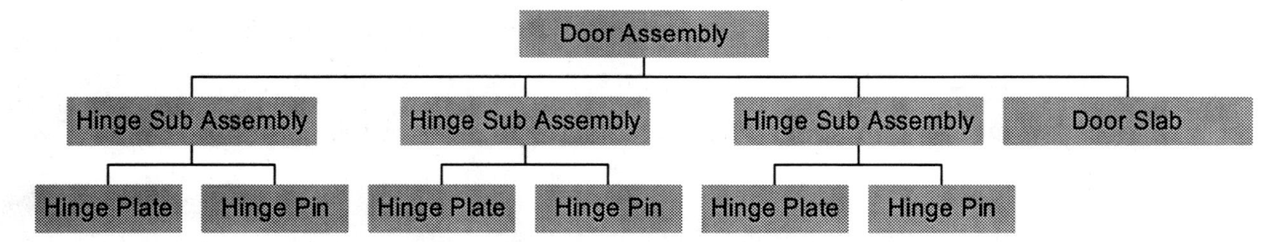

In this case, we would model everything without regard to its final location in space and locate the items (Hinge Pin Subassembly) upon assembly. While the Hinge Sub Assembly shows up three times, all three are the exact same indicating that the Sub Assembly itself is only modeled once and simply added three times to the Door Assembly.

The final approach that we will discuss is that of a mixed approach that models some items, particularly, unique items in their exact location in model space, and more common items (of which there are usually many) without regard to location. These common items would then simply be added to the assembly and positioned as needed for the design. This approach is also used in the aircraft and automotive industries where common items include standard brackets, fasteners, and clamps just to name a few.

3.2 Coordinate Systems and Right-Hand Rules

In the previous section, we discussed different approaches that objects are modeled and assembled in space. All these approaches deal with locating objects in space, thus, a coordinate system. We discuss three coordinate systems in this section. These are the absolute coordinate system, the work coordinate system, and the feature coordinate system. All these coordinate systems use right-hand rules to determine the principal axes of coordinate systems and the positive directions of revolution.

The conventional right-hand rule is that if the origin of the coordinate system is in the palm of the right fist, with the back of the hand lying on a table. The outward extension of the thumb corresponds to the positive X-axis; the outward extension of the index finger corresponds to the positive Y-axis, and the upward extension of the middle finger corresponds to the positive Z-axis.

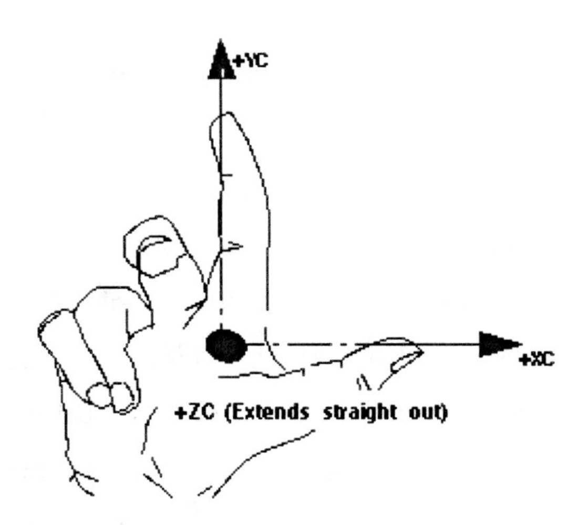

The Conventional Right-Hand Rule

The right-hand rule for rotation is used to associate vectors with directions of rotation. When the thumb is extended and aligned with a given vector, the curled fingers determine the associated direction of rotation. Conversely, when the curled fingers are held so as to indicate a given direction of rotation, the extended thumb determines the associated vector. For example, to determine the counterclockwise direction of rotation for a given coordinate system, the thumb is aligned with the ZC axis, pointing in the positive Z direction. Counterclockwise is defined as the direction the fingers would move from the positive X to the positive Y-axis.

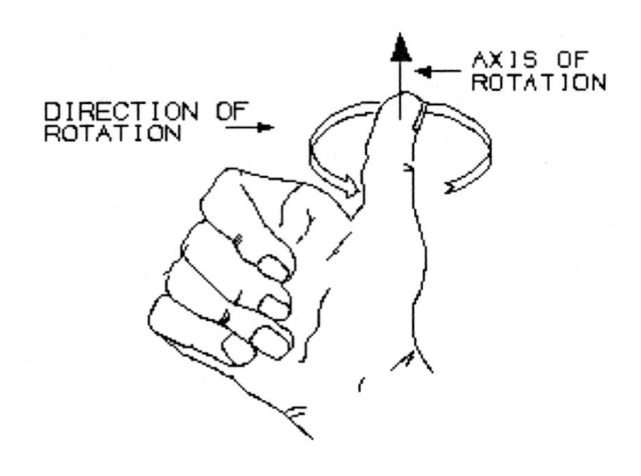

The Right-Hand Rule for Rotation

3.3 The Absolute Coordinate System (ABS)

The Absolute Coordinate System defines fixed origin (0, 0, 0) and fixed directions in X, Y, and Z axes in model space. This is fixed in that it is not a mobile coordinate system and therefore provides a 100% consistent reference point for everything from part to part as well as within an assembly. It is commonly used as the general reference for modeling and large complex assemblies when employing the “in correct space” approach to modeling a product assembly. You cannot physically see the Absolute Coordinate System (ABS) unless you superimpose (in origin and orientation) a mobile coordinate system over top of ABS.

With all of this in mind, you may be thinking that having a coordinate system that is fixed is very limiting and that it will be difficult to do anything away from it.

Well, that aspect is covered by a mobile, construction coordinate system called the Work Coordinate System or the WCS. The WCS is the system that may be placed on top of the Absolute Coordinate System to provide visualization of the ABS.

3.4 The Work Coordinate System (WCS)

As mentioned above, the WCS is a mobile coordinate system that may be moved around the model space to facilitate the construction of objects. It may also be used for determining specific points in a model and even for establishing direction vectors (for extrusions and revolutions for example). The WCS is also useful for helping to model parts in their correct origin and orientation relative to the ABS.

Generally speaking, most of the operations you will perform in Unigraphics will not require you to manipulate the WCS as features are generally built and referenced from existing geometry. However, you will often need to manipulate the WCS in order to create initial geometry such as primitive features. Primitive features include blocks and cylinders, and will be discussed in Chapter 4. Creation of primitive features depends on the various axes of the WCS for direction. The WCS is readily identifiable by a "C" next to each of the axes (XC, YC, and ZC) and is shown in red, green and blue, respectively for the axes.

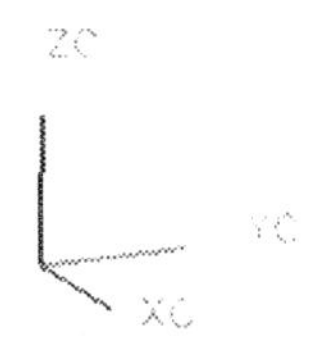

It may be helpful for you to remember the "C" as standing for Construction as this is the primary purpose of the WCS. You can display or hide the WCS by toggling **Format → WCS → Display**.

3.5 Manipulating the WCS

There are three common ways to manipulate the WCS. You can change its location or its **Origin**, you can **Rotate** the system (which does not alter its location but its orientation only), or you can **Orient** the WCS, which results in not only a rotational change to the WCS but a location change as well. All of these options may be found under the **Format → WCS** pulldown from the Main Menu as shown below.

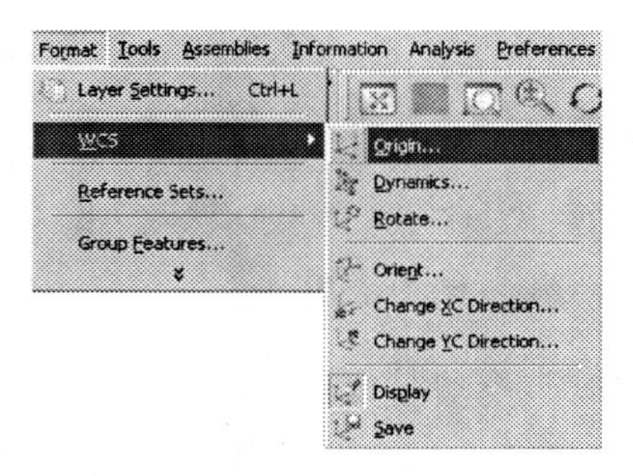

Format → WCS → Origin...

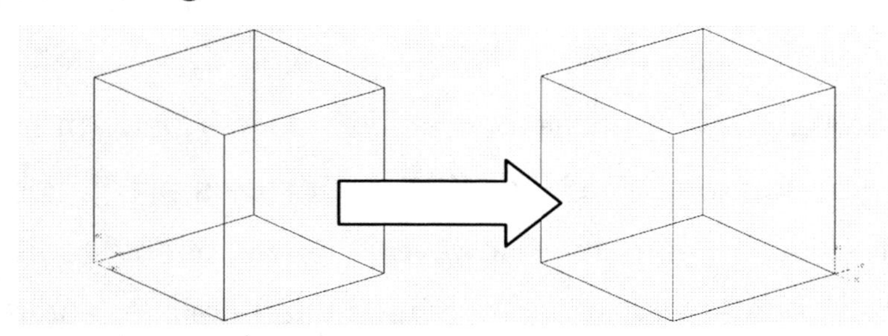

In the above illustration, the WCS simply moves from one corner to the other. In order to specify the new origin, you have to pick a point in the desired location using the Point Constructor Dialog (discussed in Section 3.5). Notice that the general orientation of the WCS remained the same and that only the location was changed.

Format → WCS→Rotate ...

This menu selection opens the below dialog. If you click OK or Apply with the current setting as shown in the dialog, the WCS rotates about the positive ZC axis by 90 degrees, turning XC toward YC. The difference between OK and Apply is that OK closes the dialog after execution while Apply leaves it still open.

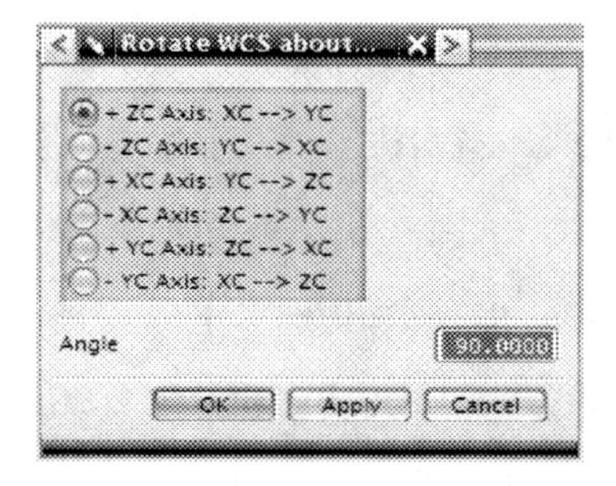

The below figure shows the WCS before and after executing the above rotation.

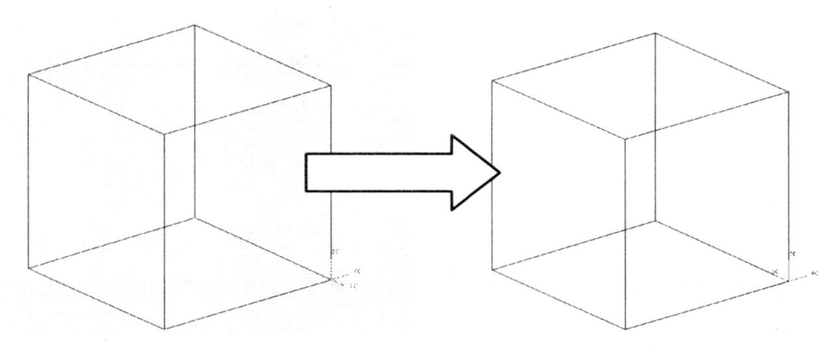

Format → WCS→Orient ...

This will allow you to change the location and the orientation of the WCS simultaneously. There are many options to choose from as shown below.

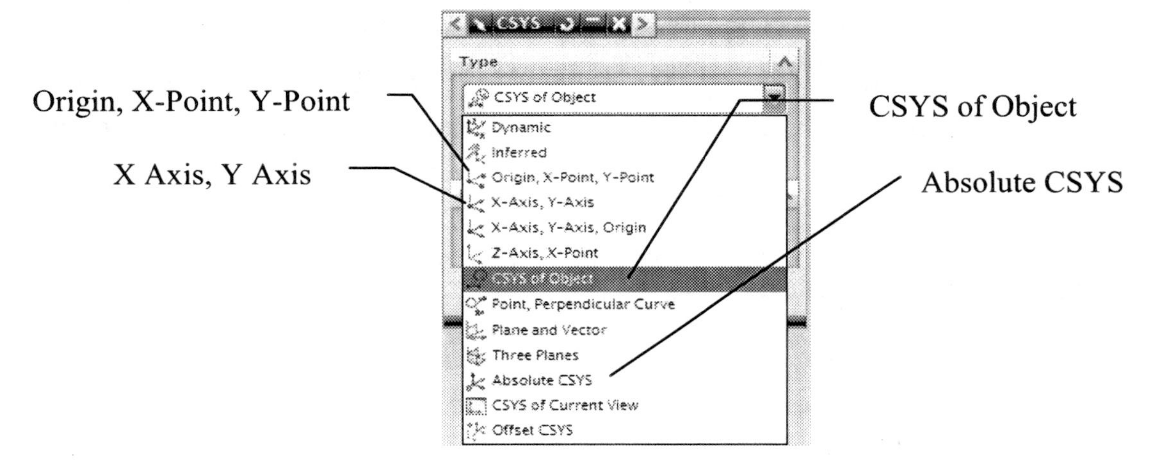

The icons represent various methods to select objects to locate the WCS aligned with or coincident with those objects. The options most used are the icons labeled above. Notice also the option for Absolute CSYS. This is the one way that you will be able to visualize the Absolute Coordinate System by moving the WCS to its location. Activity 3-1 shows how to use some of these options.

3.6 The Feature Coordinate System

Often times during the creation of various features (such as holes and pockets that will be discussed in Chapter 5) within Unigraphics, you will notice that the system temporarily moves the WCS for you. After the feature has been created, you will then notice that the WCS goes back to its previous location. What you are actually witnessing when this happens is the system created Feature Coordinate System. This coordinate system is stored with the feature and is called upon by the software mainly when the feature is being edited by the user.

3.7 Defining Points for Reference and Location in Unigraphics

Many times you will be asked by the system to provide point information so that you may locate certain items of interest. For example, when changing the origin of the WCS, you will have to tell the system where you would like to move the WCS origin on the model or in the model space. In another example, if you are constructing a block, you will need to tell the system where in the model space the lower left corner of the block, called the block origin, is. The way this is accomplished in Unigraphics is with the Point Constructor dialog. We will go over key issues frequently used. Refer to the UG help documentation for detail on other options.

The Point Constructor Dialog

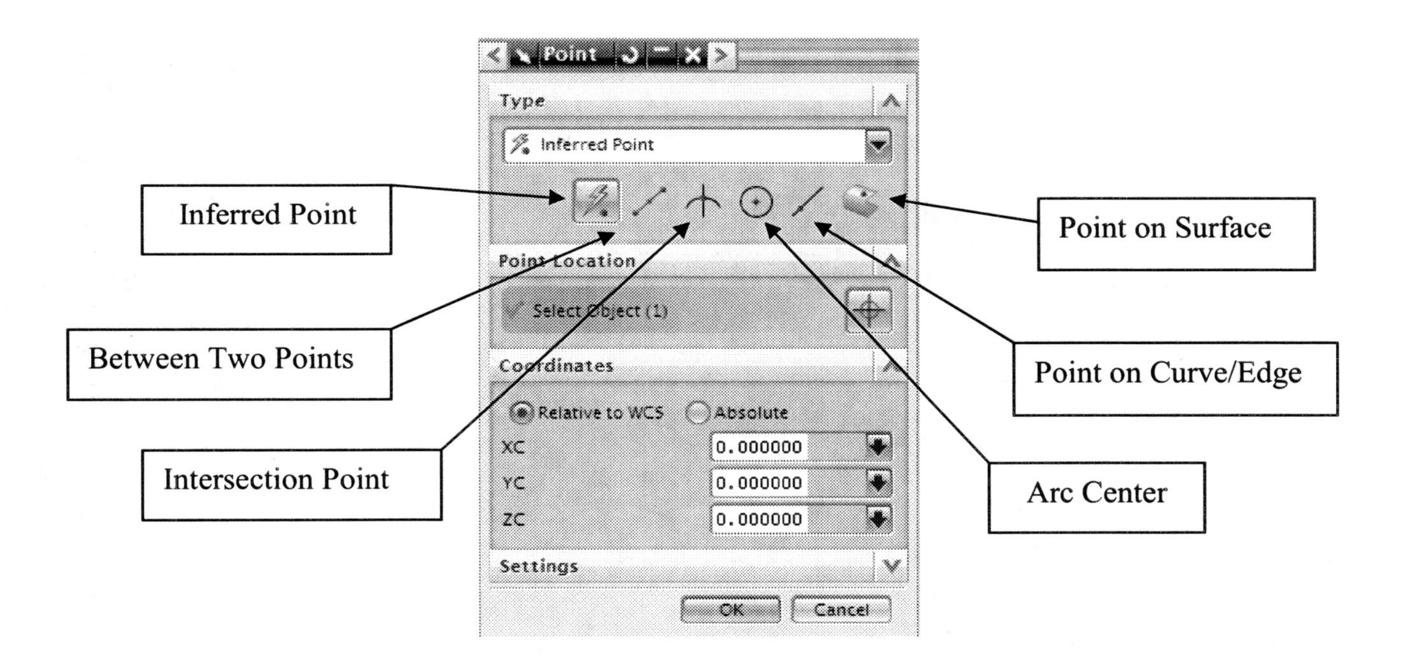

The dialog shown above is the Point Constructor Dialog and in this case was the result of a desire to change the location of the WCS origin. From this dialog, you are able to specify points relative to model space or even relative to existing objects for many varied purposes. In this instance, the Inferred button has been selected (which it is by default) as its icon is highlighted. The Inferred selection button has been put in place to facilitate selection by having the system attempt to interpret the user's selection. For example when you pick a *near* end point of a line in the inferred selection mode, the system selects its end point near your pick. When you pick an arc, the system selects its arc center.

The other options in the dialog, with the exception of the Cursor Location, require the selection of existing modeling objects from the graphics screen. Below is a sample of some of the points that are selectable from the screen.

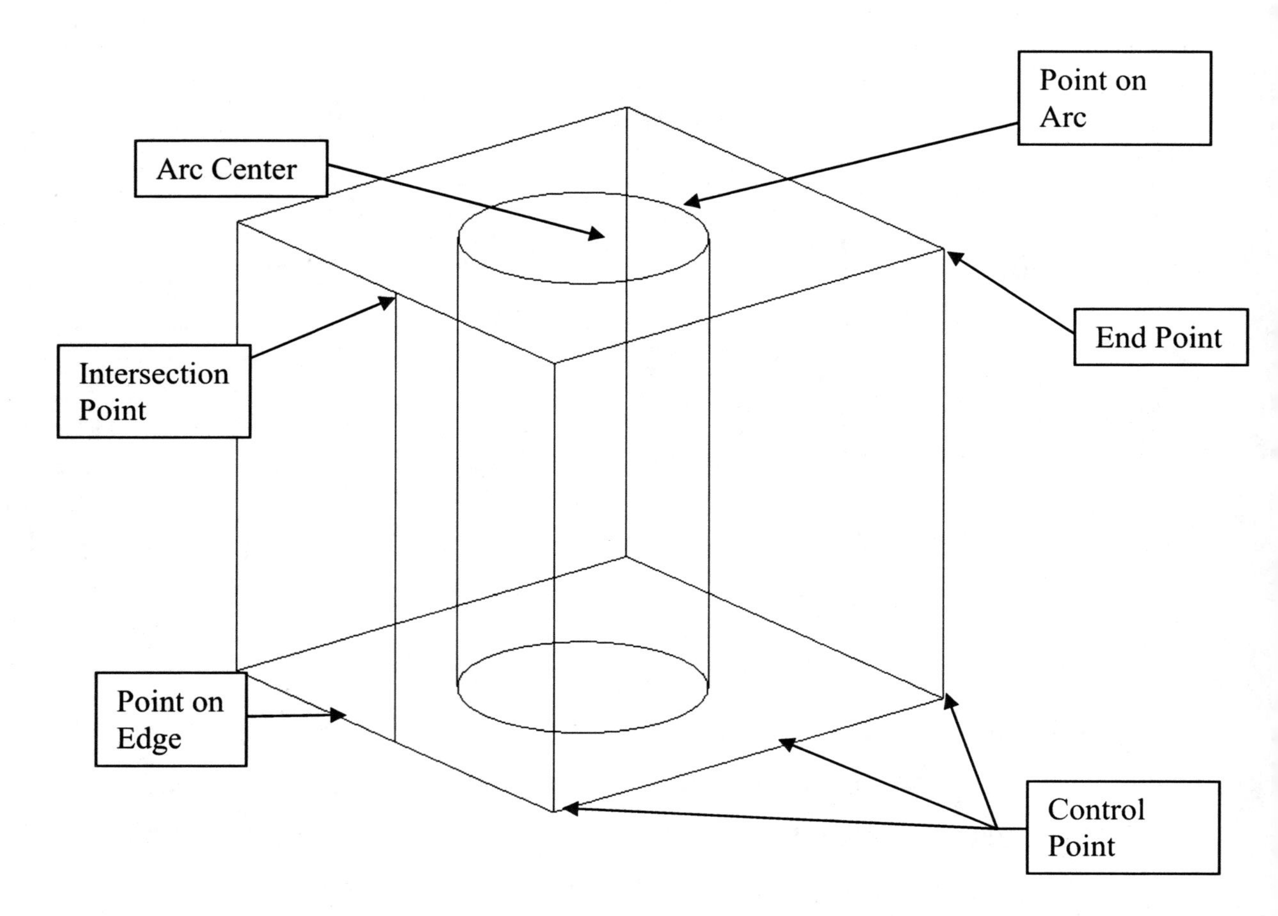

At the bottom of the dialog, you can specify if a point selected is relative to the WCS or ABS depending on which toggle you select (the WCS is selected by default).

Snap Point Tool Bar

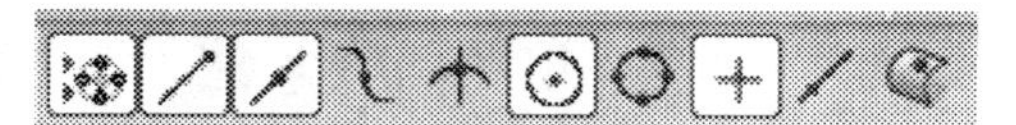

There are occasions when you want to select a certain point such as arc center or end point but the point constructor dialog does not appear. In this case, you can look for the Snap Point tool bar (usually located near the bottom area of the graphics window) as shown above. This tool bar usually appears above the cue line when needed.

In the activity that follows, you will practice manipulating the WCS and selecting different points using the point constructor dialog. Try exploring different options and ways of doing them.

Activity 3-1. Working with the WCS

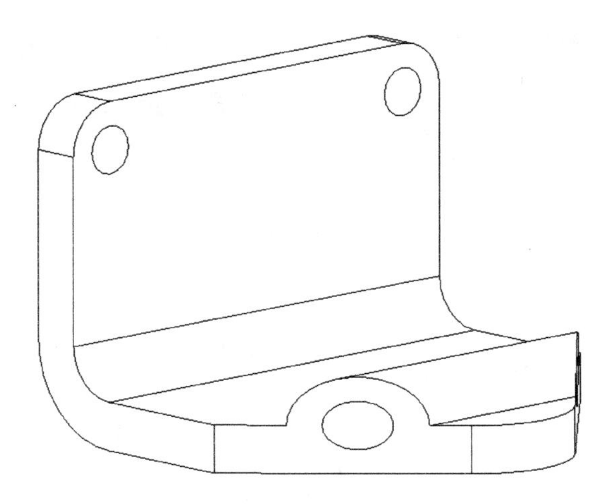

Step 1. Open part file **activity3-1.prt** by selecting **File → Open** (or Ctrl + O).

Step 2. Select **Format → WCS → Origin** and notice the **Point Constructor** dialog has been opened with **Inferred** point selection as the default.

Step 3. Select the points shown below. Notice the movement of the WCS. Does it change orientation? Should it be?

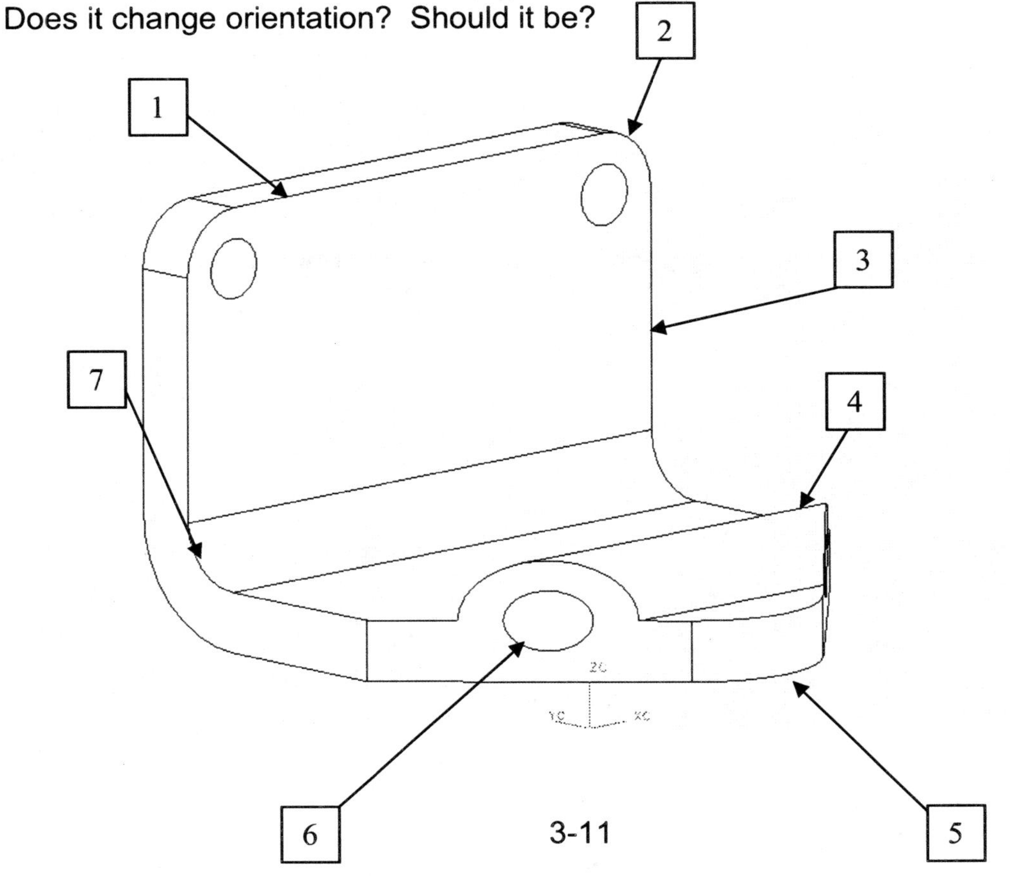

Step 4. Select **Cancel**.

Step 5. Select **Format → WCS → Orient** and the icon for X Axis, Y Axis and select the lines as shown below to re-orient the WCS. Be careful to select the lines where shown!

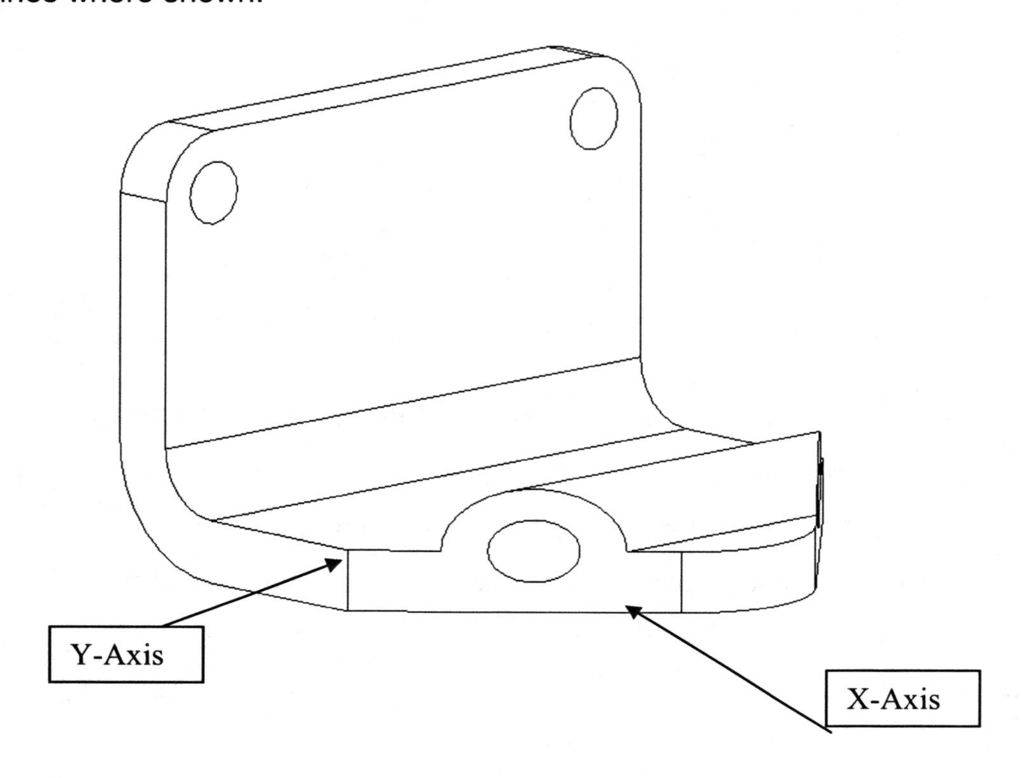

Step 6. Now, do the same thing again for the locations shown below.

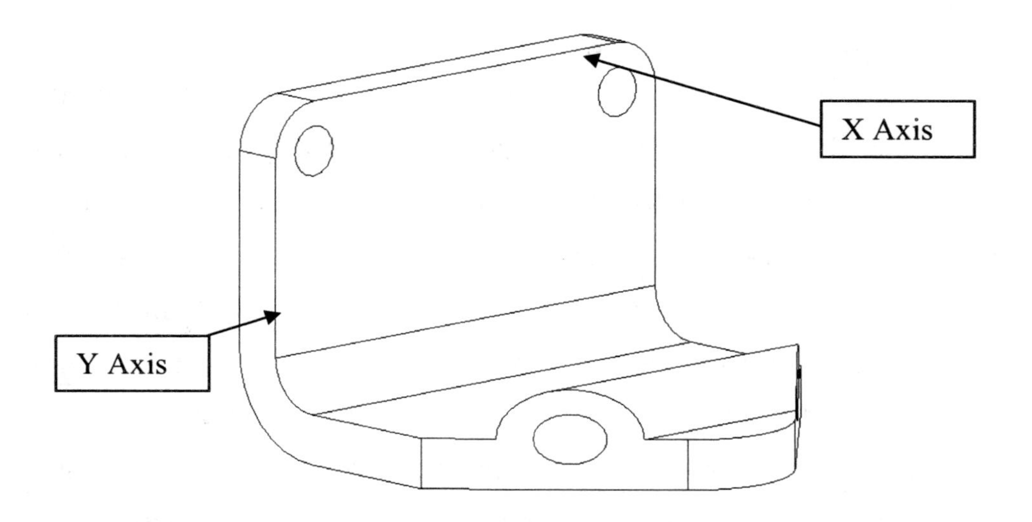

Step 7. Rotate the WCS by selecting Format → **WCS → Rotate … → +ZC Axis: XC → YC**.

Step 8. Select **Apply**.

Step 9. Select **Cancel**. Your screen should look as shown below.

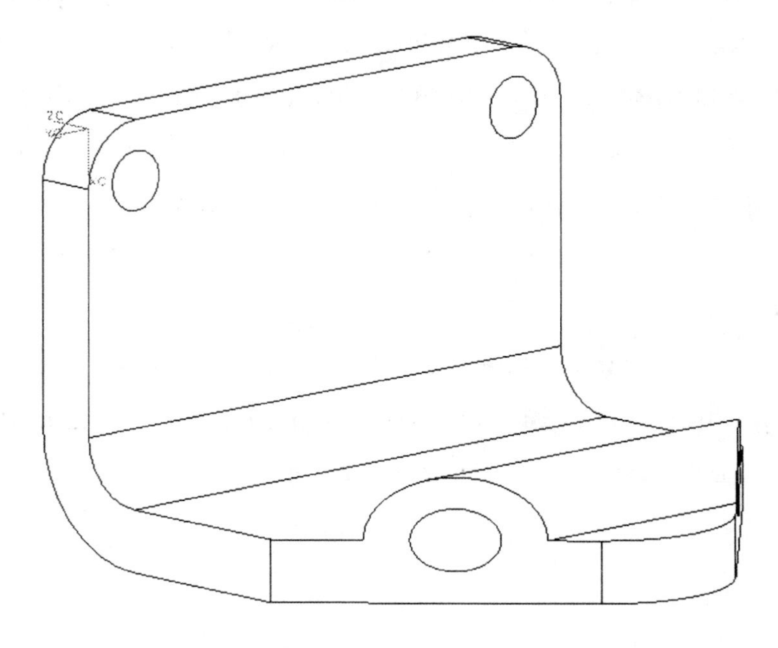

3.8 Controlling the Visibility of Objects

As you will see, there can be many reference data created throughout your use of Unigraphics when you build a model. Like many CAD systems, Unigraphics allows the users to organize their modeling file by using what are known as layers. You can think of a layer as a single piece of transparent film that has information on it. If you view only a single piece of film, then you will only see what is on that piece. In order to view other pieces of film at the same time, you would then simply lay the others overtop of your original. Also, you may desire to create new objects and as a matter of convention, they must reside on some piece of film (maybe call it your Work Film). Some of the pieces of film you will want to only see (not write on, visible only) and some of the pieces you will actually want to work with (make selectable).

Unigraphics' Layers work in much the same way. There are a total of 256 layers (layers of film) at your disposal in every single part file. At any one given time, you can only create objects on a single layer, the Work Layer. You also have the flexibility to make any or all of them **Selectable** (so that you can pick and work with the objects on a layer) or **Visible** (so that you can see, but not pick nor modify the items on that layer). Now that you have the background information, let us look at some of the mechanics of Layers from a User Interface standpoint.

To get to the **Layer Settings Dialog**, go to **Format** → **Layer Settings** to bring up the following dialog. The following bulleted items are intended to be a straightforward reference to help you in using the dialog.

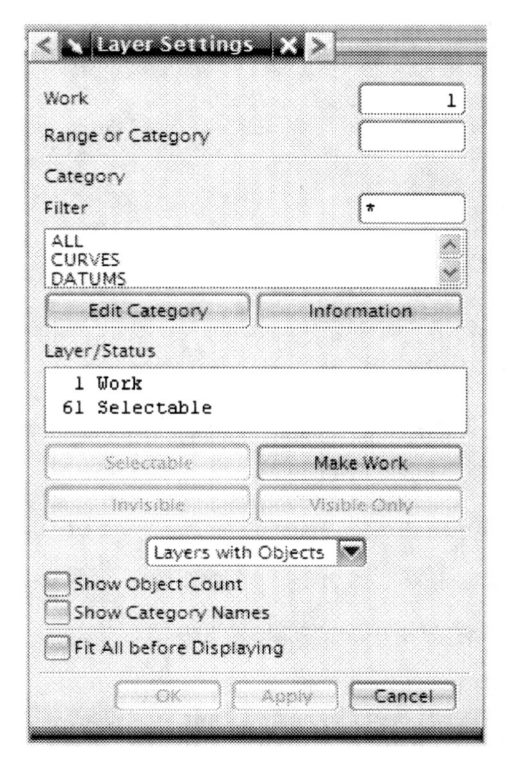

- At the very top of the dialog, you may directly change the Work Layer by entering in a layer number (between 1 and 256) followed by the enter key on your keyboard. Note that you must press enter for this to work correctly.
- To make a layer **Selectable**, you can
  - pick the layer number from the listing window in the middle of the dialog and click the **Selectable** button or
  - double-click the layer number directly on the list.
- To make multiple layers **Selectable**, you can drag over the layers and then click the **Selectable** button.
- For any of your changes to take effect, you must select **OK** or **Apply**.

Many companies have rigid standards in place when it comes to layering. These standards are typically referred to as layering standards. These standards basically tell the modeler what objects should be placed on what layer. Implementation of layering standards is a key way that companies achieve consistency in the models they produce. Adherence to these standards helps companies to communicate and interrogate the model definition. We give some of the layering standards commonly used by companies like Boeing that will be also adopted in this book as follows:

Layers 1 to 10 for Solid bodies
Layers 11 to 20 for Sheet bodies
Layers 21 to 40 for Sketch objects
Layers 41 to 60 for Curve objects
Layers 61 to 80 for Reference Datum objects

One way that companies implement these standards is through the use of a seed part file (sometimes referred to as a null file). These empty seed part files have the layers and corporate standards prearranged to make the adherence to the standard easier. The basic process is that when a user begins a new model, he or she opens the appropriate seed part and performs a **Save As** operation to save it with the correct file name for the model. With this practice, the new part file has the standards built in. While this does not guarantee compliance with the standards, it does facilitate the process.

3.9 Selecting Objects Using Quick Pick Selection

Beginning in V16, the software has employed a quick pick method for selection. This method makes the selection of an object an easier task with usage of the mouse buttons or the cursor. In addition to this, the graphical feedback helps you to determine the current selection.

When first moving your cursor to an object you want to select, your cursor will begin looking like a normal crosshair. However, if you let the cursor dwell over the current location for a few seconds without clicking any mouse button, the cursor may change to look like . Notice the three periods or small rectangles after the crosshair cursor. If this is the case, UG signals you that there is more than one object "behind" the current cursor location that is available for selection.

Upon selecting **MB1** at that point, you will be presented with another dialog box (generally to the right of where you selected) filled with numbers, one for each potentially selectable object. Also providing graphical feedback is the **Cue Line** and **Status Line** which give you hints on how to accelerate the process as well as what type of object the current selection represents. A sample example is illustrated below.

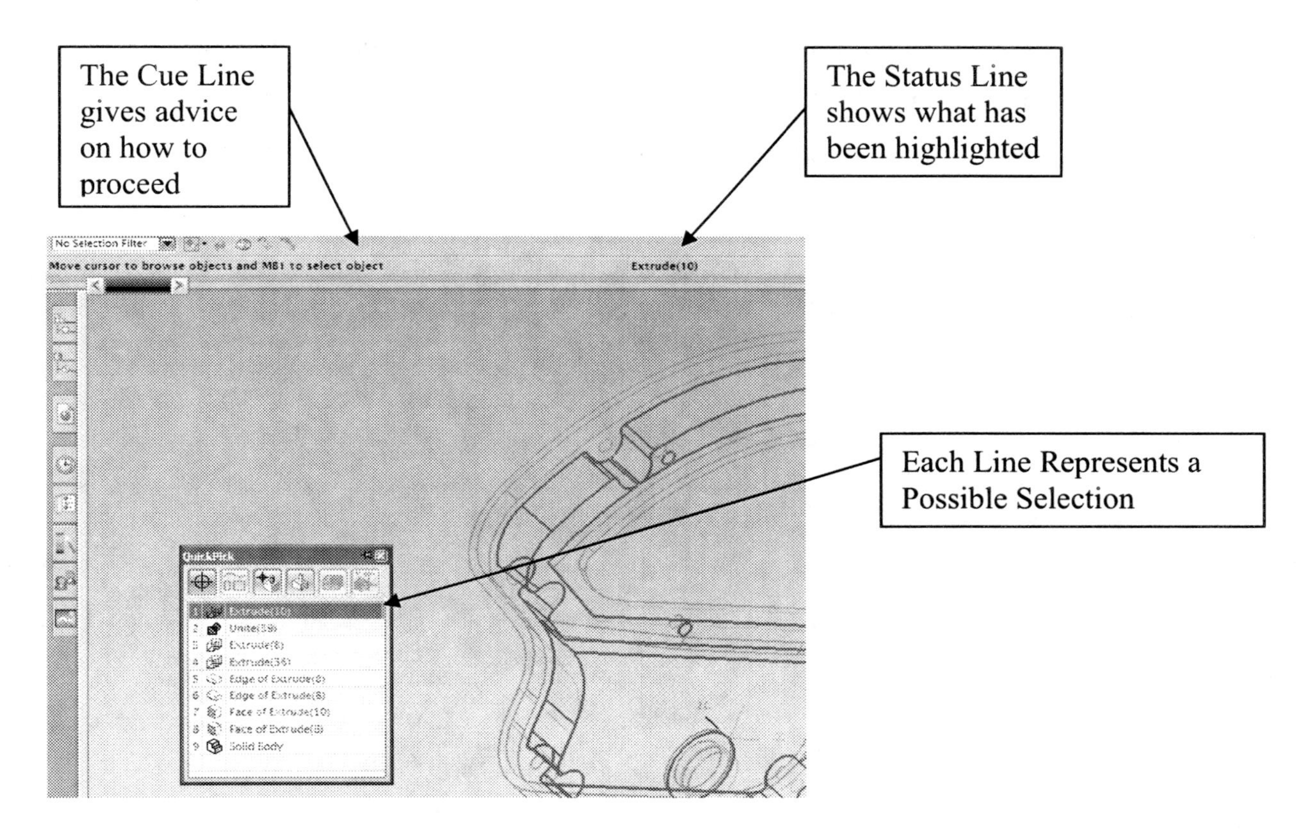

At this point, you have two options available to you as bulleted below:

- Move your cursor along the numbers while watching the screen and **Status Line** until the system has found your desired selection and pick the number with **MB1** (or press enter on your keyboard) to select the object.
- Use the right arrow key to run along the numbered selections and pick the resulting number with **MB1** (again, you may also simply press enter on the keyboard) to select the object.

3.10 Utilizing the Class Selection Dialog

There are many situations where you will encounter the **Class Selection** dialog. This dialog is little more than a global filter that allows you to decide what will be selectable and what is not. The dialog is shown below and is a result of following the selection, **Information → Object**. The small dialog comes up and objects may be selected directly. However, if help is needed then select the **Class Selection** dialog from this menu.

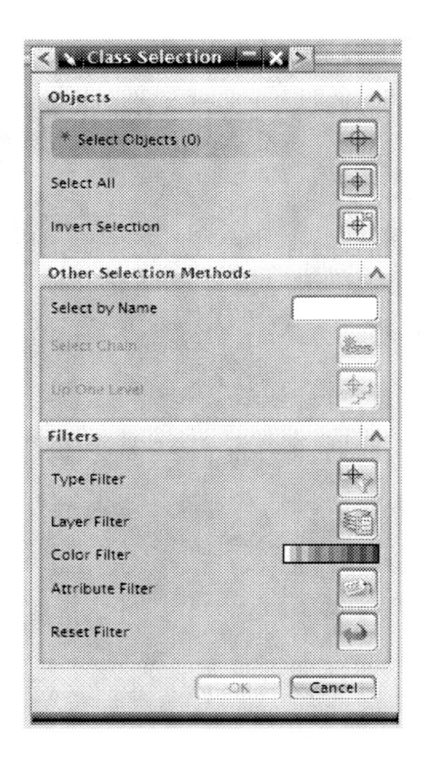

This dialog offers more flexible selection than a global selection filter. By default, all objects are available for selection when first invoked. From this point, you are able to determine what can be selected. You may choose from the types of objects, the colors of objects, the layer in which the objects are on, or various other items (line font for example) by selecting the Type, Layer, Color, or Attribute buttons, respectively. For example, if you want to select only the White, Solid Bodies on Layer 12, you would follow the selection process as depicted below. The selection process of this example may look somewhat involved. However, you will only employ such a selection process when you NEED that level of control (when there is a lot of object clutter in the graphics screen perhaps). Remember that, by default, all these items are available for selection, and, with the help of **Quickpick**, in most cases you will be able to pick them directly from the screen.

Step 1. Choose the **Type** button meaning to select objects by type. The **Select by Type** dialog appears. Select the Solid Body selection and choose **OK**.

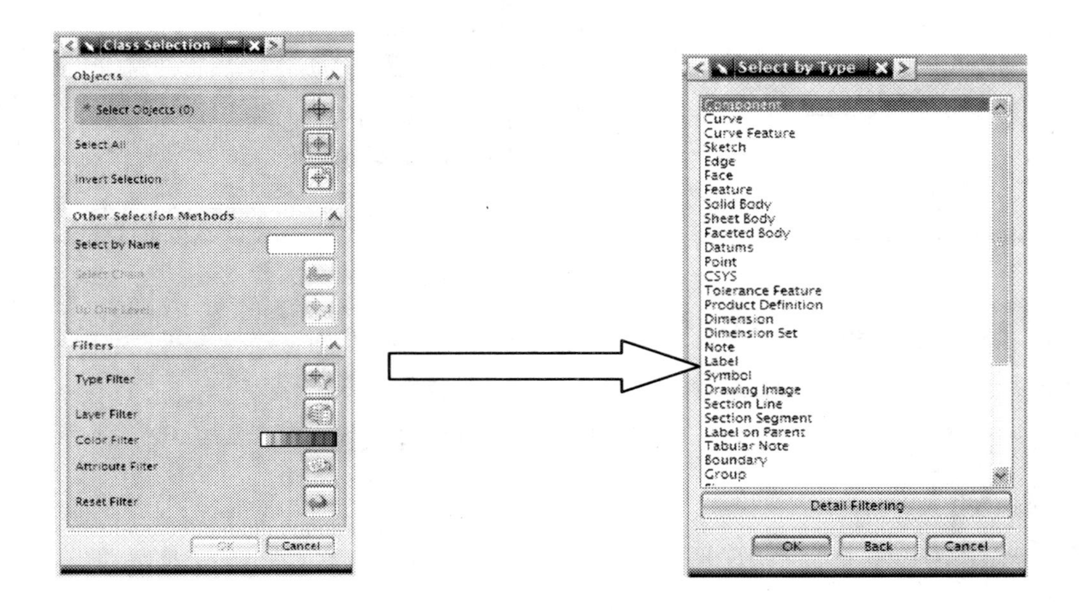

Step 2. Choose the color window and the Color map dialog comes up. Choose the color White in the color pallet and choose **OK**.

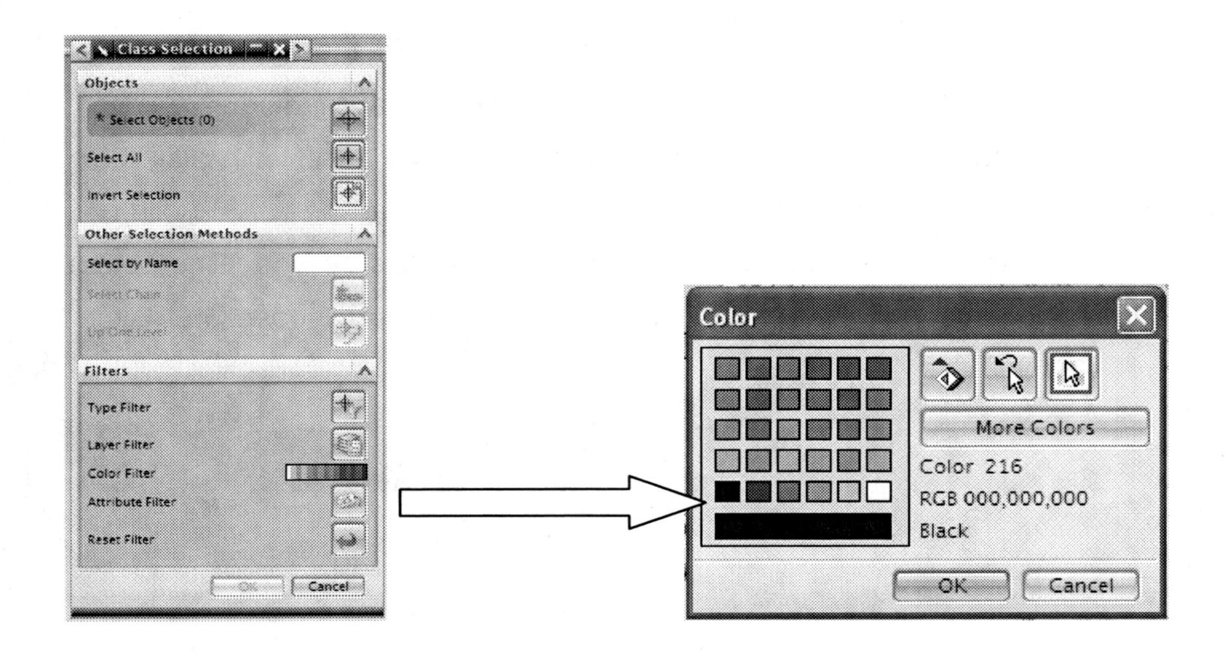

Step 3. Choose the **Layer** button and the Select by Layer dialog comes up. Select the layer number you want to use and choose **OK**.

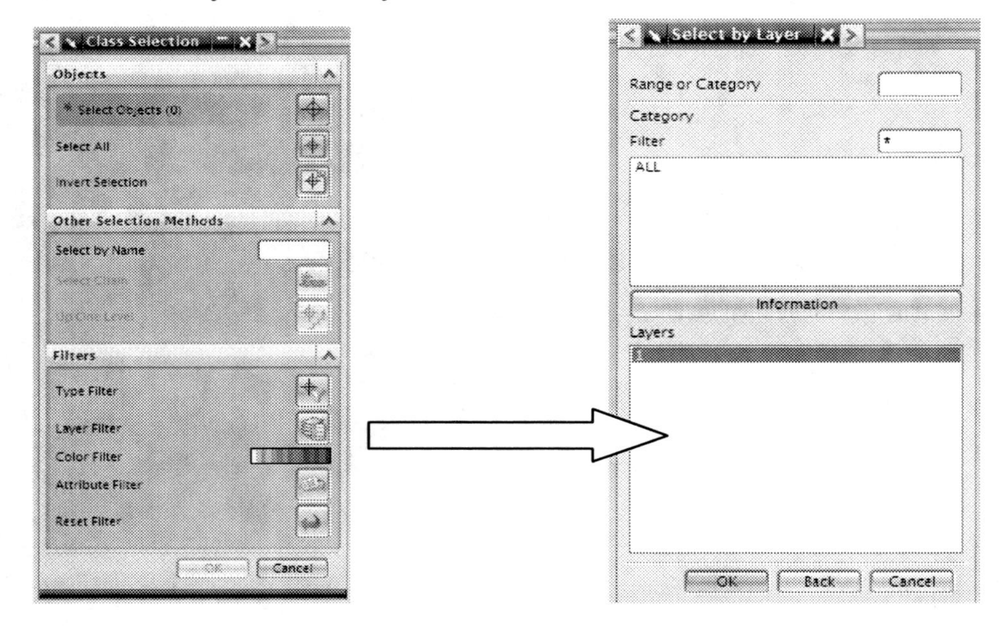

Step 4. Now the filters have been set to select only a solid body, of white color and on layer x, we are now ready to select.

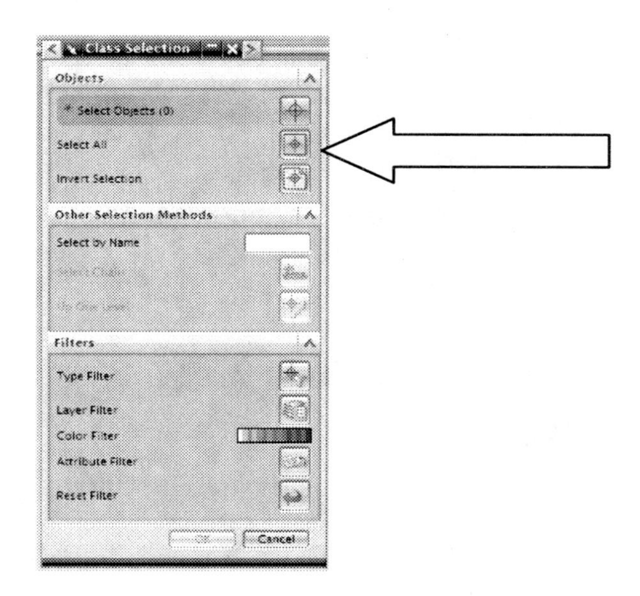

Step 5. Choose the **Select All** button and all objects fitting the criteria of Step 4 will be chosen. Choose **OK** to complete the selection process.

Drag Rectangle Selection

It is often necessary to select many objects at once. To select a group of objects using a rectangle, press and hold **MB1** and drag the cursor. When the rectangle expands to enclose the desired objects, release the mouse button to complete the selection.

The rectangle (and polygon for that matter) has five selection methods:

- "Inside" only selects items within the rectangle.
- "Inside/Crossing" selects items within the rectangle or crossing its boundaries.
- "Outside" only selects items outside of the rectangle.
- "Outside/Crossing" selects items outside the rectangle or crossing its boundaries.
- "Crossing" selects only items crossing the rectangle boundary.

To set the default rectangle selection method, use **Preferences → Selection → Rectangle Method**. You can set the method on a case by case basis using the Class Selection Tool's Rectangle/Polygon Method option. Note that if you specify a method from the Class Selection dialog, that method will be retained by the system until you change it again in the Class Selection dialog, or reset it in **Preferences → Selection**.

Chaining

This utility is used to select *connected* curves or edges. Chain is available as a button on the Class Selection dialog. To chain curves that form a closed boundary or an open section of closed curves, select the Chain button then select a starting curve followed by the ending curve. Take care in selecting the end of the starting curve that is pointing towards the ending curve as this dictates the chain direction. Next you can pick a curve that ends the chain. Alternatively, in either case of a closed boundary or an open boundary you can select only the starting curve of the chain and select End Chain on the dialog. This will chain all the curves that are connected from the starting curve toward the chain direction.

Polygon Method

The polygon method is used in situations where a rectangle is not feasible. To use this option, choose the **Polygon** button followed by screen picks that form a polygon. Once you have selected the screen picks and are satisfied, select **OK** to close the polygon off and have the system select the objects for you.

3.11 User Interface Shortcuts

In this section, we introduce some of the shortcuts that exist in the software to improve your productivity. Moving a mouse around a computer screen can take time, causing loss of productivity for repetitive operations. To the right of most entries in the pulldown from the main menu, there are key combinations listed which are known as hot keys. When pressed in the correct combination on the keyboard, these hot keys simply activate the given menu item. For example, to create a new file, you may use either the method of moving your mouse to File, then find New and select it from the pulldown menu or you may simply press Ctrl+N to accomplish the same task. Below is an abbreviated table of some of the more commonly used hot keys for your reference.

Hot Key Combination	Action Performed
Ctrl + N	File→New
Ctrl + D	Edit→Delete
Ctrl + B	Edit→Show and Hide→Hide
Ctrl + S	File→Save
Ctrl + Shift + A	File→Save as
Ctrl + A	Select All
Ctrl + T	Edit→Transform
F4	Hide/Retrieve Information Window
Ctrl + J	Edit→Object Display

While the above is not a complete list, it does represent some of the more common items.

3.12 Default Files in Unigraphics

Unigraphics offers various options and features, and has many defaults set up for users. These default settings range from default values of certain variables to user interface settings all the way to where certain files are located. Typically, companies assign system administrators to the task of configuring these default settings to comply with the company standards.

Two files, ug_english.def (for English units) and ug_metric.def (for metric units), that are usually located under C:\Program Files\UGS\NX\UGII for UG NX (C:\UGS180\ugii for V.18) are such default files that determine the preset values or interface used in the software. These are text files which users are able to edit in order to change default values. However, these defaults files are no longer used from NX3.0 on. The customer defaults mechanism has changed to use an interactive dialog instead to allow customizations of various default settings. The new dialog is found at **File → Utilities → Customer Defaults**.

The ugii_env.dat (.ugii_env on Unix) file mainly sets the system environment variables when you start a session of the software. Many companies differ in where certain files or directories are on the system. This file also tells the system which default file to use between the two ug_english.def and ug_metric.def.

3.13 Obtaining Help

You may have questions on Unigraphics that this book does not address. A very good source for assistance with your questions is contained within the online help that should be installed with your software. To activate the online help, simply use your mouse and point to **Help → On Context or Help → Documentation**. Once in the help, you have the ability to search on topics or simply browse through several aspects of the software.

3.14 Unit System and Unit Setting

When you create a new file by **File → New**, you can specify the unit system either USCS (United States Customary System) or SI (System International) by choosing either Inches or Millimeters as shown below.

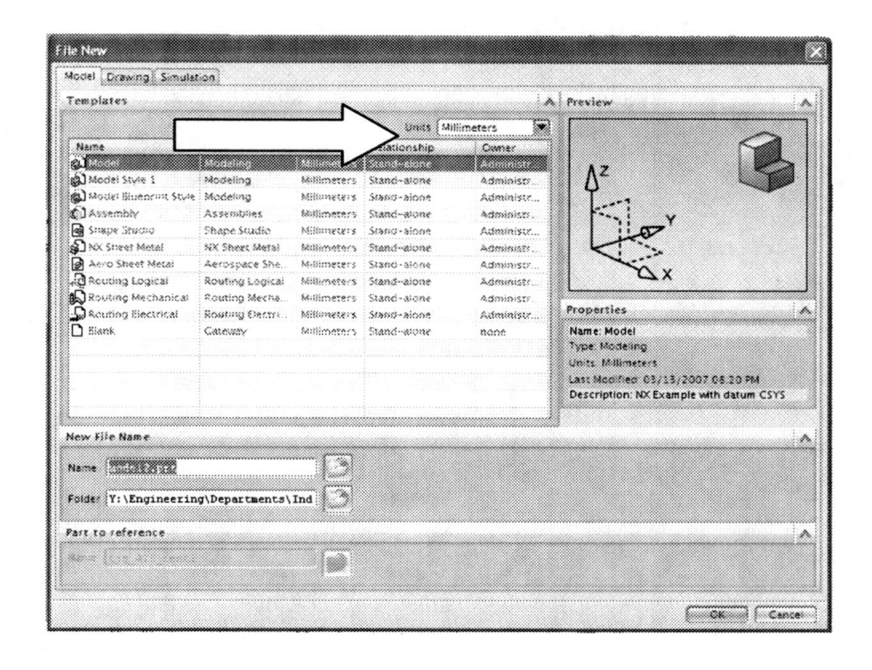

Your choice of the unit system will determine the default unit system when you enter parameter values for length, force, or such. If you want to enter these values in the other unit system or a different unit within the same unit system (for example, from millimeters to meters), you can do so by simply choosing **Analysis → Units** and then selecting the unit of your choice as shown below.

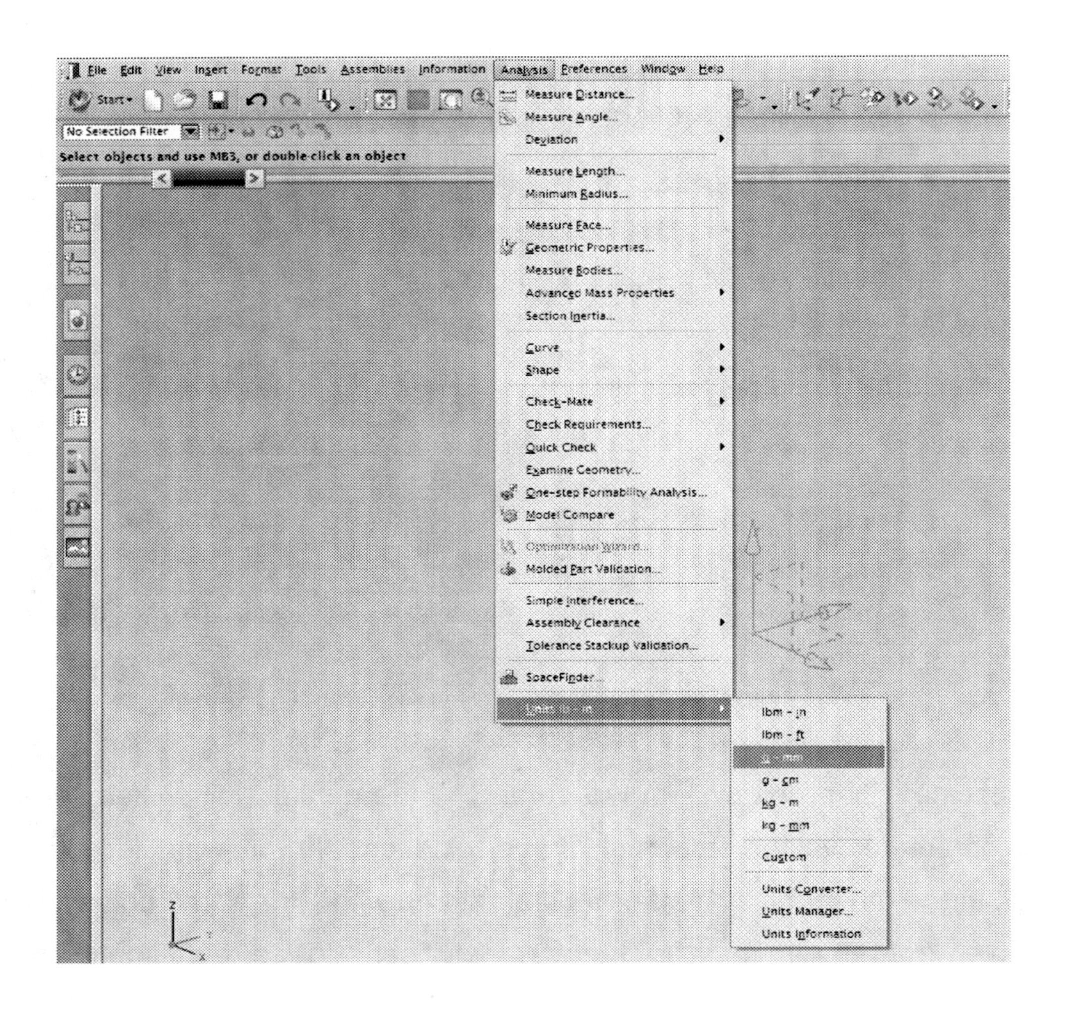
File Edit View Insert Format Tools Assemblies Information Analysis Preferences Window Help
Start
No Selection Filter
Select objects and use MB3, or double-click an object
Measure Distance...
Measure Angle...
Deviation
Measure Length...
Minimum Radius...
Measure Face...
Geometric Properties...
Measure Bodies...
Advanced Mass Properties
Section Inertia...
Curve
Shape
Check-Mate
Check Requirements...
Quick Check
Examine Geometry...
One-step Formability Analysis...
Model Compare
Molded Part Validation...
Simple Interference...
Assembly Clearance
Tolerance Stackup Validation...
SpaceFinder...
lbm - in
lbm - ft
g - cm
kg - m
kg - mm
Custom
Units Converter...
Units Manager...
Units Information

Exercise Problems

3.1. Outline and discuss the two main approaches to modeling (spatially) in Unigraphics and the implications these two methods have on using the software.

3.2. Discuss the three coordinate systems presented in this chapter. Be sure to include their purposes.

3.3. What situations might cause the manipulation of the WCS?

3.4. Who sets up the default files in Unigraphics? Why are they available for modification?

3.5. What is the purpose of Hot Keys?

3.6. What does CAX stand for?

3.7. What is the Point Constructor used for?

3.8. What is the purpose of the Class Selection Dialog?

3.9. Explain how you would select green lines on layer 11 using the Class Selection Dialog.

3.10. Explain the preferred method for confirming a selection.

3.11. How many layers does Unigraphics have?

3.12. Name two options on Layer Settings that can make layers easier to manage.

3.13. What is the difference between a layer being Selectable and Visible?

Chapter 4. Primitives

Primitive features are base features to which other features are added. UG uses five primitives, which take on the following basic analytic shapes: **Blocks, Cylinders, Cones, Spheres, and Tubes.** Primitives are non-associative, meaning they are not associated to geometry used to create them. However, you can edit certain parameters.

Section 4.1 discusses a common procedure to create a primitive. Sections 4.2 to 4.4 describe the first three primitives, Blocks, Cylinders, and Cones respectively that will be used in this book.

4.1 Common Procedure to Create a Primitive

There is a general procedure that can be followed when creating a primitive:

1. Select the type of primitive (block, cylinder, etc.) you want to create.
2. Choose the creation method.
3. Enter the creation parameter values.

After performing step 3 above, if a solid body already exists, you need to specify how the new primitive will be used with the existing solid body. In this case, the following Boolean options are displayed:

Create	To create a new solid body.
Unite	To add the primitive to the current target solid.
Subtract	To subtract the primitive from the current target solid.
Intersect	To intersect the primitive with the current target solid.

For the primitives with curved edges (all except Block), if you do not have silhouette edges displayed, it can be hard to see them. Therefore, it is recommended that you have silhouette edges displayed when creating these features, which is the default. If silhouette edges are not displayed for some reason, you can do so by using **Preferences → Visualization → Visual** and selecting **Edge Display Settings** and putting a check mark for the option. Also, after you have finished performing the Boolean operation, you can use **Edit → Object Display** and select the newly formed solid body to add U-V grid lines to make the shape of the resulting body more obvious.

You cannot position solid primitives using positioning dimensions as you can most other features. Positioning methods will be discussed in Chapter 5. During the creation of the primitive, you supply its location either using the *Point Constructor* that was discussed in Section 3.7, or by selecting geometry. We will give the examples in the subsequent sections.

4.2 Block 

There are three ways that you can create block primitives by selecting proper options as shown in the table below.

Origin, Edge Lengths	Define the length of each edge and a corner point.
Two Points, Height	Define the block height and two diagonal points of the block base.
Two Diagonal Points	Define two 3D diagonal points representing opposite corners of the block.

.

All of these options create the block such that its faces align with the axes of the WCS. Below we discuss each option in detail.

Origin, Edge Lengths

This option creates a block by defining a corner point and the length of each edge.

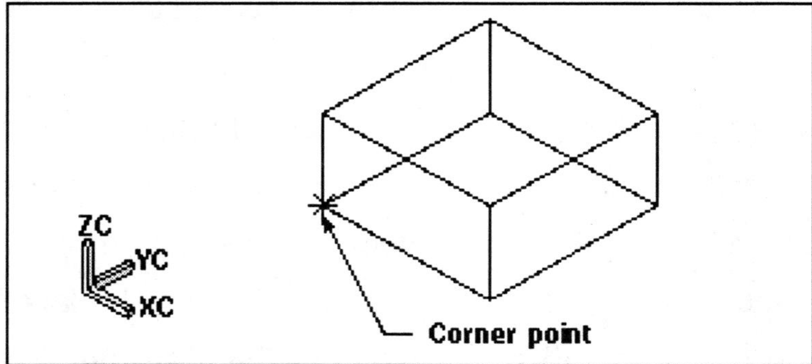

After choosing this method, you must:

1. Define the corner point of the block, which is the origin of the block.
2. Enter the XC, YC, ZC edge length values which must be positive.
   Use the Tab key to move from window to window.
3. Choose OK to accept the choices and create the block.

The system creates the block using the input values, starting from the specified point. The edges of the block are parallel to the axes of the WCS.

A special case for the block exists such that if you want the origin to be at 0, 0, 0 you do not need to select it—just enter the edge lengths and choose OK.

Two Points, Height

This option creates a block by defining the block height and two diagonal points of the block base.

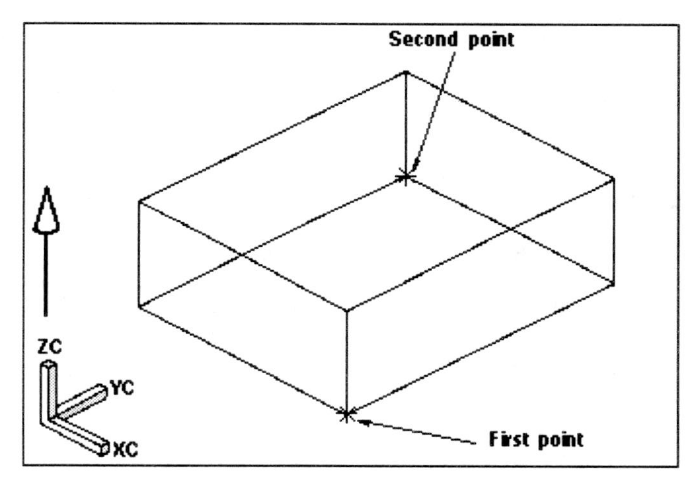

In order to use this method, in the Block dialog (**Insert → Design Feature → Block** or select Block icon), choose the middle icon from the top row Type options as shown highlighted below.

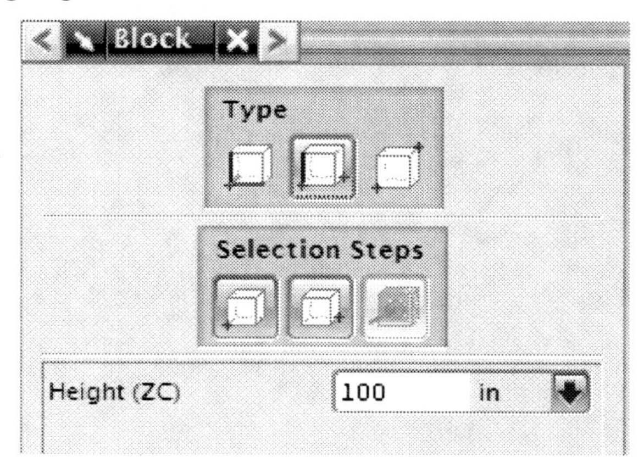

And then you must:

1. Enter the positive block height.
2. Define the diagonal points of the block base.

The value entered causes the system to create the block of that height in the direction of the +ZC axis. Define the opposite corners of the block base by specifying two diagonal points. To define the points you may have change the setting of the Point Method menu or utilize the Point Constructor dialog. The first corner point also determines the plane of the block base. This plane is parallel to the XC-YC plane of the WCS. The second point defines the opposite corner of the block base. If you specify the second point on a different plane (different Z values) than the first, the system projects the point normal to the plane of the first point, to define the opposite corner. The edges of the block are parallel to the axes of the WCS.

Two Diagonal Points

This option creates a block by defining two 3D diagonal points representing opposite corners of the block. And you may need to change the Point Method menu and utilize the Point Constructor dialog.

4.3 Cylinder 

You can create cylinder primitives by specifying an orientation, size and location, using the following options:

Axis, Diameter, Height	Axis direction, Define diameter and height values.
Arc and Height	Creates a cylinder by selecting an arc and entering a height value.

Diameter, Height

This option creates a solid body cylinder by defining diameter and height values.

To create a cylinder using this method you must:

1. Define the cylinder direction vector.
2. Enter the diameter and height values.
3. Define the cylinder origin.

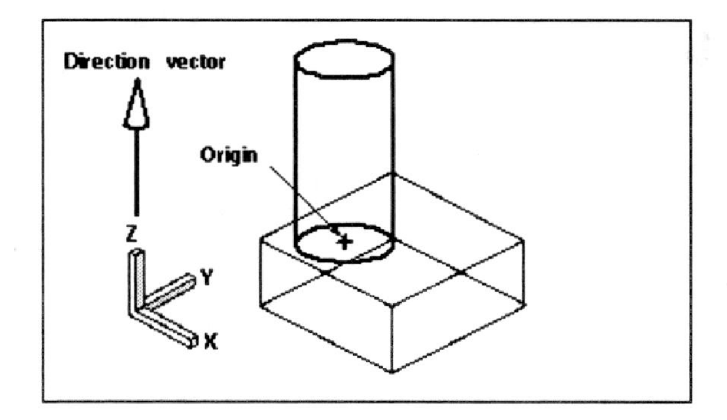

Define the cylinder direction using the Specify Vector pull down menu. The direction you specify also determines the cylinder orientation. The cylinder is created along the axis with one circular face passing through the axis origin and the other circular face at a distance of the input height value.

Height, Arc

This option creates a cylinder by selecting an arc and entering a height value.

To create a cylinder using this method you must:

1. Enter the height value.

2. Select an arc.
3. Confirm the cylinder axis direction. You can reverse the direction.

The system derives the orientation of the cylinder from the arc you select. The axis of the cylinder is normal to the plane of the arc and passes through the arc center. A vector indicates this orientation. The arc you select does not have to be a complete circle. The system creates a complete cylinder based on any arc objects. Note that the cylinder is not associative with the arc. You may create the cylinder along the displayed axis, or reverse the creation direction. The system creates a complete cylinder of the specified height and direction initiating from the selected arc. The figure below illustrates the results of the height value and direction reversals.

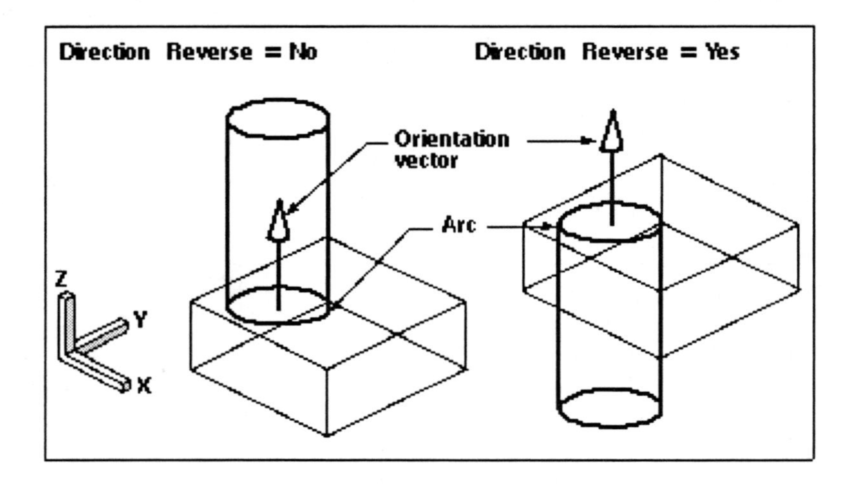

4.4 Cone  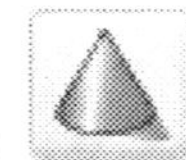

You can create cone primitives by specifying an orientation, size and location, using the following options:

Diameters, Height	Define the base diameter, top diameter, and height values.
Diameters, Half Angle	Define the base diameter, top diameter, and half angle values.
Base Diameter, Height, Half Angle	Define the base diameter, height, and half vertex angle values.
Top Diameter, Height, Half Angle	Define the top diameter, height, and half vertex angle values.
Two Coaxial Arcs	Select two arcs, which do not need to be parallel.

For these options, the cone direction (its central axis) is defined using the pull down menu. The origin of the base of the cone is defined using the Point dialog and the pull down menu. We will discuss below only the most commonly used option, which is, the first option in detail.

Diameters, Height

To create a solid body cone using this method, you must

1. Define the cone direction.
2. Enter the base and top diameters and the cone height.
3. Define the origin.

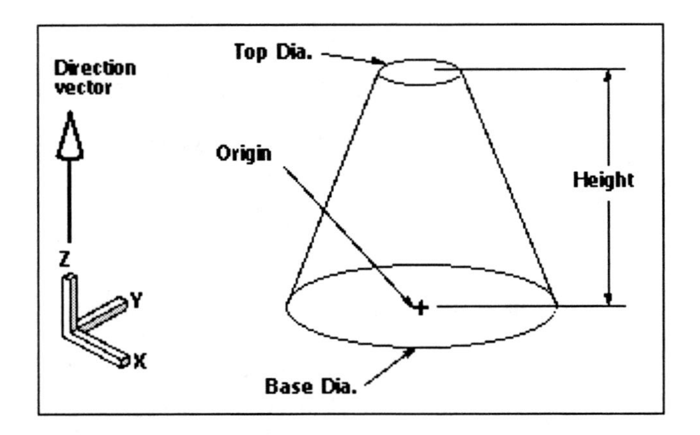

The base diameter is defined as the starting plane and origin for the cone. The top diameter is merely the diameter the height distance away. One diameter may be larger than the other—there is no rule here. To obtain a pointed cone, the top diameter must be defined as 0. The base diameter must be a positive integer.

Activity 4-1. Creating a Model with Two Blocks

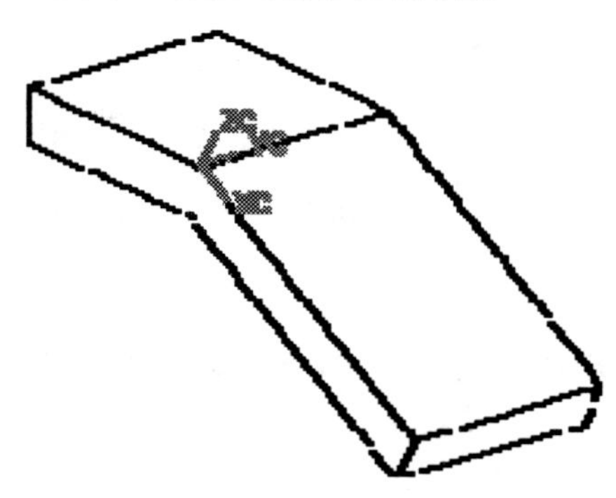

Step 1. Create a new part file **(unit: inches)** by choosing **File → New** and name the file **xxx_activity4-1.prt**. Start the Modeling Application by **Start → Modeling**.

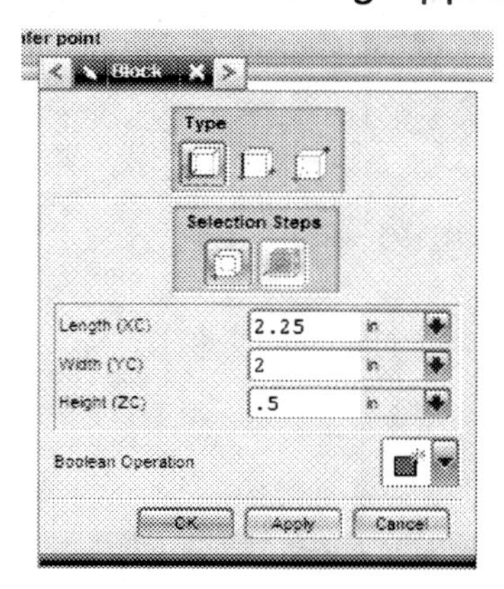

Step 2. The first block is created with the first option, edge lengths (2.25, 2, 0.5) and a corner (if you need help for this step, refer to Section 2.1 steps 1 for detail). Note that these edge lengths are in inches. If you created a part in the mm unit by mistake, you could change the unit system as described in Section 3.14.

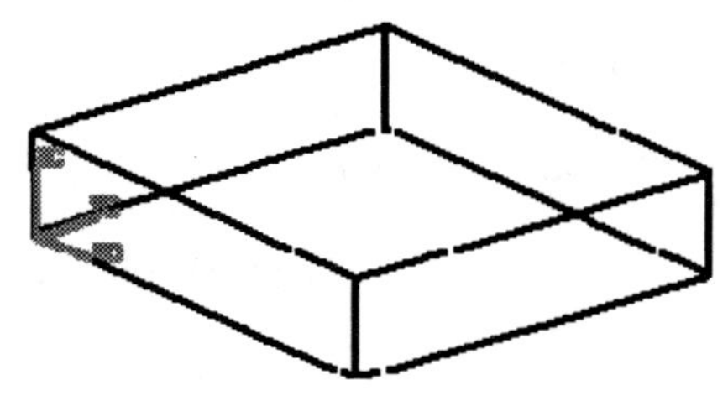

Step 3. Move the **WCS**

You will place the second block at the end of the first block, and angled from it 30 degrees. One way to do this is to change the position of the **WCS**, and then use that orientation to create the second block. First, you will move the WCS to a different location.

3.1. Choose **Format → WCS → Display.**

This allows you to see the **WCS**.

3.2. Choose **Format → WCS → Origin**.

You can indicate a new origin. The **Point** dialog displays to assist you with methods to select a point, or you can simply select a point in the view as seen in the Cue Line.

3.3. Select this corner (end point of an edge) and select **OK**.

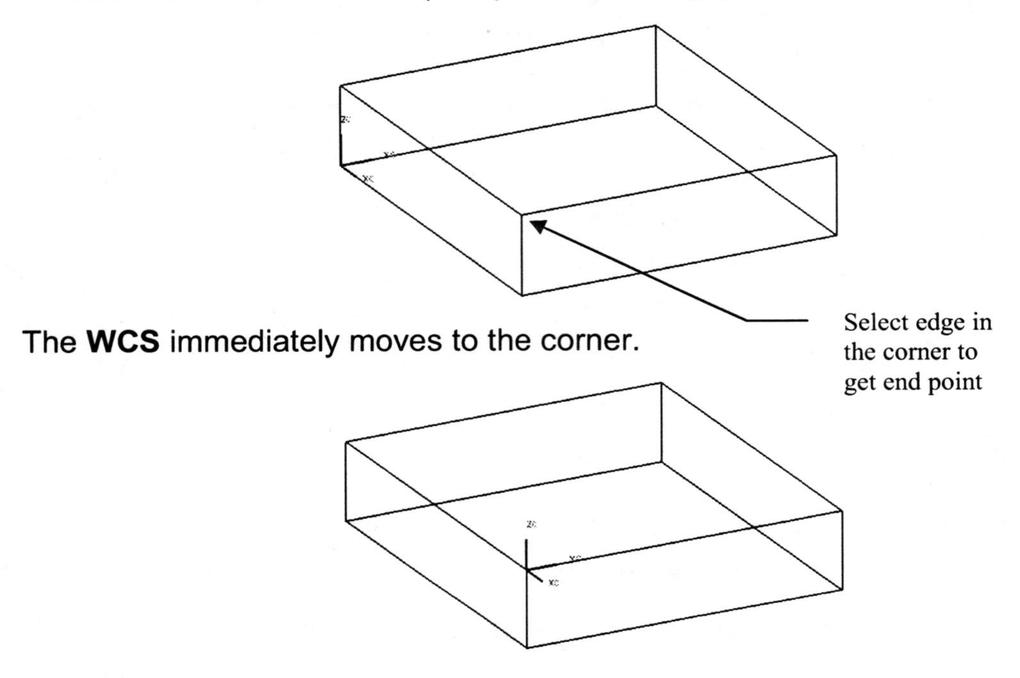

The **WCS** immediately moves to the corner.

Step 4. Rotate the **WCS**

In order to create the second block angled from the first, you must rotate the **WCS** to make its **XC** axis point in the correct direction.

4.1. Choose **Format → WCS → Rotate**.

The **Rotate WCS about...** dialog lets you specify how you want to rotate the WCS.

4.2. Click on the button by **+YC Axis: ZC → XC**.

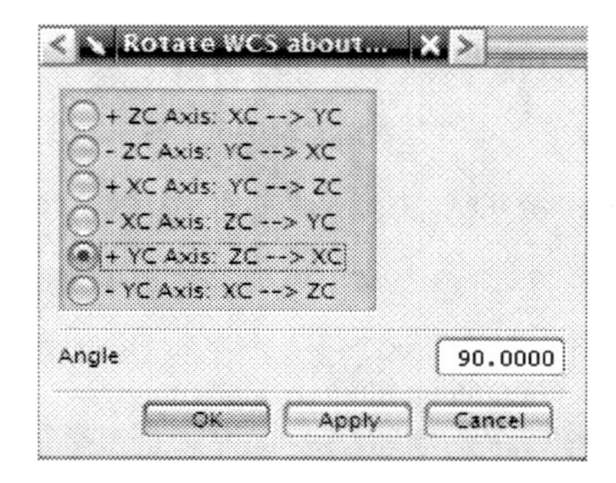

The **X length** of this second block needs to slope downward at a 30 degree angle from the first block.

4.3. Double-click in the **Angle** field of the dialog and key in an angle value of 30 (degrees). Choose **OK**. The **WCS** should be oriented like this.

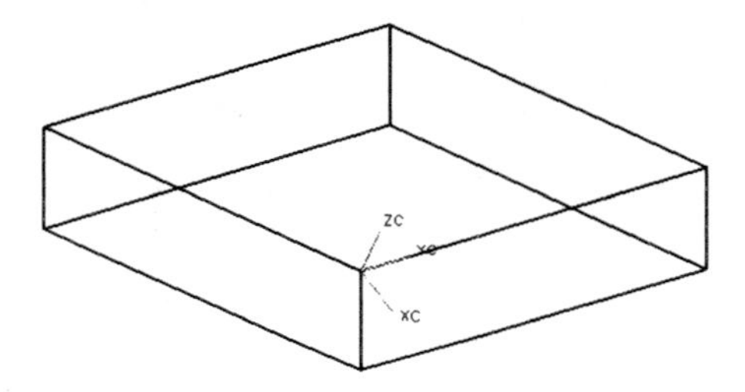

Step 5. Create the second **Block**

You can create the second block using the **Two Diagonal Points** method, which requires you to define the two diagonal points that will represent the opposite corners of the block. The dialog of three block methods should still be displayed.

5.1. If not, **Insert → Design Feature → Block**.

5.2. Choose **Two Diagonal Points** icon from the top row of icons.

Step 6. Specify the two diagonal points.

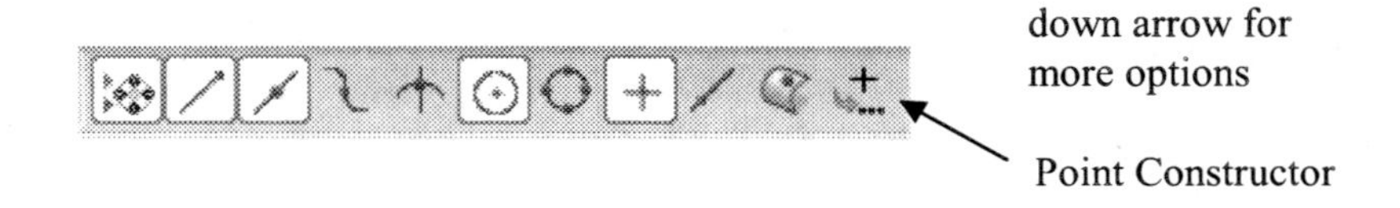

If the point constructor icon is not available in the tool bar, select the down arrow for a pulldown menu, scroll down and check the point constructor item.

If the Selection tool bar is not open, move your cursor to the tool bar area and click MB3 once. You will see a list of tool bar names appear. Check item Selection Bar. A new bar will be added just above the graphic window. At the far right is the pulldown arrow, click it, select Add or Remove Buttons, then select Selection Bar. A new list opens, scroll down to select the point constructor item. You may have to select the arrow at the bottom of the list. Close the list. If the Block dialog is still up the command will appear immediately; otherwise, the point constructor will not appear on the tool bar until a point is required by a dialog.

This block needs to angle off the first, but have the same height and width. One easy way to do this is to place the first diagonal corner point a half inch below the **WCS** along the **ZC** axis.

6.1. Toggle on the **WCS** button on the Point Constructor dialog, if necessary.

6.2. Choose the **Reset** option to change the Base Point values to zero.

6.3. Key in the following values for the first diagonal point:

XC field= 0, **YC** field 0, **ZC** field= -0.5 and choose **OK**.

In the graphics area, an asterisk displays the location for the first diagonal point.

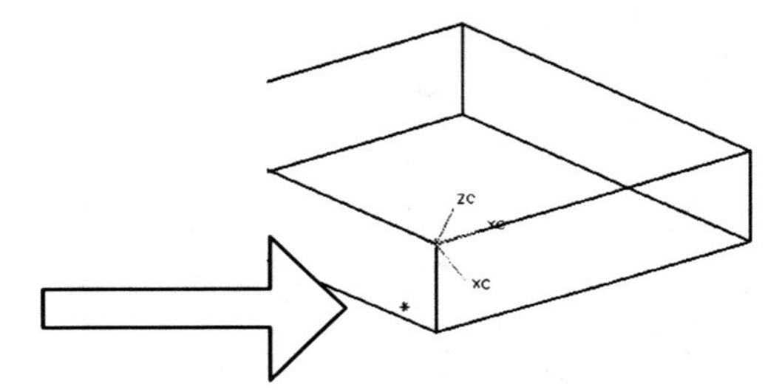

This block needs to be 3.25 inches long (along the **XC** axis), 2 inches wide (along the **YC** axis), and a half inch high. You need to specify the location of the second diagonal point.

6.4. Open the Point Constructor dialog again by choosing the last icon on the Selection tool bar as before. Key in the following values for the second diagonal point:

XC field = 3.25, **YC** field = 2.0, **ZC** field = 0 and choose **OK**.

An asterisk displays the second point. These two asterisks define where the block will be created.

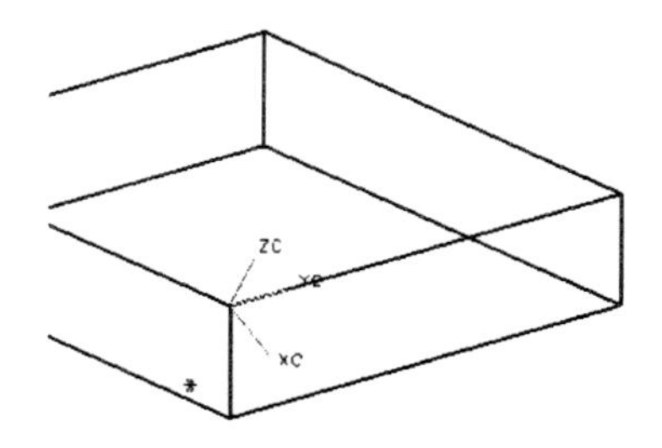

The block is not created yet, because the system knows that two solids will occupy some of the same model space, so you must tell the system that you want to use a Boolean operation. Select the Boolean Operation pull-down menu and a dialog of Boolean operations displays.

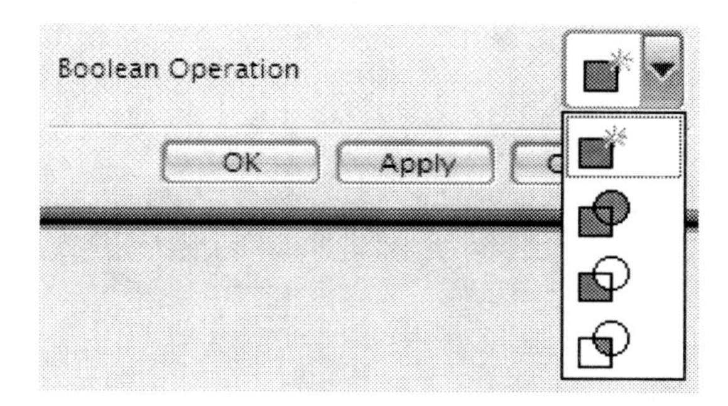

You can **Create** the block as a separate solid body, **Unite** the two solid bodies, **Subtract** the new solid body from the old, or find a solid body from the **Intersection** of the two solid bodies. You will **Unite** the two blocks into one solid body.

6.5. Choose **Unite** to join the two blocks and choose **OK**.

6.6. Use **MB3 → Fit** to fit the view.

If your part does not look correct you can use **Edit → Undo List** and choose 1 Block, or use **Ctrl+Z,** or **MB3 → Undo** to restore the part to the prior state.

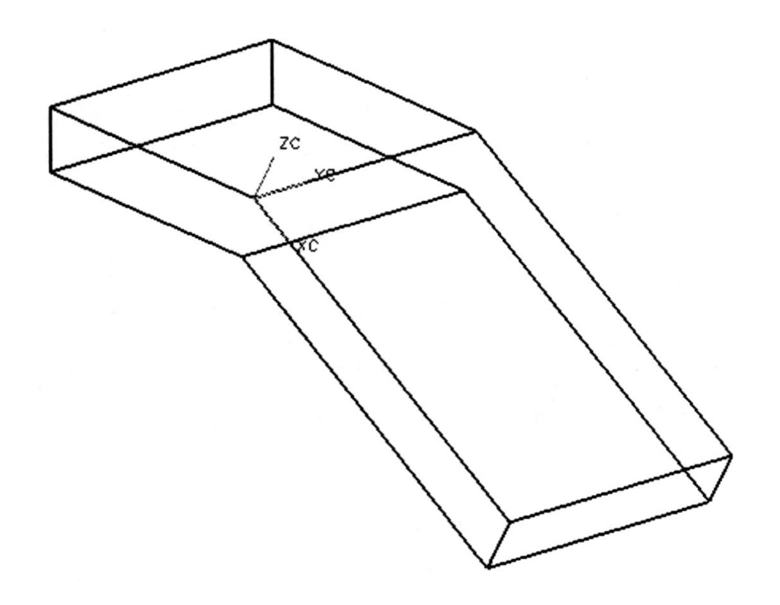
ZC
YC
XC

Activity 4-2. Creating a cylinder

You will add a cylinder to the end of the angled block.

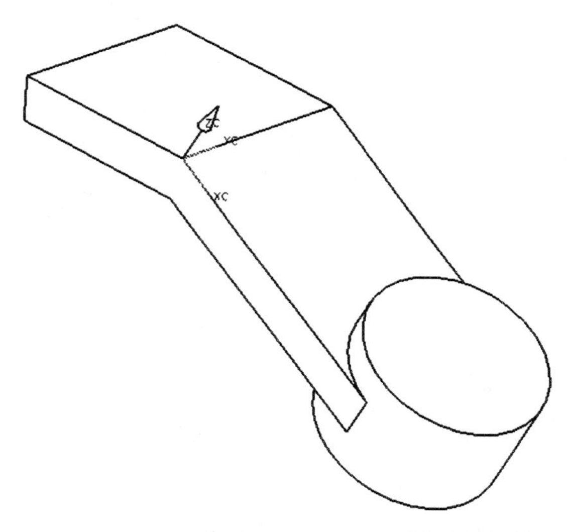

Step 1. Choose **Insert → Design Feature → Cylinder** or choose **Cylinder** icon. The **Cylinder** dialog lists two creation options. The **Cylinder** option lets you create a primitive cylinder by specifying its orientation, size, and location.

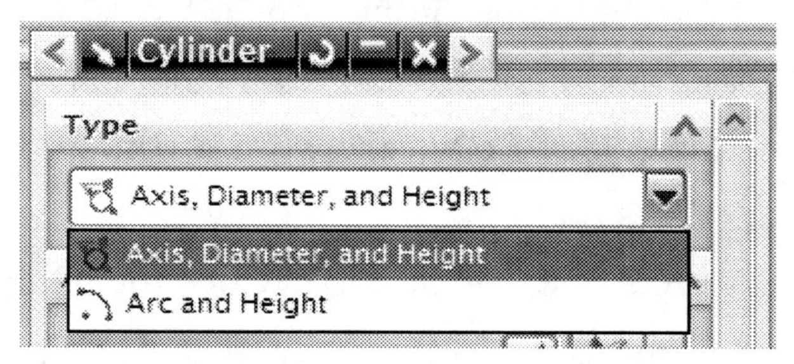

1.1. Choose the Axis, Diameter, and Height option.

Step 2. Specify the Direction.

The **Vector Constructor** dialog displays methods to define the direction of the cylinder (vector direction). The positive **ZC** axis of your **WCS** is already oriented in the correct direction for the cylinder. That is, the **ZC** axis is pointing normal to the top face of the angled block, which is the direction in which you will want to create the cylinder.

2.1. Choose **ZC Axis** on the dialog ZC.

A direction vector displays the direction in which the new primitive cylinder will be created.

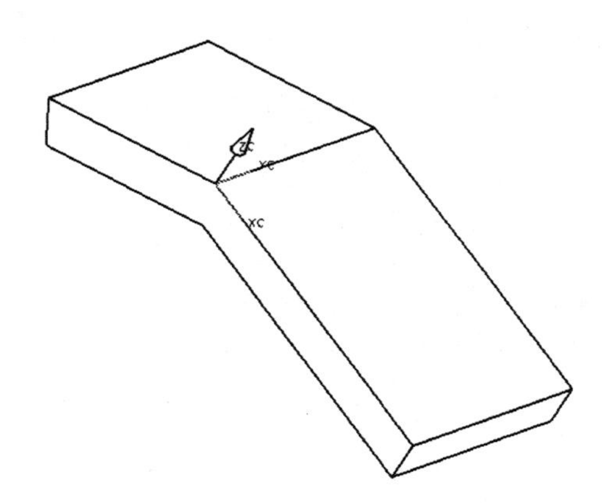

Step 3. Specify the Parameters.

In the next dialog that displays, you must specify positive diameter and height values for the cylinder.

3.1. Key in the parameters: Diameter = 2.0, Height = 1.25.

To display the **Point Constructor** dialog click next to specify point. To position the cylinder correctly, you need to place its origin (base point) directly below the midpoint of the end edge of the angled block. The coordinate of this point is XC=3.25, YC=1.0 and ZC=-.875 based on the current WCS. Make sure to turn off associative and select "Relative to WCS" in order for the points entered to be based on current WCS.

3.2. Toggle on the **WCS** option in the Point Constructor dialog if not.

3.3. In the **Point Constructor** dialog, key in the parameters for the **Base Point** as follows:

XC = 3.25, **YC** = 1.0, **ZC** = -0.875 and choose **OK**.

In the view, an asterisk displays where the base point of the cylinder will be located. If you did not like this location, you could choose **Back** on the Boolean operation dialog and respecify the cylinder location. (If it's hard to visualize the location of the asterisk in 3D space, rotate the part around to see where it is located relative to the block.)

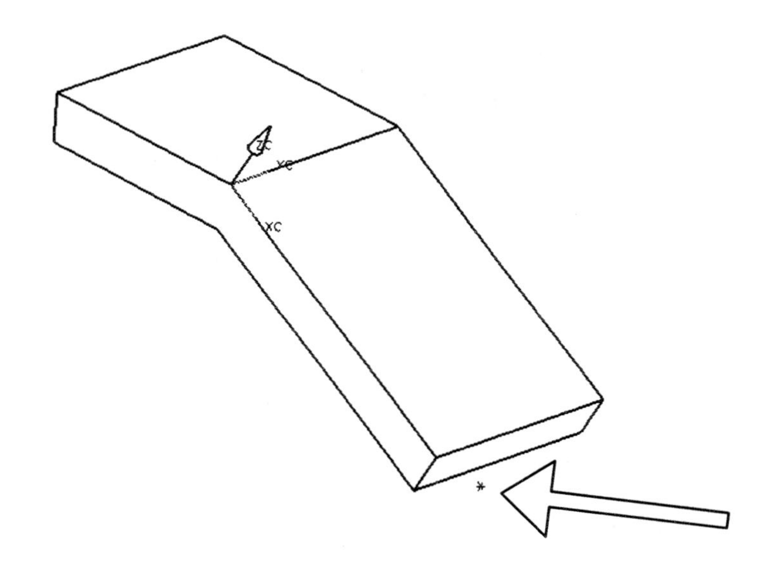

Step 4. **Unite** the Cylinder with the Block. In the four Boolean options,

4.1. Choose **Unite**.

4.2. **Cancel** the **Vector** dialog.

4.3. **Save and close** all files.

Activity 4-3. Creating a cone

You will create this truncated cone.

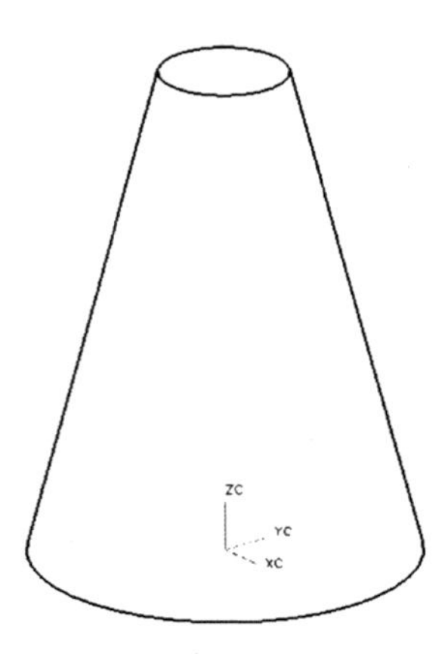

Step 1. Create a new part file **(unit: inches)** named **xxx_activity4-3.prt**.

Step 2. Specify the **Cone** creation method.

2.1. Choose **Insert → Design Feature → Cone** or choose the **Cone** icon .

The cone creation methods are available on the **Cone** dialog.

2.2. Choose the **Diameters, Height** option.

Step 3. Specify the direction for the cone.

3.1. Choose **ZC Axis** option ZC .

The direction vector points ZC positive from the **WCS** origin.

Step 4. Specify the Parameters.

The **Base Diameter** must be greater than zero. The **Top Diameter** can be zero, which will create a pointed cone, or greater than zero for a truncated cone. (This value can be larger than the base diameter value.) For this cone, you can use these values:

Base Diameter = 3

Top Diameter = 1

Height = 4

Choose **OK**.

Step 5. Specify the **Origin**.

The dialog displays the WCS coordinates by default to specify the origin (arc center) of the base of the cone. For this cone, you can use the following origin in the **WCS**:

5.1. Enter **XC** = 2, **YC** = 3, **ZC** = 0 and choose **OK**.

The truncated cone is created slightly to the right of the WCS.

5.2. If necessary, fit the view by choosing in the graphic window MB3 → Fit.

5.3. Cancel the **Vector** dialog.

As with cylinders and blocks, you can unite, subtract, or intersect multiple bodies in the part file.

Step 6. Save and close this file.

Project 4-1. Primitives

Using the primitives that you learned in this chapter, create a solid model that is depicted in the drawing below (units: inches). Note that the cone and the cylinder are concentric and are centered in the block. Save it with filename **xxx_project4-1.prt**, where **xxx** is your three-letter name initials. You will use it again in Chapter 5.

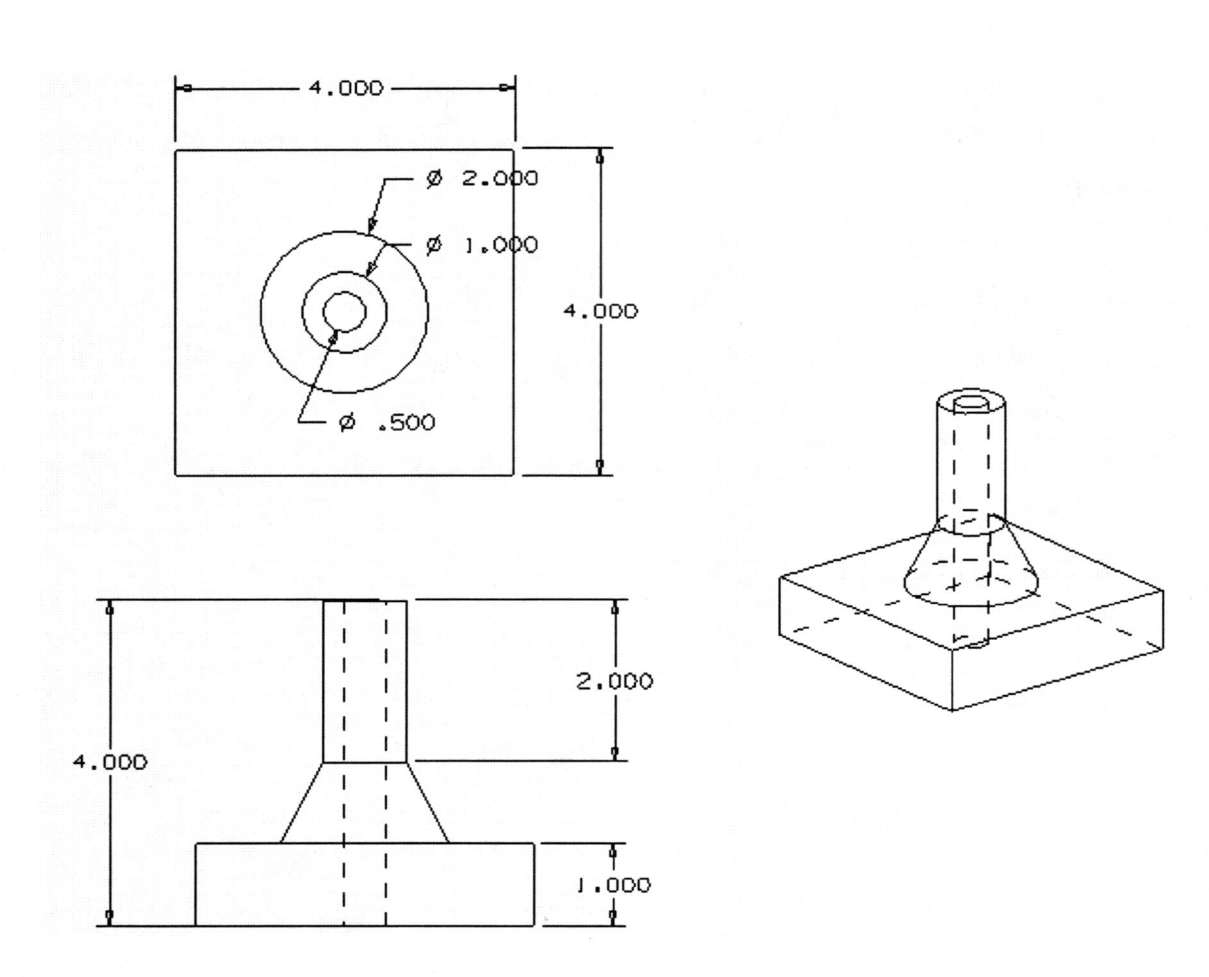

It is important to note that generally it is not a recommended design practice to use more than one primitive to model a part. In Chapter 5, you will model this same part using only one primitive, adding form features to it. You will compare these two modeling approaches and learn why using more than one primitive to model a part is not a recommended design practice.

Exercise Problems

4.1 What are primitives?

4.2 State five primitives used by Unigraphics.

4.3 Write a common procedure to create a primitive.

4.4 Name three ways to create a block.

4.5 Name two ways to create a cylinder.

4.6 There are several ways to create a cone. Name one frequently used.

Chapter 5. Basic Form Features

Form features are features that are added to or subtracted from a base feature. A base feature can be started with either a primitive discussed in Chapter 4 or a swept feature discussed in Chapter 9. Thus, form features supply detail to your model. In this chapter, we introduce five form features that are most commonly used. These are hole, boss, pocket, pad, and slot. Section 5.1 discusses a common procedure to create a form feature and Section 5.2 discusses positioning methods to locate form features on the existing base feature. Sections 5.3 to 5.7 describe the five form features in order.

5.1 Common Procedure to Create a Feature

The basic procedure to add a form feature to your part is:

1. Select the **Planar Placement** face.
2. Select a **Horizontal Reference** if required.
3. Select one or more **Thru Faces** if required.
4. Enter values for the feature parameters.
5. **Position** the feature.

Planar Placement Face

Most form features require a *planar placement face*, which the form feature is created on and is associative with. The planar placement face can be a datum plane or a planar face on the target solid. The feature is created normal to the face and near the location where the face was selected. At a later time it can be more precisely located using positioning methods of Step 5. The feature is automatically linked to the selected face, so that when the face is translated or rotated, the feature remains perpendicular to the face, and the height of the feature remains constant with respect to the face.

If you select a datum plane as the planar placement face, a direction vector is displayed, showing the side of the datum plane on which the feature will be created. You are given the option to accept this default side or to flip the vector to the opposite side. We will discuss the datum plane in Chapter 8.

Horizontal Reference

Some form features require a horizontal reference, which defines the XC direction of the feature coordinate system. You can select an edge, face, datum axis, or datum plane as the horizontal reference. The horizontal reference defines the length direction of those form features that require a length, including Slot, Pocket and Pad. For these options, you are not required to select a Horizontal or Vertical Reference during the positioning of the feature since it is already defined during feature creation.

Thru Faces

Holes and slots can be created to always go completely through selected faces, regardless of modifications to the solid body, using the Thru option. When a thru face is selected, the hole or slot feature is considered to extend completely through the solid limited only by the placement face and the thru face(s).

Parameter Values

Each feature type requires you to enter values, which define the dimensions necessary for that particular feature. These are sometimes referred to as the feature's *parameters*. These, for example, include length, diameter, or depth.

5.2 Positioning Form Features

You can position a feature or a sketch relative to existing solid body geometry or datum planes. Sketch will be discussed in Chapter 10, and in this chapter we will focus on feature positioning. We can position a form feature by creating positioning dimensions called dimensional constraints. These constraints control the location of a feature relative to some existing solid body geometry or datum planes. For example, suppose that we want to create a hole feature in a block with the location of the hole shown in the below figure. Two positioning dimensions are used to locate the hole.

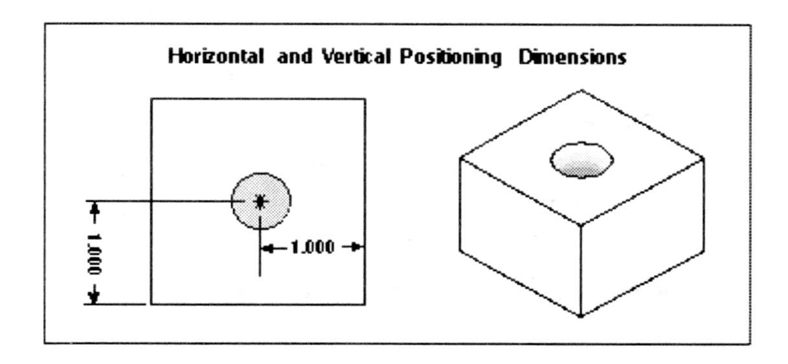

You can complete creating a form feature without constraining its location by choosing OK before choosing a dimension type. The feature can later be positioned or moved using options found under **Edit→Feature→Edit Positioning**. Depending on the number and type of positioning dimensions specified, the UG Status line indicates that the feature is fully specified, underspecified, or overspecified. The feature is fully specified when it is uniquely located by the positioning dimensions specified. It is underspecified when it is not yet uniquely located. It is overspecified when it has more positioning dimensions applied to it than are necessary.

To position a feature, you must:

1. Choose the positioning dimension type.
2. Select the objects to dimension.
3. Enter the new value for positioning.
4. Choose OK.

There are up to nine dimension types that can be used to constrain the location of a feature. These are Horizontal, Vertical, Parallel, Perpendicular, Parallel at a Distance, Angular, Point onto Point, Point onto Line, and Line onto Line, and their icons are shown in the figure below. Some features do not require all nine and only those will be shown that are applicable.

When selecting the objects to dimension, you must select one of the objects from the target solid, curve, or datum and the other from the feature to be positioned. When you position a boss or hole, the system automatically selects the object on the feature (the arc center) for you. Selection of objects varies with the type of dimension you wish to create. For example, if you want to use edges for a horizontal dimension, you must select their endpoints. For those positioning dimensions that require the selection of points, the points must be part of the solid body (i.e., midpoints, endpoints, arc centers, tangency points).

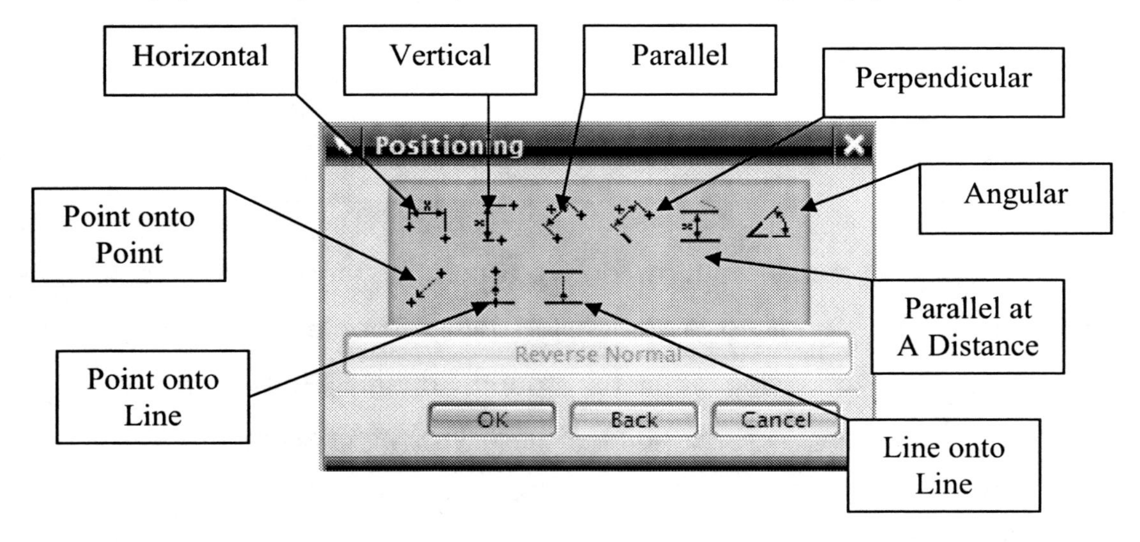

In the following section, we provide more detail of these nine positioning methods.

Nine Positioning Methods

(1) **Horizontal**: This creates a positioning dimension between two points.

A horizontal dimension is aligned with the Horizontal Reference, or is 90 degrees from the Vertical Reference

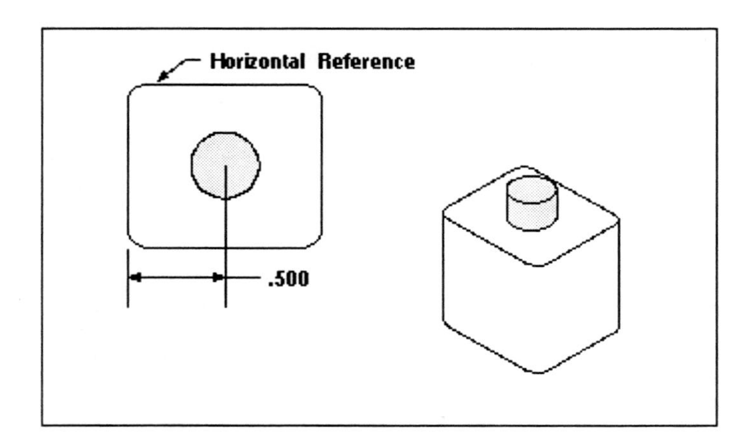

(2) **Vertical**: This creates a positioning dimension between two points. A vertical dimension is aligned with the Vertical Reference, or is 90 degrees from the Horizontal Reference.

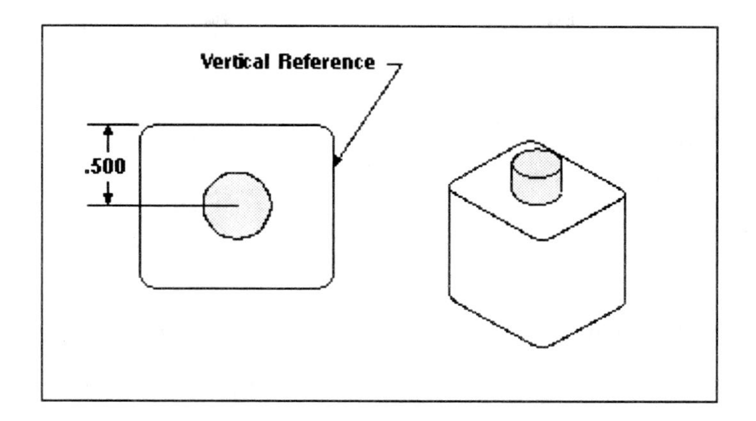

When creating horizontal and vertical dimensions, you may be required to define a Horizontal or Vertical Reference by selecting a linear edge, a solid face, or a datum axis or datum plane. A Horizontal (Vertical) Reference defines the horizontal (vertical) direction for dimensioning the feature's location. When using a face or datum plane for a Horizontal or Vertical Reference, the reference is formed by the intersection line of the selected face or datum plane and the selected planar placement face.

Positioning dimensions are associated to the geometry used to create them. If you move or delete geometry, the associated positioning dimension is also moved or deleted. Whenever a feature whose edge/face was used as a horizontal or vertical reference is deleted, any Horizontal or Vertical positioning constraints associated with the reference are also deleted.

(3) **Parallel**: This option creates a positioning dimension which constrains the distance between two points (e.g., existing points, entity endpoints, arc center points, or arc tangent points) and is measured parallel to a straight line between the two points. In the figure below, a pad is dimensionally constrained on a block. You can imagine a parallel dimension as a rope joining two points at a specified distance. It takes 3 "ropes" to locate this feature.

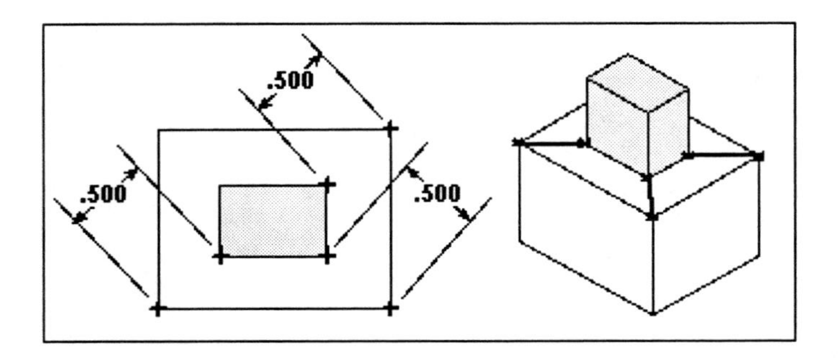

When you create a parallel or any other linear type dimension to a tangent point on an arc, there are two possible tangency points. You must select the arc near the desired point of tangency.

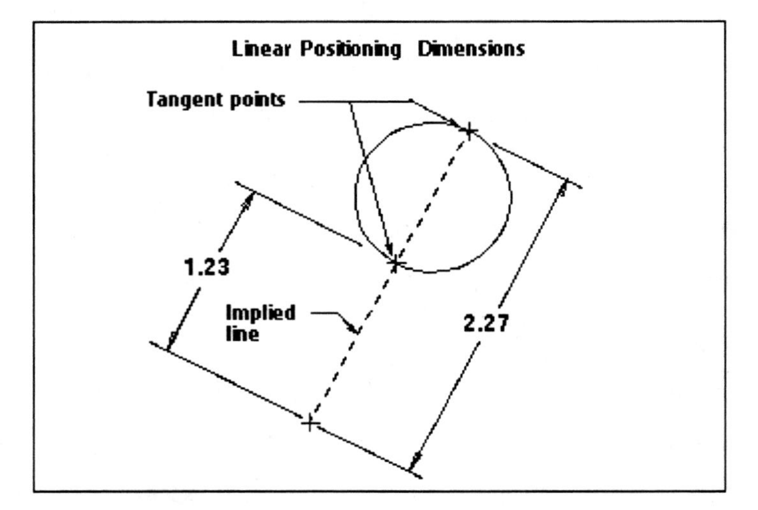

(4) **Perpendicular**: This creates a positioning dimension, which constrains the perpendicular distance between an *edge* of the target solid and a *point* on the feature. You can also position to a datum by selecting a datum plane or datum axis as the target edge, or any existing curve (which need not be on the target solid). This constraint is used to dimension linear distances that are not parallel to the XC or YC axis. It only locks the point on the feature to the edge on the

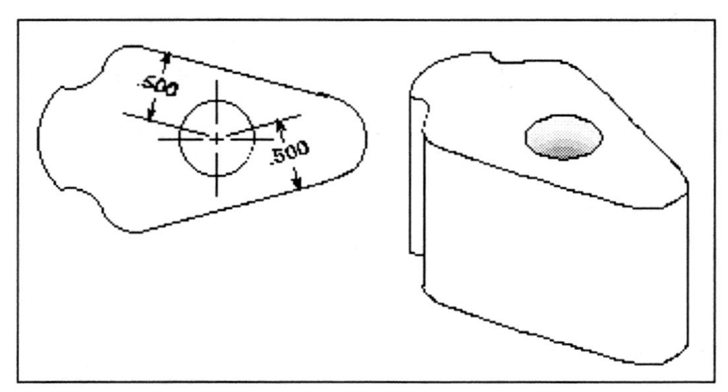

target solid, or to the curve, at the specified distance.

(5) **Parallel at a Distance**: This option creates a positioning dimension, which constrains a linear edge of the feature and a linear edge of the target solid (or any existing curve, on or off the target solid) to be parallel and at a fixed distance apart. This constraint only locks the edge on the feature to the edge on the target solid or the curve at the specified distance.

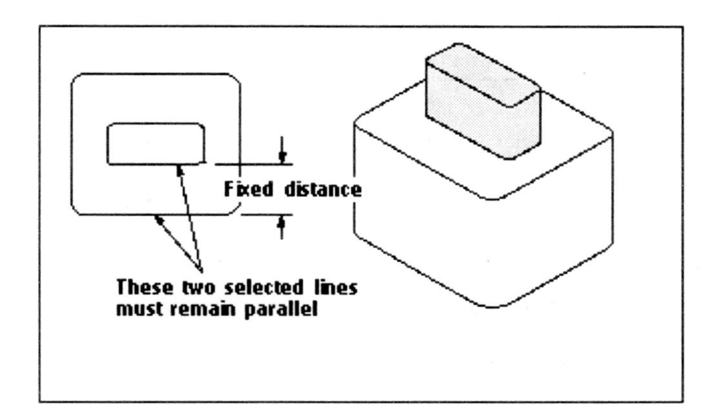

(6) **Angular**: This option creates a positioning constraint dimension between a linear edge of the feature and a linear reference edge/curve at a given angle. Be sure to select the lines to dimension at the proper location. Each line has three control points, one at each end and one at the exact center. The angle created depends on which side of the center control point you select.

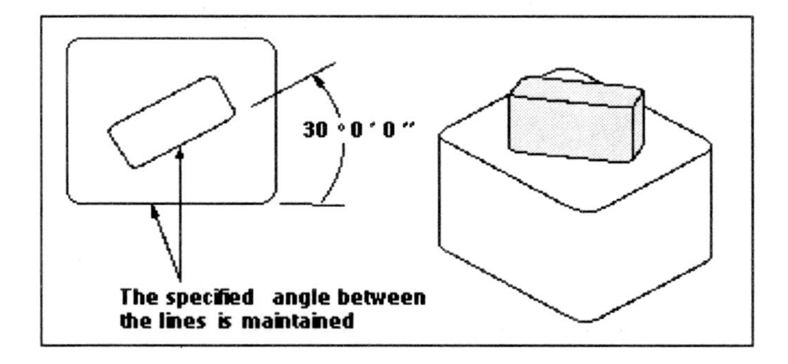

The figure below shows two lines with their center control points highlighted by asterisks (*). The 33°25' angle was created by selecting the lines at the positions indicated by A. The 146°36' angle was created by selecting the lines at the positions marked B. As you can see, selecting the smaller line at a position left of the center control point creates the complementary angle.

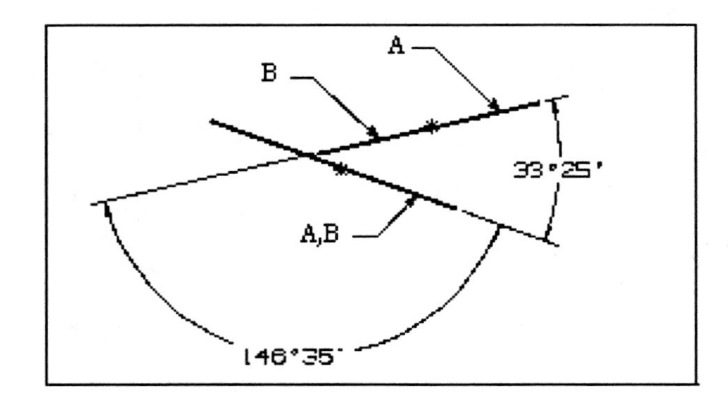

(7) **Point onto Point**: This option creates a positioning dimension the same as the Parallel option, but with the fixed distance between the two points set to zero. This positioning dimension causes the feature to move so that its selected point is on top of the point selected on the target solid.

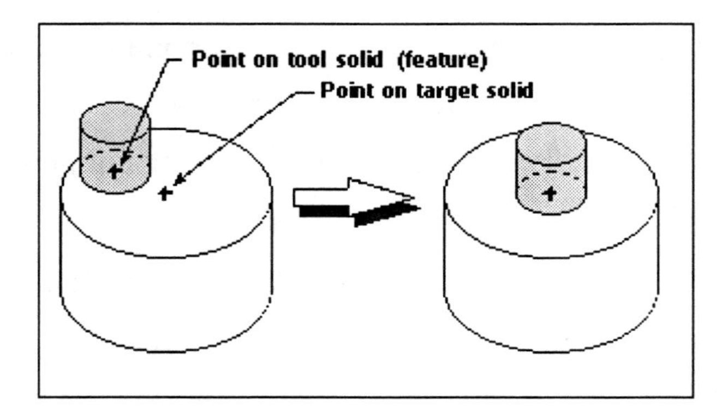

(8) **Point onto Line**: This option creates a positioning constraint dimension the same as the Perpendicular option, but with the distance between the edge or curve and point set to zero. This positioning dimension causes the feature to move from its selected point, normal to the edge or curve selected on the target solid, until the point is on the edge. This constraint only locks the point on the feature to the edge on the target solid.

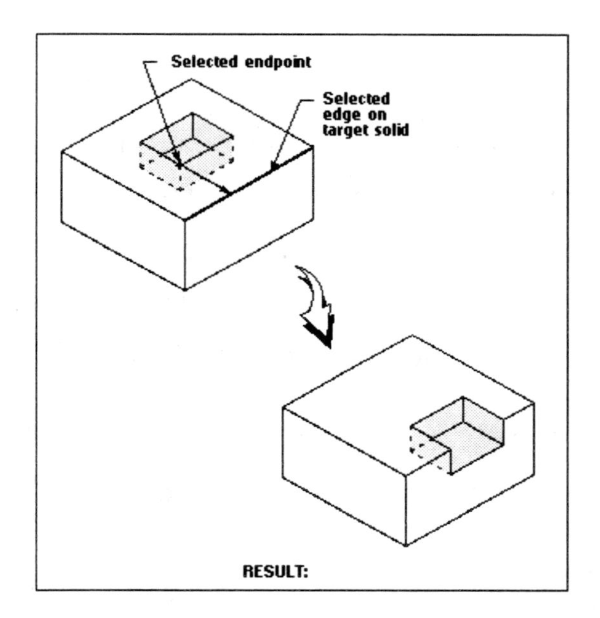

(9) **Line onto Line**: This option creates a positioning constraint dimension the same as the Parallel at a Distance option, but with the distance between the linear edge of the feature and the linear edge or curve on the target solid set to zero. This positioning dimension causes the feature to move from its selected edge perpendicularly to the edge or curve selected on the target solid. This constraint only locks the edge on the feature to the edge/curve on the target solid.

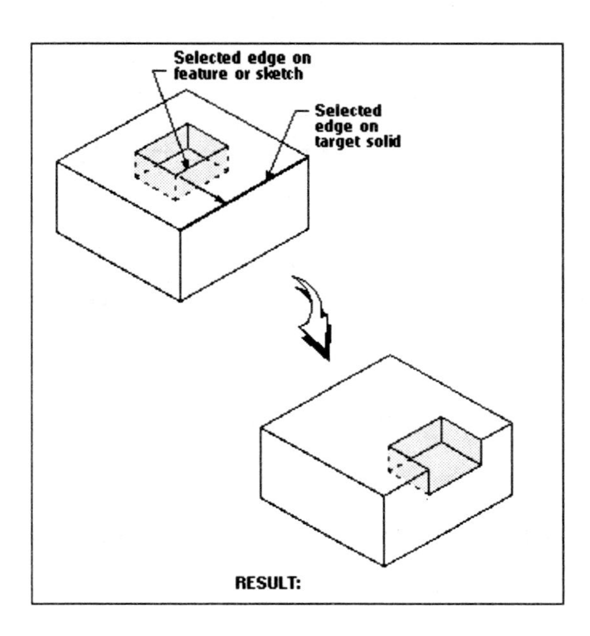

5.3 Hole or

This feature lets you create a simple hole, a counterbore hole or a countersunk hole in a solid body. For all hole creation options, the depth values must be positive. NX5 provides two different methods to create a hole: form-feature based and sketch-based. In this section, we will discuss the first method. Later in Chapter 10, we will discuss the second method. In order to create a hole using the form-feature based method, first add the icon Pre-Nx5 Hole into a toolbar as follows: **Tools → Customize →Commands tab**. Choose Feature under Categories and choose Pre-NX5 Hole under Commands (you may need to scroll down for each) as shown below.

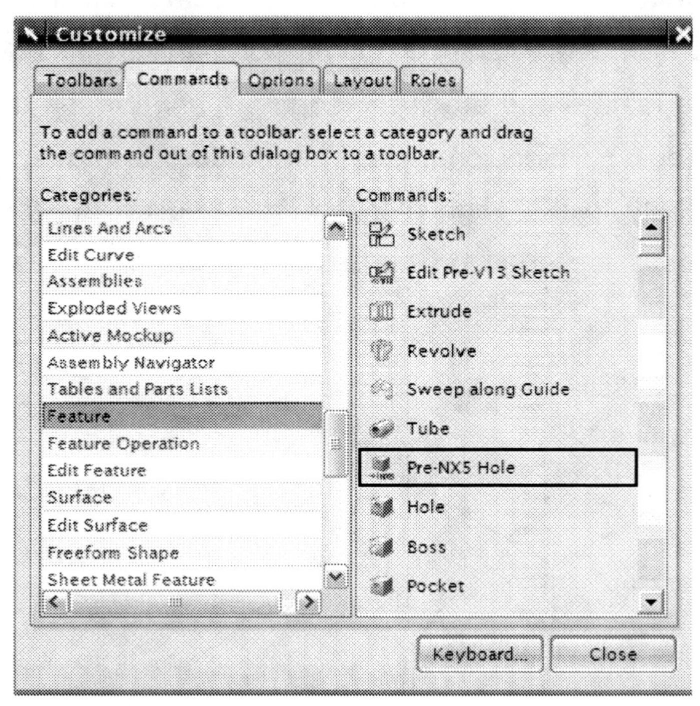

Drag and drop the Pre-NX5 Hole into a toolbar (any toolbar below the UG Main Menu into which you want to add the icon would work, but typically the feature operation toolbar is chosen). Choosing this Icon presents the Hole dialog for the hole creation options as follows:

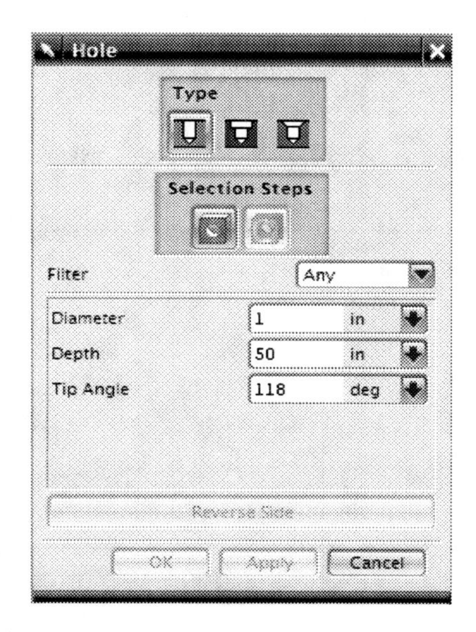

Thru Hole

If you want any of the hole types to go through the target solid, you merely select the Thru face when prompted for it on the Cue Line. Thru holes do not stop at the first contact with the thru face, but continue completely through the thru face. If the hole intersects the thru face more than once, the hole will continue to the last intersection. If a Thru hole is not required, then set the parameters in the dialog and choose OK.

Simple Hole

To create a simple hole, you must specify a placement planar face/plane, diameter, depth and tip angle. Tip Angle lets you create either a flat or pointed end hole. A zero tip angle value results in a flat end (blind) hole. A positive tip angle value creates an angled tip, which is added to the depth of the hole. The tip angle must be greater than or equal to 0 and less than 180. If you choose to create a Thru Hole, the Tip Angle window and the Depth window are grayed out.

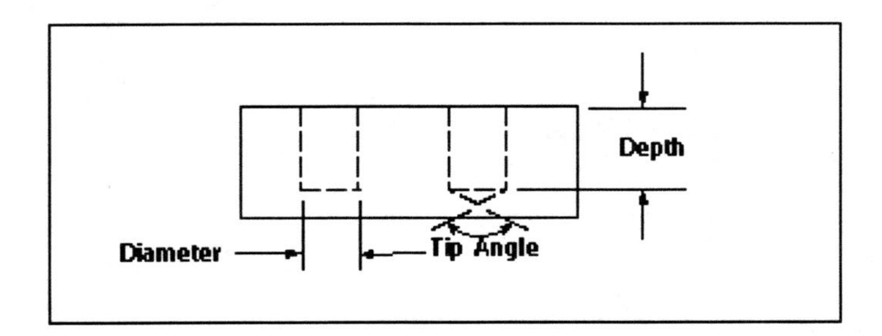

Counterbore Hole

To create a counterbore hole, you must specify two more parameter values than a simple hole, which are C-Bore Diameter and C-Bore Depth.

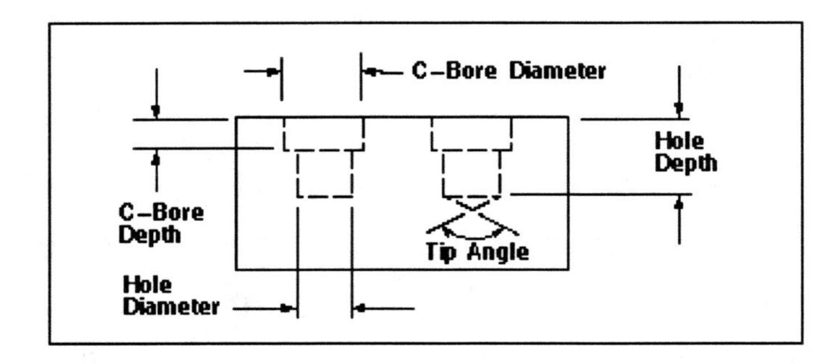

Countersink Hole

A countersink also requires two more parameter values than a simple hole, which are C-Sink Diameter and C-Sink Angle.

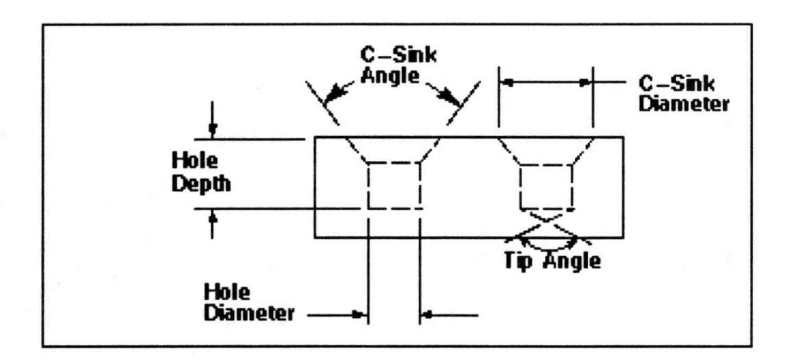

5.4 Boss 

Feature Boss can be viewed as opposite to a hole, that is, adding a cylinder on an existing solid instead of subtracting it from. To create a boss you must specify three parameters, Diameter, Height, and Taper Angle and a planar placement face/plane. Taper Angle is the angle at which the cylinder wall of the boss inclines inward. This value can be negative or positive. A zero value results in a vertical cylinder wall with no taper.

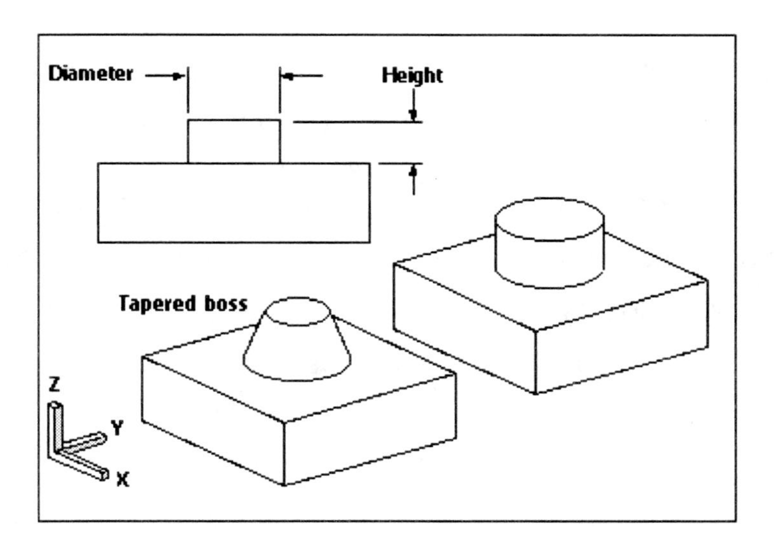

Activity 5-1. Creating holes

In this activity, you will create a simple thru hole and a countersink hole as shown below.

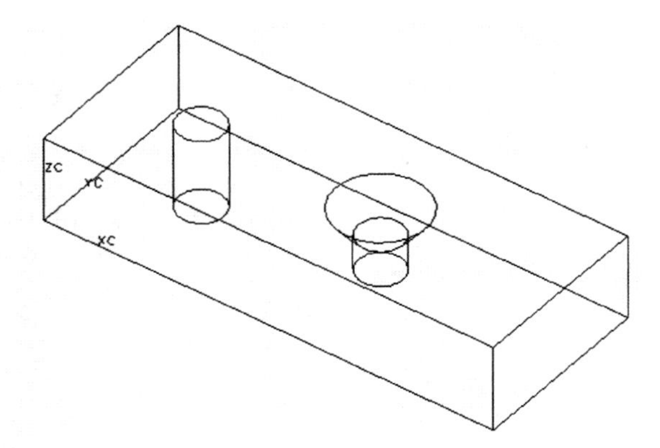

Step 1. **Open** part file **activity5-1.prt,** and save it as **xxx_activity5-1.prt** where xxx is your three letter name initials. Start the **Modeling** application.

1.1. Choose the Pre-NX5 Hole icon from the toolbar (refer to Section 5.3 Hole for the instruction on how to add this icon into a toolbar). The **Hole** dialog is displayed.

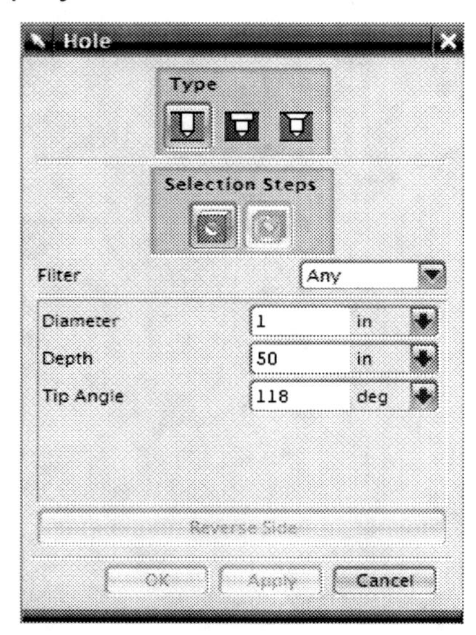

Step 2. Create a thru hole.

2.1. Specify **Planar Placement Face** for the hole by selecting the top face of the block as near where you want the hole to be created.

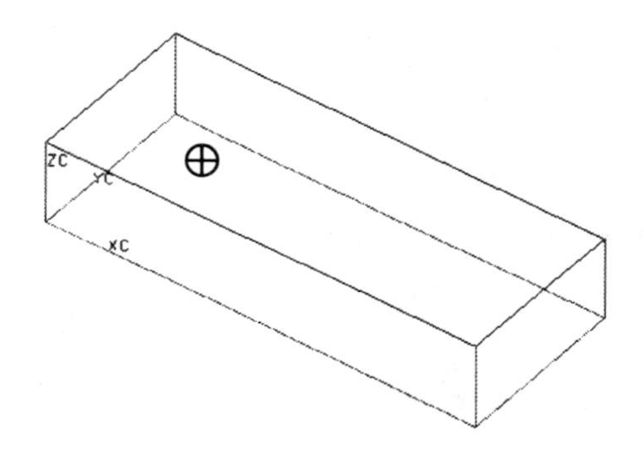

The **Cue Line** is prompting for a thru face.

2.2. Specify the **Through Face**. All form features, except grooves, are created normal to the selected planar face. If you select where the top and bottom faces overlap in the view, the system will automatically select the bottom face. Select the bottom face like this.

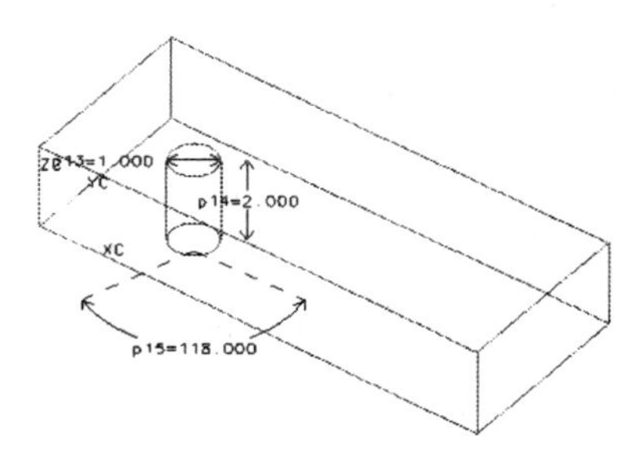

Step 3. The Hole dialog is still up. In the Parameters windows, key in the diameter of 1.0 and choose **OK**. The hole tool solid temporarily displays as a long green cylinder that extends beyond the thru face. The **WCS** temporarily moves to represent the feature coordinate system at the top center of the hole where the feature is first displayed.

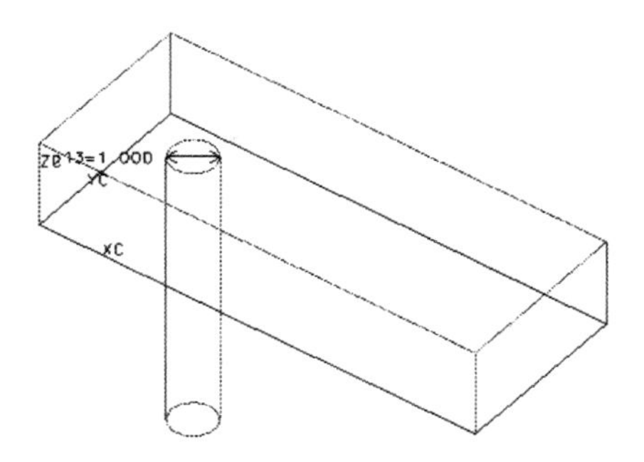

The feature coordinate system shows the XC-axis and YC-axis that represent the horizontal and vertical references that will be used by the system to establish the relationship for positioning the hole.

The **Positioning** dialog displays positioning methods. Note for a hole that only 6 positioning methods are available and the Perpendicular method is the default.

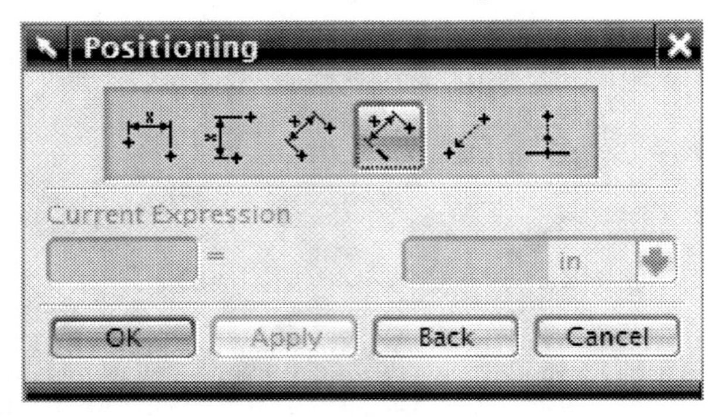

Step 4. Position the hole. One way to position it is to provide **Horizontal** and **Vertical** positioning dimensions to constrain the hole to the solid.

4.1. Choose the **Horizontal** icon . A **Horizontal** positioning dimension is the shortest distance along the X-axis between a target edge and in this case, a cylindrical feature like this hole, the dimension will be measured by default to the center of the hole (tool solid).

4.2. Specify a **Horizontal Reference**. The Horizontal Reference defines the direction of the positioning dimension. A horizontal positioning dimension will run horizontally with this reference. A vertical positioning dimension will run 90 degrees to it. Select the front face of the block.

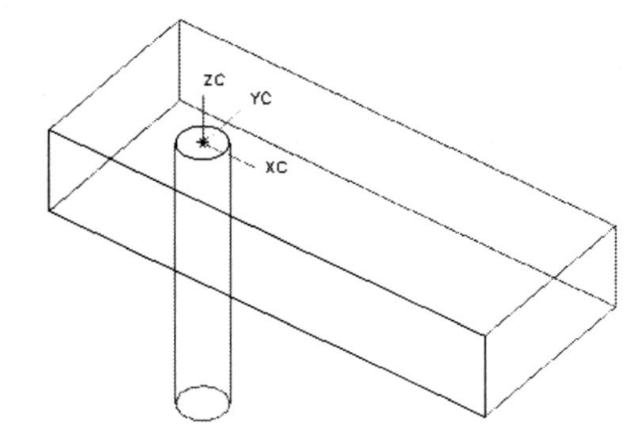

A temporary direction vector shows the horizontal reference.

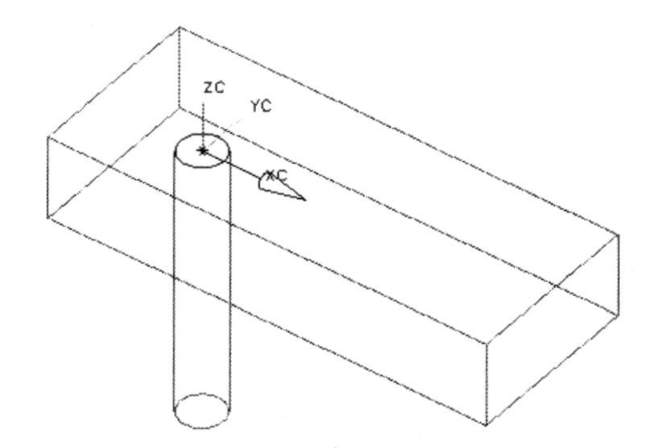

4.3. Specify a Target. Select the left edge of the top face as the target edge for the horizontal dimension, which aligns in the XC direction.

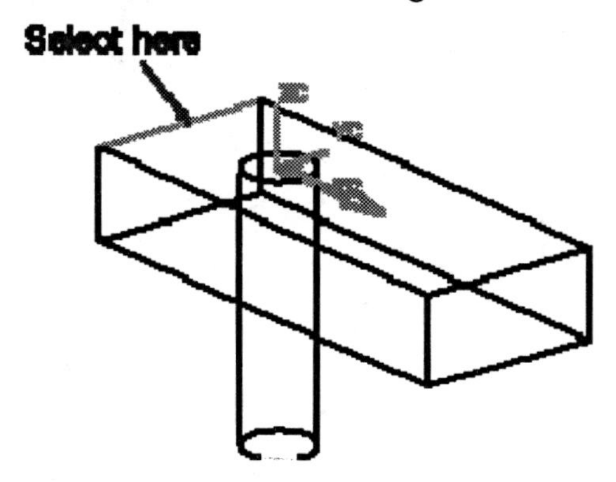

4.4. The temporary horizontal dimension displays from the target edge to the center of the hole. The Position dialog displays with the temporary value, so key in the distance of 2 and choose **Apply** (because we are not finished with the Positioning dialog). This places the hole two inches from the left edge. This horizontal positioning dimension constrains the simple thru hole in one direction.

4.5. With the **Positioning** dialog still displayed, you can define the second constraint, so that the hole is fully specified (constrained). Choose the **Vertical** icon .

4.6. Select the front edge of the top face as the target edge for the vertical dimension.

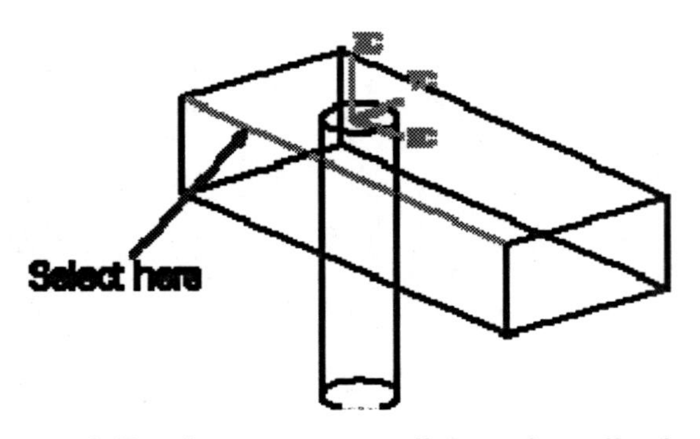

4.7. A temporary dimension displays from the edge to the center of the hole. The **Position** dialog displays the current distance value in the **Current Expression** window. Key in 2 and choose **OK** (because we are finished with the Position dialog). The hole is now positioned.

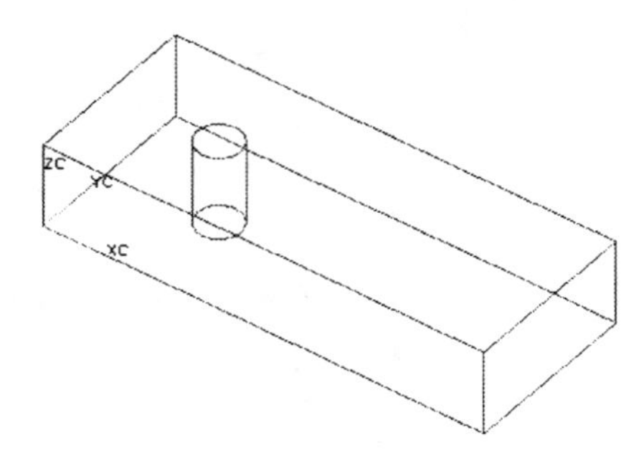

The dimensions are no longer displayed, but you can redisplay the feature dimensions by choosing **Information** → **Feature** to open the Feature Browser dialog, toggling on **Display Dimensions**, and choosing SIMPLE_HOLE(1). Any feature created using the thru option will be associated to the thru faces. If you were to change the size of this block, the hole would remain a through hole positioned two inches from each of the two target edges. Activity 5-2 will address this associativity issue more in detail.

Step 5. Create a countersink hole on the top face of the block with the following parameters:

C-Sink Diameter = 2.

C-Sink Angle = 75

Hole Diameter = 1.

Hole Depth = 1.5

Tip Angle = 0

Step 6. Create the hole without positioning it. The WCS is temporarily displayed at the top center of the countersink hole. Choose **OK** in the **Positioning** dialog to accept the present location. The countersink hole is complete even though the location is unconstrained, i.e., the hole is underspecified.

Step 7. **Undo** Step 5 by **MB3→ Undo.** Position the countersink hole such that its center is located with 2 inch away from the front face of the block and with 4 inch away from the right edge of the top face of the block as shown below. If you have difficulty in positioning, redo **Step 4**.

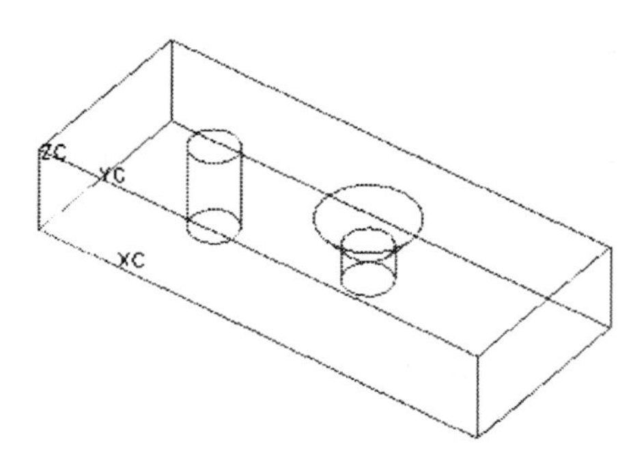

Step 8. Save and close the file.

5.5 Pocket 

You can use the Pocket option to create a cavity in an existing body, using one of the following three methods:

Cylindrical	Lets you define a circular pocket, to a specific depth, with or without a blended floor, having straight or tapered sides.
Rectangular	Lets you define a rectangular pocket, to a specific length, width, and depth, with specific radii in the corners and on the floor, having straight or tapered sides.
General	Lets you define a pocket with much greater flexibility than the cylindrical and rectangular pocket options.

Below we will give more detail description on the Rectangular Pocket. Specific parameters are described below:

Length	The length of the pocket parallel to Horizontal.
Width	The width of the pocket perpendicular to Horizontal
Depth	The depth of the pocket.
Corner Radius	The rounded radius (zero or greater) for the vertical edges of the pocket.
Floor Radius	The rounded radius (zero or greater) for the bottom edges of the pocket. The Corner Radius must be greater than or equal to the Floor Radius.
Taper Angle	The angle at which the four walls of the pocket incline inward. This value cannot be negative. A value of zero results in vertical walls.

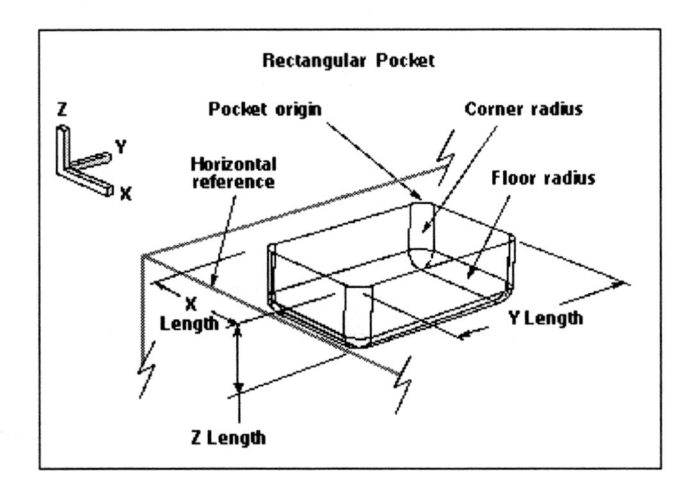

5.6 Pad 

This feature creates a pad on an existing solid body. This is somewhat opposite to the Pocket feature as Boss is to Hole. There are two methods of creating a pad:

Rectangular	Lets you define a pad to a specific length, width, and depth, with specific radii in the corners, having straight or tapered sides.
General	Lets you define a pad with greater flexibility than the rectangular pad option.

Below we will give more detail description on the Rectangular Pad. Specific parameters are described below:

Length	The length of the pad in the direction of Horizontal.
Width	The width of the pad.
Height	The height of the pad.
Corner Radius	The rounded radius for the vertical edges of the pad. The radius specified must be a positive or zero. (A zero radius results in a sharp edged pad.)
Taper Angle	The angle at which the four walls of the pad incline inward. This value cannot be negative. (A zero value results in vertical walls.)

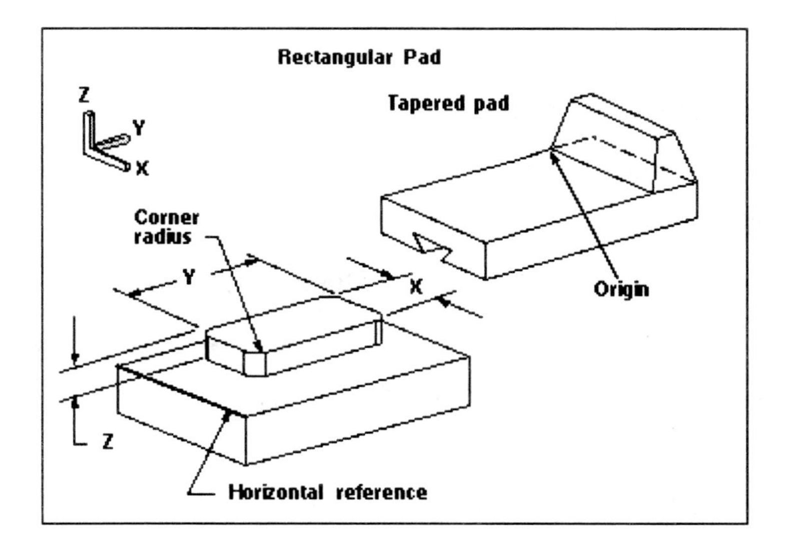

5.7 Slot 

This feature creates a passage through or into a solid body in the shape of a straight slot as if cut by a milling machine tool. The slot feature will be created so that the axis of the imaginary cutter tool is normal to the planar placement face or datum plane selected. The path of the slot is parallel to the horizontal reference selected. There are five types of slots to choose from and the parameters for each type of slot are described below. The depth value for all slot types is measured normal to the planar placement face.

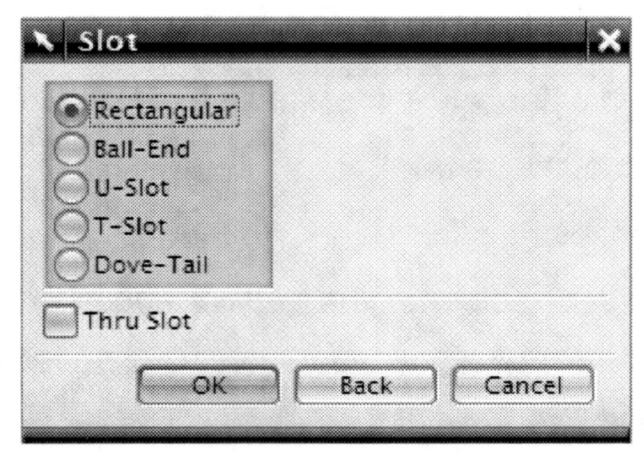

Rectangular	Lets you create a slot with sharp edges along the bottom.
Ball-End	Lets you create a slot with a full radius bottom and corners.
U-Slot	Lets you create a slot with a "U" shape (rounded corners and floor radii).
T-Slot	Lets you create a slot whose cross section is an inverted T.
Dove-Tail	Lets you create a slot with a "dove-tail" shape (sharp corners and angled walls).

For all types, the **Thru Slot** option extends the slot length along the placement face in the direction of the horizontal reference thru 2 other faces that you select. These faces are the **starting thru face** and the **ending thru face**, as shown in the figure below.

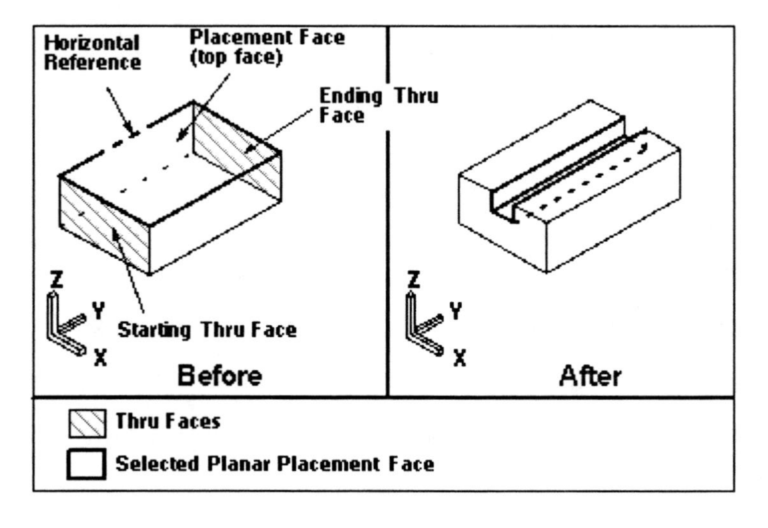

When positioning a thru slot, a single constraint that locates the slot perpendicular to its length is all you need. Do not dimension to the end arcs of the slot. Thru slots intersecting a thru face more than once may result in multiple solutions as shown below.

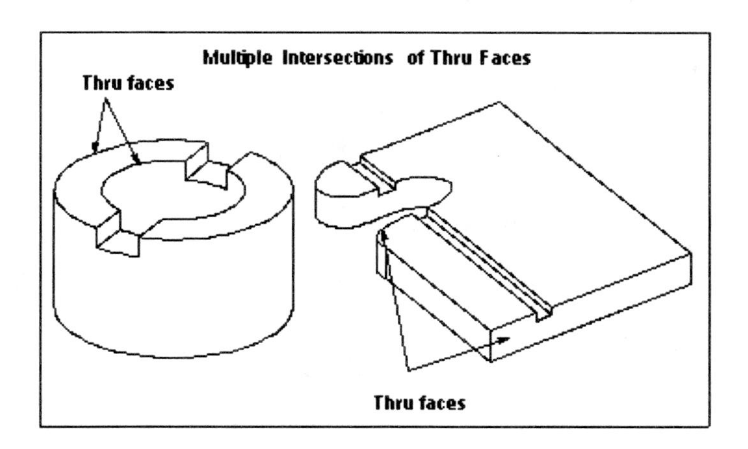

Rectangular Slot

This option creates a slot using a tool that has cylindrical end faces and leaves sharp edges along the bottom. Its parameters are

Width	The width (diameter) of the tool that forms the slot.
Depth	The depth of the slot is the distance from the origin point of the slot to the bottom of the slot.
Length	The length of the slot, measured in a direction parallel to the horizontal reference.

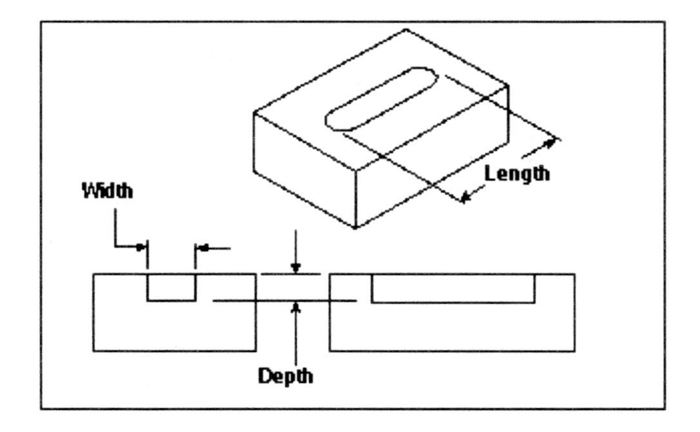

Ball-End Slot

A ball-end slot leaves a full radius bottom and corner. Its parameters are Ball Diameter (i.e., the width of the slot resulting from the tool diameter), Depth, and Length.

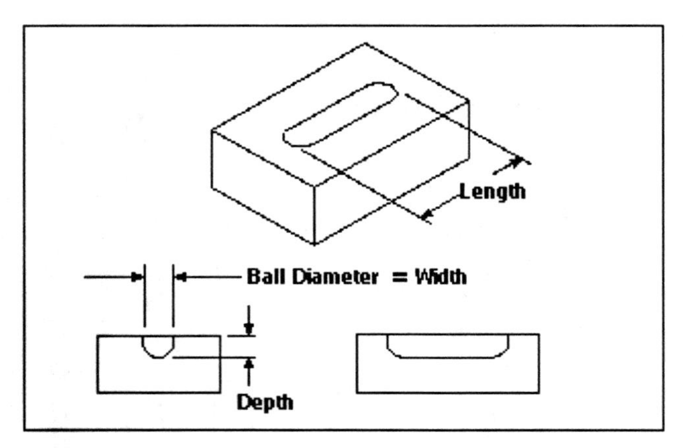

U-Slot

This option creates a slot with a "U" shape. This type of slot leaves rounded corner and floor radii. You must specify the following parameters: Width, Depth, Corner Radius (i.e., the cutting tool edge radius), Length. The Depth value must be greater than the Corner Radius value).

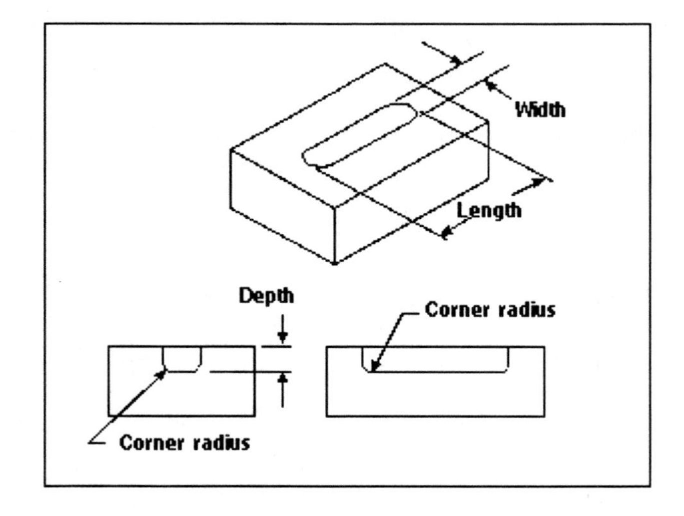

T-Slot

This option creates a slot whose cross section is an inverted T. You must specify the following parameters: Top Width, Top Depth, Bottom Width, and Bottom Depth, and Length.

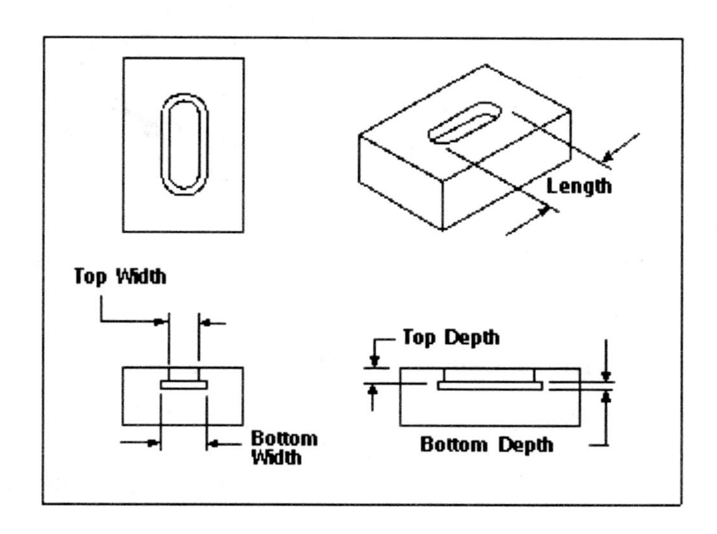

Dove-Tail

This option creates a slot with a "dove-tail" shape. This type of slot leaves sharp corners and angled walls. You must specify the following parameters: Width, Depth, Angle, and Length.

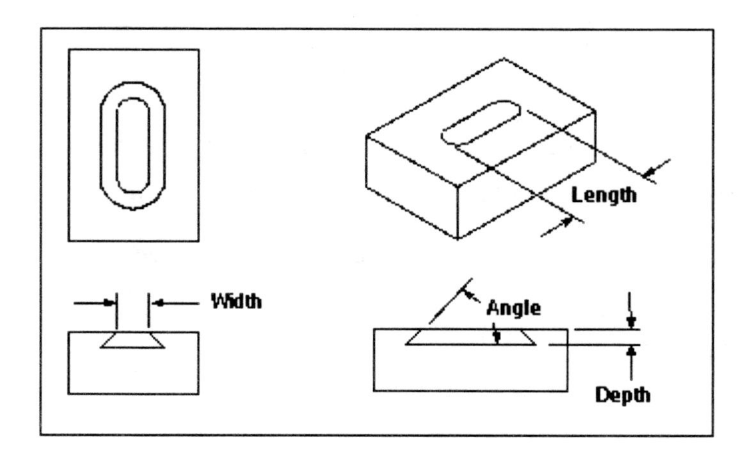

Activity 5-2. Creating a thru T-slot

The design intent here is that a T-slot be created and located so that it aligns with a pad near the back of the block as shown in the following figure.

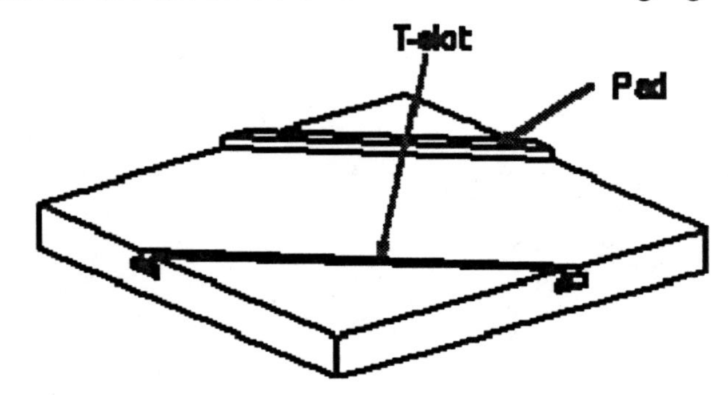

Step 1. **Open activity5-2.prt** and **Save** it as **xxx_activity5-2.prt** and start the **Modeling** application.

Step 2. Choose **Insert** → **Design Feature** → **Slot** (if you do not see the slot option, change your Role setting to Advanced with full menus as described on page 1-16) or choose the **Slot** icon . The **Slot** dialog displays.

2.1. Toggle **Thru Slot** to on.

2.2. Choose the **T-Slot** type. Planar face menu comes up.

Step 3. Select the top face as the planar placement face.

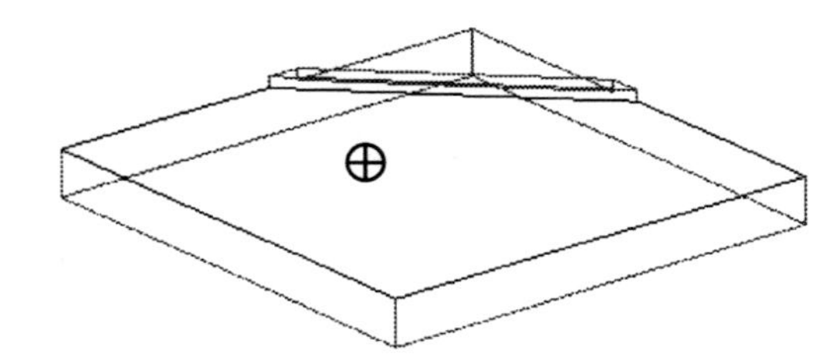

Step 4. Specify a horizontal reference to define the direction of the slot.

4.1. For the horizontal reference, choose **Solid Face** on the dialog.

4.2. Select the closest face of the pad. You may need to zoom in on the pad for careful selection.

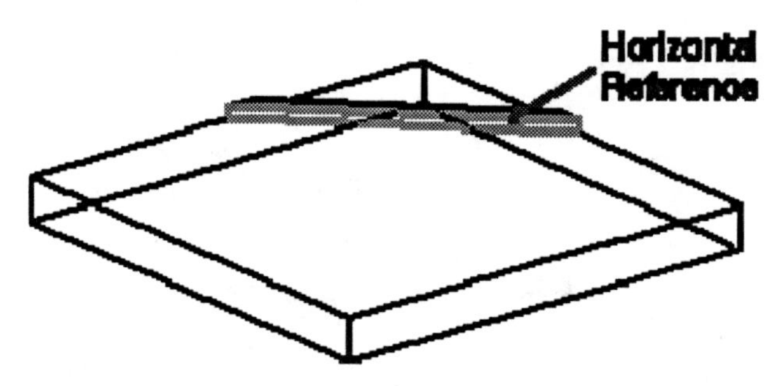

A vector displays the X direction (length/path) of the slot.

Step 5. Specify the thru faces. Select these two thru faces.

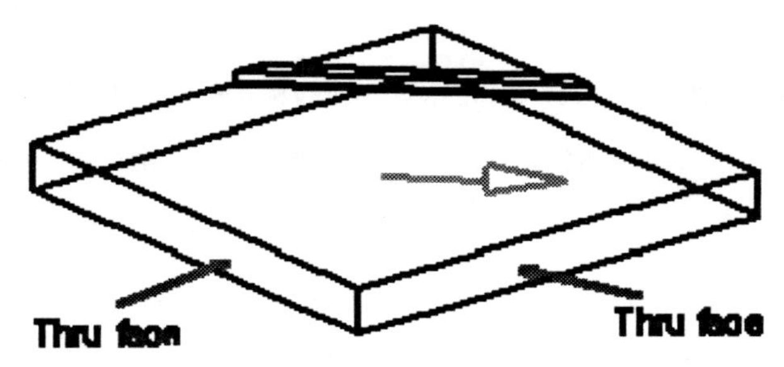

Step 6. Enter the following parameters for the thru **T-slot**:

Top Width = 9/32

Top Depth = 1/4

Bottom Width = 0.5

Bottom Depth = 7/32 and choose **OK**.

A temporary tool solid displays on the part.

Step 7. Position the slot using the **Parallel at a Distance** option. This option constrains the linear edge of the feature to the linear edge of the target solid to be parallel at a specified distance.

7.1. Choose the **Parallel at a Distance** icon .

7.2. Select the target edge of the pad. You may need to zoom in to make selection easier.

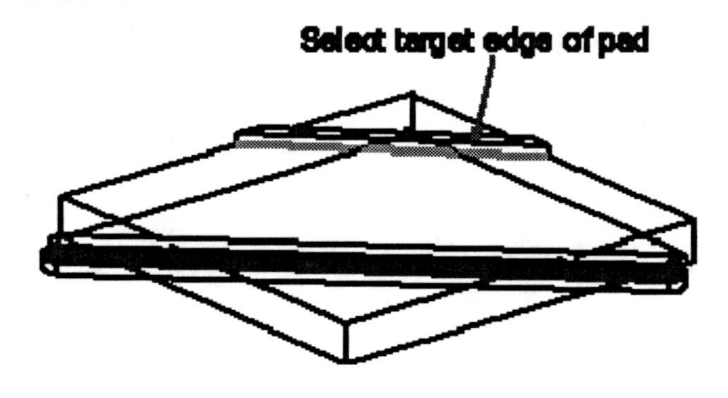

7.3. For a tool edge, select the top edge of the T-Slot closest to the pad feature.

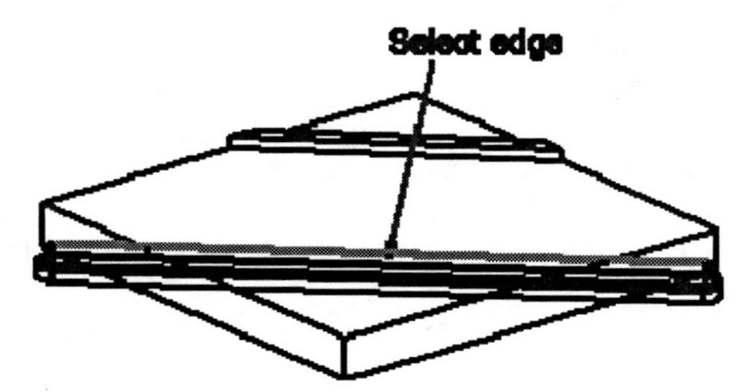

7.4. The **Create Expression** dialog displays the current distance. Key **6.0** into the value field, and choose **OK** twice to complete the positioning of the slot.

7.5. Cancel the dialog.

7.6. Fit the view by **MB3** → **Fit**, if needed.

Do not close the file since you will continue to use it in the following activity.

Activity 5-3. Editing an existing slot and creating a pad

In this step, you will change the position of the thru T-slot that was just created, in the previous activity and then add a pad such that the resulting part looks as follows:

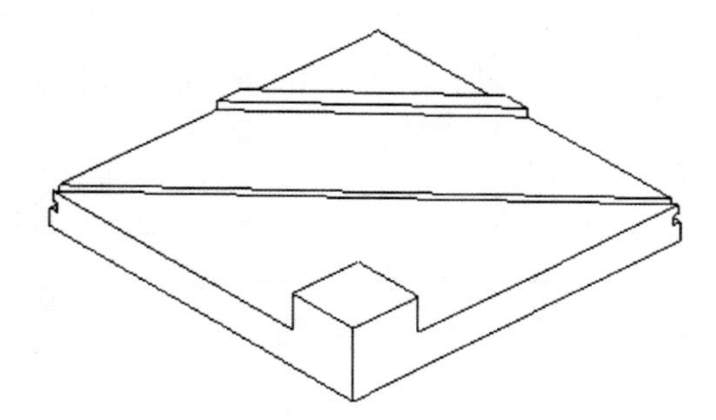

Step 1. Choose **Edit** → **Feature** → **Edit Positioning** or choose the **Edit Positioning** icon in the Edit Feature toolbar. The **Edit Positioning** dialog displays all features.

Step 2. Select **T_Slot** on the dialog, and choose **OK**. Three choices are available to modify the positioning.

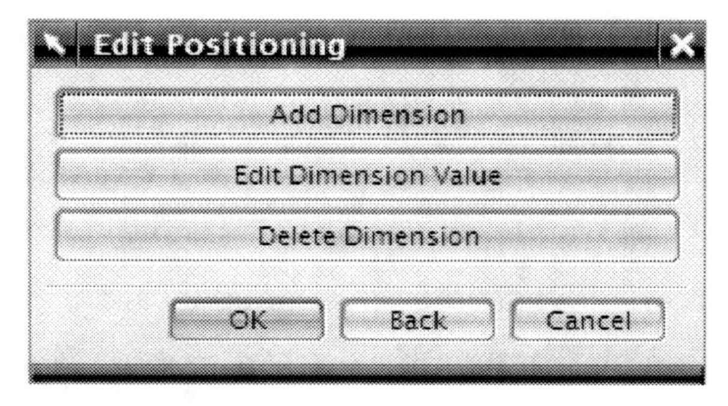

Step 3. Choose **Edit Dimension Value**.

The current positioning dimensions are displayed on the part and the Cue Line prompts to enter a new value for positioning. The Edit Expression dialog displays the current distance 6.0, so change **6** into **3.5.**

Choose **OK**. Choose **OK** until the slot changes (that's 3 times to choose OK). The model updates. Notice that the slot now pierces four walls.

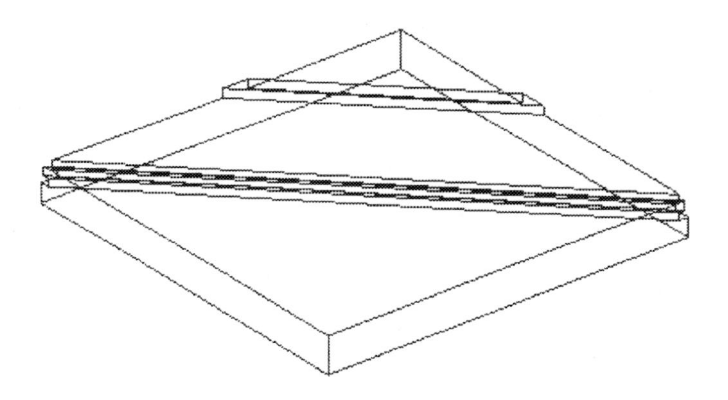

Step 4. Add a design feature **Rectangular Pad** on the top face of the block with the following parameters: Length=2, Width=2, and Height=1. Position it using two **Line onto Line** option such that the resulting figure looks as shown in the beginning of the activity.

Step 5. **Save and close** all part files.

Activity 5-4. Comparison of Modeling Approaches: Primitives vs. Design Features

Design features are associative - if you change the geometry used to create a feature, the feature is also updated. Primitives such as Block and Cylinder are not associative. You will investigate this concept using this project. Use the same drawing of Project 4-1 and create a solid model using only one primitive. Note that in Project 4-1 you used four primitives to model the part. Save the new part file as **xxx_activity5-4.prt**.

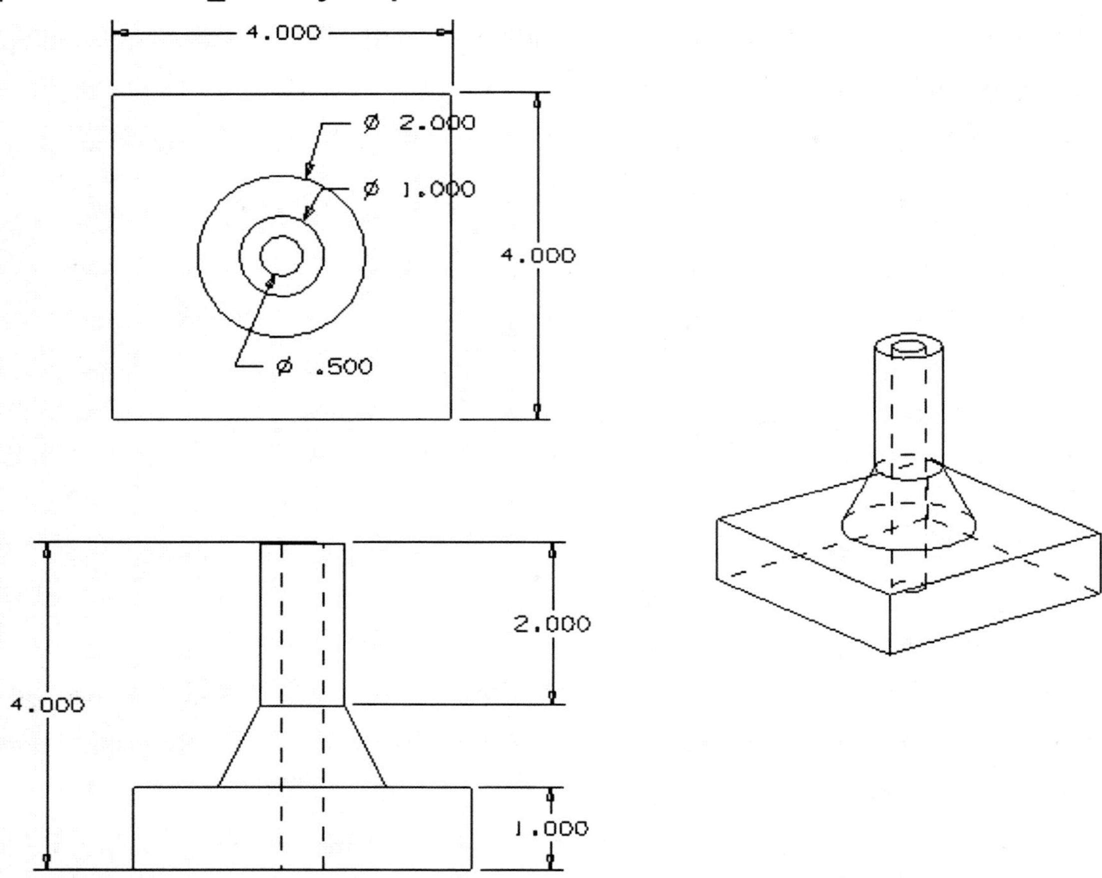

Step 1. Create a model starting with a block primitive.

Step 2. Add a tapered boss using the top face of a block as the placement face.

- **2.1.** Enter the taper angle as the following trigonometry function: arctan(0.5/1). Alternatively, you can simply enter the resulting value 26.565 (degrees).
- **2.2.** Use two **Perpendicular** positioning dimensions to locate the boss.

Step 3. Add the second boss using the top face of the first boss as the placement face. Its taper angle is zero. Use the **Point onto Point** positioning dimension to align the circle center points of these two bosses.

Step 4. Add a thru hole using the top face of the second boss as the placement face. The thru face is the bottom face of the block. Use the **Point onto Point** positioning dimension to align the circle center of the hole to the circle center of the second boss.

In the following Steps 5 and 6, you are going to change the height of the block for this model and that of the block that you created in Project 4-1, and then to observe the difference of the two models.

Step 5. Choose **Edit → Feature → Edit Parameters**. "**Edit Parameters**" dialog appears.

5.1. Select **Block(0)** and **OK**.

5.2. Click "**Feature Dialog**" and Change **Z length** from 1 to 2 and then click **OK** or MB2 three times.

You will see the block height increase from 1 to 2 and all the form features move upward as the block top face rises due to the increased block height.

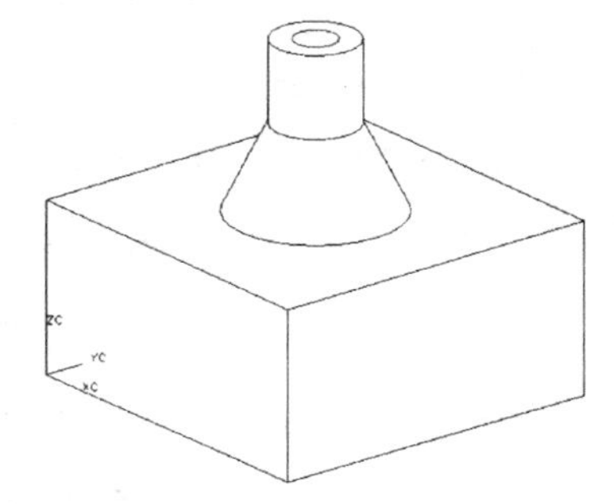

Step 6. Now open **xxx_project4-1.prt.** Repeat **Step 5**. As the block height increase from 1 to 2, you will see the cone primitive disappear (in fact, it is buried inside the block).

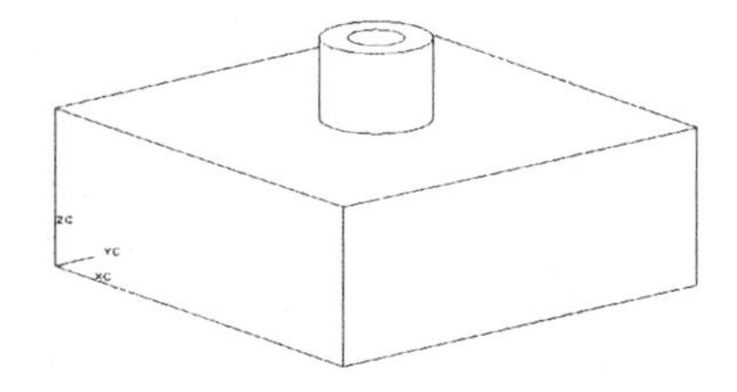

Why do the resulting models of the above two steps differ? This is because design features are associative but primitives are not. Consider how design features are constructed. The first tapered boss is associated with the block in that the block top face was used as the placement face of the boss. As this face rises, the boss on the face also rises. The second boss is associated with the first boss through the placement face (top face of the first boss), and the hole associated with the second boss through the placement face (top face of the second boss). As a result, these two features move upward as the first boss moves upward. On the other hand, the cone or cylinder primitives are placed in space of WCS with no association (relation) with the existing solid body, which is a block. The next question you may ask is which approach is right between the two. This is determined by the design intent of a designer. Therefore, modelers need carefully interpret the design intent when they model parts.

Design Intent

An important aspect of the design process is the conditions under which the designer must operate as defined by marketing, manufacturing and sales considerations that will affect the final design. All of these considerations are contained in the term design intent. More formally stated, design intent is the ability to capture geometric conditions, dimensional relationships, and relative positions between features on solid models. A CAD software application that can capture design intent is called a parametric modeler. Unigraphics has powerful parametric modeling functions. The program gives you a choice to add, change, or later remove, design intent. At the start of the design, you may not know what sort of design intent is necessary. Using expressions, positional dimensions, instances and feature editing, you can change the model easily as design conditions and intent change. In the following chapter, we will discuss editing models.

Project 5-1. Shaft Support

Create a model of the part using the drawings and dimensions (in inches) given below. Name your part **as xxx_project5-1.prt** where xxx is three-letter initials of your name.

Hint: One approach to model this part is to start with a block and add the following form features: two bosses and 6 simple thru holes.

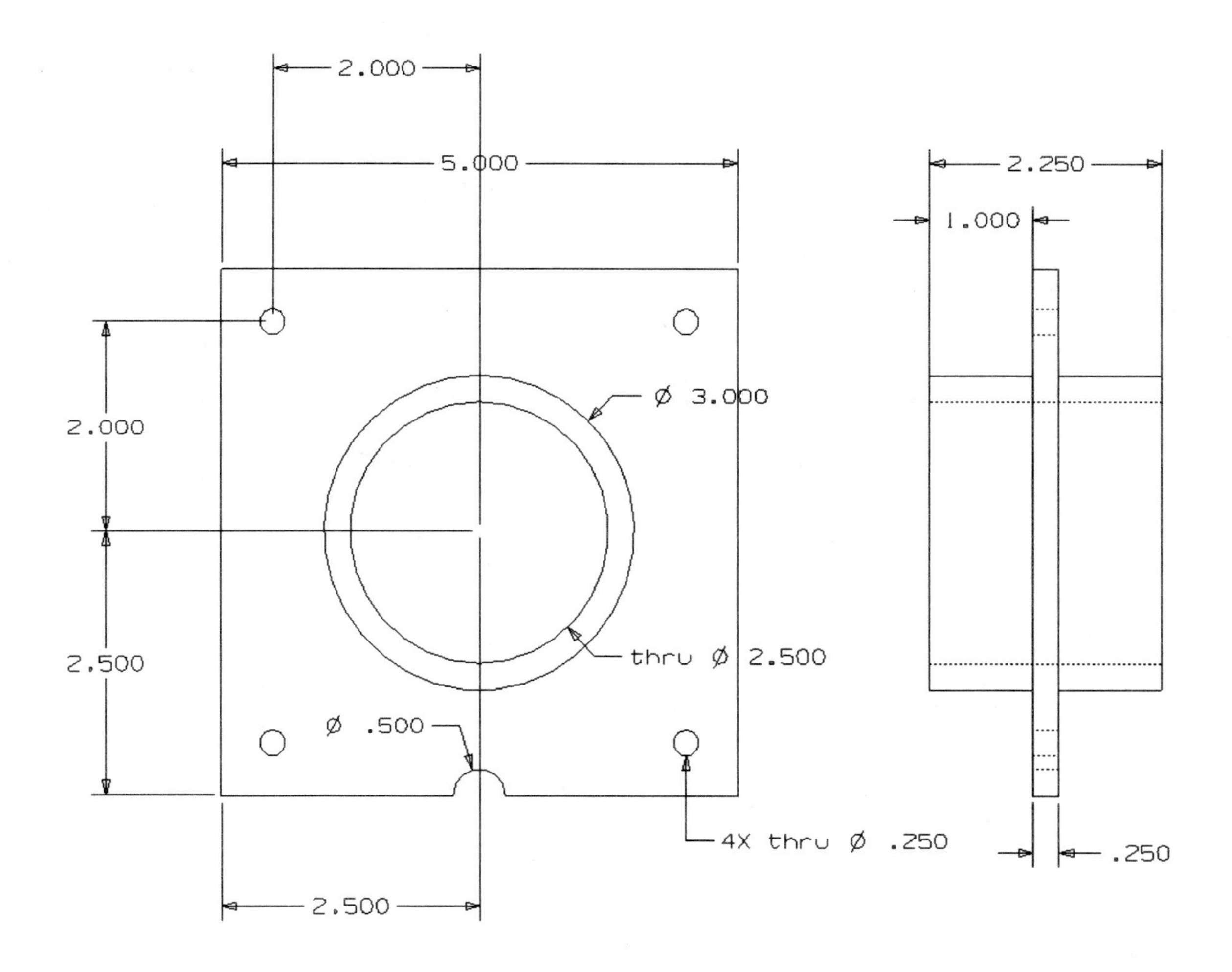

Project 5-2. Bracket

Create a model of the part using the drawings and dimensions (in inches) given below. Name your part as **xxx_project5-2.prt** where xxx is three-letter initials of your name.

Hint: One approach to model this part is to start with a block and add the following features: four simple thru holes, two bosses, two pads, three rectangular slots, and another two simple thru holes.

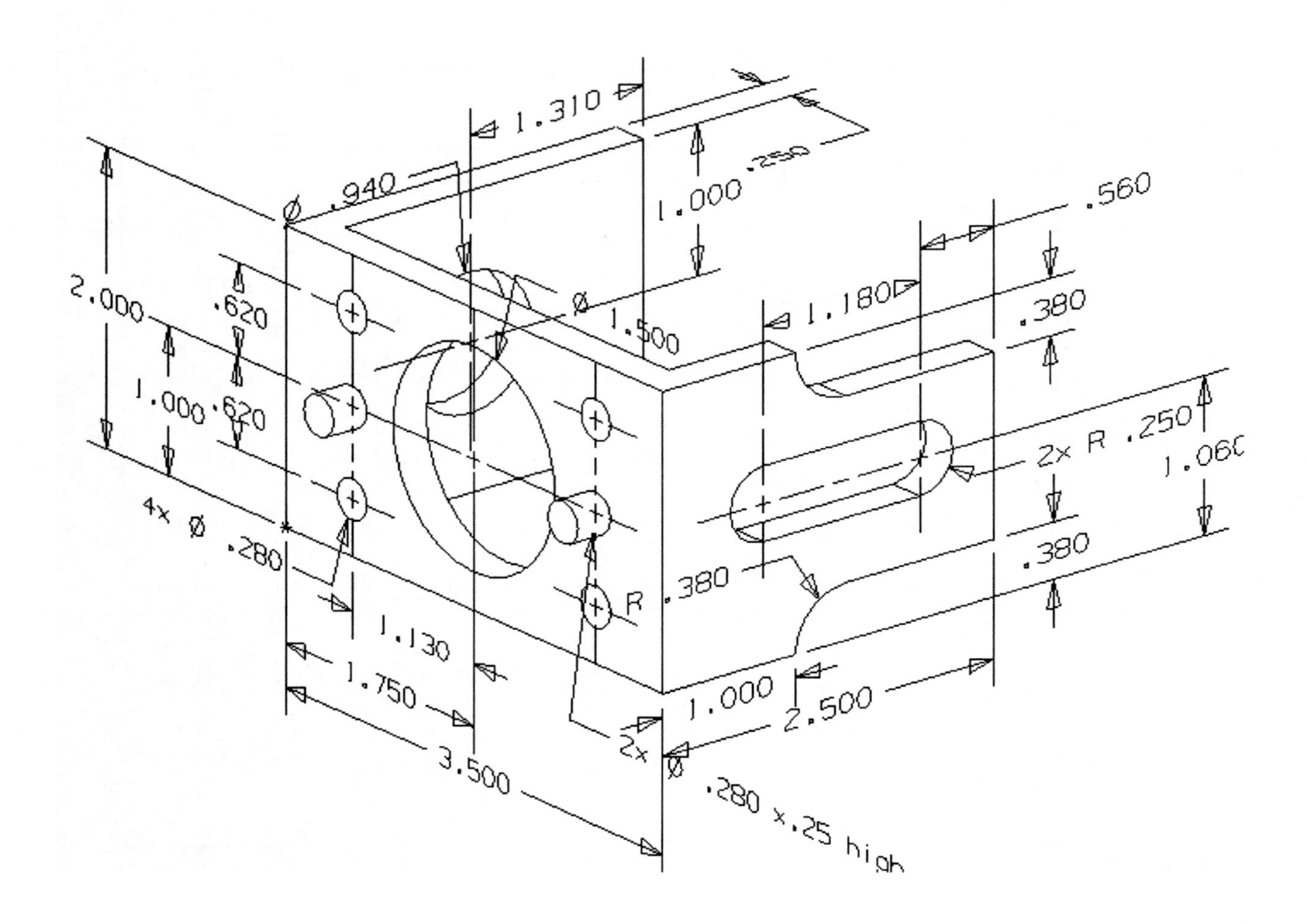

Project 5-3. Mounting Plate

Create a model of the part using the drawings and dimensions (in inches) given below. Name your part **as xxx_project5-3.prt** where xxx is three-letter initials of your name.

Hint: One approach to create a model is to start with a block and add the following features: 4 pads, 1 rectangular pocket, 1 boss, 4 simple thru holes, 1 counter bore thru hole, and 1 rectangular slot. You need to apply various positioning methods to locate the form features.

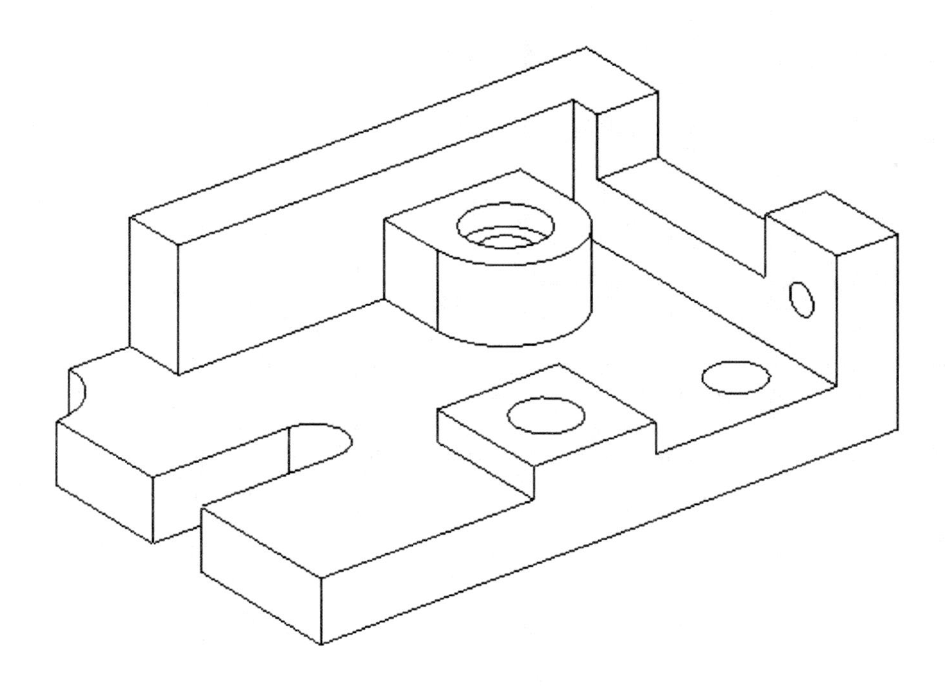

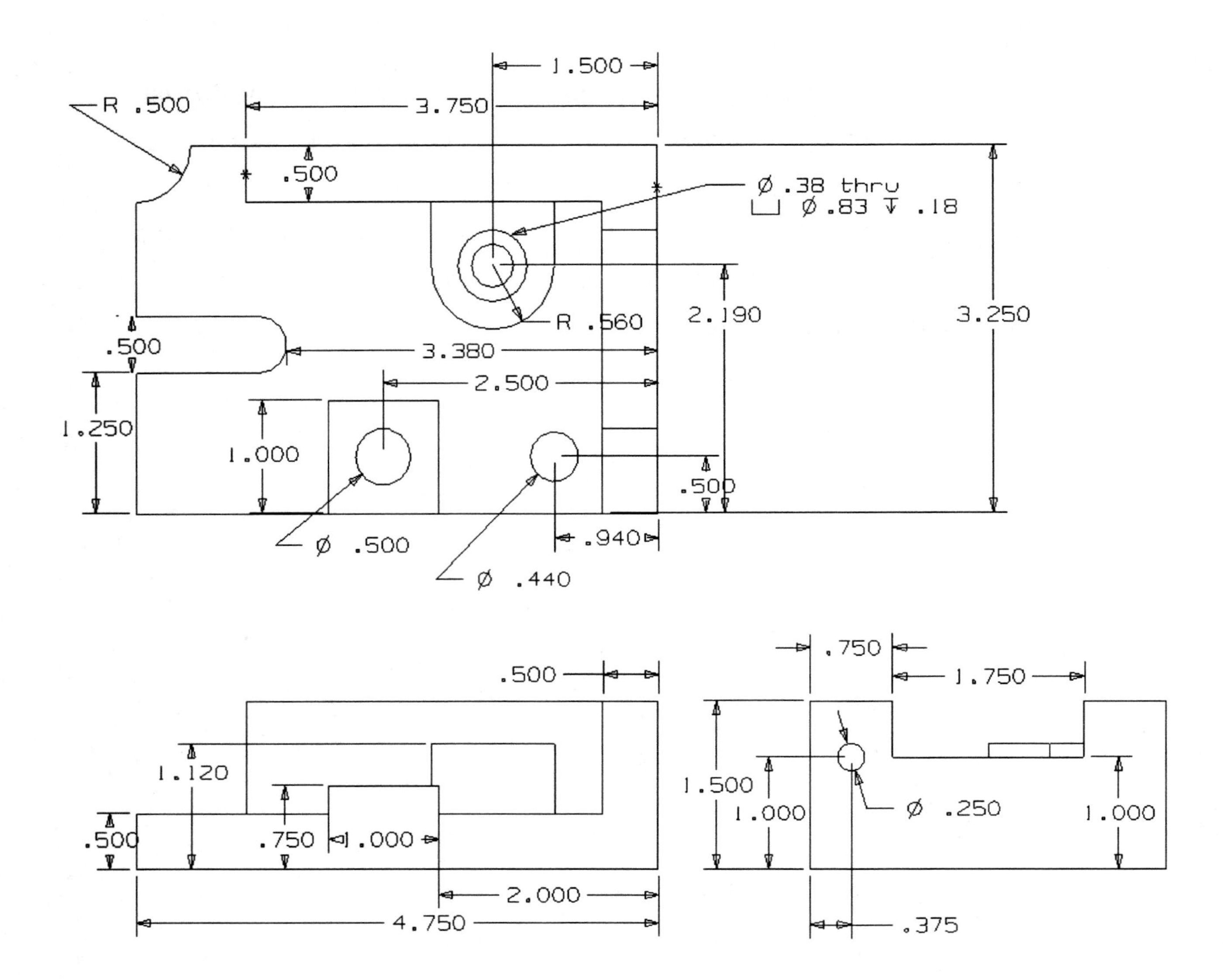

1.500
R .500
3.750
.500
Ø .38 thru
⌴ Ø .83 ↧ .18
R .560
2.190
3.250
.500
3.380
2.500
1.250
1.000
.500
.940
Ø .500
Ø .440
.750
.500
1.750
1.120
.750
1.000
.500
1.500
1.000
Ø .250
1.000
2.000
4.750
.375

Exercise Problems

5.1 Write a common procedure to create a design feature.

5.2 What is the planar placement face? Give me three design features that require the planar placement face.

5.3 What is the horizontal reference? Give me three design features that require it.

5.4 Name the nine positioning methods.

5.5 What is the advantage of using the Perpendicular positioning method over the Horizontal or Vertical positioning methods?

5.6 Is the Point onto Point positioning method a special case of the Parallel one?

5.7 What happens to the positioning dimensions when their reference objects or geometry used for positioning are deleted?

5.8 Suppose that you need to add a cylinder shape of a solid to an existing solid body. You can do so by adding either a primitive Cylinder or a form feature Boss. Compare these two approaches.

5.9 What is the difference between two form features, the Rectangular Pocket and the Rectangular Slot?

5.10 The Thru Hole and the Thru Slot differ in picking "thru" faces. Discuss the difference.

5.11 What is the design intent?

Chapter 6. Querying and Editing Models

When you work on part models, particularly, ones created by others, you often need to query or interrogate a model for various pieces of information. Sometimes, this information is geometrically related such as the distance between two objects like edges or points. Other times, you need to understand how a model was constructed for editing purposes. Investigation of construction methods helps you to understand the design intent of the model. This design intent includes not only the dimension and function of the model itself, but also the intent of design change, that is, how the model is supposed to react to changes of parameter value(s).

This chapter consists of six sections. Section 6.1 discusses how to use layers to find objects. Section 6.2 addresses the Model Navigator, which helps to navigate your model easily for querying or editing. Section 6.3 discusses expressions and their relation to parametric modeling. Sections 6.4 through 6.6 discuss three other querying functions in order: (1) playing back to reconstruct a model from the beginning, (2) retrieving information on features, and (3) finding distance between objects.

6.1 Using Layers to Find Objects

In Chapter 3, you were introduced the concept of layers as well as the tool within Unigraphics to control them. In this section, the Layer Settings dialog is used to determine what layer an object is on, which is very helpful in determining the overall organization of the part file.

Most companies employ layering standards, which are company standards to dictate what layer or layers objects should be on. For example, a company may dictate that solids be placed on layer 1, and sketches on layer 21 in all part files. Obviously there are benefits of following the standards. For example, if these standards are well followed, it helps one to understand a part model developed by another by quickly identifying how many objects and what kinds of objects are on which layers.

To find the layer that an object is on, simply go to **Format → Layer Settings** to bring up the **Layer Settings** dialog shown below. Once the dialog has been opened, pick the category, **ALL** followed by **Selectable** followed by **Apply** as shown below. This will show all 256 layers including ones with no object, and make all objects selectable. Changing from **All Layers** to **Layers with Objects** (see number 4 in the below figure) will show only layers that have at least one object.

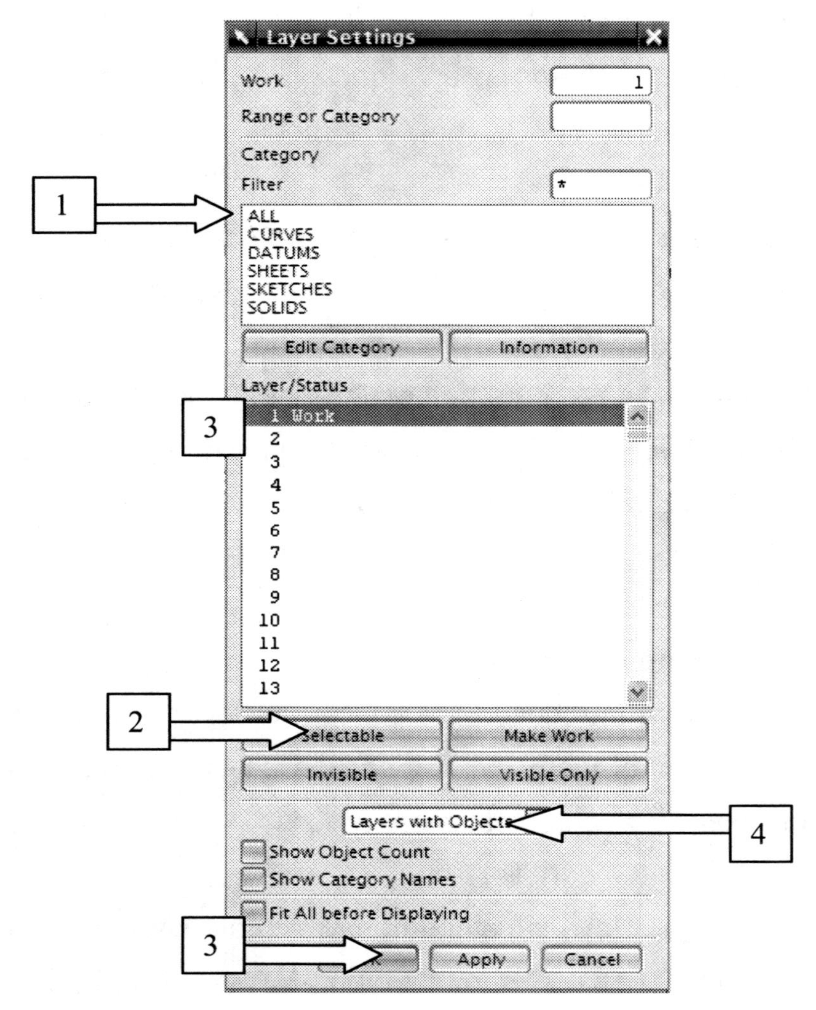

Now, with all objects being displayed on your screen, simply select them in the graphics window and look at the **Layer Settings** dialog. In the list of layers, the layer that the object you selected is on will highlight, informing you on which layer the object you selected is placed.

6.2 The Part Navigator (PN)

The Part Navigator (PN) is one of the most useful tools in modeling. It gives you direct access to the various features that have been constructed in your model. From gathering information to changing parametric values, PN gives you single point access capabilities for a variety of tasks.

Open part file Mounting_Plate.prt. To turn on PN, select the Part Navigator icon from the resource bar on the right side. A sample PN window is shown below. If you want to keep this window open, click the pin button.

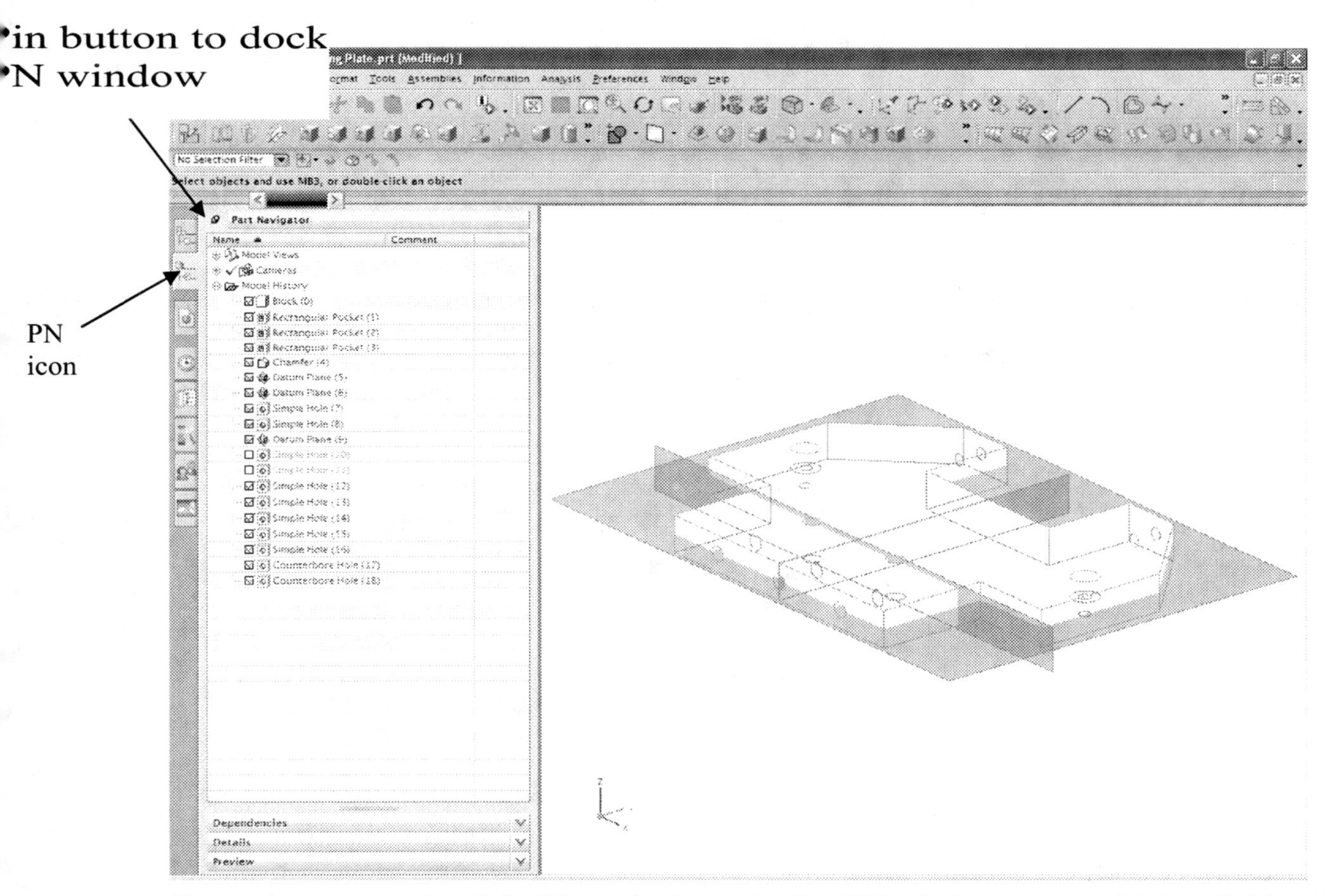

If you do not see the list of form features in the PN window, expand it by clicking + icon. Notice the correlation between the types of primitives/form features that make up the solid model and the list of items in the PN window. For each hole, there is a feature "HOLE" that shows up in the Part Navigator along. If you click, say, 'SIMPLE HOLE (16)' in PN with MB1, the corresponding feature is highlighted in the graphics window. Click + icon in front of 'SIMPLE HOLE (16)' and it will be expanded further to show how the form feature is created such as its placement face. If you click 'Face' with MB1, that face will be highlighted on the graphics window.

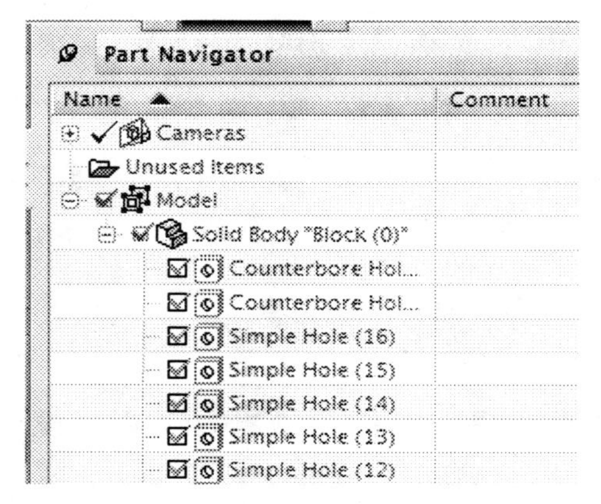

There are other features (neither form features nor primitives) that are used to create solid models but not shown in the list on the current PN window. These include reference features which will be discussed in Chapter 8. In order to see these features as well, click MB3 in the PN window once (do not click on any object) and the following pop-up menu appears.

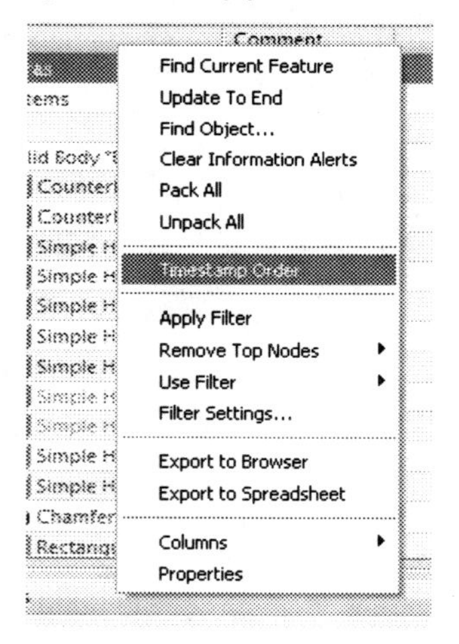

Click Time Stamp Order. The PN window now shows that this model consists of 19 features as shown below. Three of them are Datum Plane which is a reference feature. The 'Layer' column in PN shows that all the three Datum Plane features were created in layer 61. Right next to a feature name in the PN window, there is a number in the parentheses, which is referred to as the timestamp. The timestamp simply indicates the order in which features are constructed. BLOCK(0) means that the block is the first feature created and COUNTER_ BORE_HOLE(18) is the last feature created among the features in the list.

With the PN window open, simply pick the feature from the graphics window [PN window] and the feature will be highlighted in the PN window [graphics window]. Once the feature of interest is highlighted, place your cursor over the highlighted item in the PN window and click MB3. A quick menu pops out in the graphics window with various options to choose from as shown below. Note that RECTANGULAR_POCKET (1) is highlighted in this picture. We explain commonly used options below.

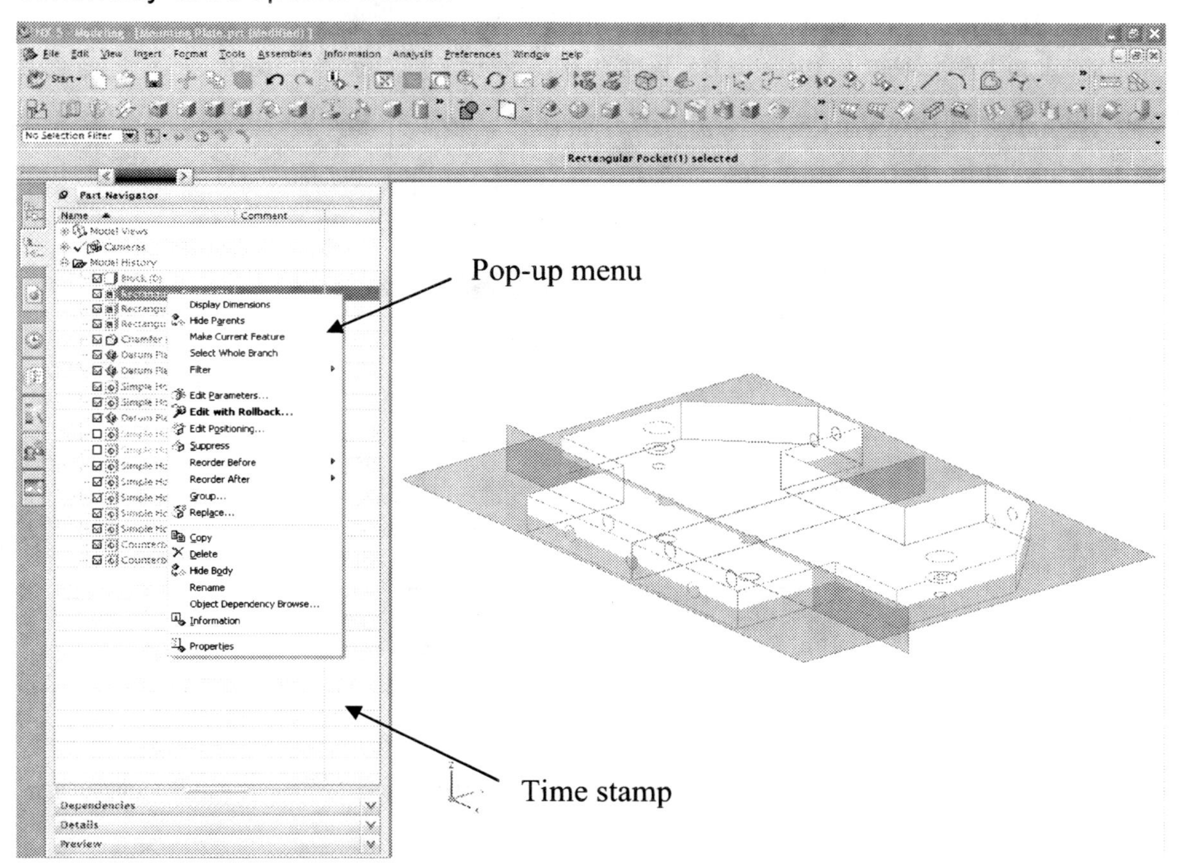

Display Dimensions will show you the parametric expressions and their associated values that control the feature you are working with. This is very helpful for investigating a model.

Suppress will hide the feature from view. Note that this does not delete the feature but simply does not display it. Alternatively, you can uncheck a small square box in front of the feature name by clicking it with MB1. Click it again to Unsuppress.

Edit Parameters will allow you to modify the controlling expressions for that feature (this would be a parametric modification). Selecting this option will result in the dialog shown below as for the Rectangular Pocket feature. Note that this

dialog is similar to the one used when the feature was created, and thus the dialog format is feature-dependent. At this point, you may either select one of the temporary dimensions that are shown on the graphics window or simply select the Feature Dialog button. Selecting this button will bring up the following dialog.

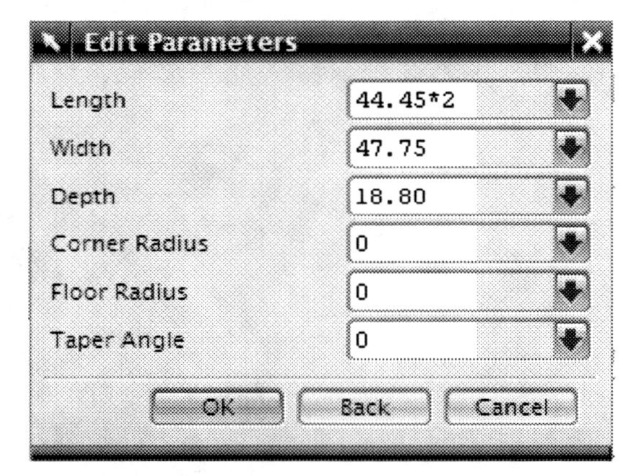

Here, simply modify the values that you would like to change and select **OK** twice. The feature is updated with the new parameter values.

Edit Positioning (only available when dealing with form features) will allow you to modify a located feature's position relative to the model.

Replace Feature allows certain types of features to be replaced by another feature. This option is limited and must be used carefully.

Reorder Before/After modifies the timestamp of the given feature to show up earlier or later.

Make Current Feature allows the insertion of new features after the current feature highlighted in the PN window.

Delete will remove the feature (and any of its dependents) from the model.

Rename allows you to rename the feature with a user-defined name.

Object Dependency Browser will show you what objects (curves, edges, faces, bodies) this feature is dependent upon for its existence. This can be rather helpful for investigative purposes and will help to establish the design intent mentioned earlier. Similar info can be obtained by clicking "Dependencies" near the bottom of the PN window.

Group lets you create a new feature in PN that represents a group (subset) of features in the model that are selected by you. This group feature created then acts as one feature for manipulation.

Information will open a window with information on the feature. This option is helpful for querying and investigating models. The sample list information is given below for feature RECTANGULAR_POCKET(1).

```
============================================================
Information listing created by Joe Miller
........
RECTANGULAR_POCKET(1)
======================================================

p3=44.45*2                                    88.9
Length X

p4=47.75                                     47.75
Length Y

p5=18.80                                      18.8
Length Z

p6=0                                          0
Corner Radius

p7=0                                          0
Floor Radius

p8=0                                          0
Taper Angle

p10=120.65                                    120.7
Positioning Dimension Horizontal Distance

p9=0.0                                        0
Positioning Dimension Parallel To Edge Distance

Feature Parameters for:RECTANGULAR_POCKET(1)
========================================================================
```

```
FEATURE TYPE - RECTANGULAR_POCKET(1)
X Length        =    88.90
Y Length        =    47.750
Z Length        =    18.80
Corner Radius     =    0.0
Floor Radius      =    0.0
Taper Angle       =    0.0

Feature Associativity for:RECTANGULAR_POCKET(1)
==========================================================================
 Parent(s):
   BLOCK(0)
```

Portion of this info can be obtained by clicking "Details" near the bottom of the PN window.

Properties option allows you to add and edit non-graphics comments and attributes to the feature selected. It allows renaming of a feature also.

Activity 6-1. Practice with Layers and Part Navigator (PN)

Step 1. Open file Mounting Plate.prt.

Step 2. Use the Layer Settings dialog (**Format → Layer Settings**) to find how many objects are on which layers and what type of objects (solid, curve, sketch, datum) they are.

Step 3. Use PN to find all the parameter values for RECTANGULAR_POCKET(3) including positioning dimensions. Edit this feature to reduce its length by 25%.

Step 4. Suppress and unsuppress RECTANGULAR_POCKET(3).

Step 5. Close the file without saving.

6.3 Expressions

Expressions are algebraic or arithmetic statements used to control the characteristics of a part such as dimensions and relationships of a model. Expressions are created anytime you enter information into the system to dimension a feature or its location. For example, creating a hole feature leads to creation of expression of its diameter. Expressions are also created when a sketch is dimensioned or positioned. Expressions are considered as driving force behind parametric modeling and can easily be edited for model modification.

Expressions have their own grammar, which generally mimics the C programming language. An expression consists of both a name and a value. The name of the expression always appears on the left side of the equal sign and may be comprised of alphabetic and numeric characters provided that you begin with an alpha character. No spaces are allowed in expression names and expression names are *not* case sensitive. There are three commonly used categories of expression names: (1) system defined expression starting with letter p, (2) named expression (system defined one is renamed as another name by a user), and (3) user defined expression (one created by a user that may not refer to any geometry object). An expression name is associated with a formula and its value that appear as columns in the Expression dialog. The expression formula may be a simple numerical value, an equation, or even a logical/conditional statement. You can use constants, multipliers, and even trigonometry and other built-in functions in the expression. A sample set of the functions are listed in the table on the next page and more can be found when you click the function button f(x). An expression formula example is shown below:

6.8*sin(30) + p6 + 9

Notice that this particular expression references another expression, p6. Therefore, the Width of this part (or solution of this expression) is directly related to the solution of p6. If you rename p6 to something else, it changes not only its original location name and also any other place where it is referenced.

To open the expression dialog, go to **Tools → Expression** (or simply press Control + E) inside Application Modeling. Its dialog shows below. Three common operations done through the dialog are as follows. To create new expressions, type in the name and the associated value in the editing area, and hit "Enter" key. To edit an existing expression, select it from the expression list and modify it in the editing area, and hit "Enter" key. To rename an expression, select it from the expression list, and enter a new name and hit "Ok". For all of these three operations, you must click OK or Apply after editing or creating, in order to have your changes saved and reflected in the model.

On the top of the Expression dialog, there is a pull-down menu called "Listed Expressions". With this, you can control the display of expressions in nine ways: Filter by Name, value or formula, Display User defined expressions, Display all the expressions, and Display unused expressions etc.

If desired, you can associate an expression with an engineering dimension type (length, mass, etc) by choosing it from the pull down menu in the expression dialog. Its corresponding unit (inches, Kg, etc) is shown accordingly right below. You can change it into a different unit from another pull down menu.

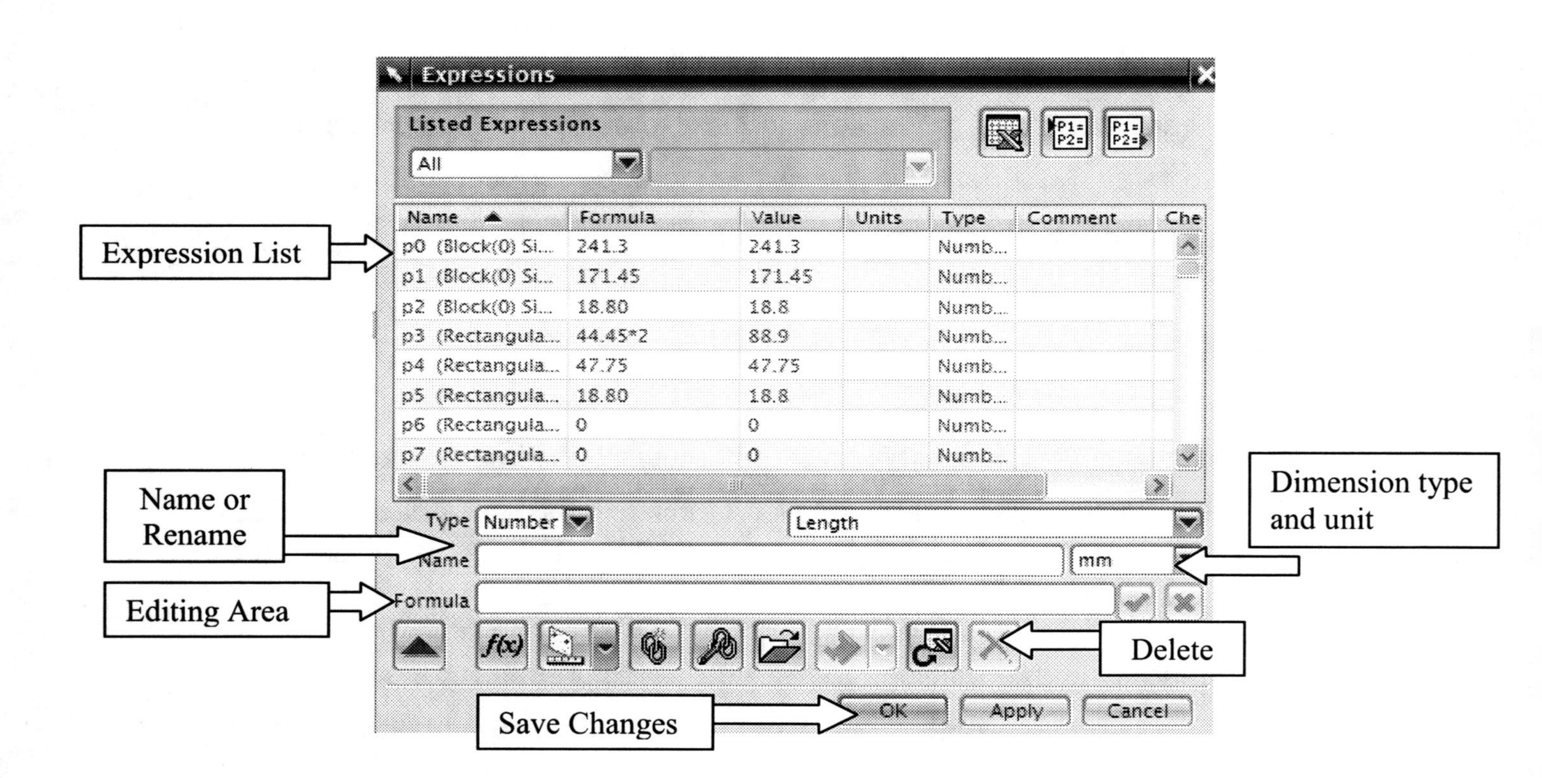

Table for Built-In Functions

Name	Description
abs	Absolute Value, abs(x) = |x|
arccos	Arc Cosine, arccos (x) = arc cos(x), (result in radians)
arcsin	Arc Sine, arcsin (x) = arc sin(x), (result in radians)
arctan	Arc Tangent, arctan (x) = arc tan(x), (result in radians)
arctan2	Arc Tangent, arctan2 (x, y) = arc tan(x / y), (result in radians)
ceiling	Ceiling, ceiling (x) = x is a real number, function returns closest integer greater than or equal to x if a= ceiling (.1), a evaluates to 1
cos	Cosine, cos(x) = cos(x), (x must be in degrees)
hypcos	Hyperbolic Cosine, hypcos (x) = cosh(x)
equal	Compares two expressions, equal(x, y) and returns a Boolean value
floor	Floor, floor(x) = x is a real number, function returns closest integer less than or equal to x if a=floor(.1), a evaluates to 0
log	Natural Logarithm, log(x) = ln (x) = loge (x)
log10	Common Logarithm, log10(x) = log10(x)
max	Returns the maximum of given numbers max(x, y, z)
Radians	Radian Conversion, Radians(x) converts degrees to radians
sin	Sine, sin(x) = sin(x), (x must be in degrees)
hypsin	Hyperbolic Sine, hypsin (x) = sinh (x)
sqrt	Square Root, sqrt (x) = /x
tan	Tangent, tan(x) = tan(x), (x must be in degrees)
hyptan	Hyperbolic Tangent, hyptan (x) = tanh (x)
pi()	Takes no arguments and returns the value of pi
ug_excel_read	Value = ug_excel_read("XXX.xls", "A5") Returns value from the specified cell in a given spread sheet

6.4 Playback a Model

Playback allows the replay of the construction of the model so you can better understand the way it was built. It is one of the Edit Feature menu choices. To run the playback tool, choose **Edit → Feature → Playback...** as shown below.

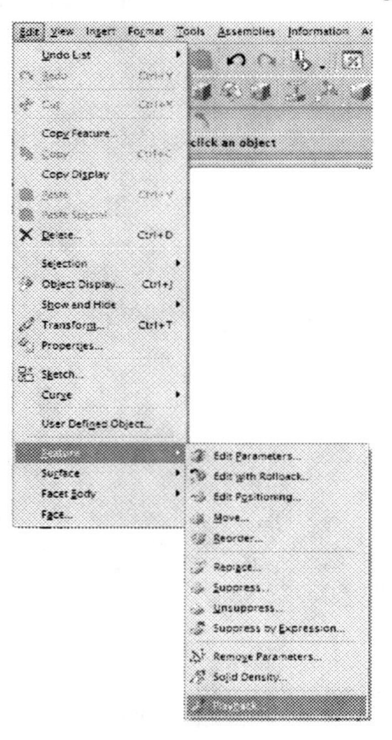

This brings up the playback dialog as shown below:

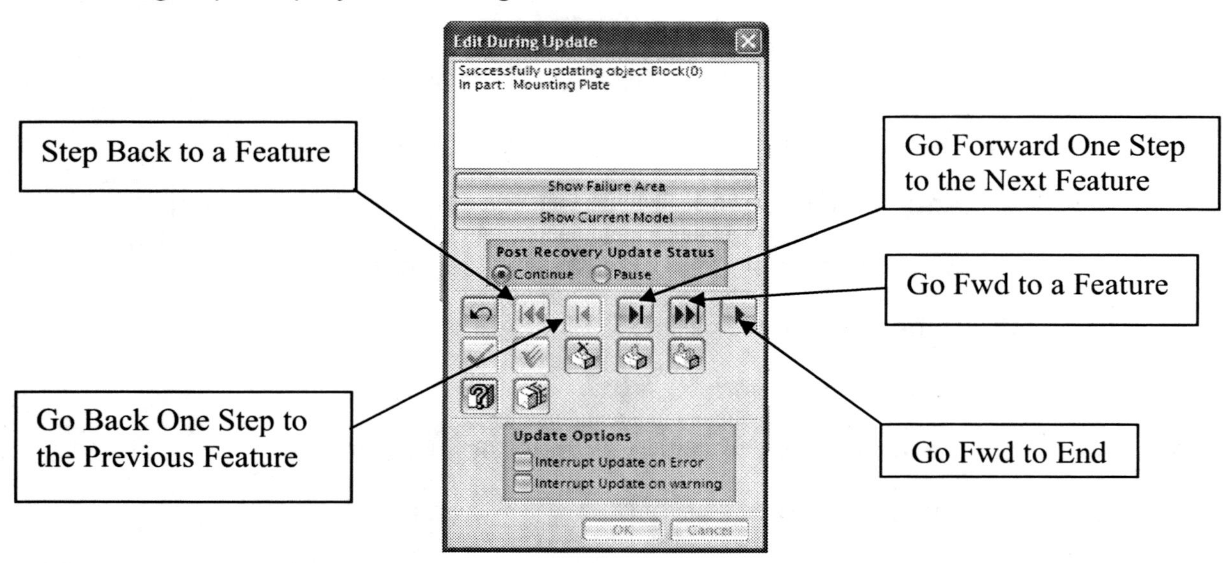

To use the dialog simply step forward or backwards through the model using the buttons. Note that some of other Edit Feature menu choices are directly accessible by the Model Navigator with the third mouse button click on the highlighted feature.

6.5 Information → Feature

Another way that you can get access to information about the features in a model is through the Information menu bar. Simply go to **Info → Feature** to bring up the dialog as shown below.

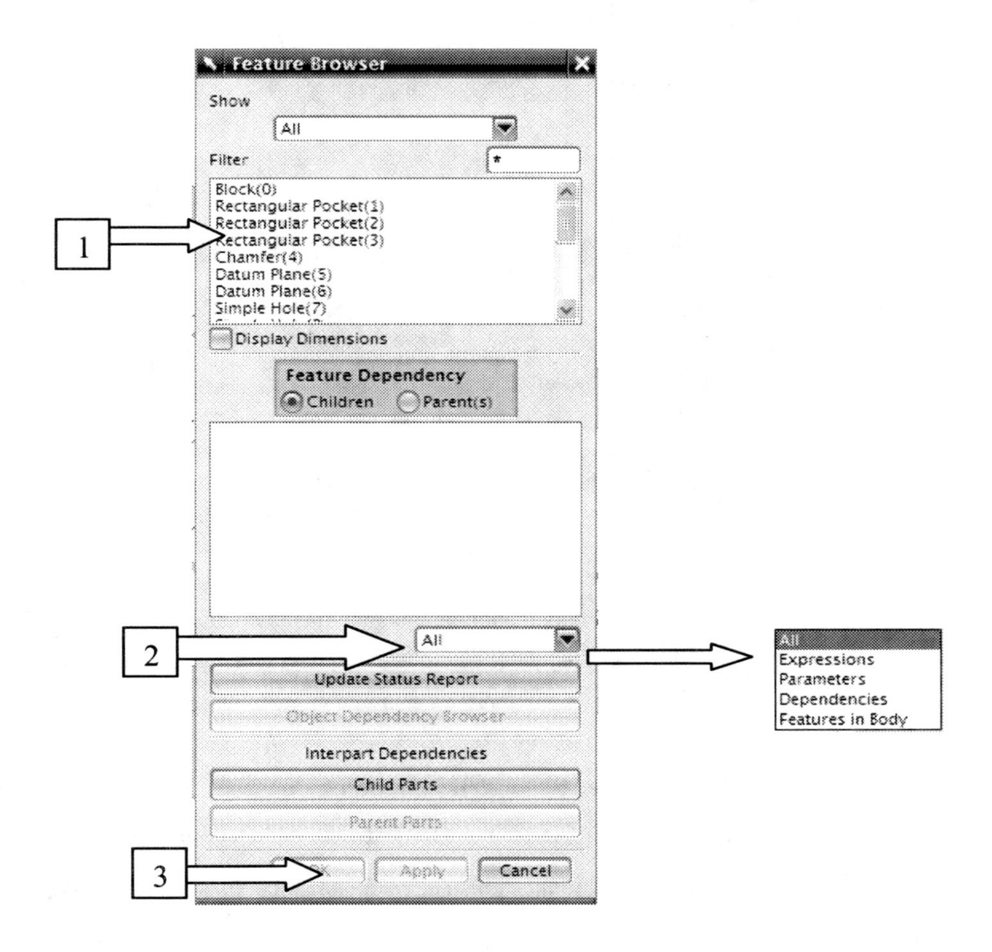

To get a listing window of information on a given feature,

1. Select the feature from the feature listing area
2. Change the **List** pulldown to reflect the information you seek
3. Select **OK**.

If you would like to have the information shown graphically as well, then you may turn on the **Display Dimensions** box in the dialog.

6.6 Distance Between Objects

In many instances, you need to know how far one object is from another. To find this distance information, go to **Analysis → Distance** to bring up the dialog shown below.

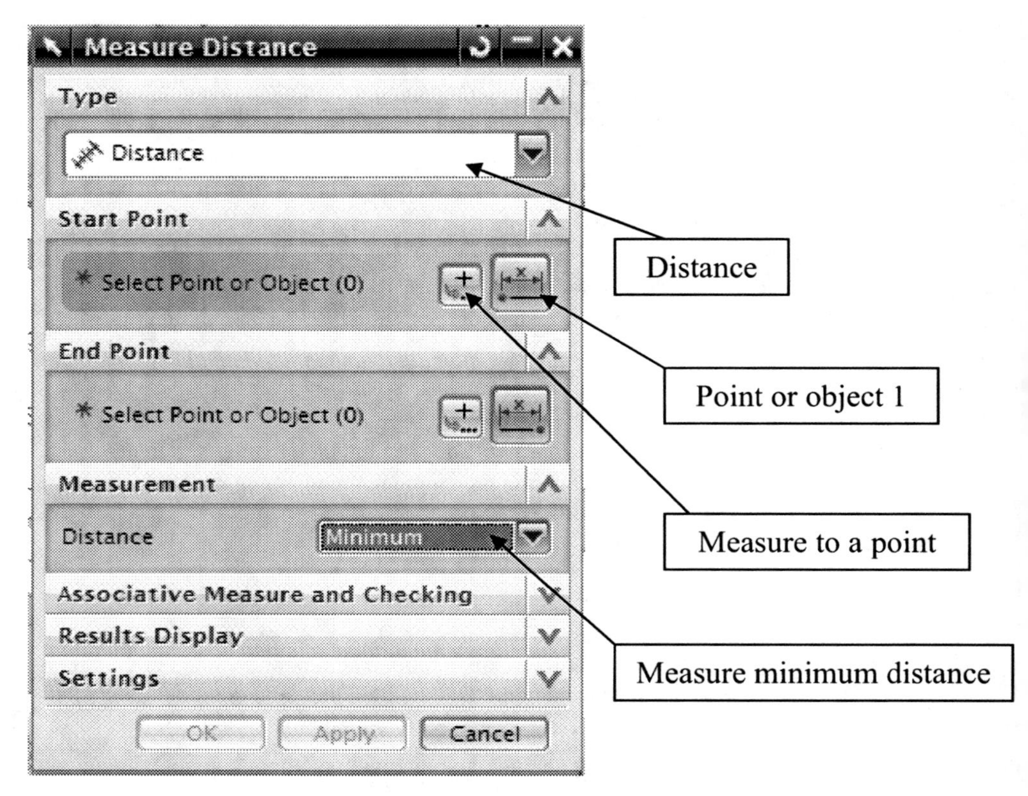

At this point, you must select two objects between which you would like the distance. The dialog shown above is little more than a selection filter to aid you in selecting objects from a crowded graphics window. For example, if you would like the minimum distance between two edges and the graphics window has too many objects, you would select the Measure minimum distance button, select the first edge and then select the second edge. At this point, a string with a scale along with the distance will be displayed on the screen. If you would like the minimum distance between two points, you click the icon to access the point constructor and then you can just pick the two point objects in the graphics window directly.

Activity 6-2. Query and Parametric Modification of a Part

In this activity, you will exercise two things: (1) how to interrogate the model to find desired information, and (2) how to modify a part using the information. You will use the shaft support part completed in Project 5-1 of Chapter 5. The figure below shows a part before and after modification.

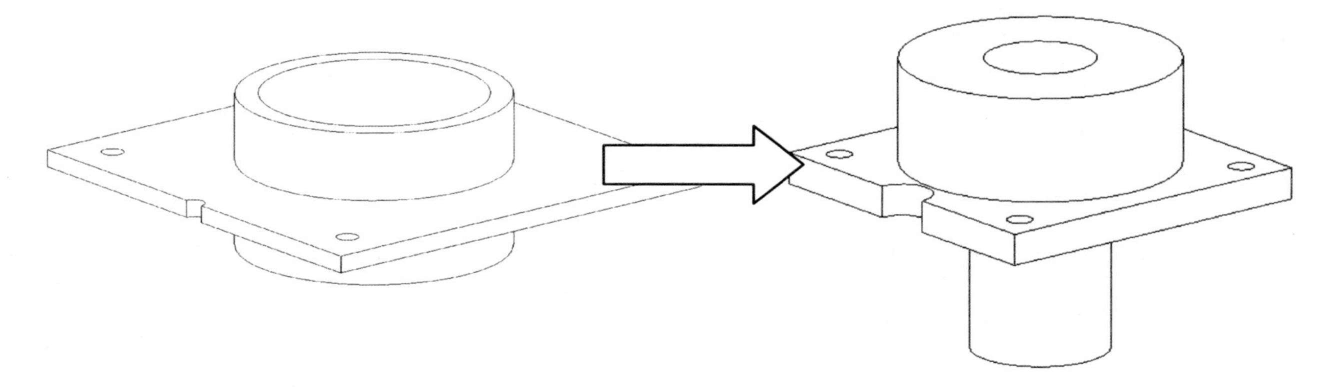

Step 1. **Open** your part **xxx_project5-1.prt** as it was completed, and save it as **xxx_activity6-2.prt**. Start the Modeling application.

Step 2. Turn on the **Part Navigator** icon and click the thumb tab to keep the PN window open.

Step 3. Shade the model.

Step 4. Review the features in the model against those shown in the Model Navigator by selecting them from your graphics window.

Step 5. Using **Analysis → Measure Distance**, find the distance between the outer holes and record it.

5.1. Go to **Analysis → Distance**. The following menu appears.

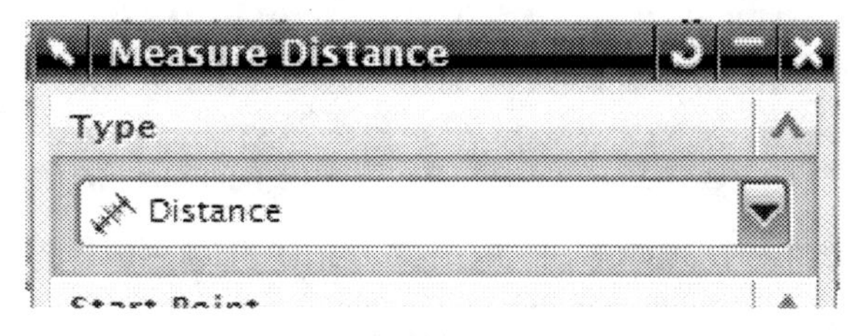

5.2. Select the **Distance** Option

5.3. Move the cursor onto the hole shown below and the status line reads Edge(Arc Center). Select the first hole. If the QuickPick dialog appears, choose the Arc Center option.

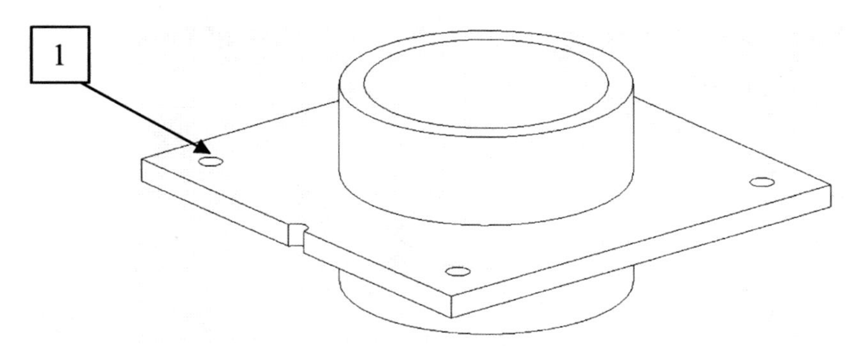

5.4. Move the cursor to the second hole as shown below and the status line again reads Edge(Arc Center). Select the second hole.

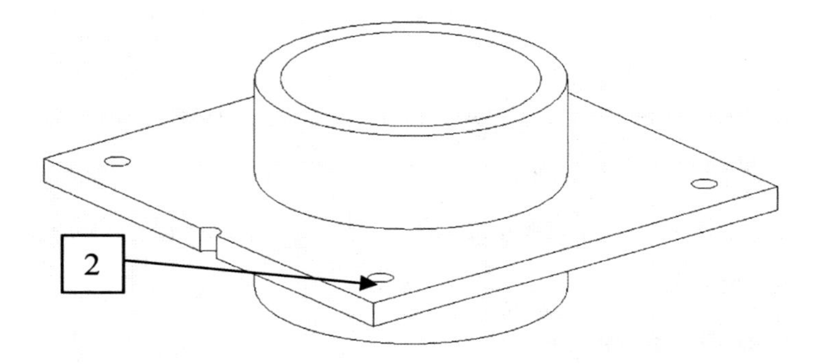

5.5. Record the **Distance** below and close the distance tool bar window.

DISTANCE = ____________

Step 6. Edit the top **Boss** from the **Part Navigator**.

6.1. Find the top **Boss** in the **Part Navigator** and select it with the left mouse button MB1.

6.2. While the cursor remains over the Boss feature in the PN window, click your right mouse button MB3 once, and choose **Edit Parameters**.

6.3. Choose **Feature Dialog**.

6.4. Change the **Diameter** to a value of **4**. (Note that this is also automatically modifying the expression that controls this value.)

6.5. Change the **Height** to a value of **2**. (Note that this too automatically changes the expression that controls this value).

6.6. Select **OK** twice. **Boss** was updated.

Step 7. Edit the centerline hole diameter from the **Expression** dialog.

7.1. Find the simple hole feature for the centerline hole in the **Part Navigator** by selecting it from the graphics window.

7.2. Place your cursor directly over the simple hole feature in the **Part Navigator**.

7.3. Click your right mouse button once and choose **Information**.

7.4. Below, write down the expression name that controls the diameter (pXX).

Diameter is Controlled By: ______________

Close the Information window.

7.5. Go to the **Expression** dialog by going to **Tools → Expression** or simply holding down Control + E. This will bring up the **Expression** dialog as shown below. Change the Listed Expressions filter to **Filter by Name**.

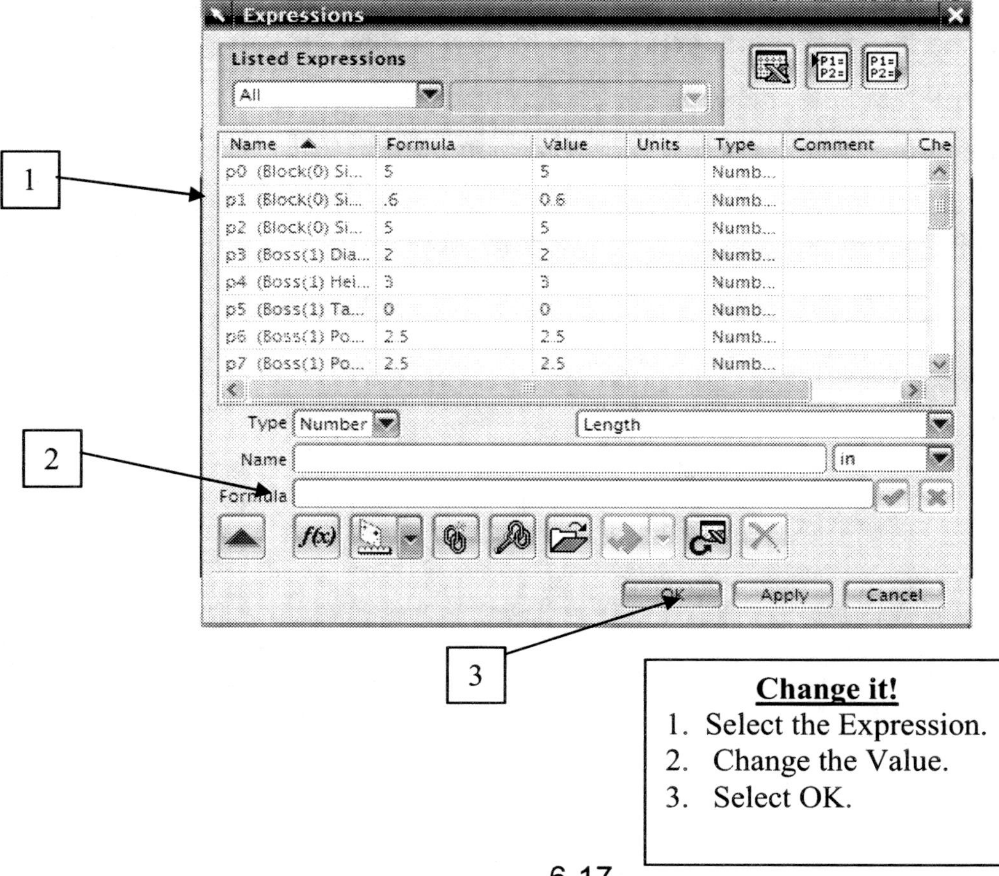

7.6. In the expression dialog, find the expression that controls the diameter (the one you wrote down above) and modify it to be 1.6. Choose OK. Your screen should look similar to one below.

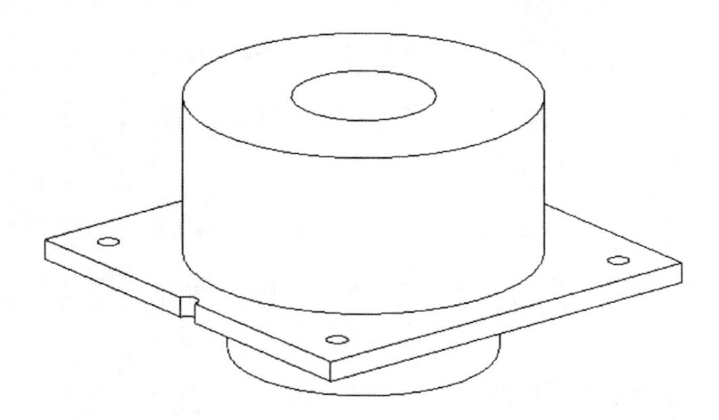

Step 8. Make the following changes to your model using the two methods discussed above. Write down the controlling expressions for each feature next to each step or on a spare piece of paper.

8.1. Change the lower boss to have a diameter of 2 and a height of 3.

8.2. Change the corner mounting holes to have a diameter of .375.

8.3. Change the plate thickness to be .6.

8.4. Change the locating notch to have a diameter of .9.

Step 9. Find the distance between the two arc centers shown below.

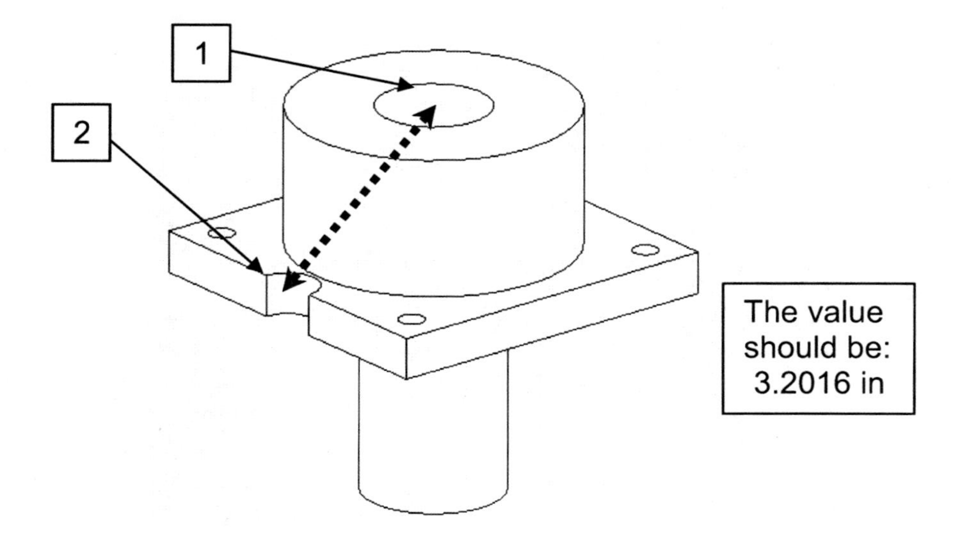

Step 10. Find the distance between two diagonal points of the block as shown below.

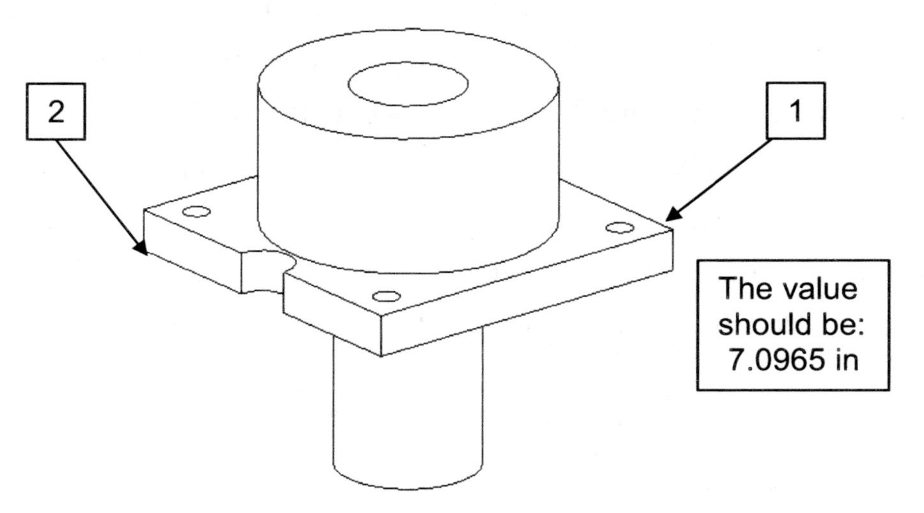

Note that you have to use the following setting (the Point Constructor) to compute the distance.

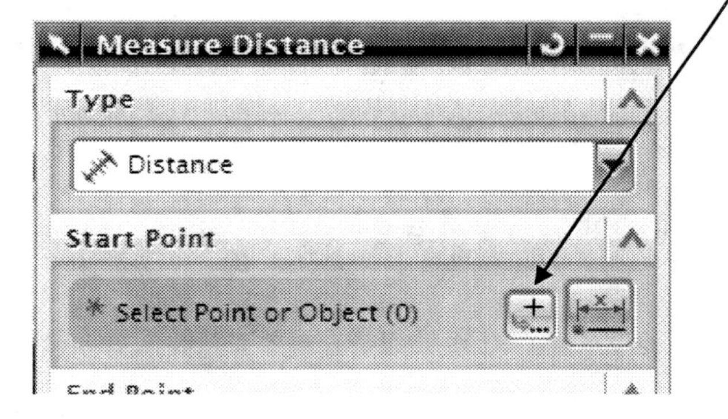

Step 11. Review the construction of your model by playing it back with the following menu picks, **Edit→ Feature→ Playback**. Dialog "Edit During Update" appears. Review the features of the model by clicking the Step button .

Step 12. **Save and close** this part.

You will use this part on **Project 6-1**.

Project 6-1. Two Changes in Design Intent

Open xxx_activity6-2.prt and save it as xxx_project6-1.prt where xxx are your three-letter name initials. There are two changes in design intent in this project and you will incorporate the changes into the model using expressions.

The first change. Until now, you have allowed the upper and lower bosses to be of a different size (diameter and height). In this project, both bosses will always have the same diameter and height. For ease of modification, use the expression names "dia" and "height" for them.

Hint:

1. Find the expressions for each boss.
2. Rename the expressions that control one of the bosses to be dia and height.
3. Edit the expressions that control the second boss to have the value of dia and height.

Verification:

Change dia and height to be of a different value and verify that both bosses update together.

The second change. There are four small holes on the plate. The diameters of all these holes are identical and are 1/8 of that of the bosses, "dia." After making proper changes in the expressions, verify if they are correct.

Exercise Problems

6.1. Describe how to find a layer on which a certain object is located.

6.2. Describe the functions of the Part Navigator.

6.3. When are expressions created?

6.4. What is Playback? Why is it useful?

6.5. Describe a procedure to find a distance between two edges.

Chapter 7. Feature Operations on Edge and Face

In this chapter, you will learn some of features that apply to edges or faces of a solid body. These include edge blend, chamfer, and hollow features. Edge blend and chamfer features apply to edges of a solid while a hollow feature apply to faces of a solid. Sections 7.1 to 7.3 discuss these three features and their options in order.

7.1 Edge Blend

An Edge Blend is a radius blend that is tangent to the blended (adjoining) face(s). This feature modifies a solid body by rounding selected edges. Blending works by rolling a spherical ball along the edge being blended (the blend radius), keeping it in contact with the two faces that meet at the edge. The ball will roll on the inside or outside of the two faces, depending on whether you are creating a fillet or a round.

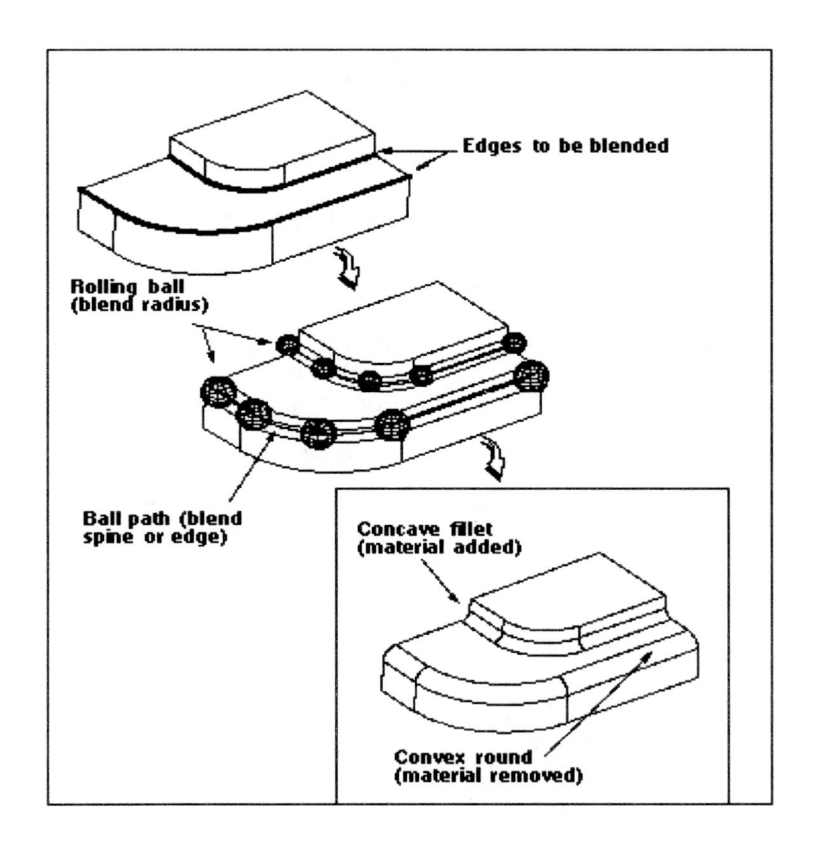

Choose **Insert → Detail Feature → Edge Blend** or choose the **Edge Blend** icon. The **Edge Blend** dialog comes up as shown below. By default the Constant Radius button is highlighted.

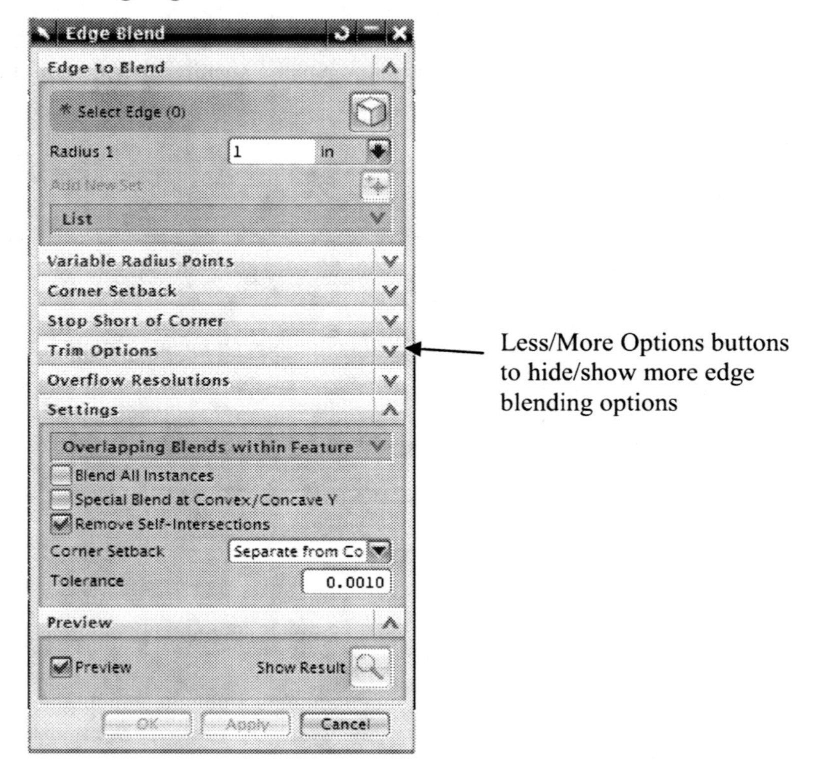

Now select the edge to apply the blend and a small pop up menu appears on the screen with a field to enter the blend radius (Default Value = 0.25 inch).

If you want to edge blend more than one edge with the same blend radius, select each of them and then choose Apply or OK. Instead of selecting edges individually, you can select a set of edges quickly by using the Selection Intent tool bar.

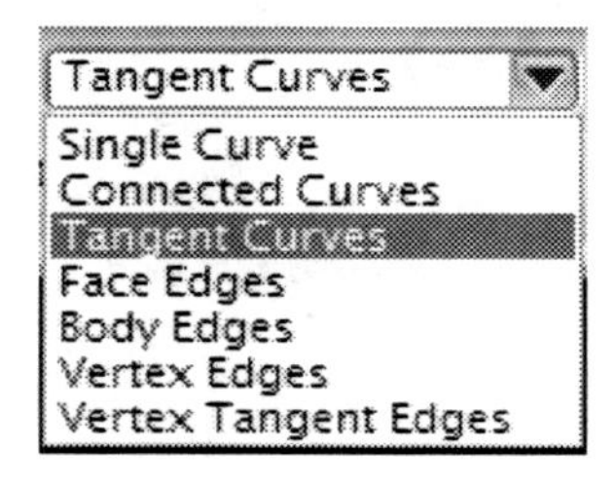

For example, when you select the “Tangent Curves ” option in the Selection Intent tool bar, picking any one edge in the graphics window causes the system to select the rest of the edges connected in the tangent string. The transition between adjacent faces of selected edges must be smooth.

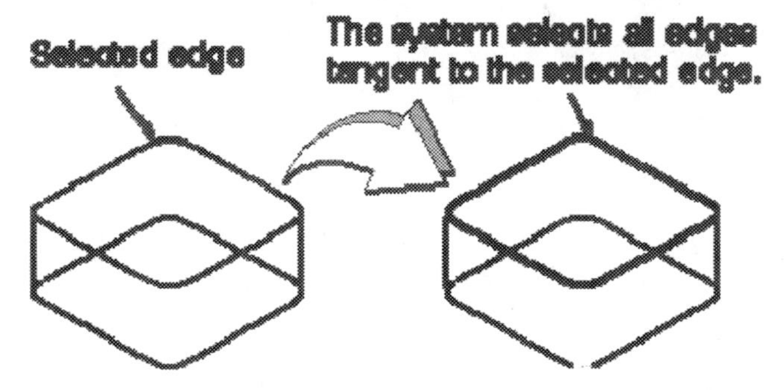

When possible, you should blend sets of edges rather than one edge at a time. When many edges are blended at one time, the result is a single blend feature that has many blended edges. However, if you blend several edges together, and later want to change the blend radius at only one edge, it will be more difficult than if you had blended them separately. Different blending sequences of the same set of edges can result in different outcomes as shown.

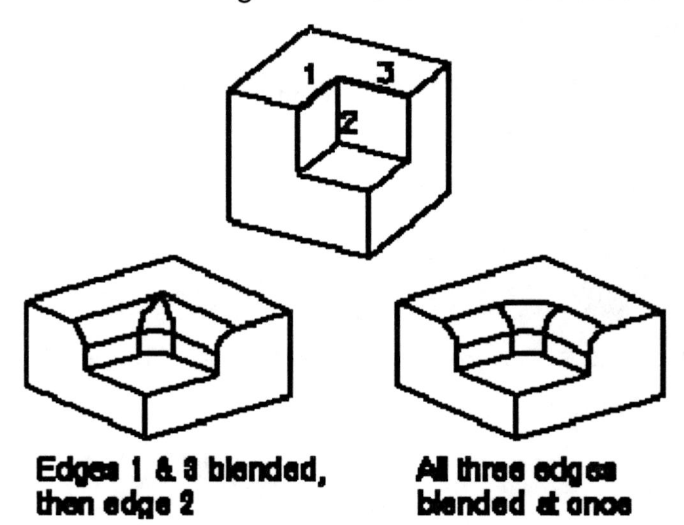

Activity 7-1. Creating Edge Blends

You will practice various edge blends in this activity.

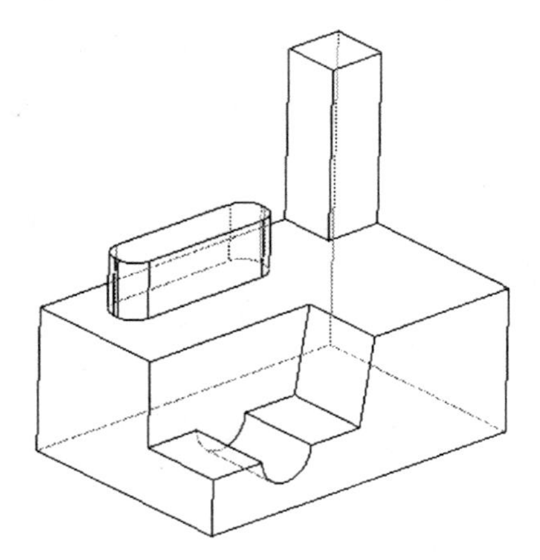

Step 1. Open the provided part file **activity7-1.prt** and save it as **xxx_activity7-1.prt** where xxx are your name initials.

Step 2. Start the **Modeling** application.

Step 3. Choose **Insert → Detail Feature → Edge Blend** or choose the **Edge Blend** icon. It brings up the **Edge Blend** dialog.

Create a radius .25-inch (default value) edge blend that removes material on edges of the tall post in the model by blending the four edges of the top face of the post and three sides of the post.

Step 4. First, change the **Curve Rule** filter in the Selection Bar into **Face Edges** as shown. If you find the Selection Bar not open, open it first by moving your cursor in any toolbar area and clicking MB3 and selecting Selection Bar in the list of toolbars.

Select the top face of the post.

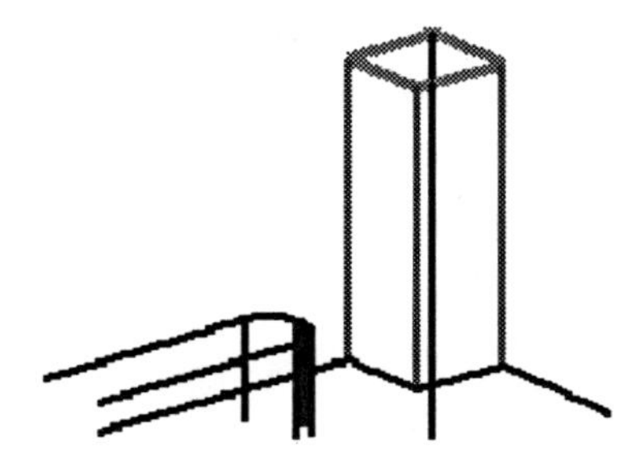

Step 5. Now, change the **Curve Rule** filter to **Single curve** and select the three vertical edges of the post as shown.

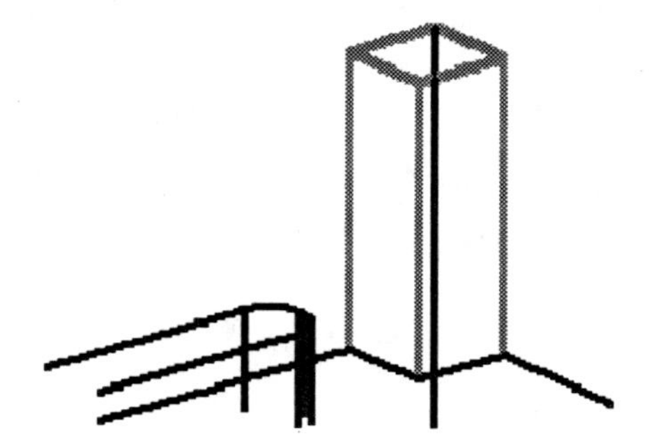

You can deselect any selected edge by pressing Shift and **MB1** on the edge you want to deselect. Set the blend radius to 0.25 inch if necessary.

Step 6. Choose **Apply** to complete the blend. Notice that blending three edges at a vertex requires the system to create an extra blended face.

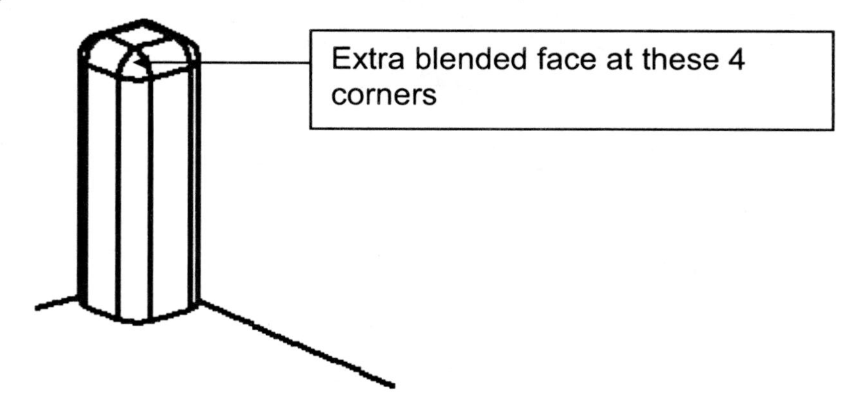

Step 7. Create edge blends with convex and concave faces.

7.1. Pick the three edges of an inside corner.

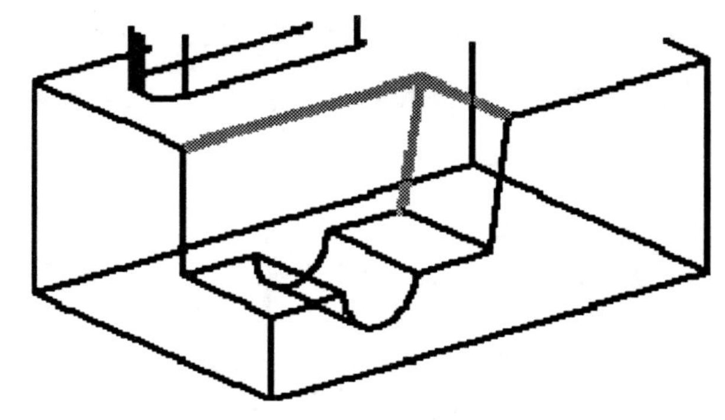

7.2. Key in 0.3125 in the **Edge Blend menu** that appears on the screen.

7.3. Choose **Apply** to create the blends that looks as follows:

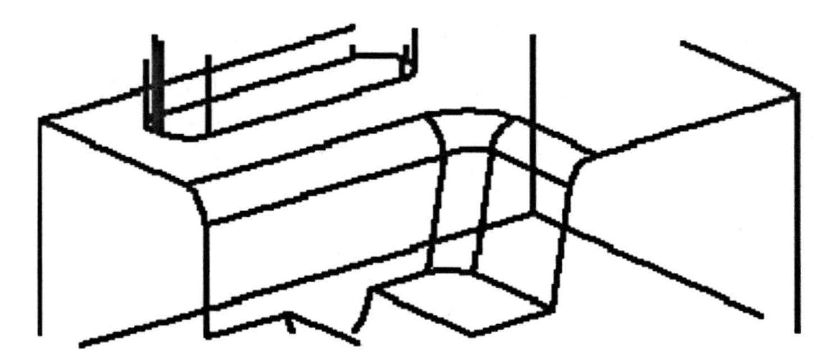

Step 8. Select the two edges as shown below and key in .375 in the blend radius field and choose **Apply**.

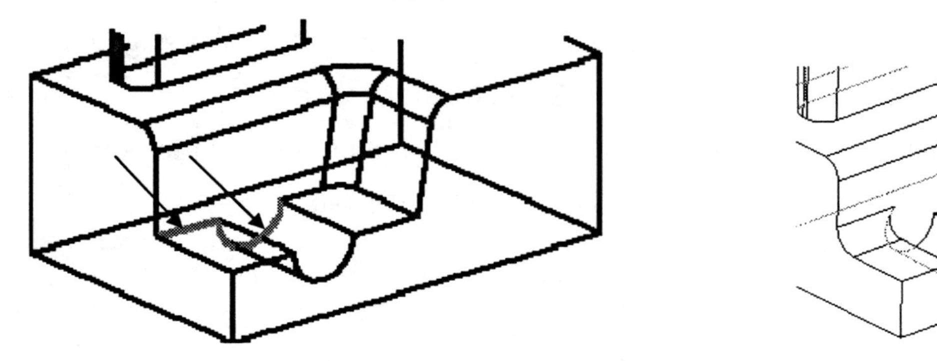

The blend was created but it obviously doesn't look correct. This is because the hole feature in this model has a radius of 0.25 and the blend radius is larger than the hole radius.

Use the popup menu and **Undo** the blend.

Step 9. Create the smaller blends on the lower edges of the rectangular pocket, using **Tangent Curves**.

9.1. Open the Blend Radius dialog, key in 0.25 in the blend radius field. Change the filter in the Selection Bar to **Tangent Curves**.

9.2. You need to select the same two edges plus the one edge of the tangent set of edges as shown below.

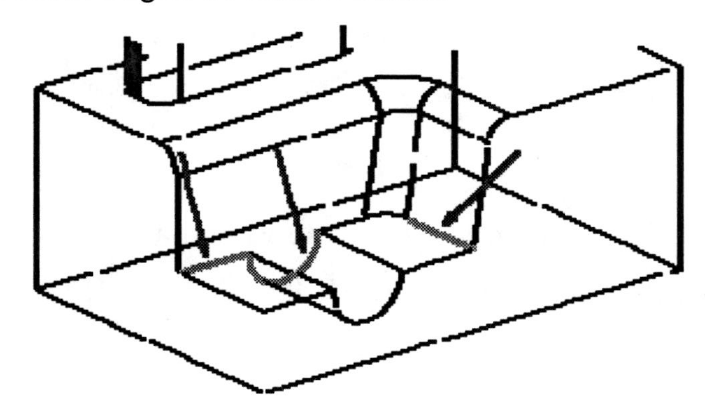

The system will select all the tangent edges. The entire string of edges highlights.

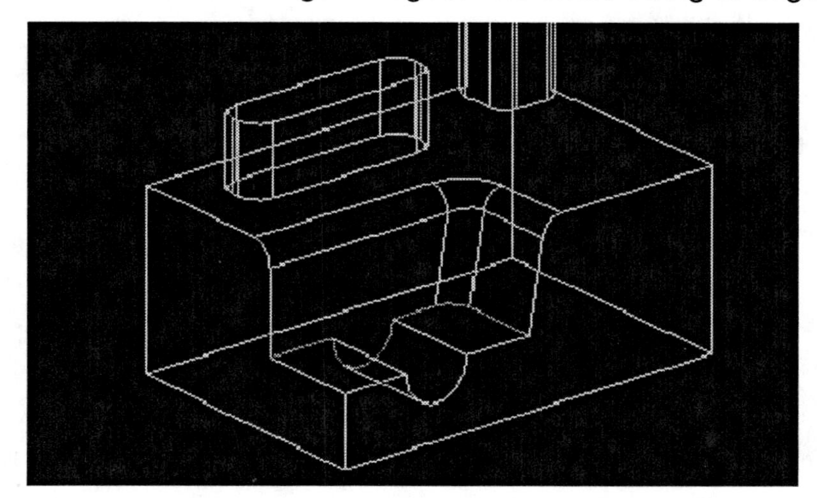

9.3. Choose **OK** to complete the blend and the resulting model looks as below.

9.4. **Save and close** the part file.

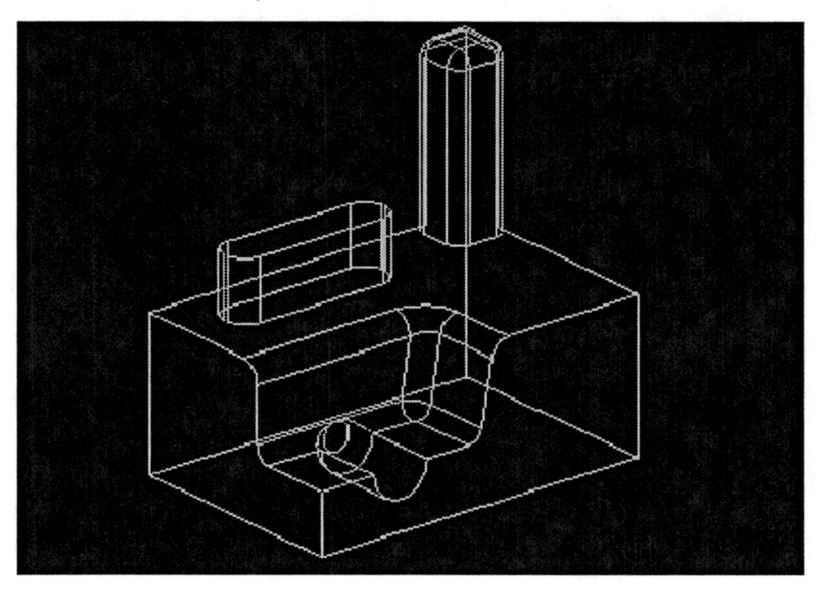

Project 7-1. Pivot Arm

Model the pivot arm part using the drawings below and save it as **xxx_project7-1.prt** in your directory.

Hint: Start out with a block and add an edge blend feature.

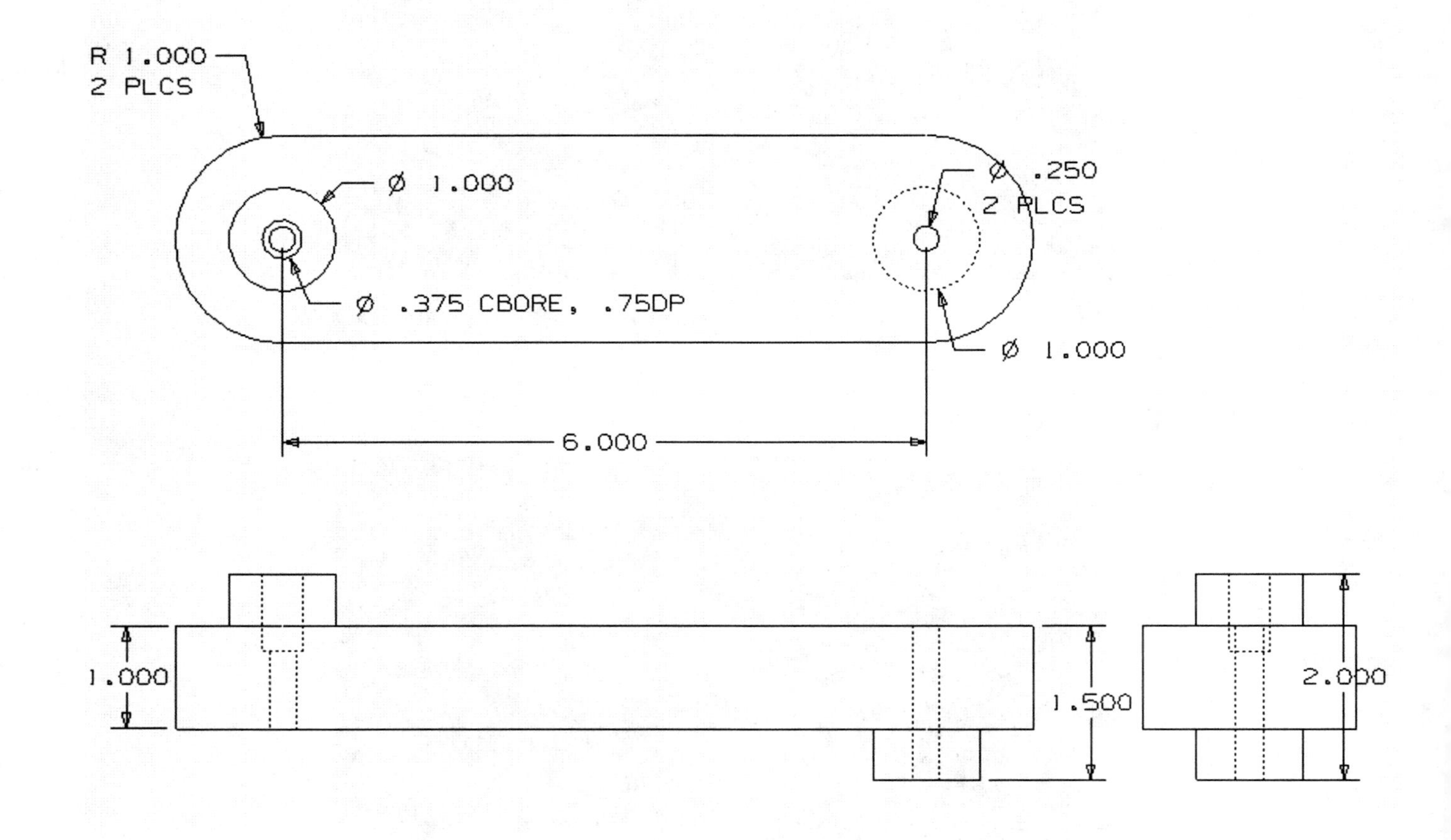

7.2 Chamfer

This option lets you bevel edges of a solid body by defining the desired chamfer dimensions. The chamfer function operates very similarly to the blend function by adding or subtracting material relative to whether the edge is an outside chamfer or an inside chamfer. Thus, the rules and restrictions for chamfers are basically the same as those for blends.

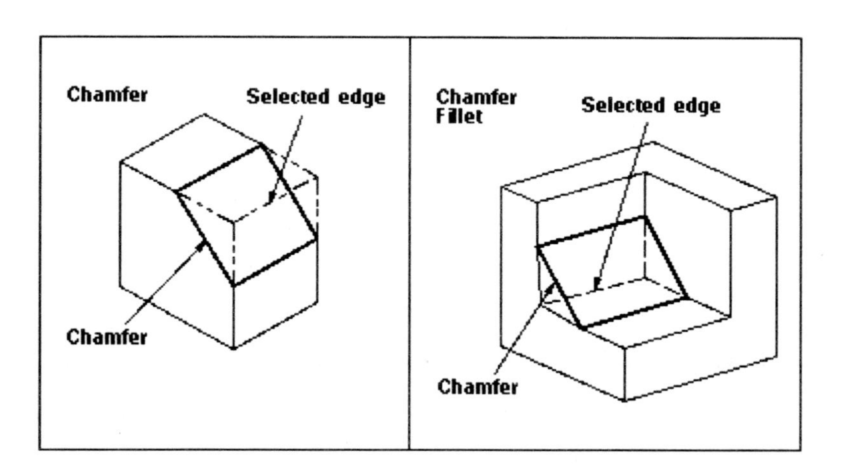

Below is given a general procedure to create a chamfer, which will be followed by more detail descriptions.

Chamfer Procedure

To create a chamfer, following these steps:

1. Choose a chamfer method.
2. Select the edges you wish to chamfer.
3. Enter the chamfer dimensions.
4. If the selected edges are part of an instance set, choose Chamfer all instances or Do not chamfer all instances.
5. (Optional) Choose Flip Last Chamfer to flip the direction of the chamfer. This step applies to Asymmetric and Offset and Angle.

You can create a chamfer using different methods. In this section, you will learn three of them. They are Symmetric Offset, Asymmetric Offset, and Offset and Angle. For these methods, the offset values are measured along the faces from the edge being chamfered.

Symmetric

This option lets you create a simple chamfer whose offset is the same along both faces. You must enter a single *positive* offset value.

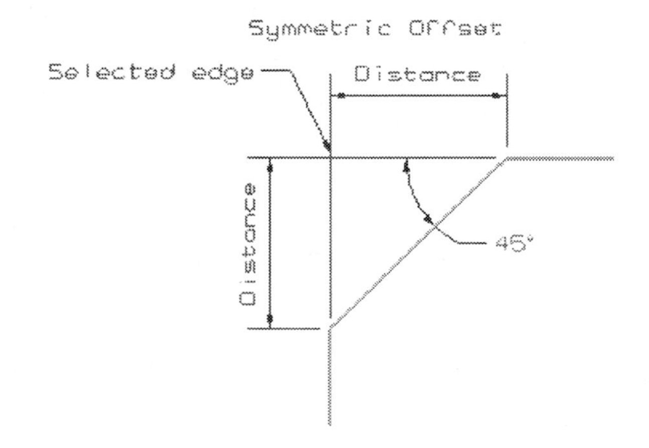

Asymmetric

For this option, you must enter values for **Distance 1** and **Distance 2**. These offsets are measured along the faces from the selected edge. Both of these values must be positive, and are applied as illustrated in the figure below. However, there is no way to assign which direction the Direction 1 and Direction 2 will be applied. Therefore, after creating the chamfer, you can choose **Flip Last Chamfer** if the chamfer's offsets were applied opposite to what you want.

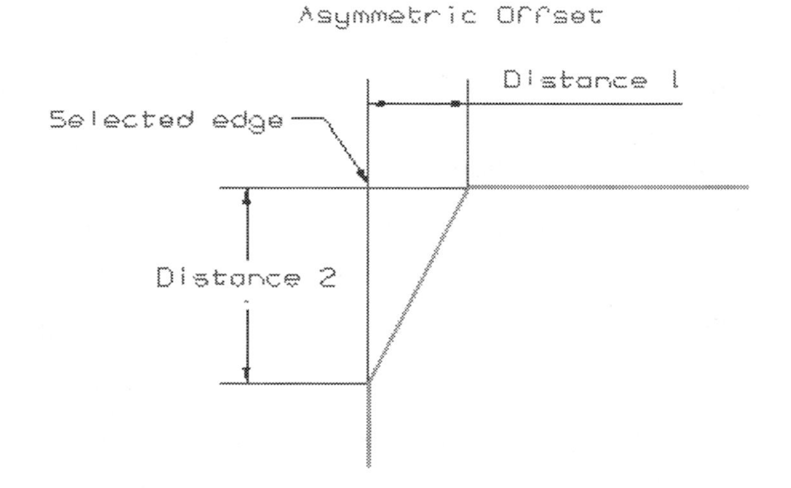

Offset and Angle

Sometimes an angle is used to define a simple chamfer. For this option, you must enter values for the **Distance** and **Angle**. The angle is measured from Face 2, as shown in the figure below. After creating the chamfer, you can choose **Flip Last Chamfer**.

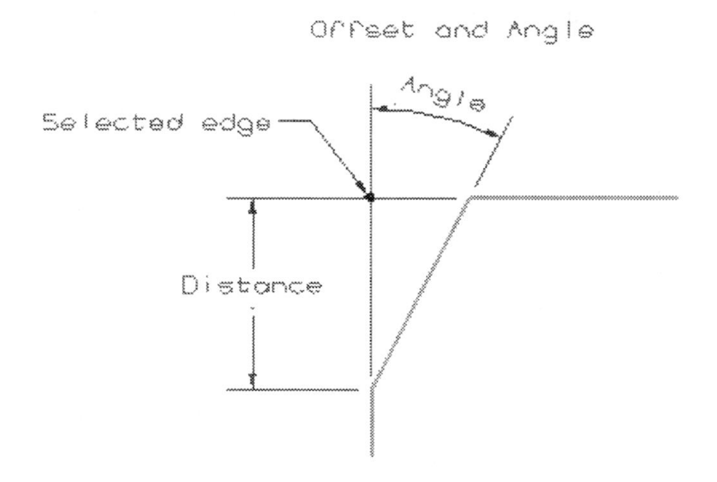

Edge Selection

All selected edges must belong to the same body, and can be of any curve type (see the figure below). The system tracks the total number of edges you have selected. Choosing **Back** during edge selection causes the system to deselect the last edge selected. Choosing **Back** repeatedly causes the system to deselect one edge at a time in the reverse order of selection. After you have selected all of the desired edges, choose **OK** to proceed with chamfering the edges you have selected.

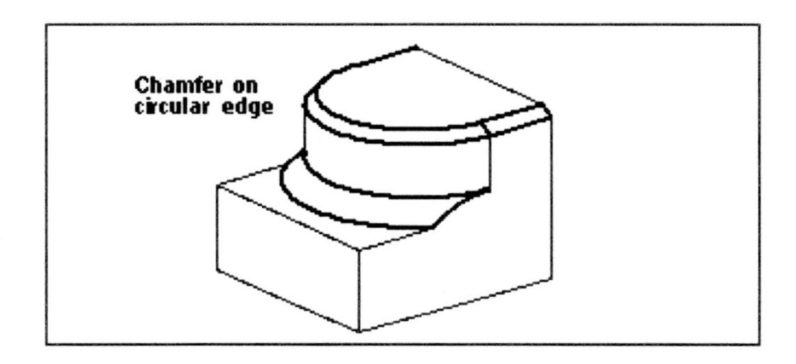

Instances

If you chamfer an edge of a feature that is a member of an instance set, a menu will be displayed at the end of the interaction, before the chamfer is created. You can choose either chamfering all instances of a feature or not. However, you cannot instance a chamfer. If you instance a feature that has a chamfered edge, the instance set will not contain the chamfer. There will be more discussion on instancing a feature in Chapter 11.

Activity 7-2. Creating Chamfers

In this activity, you will partially complete modeling the part below. You will practice chamfering edges that you learned in the above section. You will complete modeling the part in Project 7-5.

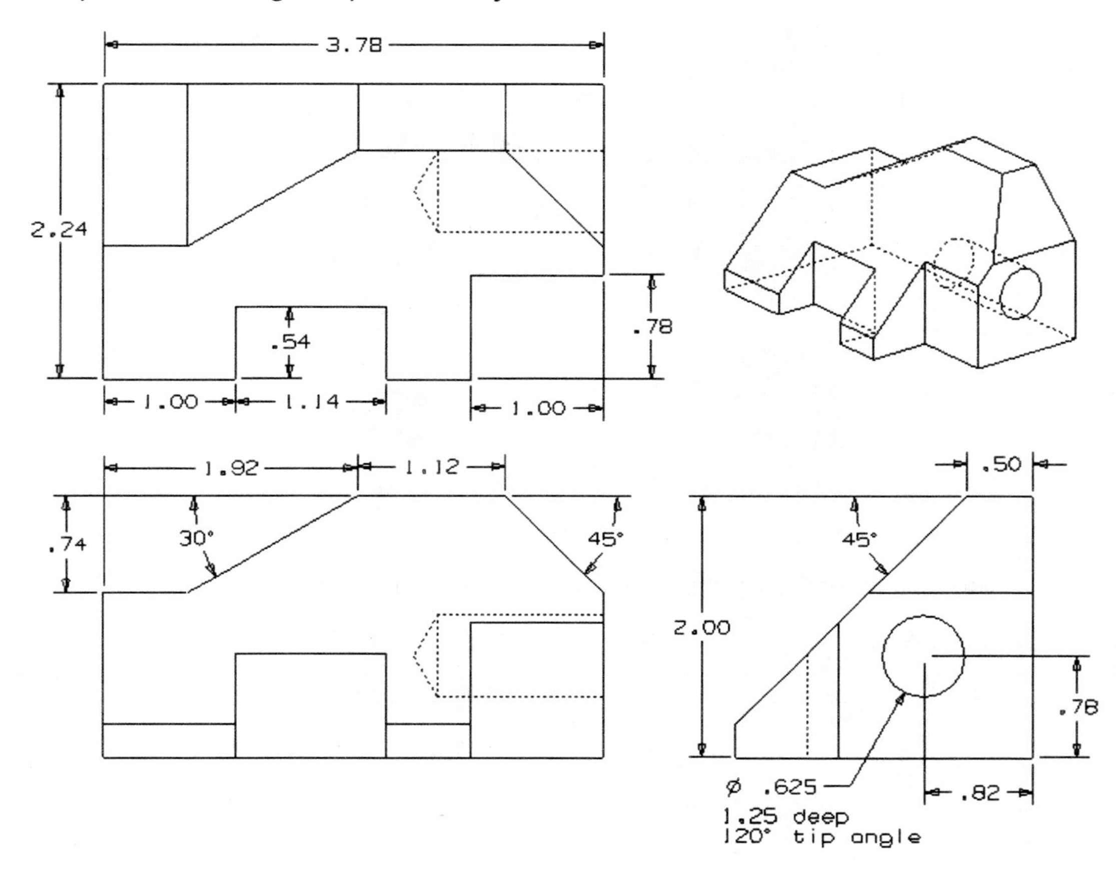

Step 1. Create a new part file (unit: inches) named **xxx_activity7-2.prt**.

Step 2. Start the **Modeling Application**.

Step 3. Replace the view to the **TFR-TRI**, Trimetric.

Step 4. Create a **Block** using the outside dimensions of the solid above and locate it at **WCS 0,0,0**.

Step 5. Create a chamfer on the front face.

5.1. Choose **Insert → Detail Feature → Chamfer...** or choose the **Chamfer** icon  .

5.2. You are creating a 45° chamfer so you may choose the **Offset and Angle** which is the third icon from the left.

5.3. Select the top edge of the front face of the block and select **OK**.

5.4. The distance of the offset is next. Make the offset to end .5 from the back face of the block.

Enter the **Offset** distance as 2.24-.5

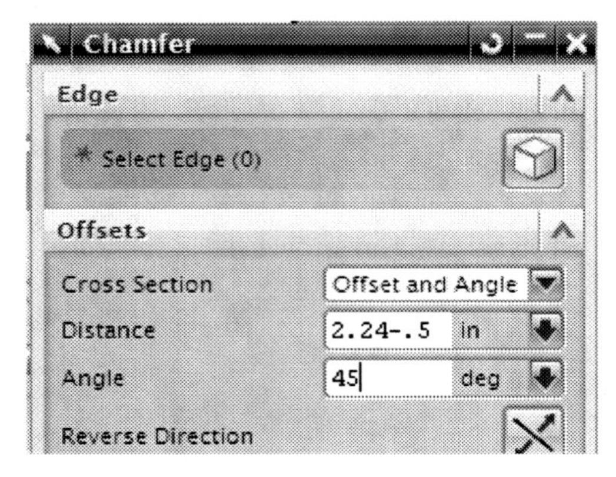

5.5. Choose **OK** and the chamfer is created.

Step 6. Create the chamfer on the right face of the block.

6.1. Select the Chamfer icon again and choose the **Offset and Angle.**

6.2. Select the top edge of the right face of the block.

6.3. This time the **Offset** distance is measured from the left face. It is a distance of 3.78-1.92-1.12 from the left edge. Enter this equation in the **Offset** distance window.

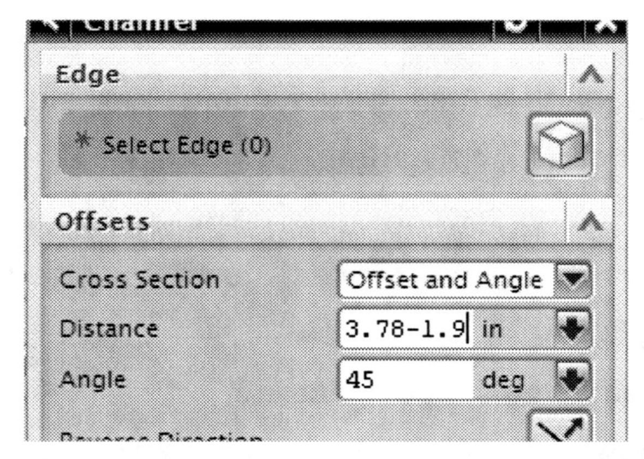

6.4. Choose **OK** and the chamfer is created. The block looks as below. **Save** and close the file.

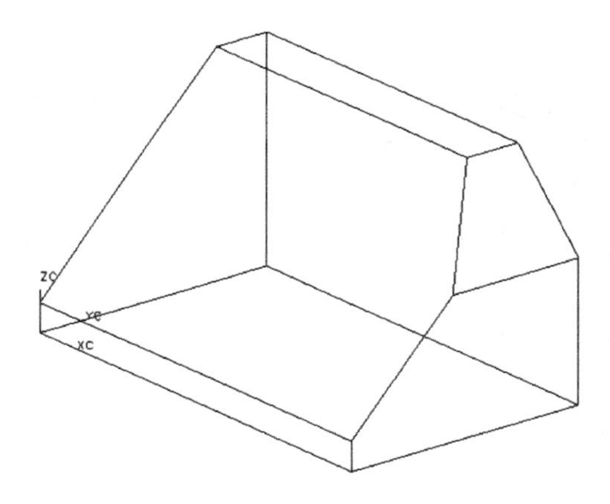

Project 7-2. Mounting Block

Open your part file **xxx_project5-3.prt** that you created from Project 3 of Chapter 5, and save it as **xxx_project7-2.prt**. Redesign the model by adding edge blend and chamfer features as shown in the bottom figure. All fillet and round R = .125.

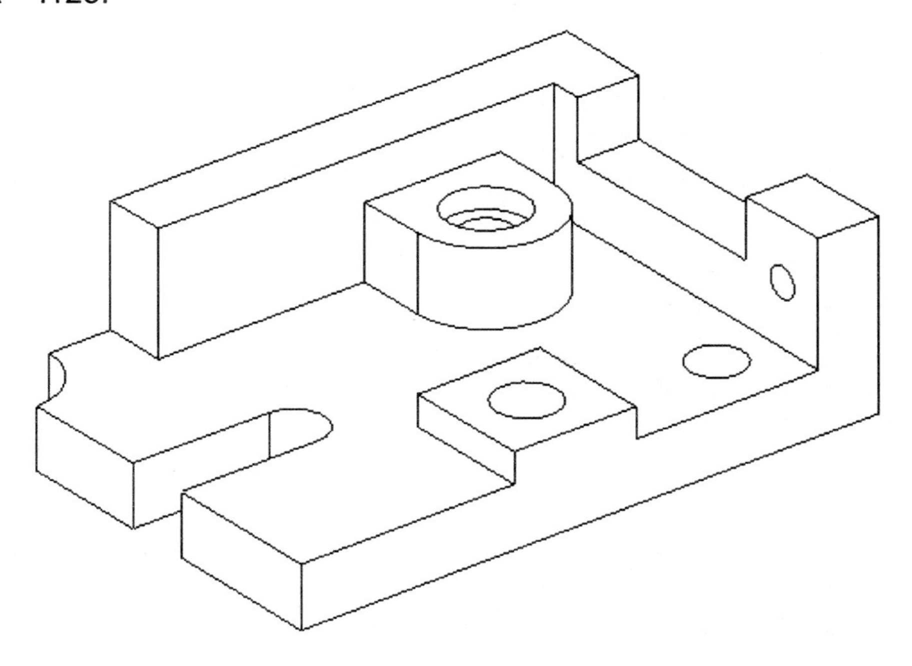

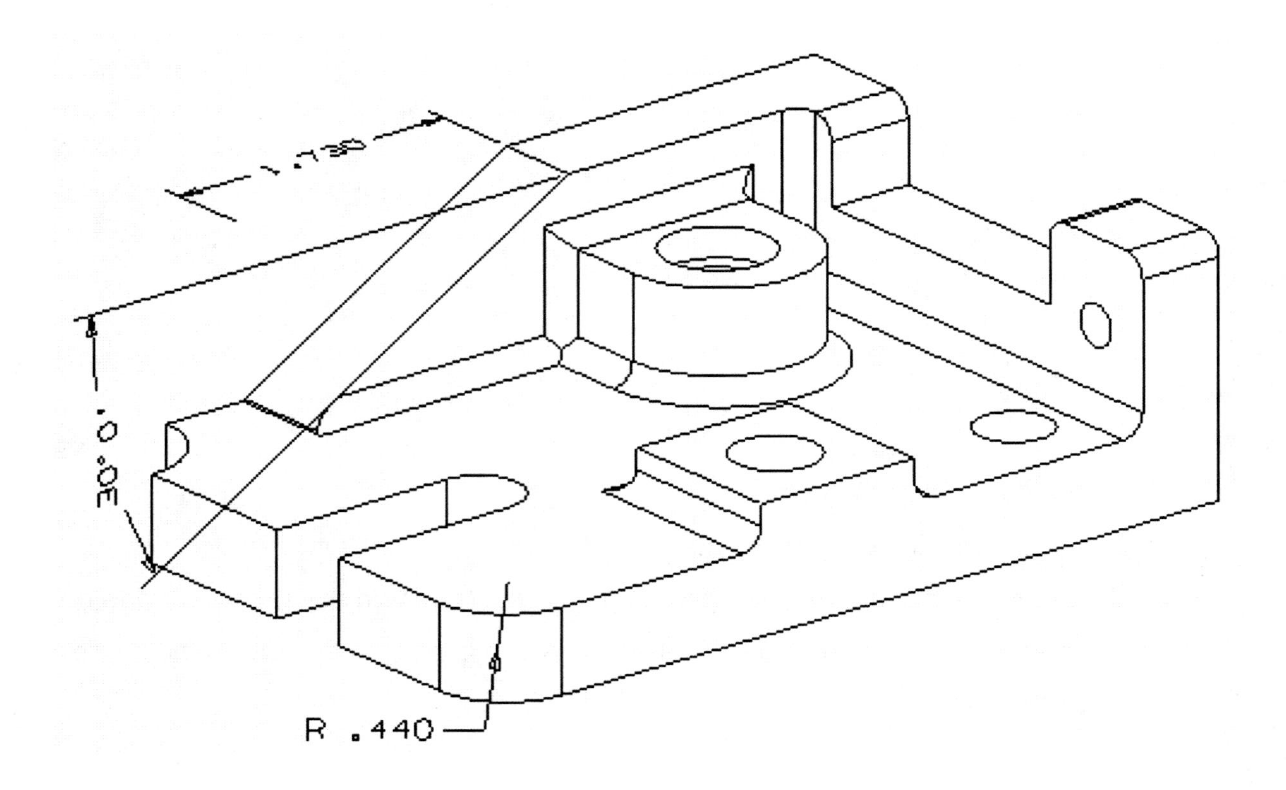
1.730
30.0°
R .440

7.3 Shell

This feature lets you hollow out or create a shell around a single solid body based on specified thickness values. You can assign individual thicknesses for faces, and select regions of faces for piercing during hollowing.

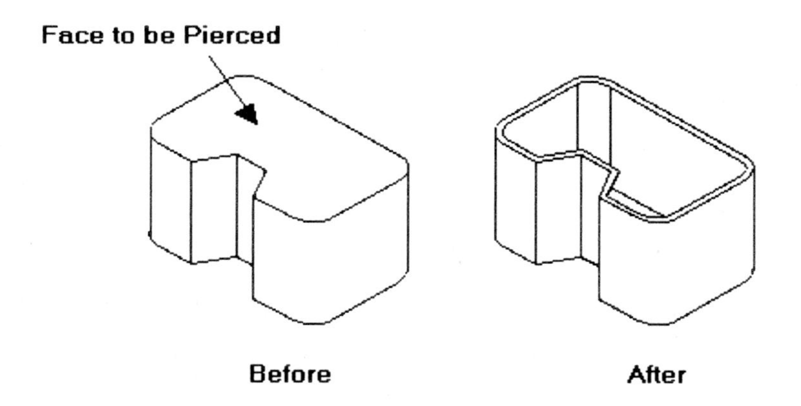

In order to open the Shell dialogue, choose **Insert → Offset / Scale → Shell**, or choose the Shell icon. The following **Shell** dialog opens.

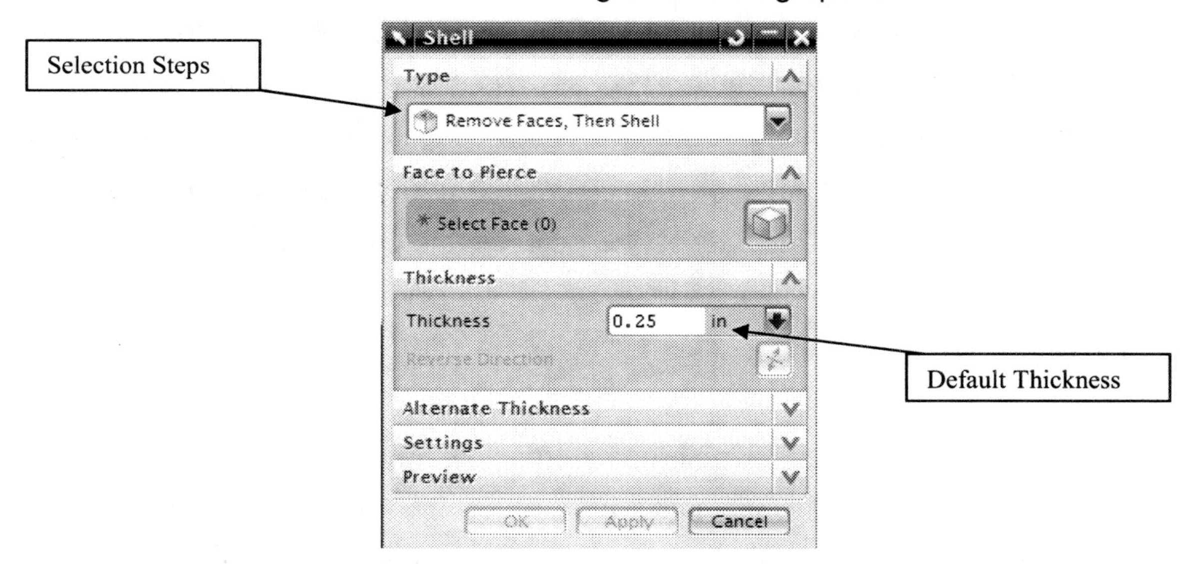

There are two types of Shell operation: **Remove Faces** and **Shell All Faces**.

In this section, we will describe two types, Remove Faces and Shell All Faces. For both types, you need to specify wall thickness. A positive thickness value hollows the existing solid so that the wall thickness is measured inward from the original faces of the solid. A negative value results in a hollow that forms a shell of the specified value around the original solid. The Variable Thickness is optional. Each selected offset face is created based on its variable thickness. Non-selected faces are created based on the default thickness (except for the face(s) selected to be pierced).

7.3.1 Remove Faces

This lets you create a hollow by collecting pierced and offset faces. You can select any faces to pierce which are removed, and you can offset the rest of the faces at different thickness values. To create a hollow using the Face Type, follow these basic steps:

1. Choose one or more faces to pierce for the hollow.
2. Choose one or more offset faces (optional). For each offset face you specify, an entry will appear in the Offset Face changeable window with an identifier and the default thickness value. You can change the thickness of individual offset faces by selecting its entry in the changeable window and specifying a new value in the Alternate Thickness field (see the figure above). Offset faces you select in the changeable window are highlighted in the graphics window in a cyan color.
3. Choose Apply or OK.

The figure below shows a shell being created on a simple block using the Remove Face type, with a single pierced face (the right face of the block). The top face gets a variable offset of 0.4. The remaining faces automatically get the default offset of 0.1 (except the pierced face).

Face Type

Selection Step: Offset Faces

Selection Step: Pierced Faces

Select single pierced face (magenta)

ZC

YC

XC

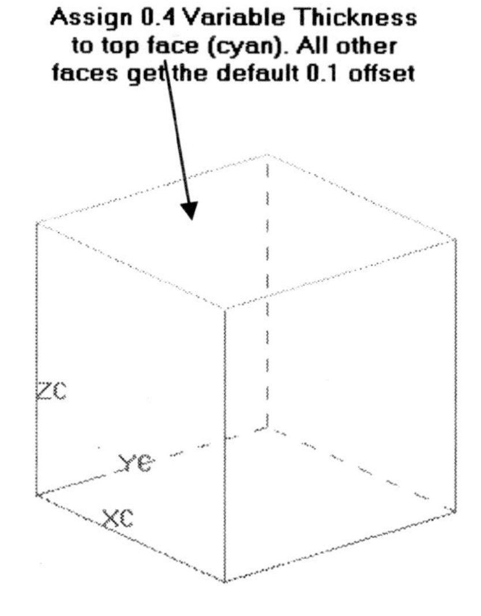

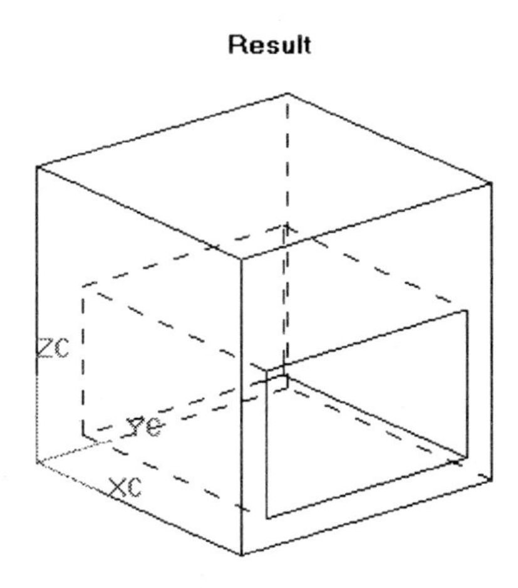

You can select any faces (more than one face) you wish to pierce before the hollowing takes place. In the figure shown below, three faces are pierced. You can use **Edit→ Feature→ Parameters** or in the Part Navigator window click MB3 on this feature → Edit Parameters in order to select additional faces to be pierced, or to deselect previously selected faces.

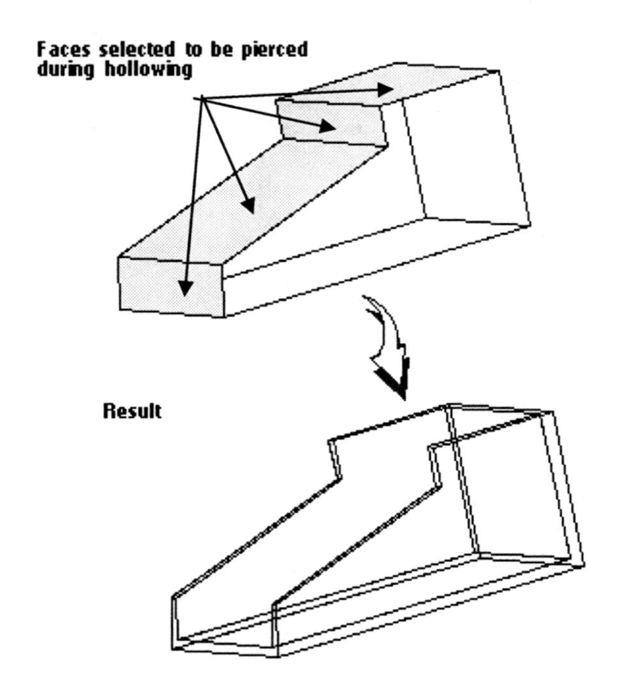

7.3.2 Shell All Faces

This option lets you create a shell around a single solid body based on a specified thickness value. You can select any faces and distance values to offset during shelling. You may not create an open face. To create a shell using the this type, follow these basic steps:

1. Choose the body to be shelled.
2. Choose one or more offset faces (optional).
3. Choose Apply or OK.

The figure below shows a shell being created on a simple block using the Shell All Faces type, with two faces selected for variable offsets (0.3 and -0.2). The remaining faces get the default 0.1 offset: The -0.2 offset has the effect of adding to the solid body, since the negative thickness value causes the wall to move outward.

Body Type

Selection Step: Solid Body

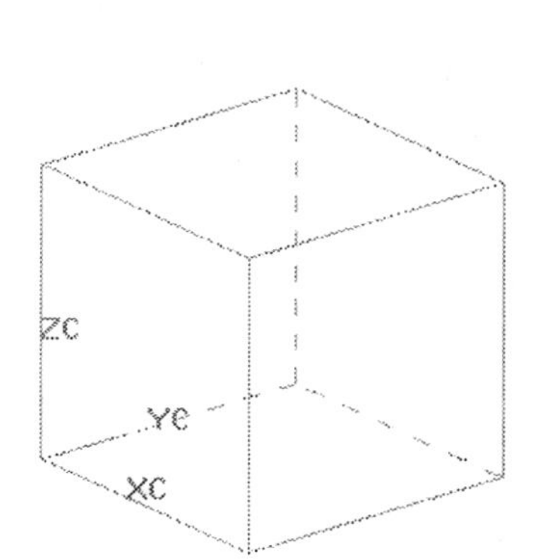

Selection Step: Offset Faces

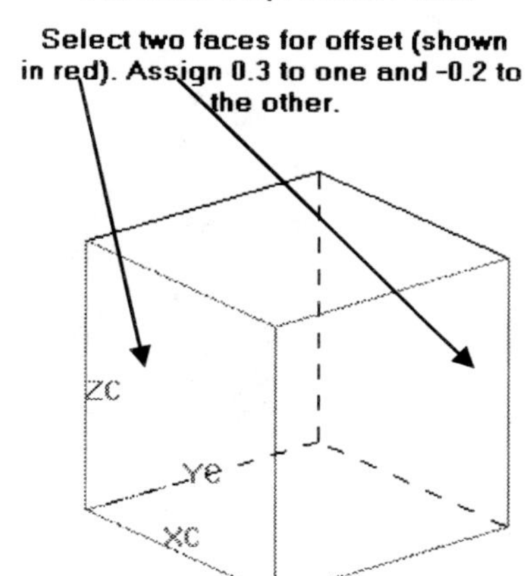

Result

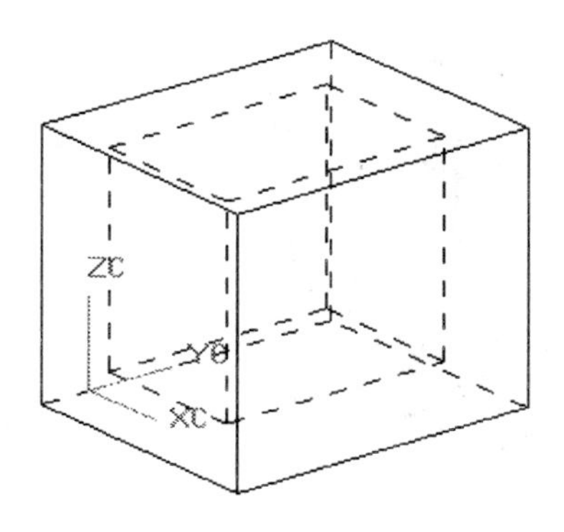

Activity 7-3. Creating Shell

You will hollow the solid to create a shell model like this.

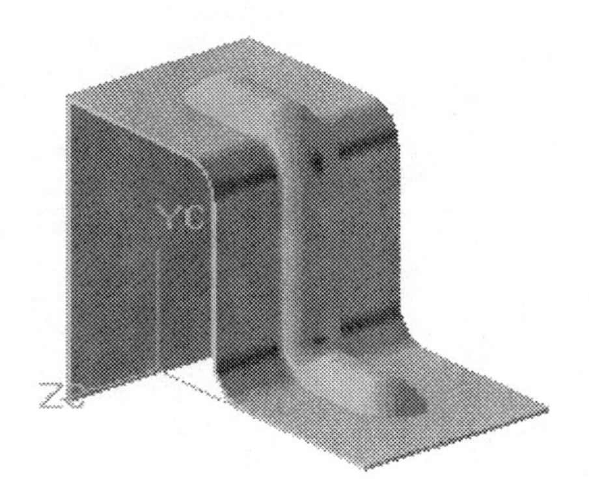

Step 1. **Open** the part file **activity7-3.prt**.

1.1. Start the **Modeling** application.

1.2. Save the part as **xxx_activity7-3.prt**.

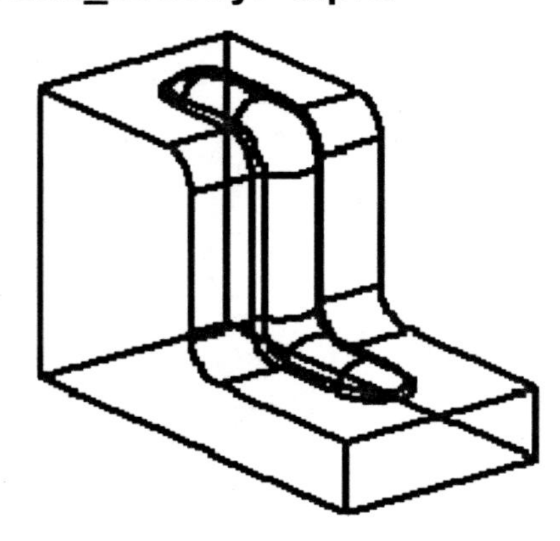

Step 2. Choose **Insert** → **Offset/Scale** → **Shell**, or choose the icon. The **Shell** dialog is displayed.

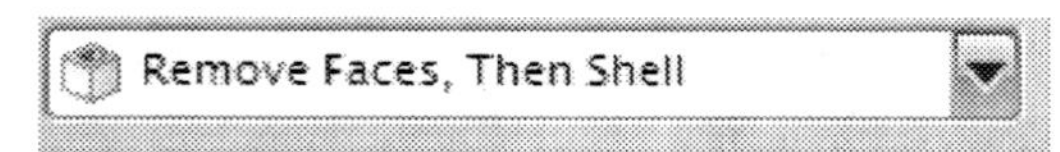

Since you will be removing faces and creating a thin walled solid body, you need to use the **Remove Faces** hollow type.

2.1. Choose the **Remove Faces** type icon, if it is not already active.

The **Preview** option of the dialogue lets you view the model before accepting it, when toggled on.

You can change the default thickness value. You can also specify a different (alternate) thickness for each wall that will be created. (You will change the alternate thickness value later.)

Step 3. Specify a wall **Default Thickness**, so key in the positive value of 0.120.

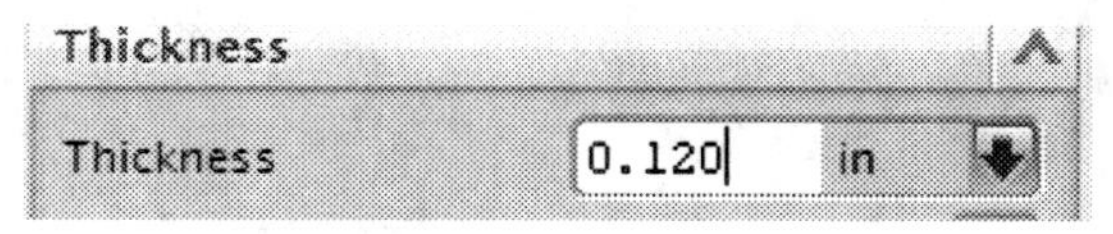

Step 4. Select faces to remove.

4.1. Select the closest two faces first. Select the front face and the right face.

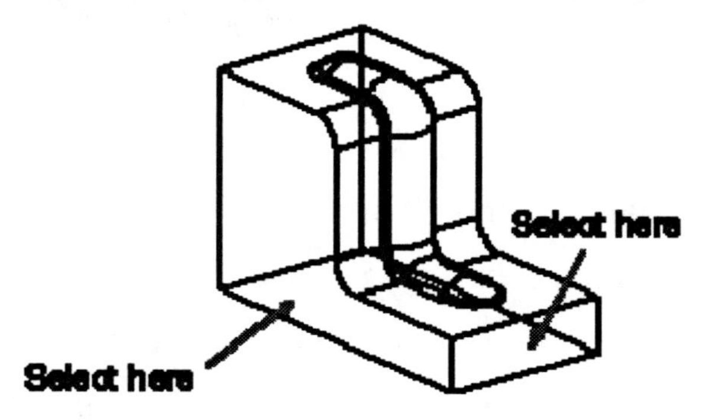

4.2. Now, you can select the last two faces to be removed. Select the bottom face and back face.

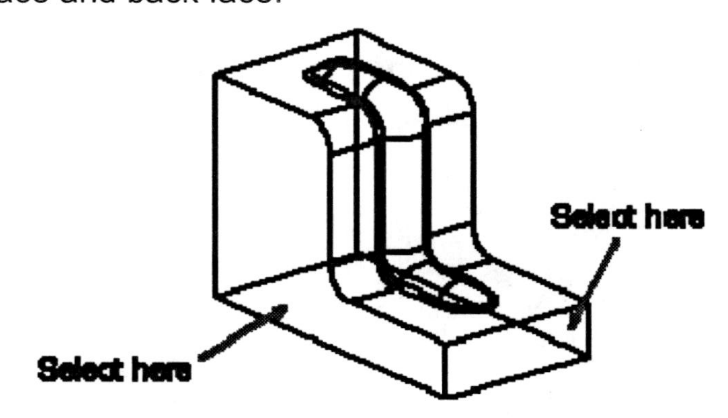

The model should look like this.

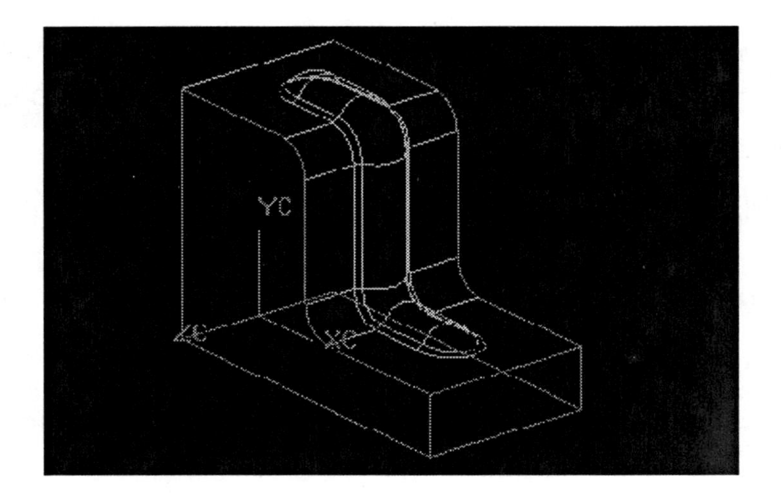

Step 5. Change one face wall thickness because the design intent specifies one face needs a thicker wall than the other faces.

The **Default Thickness** will be set to .12″ and the **Shell** operation will create all other faces (not selected to **Pierce**) to a .12 wall thickness without the need to select them. Therefore, to set a wall thickness to an alternative wall thickness we must select the face we want to be different.

You will make the left wall thicker than the other walls.

5.1. Choose the **Alternate Thickness List** icon as shown below.

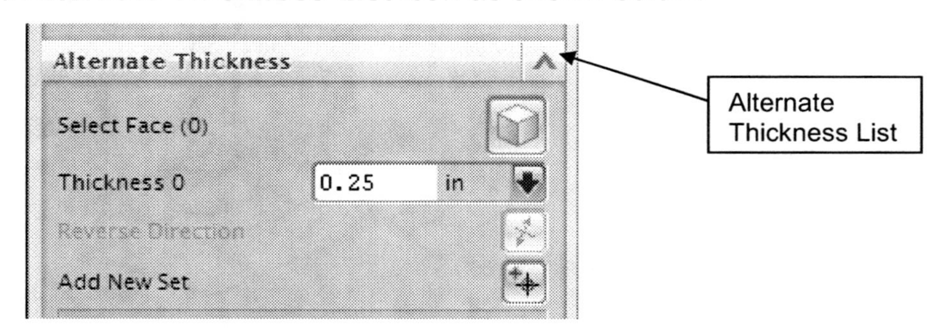

5.2. Select the left wall to receive the alternate thickness.

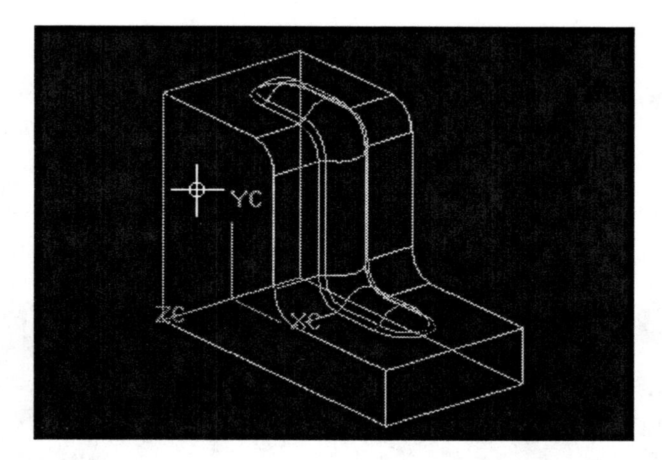

Now, the thickness parameter displays in the list box of the **Shell** dialog. Whenever a wall thickness is highlighted in the dialog, you can specify its alternate thickness value.

5.3. In the **Alternate Thickness** field, key in 0.25 as the value.

5.4. Choose **Apply** to form the hollow.

The model is temporarily built. Notice that the left wall is twice as thick as the other walls.

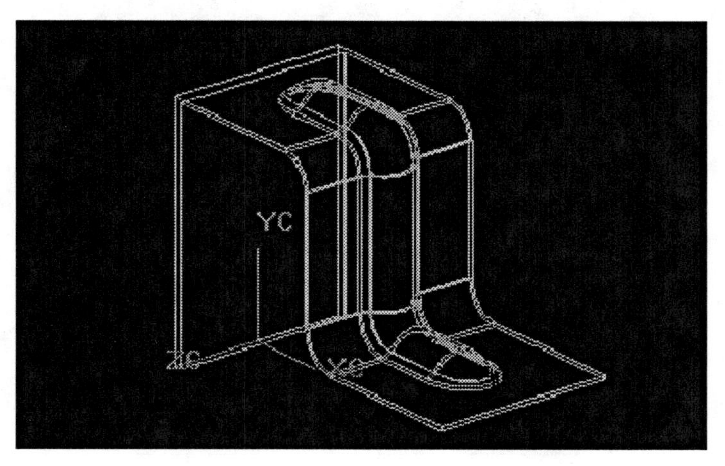

Step 6. Inspect and close out the part.

6.1. **Cancel** all dialogs.

6.2. **Shade** the model.

6.3. **Save and close** all part files.

Project 7-3. Shell Project

Create a solid model of the part as shown below and save it as **xxx_project7-3.prt** where xxx is your three letter name initials.

Hint: You may use features such as Block, Pad, Edge Blend and Hollow. Note that the part has two different wall thicknesses, .1 and .5.

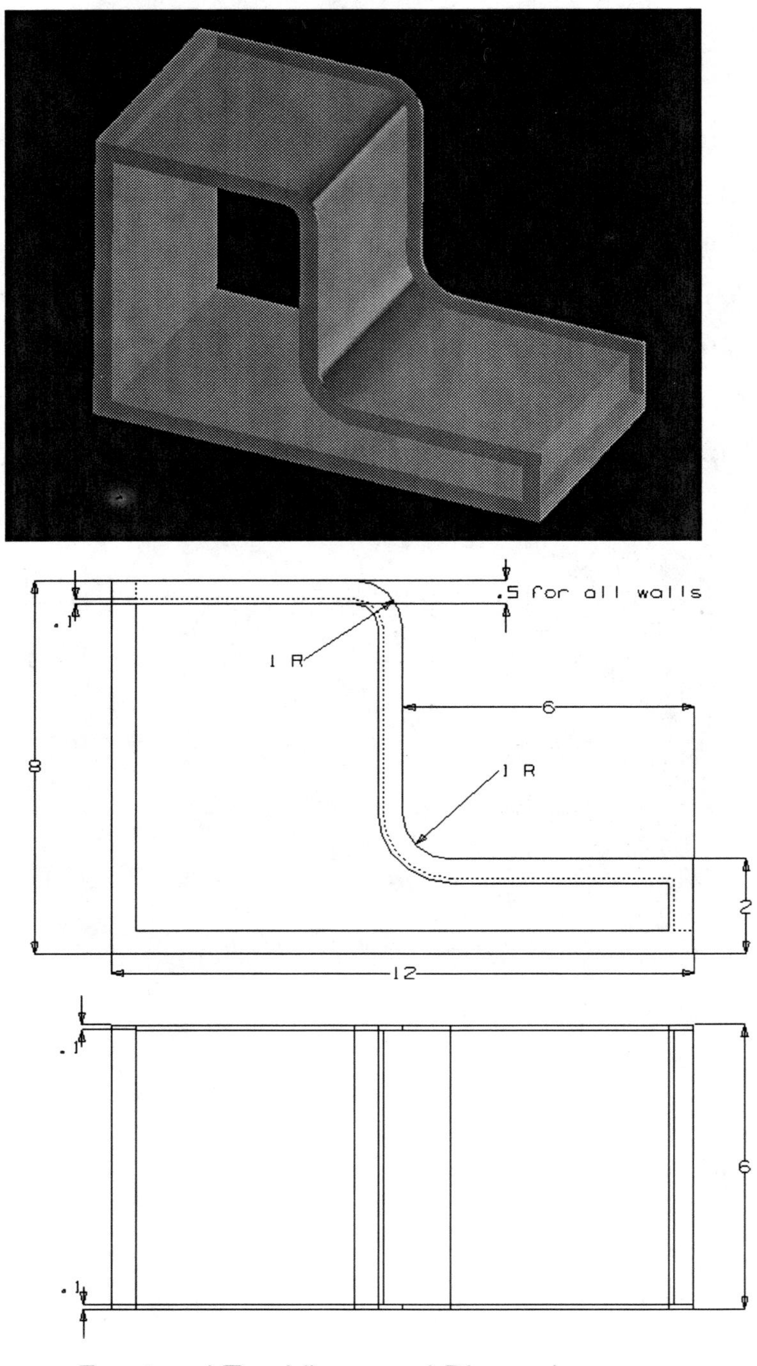

Front and Top Views and Dimensions

Project 7-4. Swivel Bracket

Create a model of the part using the drawings and dimensions (in millimeters) given below. Name your part as **xxx_project7-4.prt** where xxx is three-letter initials of your name. When you create a new part file for this project, make sure to set the unit in millimeters.

Hint: Use all of the features discussed in this chapter, edge blend, chamfer and hollow.

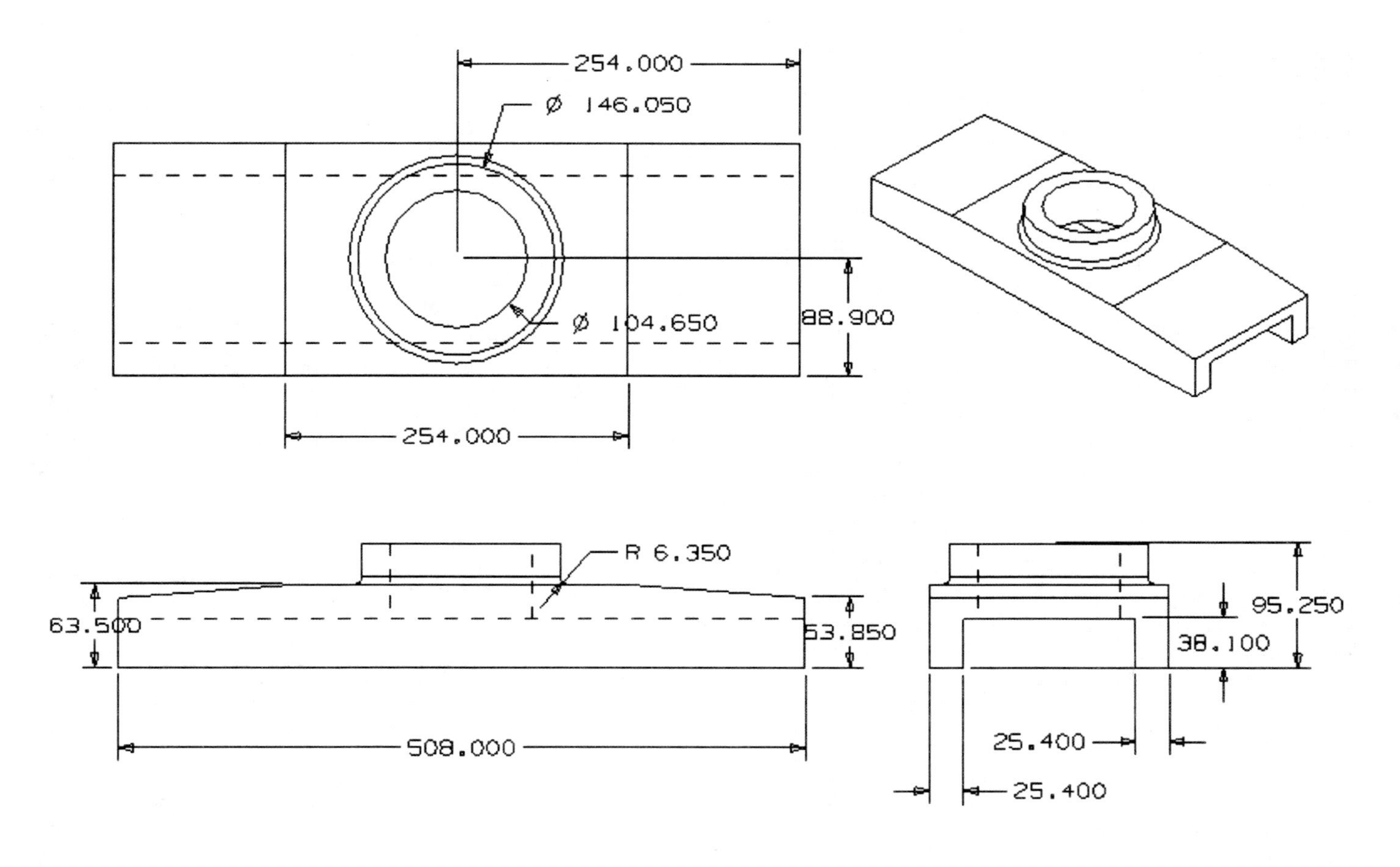

Project 7-5. Tool Holder

If you already completed Activity 7-2, open **xxx_activity7-2.prt** and save it as **xxx_project7-5.prt** where xxx is three-letter initials of your name. Finish modeling the part using the drawings and dimensions (in inches) given below, and save it into this new file.

Hint: One approach to model this part is to start with a block and add the following form features: a simple hole, chamfers, and pockets.

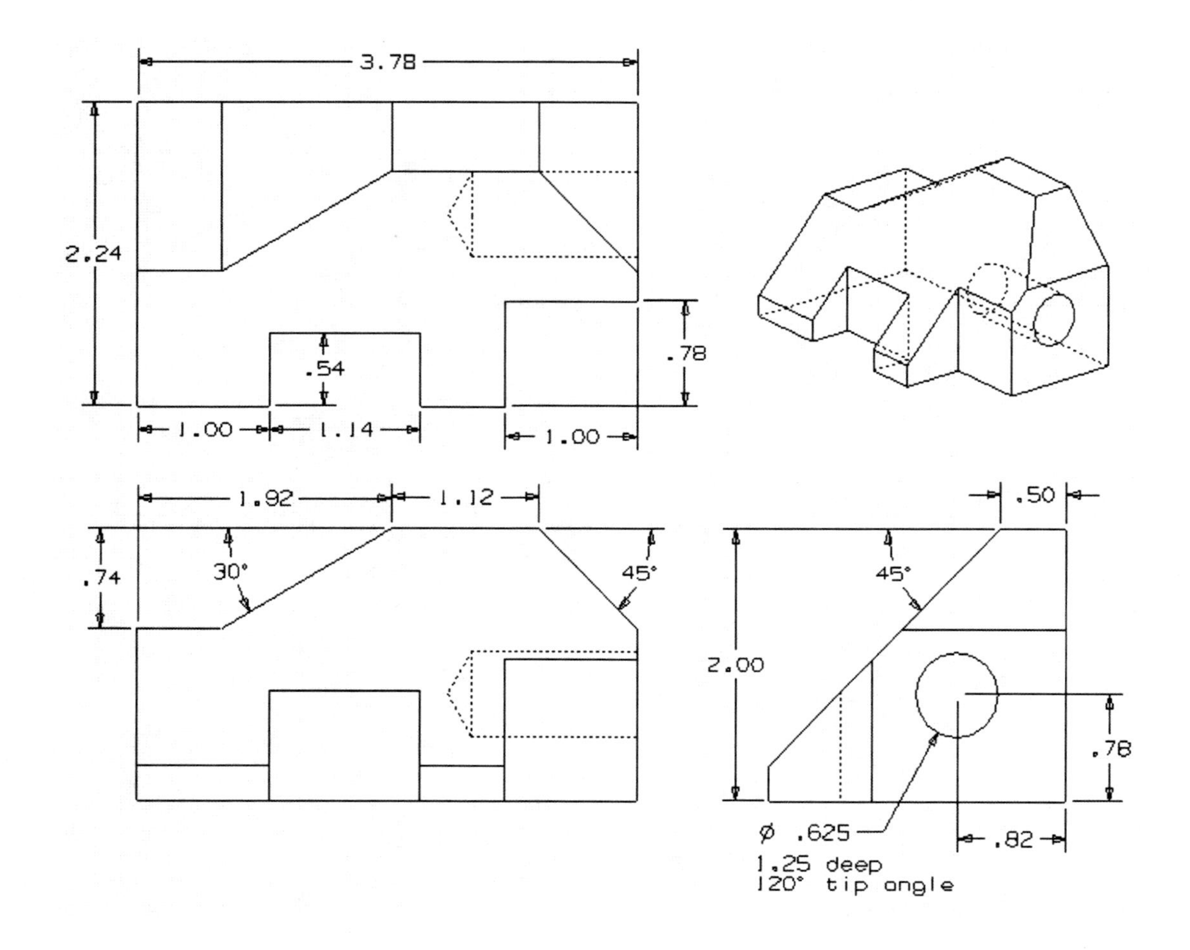

Exercise Problems

7.1 Several edges need to be blended. Is it always better to use one Edge Blend feature to blend all of them at once?

7.2 When several edges need to be blended with more than one Edge Blend feature, does the order of edges being blended matter? Can you give such an example?

7.3 Is Symmetric Chamfer a special case of Offset and Angle Chamfer?

7.4 Both Chamfer and Edge Blend can add material to or subtract material from the base solid body. How can the system know when to add and when to subtract?

7.5 Suppose that you want to create a void inside a solid body with a uniform surrounding wall thickness. What is a simple way to do so?

Chapter 8. Reference Features

Reference features are construction tools that assist you in creating other features and sketches in desired locations and orientations. You can create them either relative to an existing target solid, or fixed. Reference features are theoretically infinite in size but the display is limited to a size slightly larger than that of the objects selected to create them. There are two types of reference features: **Datum Planes** and **Datum Axes**. Section 8.1 discusses datum planes and Section 8.2 discusses datum axes.

8.1 Datum Planes

Datum Planes are reference features that can be used as construction tools in the building of a model. Such examples include (1) creating features on non-planar faces, (2) providing a planar location for a sketch (see Chapter 10), and (3) providing a planar trimming object. Suppose that you want to create a hole on a cylindrical face. A hole feature requires a planar placement face but a cylinder face is not planar. In this case, you need to create a datum plane and use it as the placement face for the hole feature. This is illustrated as follows:

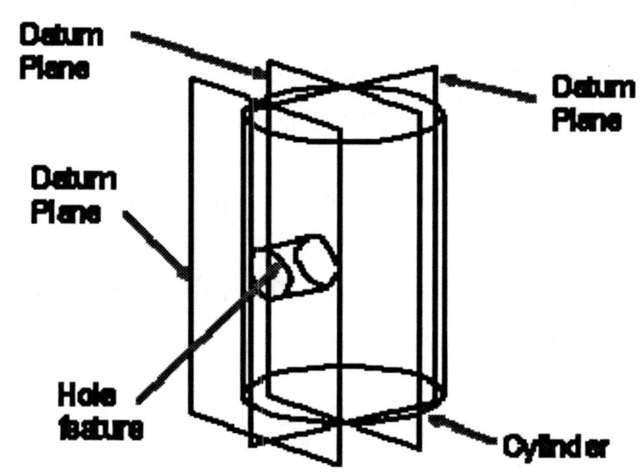

There are two types of datum planes. You can create either relative or fixed datum planes. Relative datum planes are relative (constrained) to other objects in the model such as curves, faces, or other datum. Thus relative datum planes are associated to the objects. When the objects move, the relative datum planes move according to the relationships (constraints) used to create them.

On the other hand, fixed datum planes are only related to WCS. Fixed datum planes do not reference, and are not constrained by, other geometric objects. Fixed datum planes are useful when a solid body started out from a sketch. We will focus on relative planes in this chapter.

Constructing Datum Planes

There are two basic methods you can use to construct datum planes:

- Select the edges, faces or wireframe geometry needed to specify the datum, and then choose the Datum Plane option. The system attempts to infer the best mode to use with the selected objects to successfully define a datum, and presents a preview of the datum in the graphics window.

- Invoke the Datum Plane option and then select the required objects for the datum from the graphics window. When you have selected enough valid objects to define a datum, a preview of the datum is displayed in the graphics window.

In either case, a previewed datum is inferred and then presented, based on the selected geometries and constraints. In case a datum plane cannot be created based on the selected objects, use the Datum Plane icon options to change the mode, add additional objects or change the constraints.

Datum Plane Icon Options and Inferred Method

Choosing **Insert → Datum/Point → Datum Plane** or the **Datum Plane** icon first displays the Datum Plane icon options in the upper left corner of the graphics window.

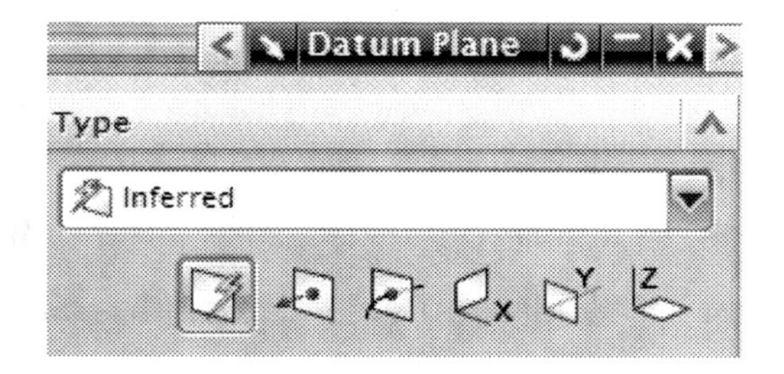

Datum Plane Icon Options

The icon options use the **Inferred method** as a default to let you quickly create a datum plane based on your object and optional constraint selections. The Datum Plane icon options include the following tools:

Inferred Plane. The constraints are inferred by the objects selected.
Point and Direction creates a datum plane that is defined by a point to locate, and a vector to orient its normal.
Plane on Curve creates a datum plane tangent to, normal or binormal to a point on a curve. You can drag the location handle for the point along the curve or edge. You can also use this method to create a datum plane coincident with a point on curve, with its direction defined by a second object. There is also an icon on the dialog — **Reverse Curve Direction** — which reverses the start point of the selected curve.
At Distance creates a datum plane that is parallel to a planar face or another datum plane at a distance you specify. Once you select the planar face or datum plane, you can specify its offset, parallel distance by dragging the offset handle, or entering an offset value in the dynamic input box or dialog Offset field. From the dialog, you can also enter the number of copies of the new plane you want to create in the Number of Planes field. Each datum plane is successively created and spaced evenly from one another using the same offset value.
At Angle creates a datum plane using a specified angle in relation to selected planar geometry. You first select a planar face, plane, or datum plane. Then you select a parallel linear curve or datum axis. You can specify the angle by either dragging the angle handle or by entering an angle value in degrees in the Angle field of the dialog or dynamic input box.
Bisector Plane create a datum mid-plane using the bisected angle between two selected planar faces or datum planes. Cycle Solution will cycle between the bisector plane and one normal to the bisector plane.

Curves and Points creates a datum plane by first specifying a point, and then a line, datum axis, curve, or edge. The new datum plane passes through both objects.
Two Lines creates a datum plane by selecting two existing lines. The resulting plane contains the first line and is parallel to the second. If the two lines are coplanar, then the plane will include both lines.
Tangent to Face At Point, Line or Face creates a datum plane tangent to a selected cylindrical face and passes through a selected point, line or face.
Plane of Object creates a datum plane based on the plane of an existing curve, edge, face, datum, or plane that you select. The curve, edge, or face can be planar or non-planar. If you select a conical face, the datum plane is created on the axis of the conic.

In this chapter, we will focus on the inferred method which is default and sufficient to create datum planes most of the time. In the Inferred method, the system infers constraints based on objects as you select them. These objects include points, edges, faces, planes, and axes (face or datum). Depending on the objects you select, constraints are used by the system as it attempts to provide a likely datum plane solution that is previewed. Before creating this previewed datum plane, you can enter a specific value for a distance or an angle to override the default value or drag the handles displayed in the graphics window. Datum planes are created with one of three constraint types: single, dual, or triple constraints. The objects you select and the sequence in which you select the objects determine a particular constraint type

(1) Single Constraints

Under this constraint type, the datum plane is constrained to the target solid by this single constraint. You can create single constraints using the following options:

- **Offset to Plane** creates a datum plane that offsets from (and is parallel to) a planar face or existing datum plane by selecting the face or plane. If the offset value is zero, the datum plane is coincident with the planar reference face/datum plane.
- **Through Face Axis** creates a datum plane that passes through the imaginary axis of a cylinder, cone, or revolved feature. Select the face of the feature.
- **Center of Face** creates a datum plane that passes through the imaginary axis of a sphere. Select the feature and choose OK.

(2) Dual Constraints

Under this constraint type, constraints must be specified in pairs to create the datum plane. Creating a Center Plane is an example of dual constraints

- **Center Plane** creates a datum plane at the center of two parallel planar faces or datum planes. When you select the first plane, the constraint applicable is **Offset to Plane**. When you select the second plane, the **Offset to Plane** constraint changes to dual constraints, **Parallel** and **Center**.

We give below two other examples of a datum plane created with dual constraints: (1) **Through Face Axis** and **Angle to Plane** and (2) **Angle to Plane** and **Through Edge (Linear Geometry)**.

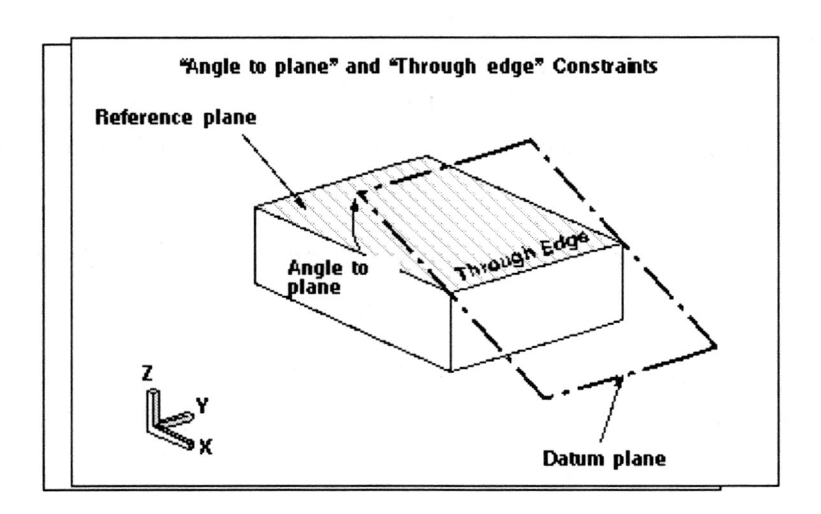

(3) Triple Constraints

There is one triple constraints method for creating datum planes: three **Through point** constraints. This creates a datum plane through three reference points, for instance, endpoints, midpoints, arc centers. The orientation of the datum plane is determined by how you select the points. The X direction of the datum plane is defined by a line drawn between the first and second selected point.

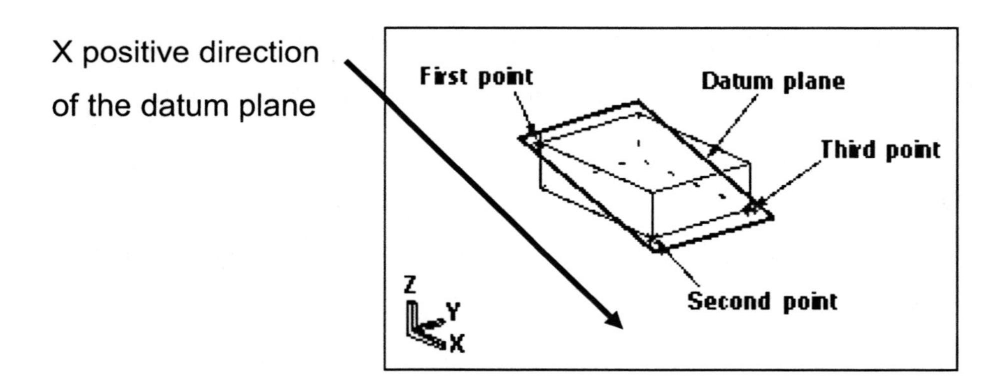

The following table summarizes supported combinations of geometries and available constraints to create relative datum planes.

Table. Supported Combinations of Geometries and Constraints to Create Relative Datum Planes

Single Constraints	Offset from Planar Face/Datum Plane Through Face Axis** Through Center of Spherical Face	
Double Constraints	First Constraint	Second Constraint
	Parallel to Planar Face/Datum Plane	Center to Planar Face/Datum Plane
	Through Point	Through Linear Geometry Through Datum Axis Through Face Axis Tangent to Cyl/Con/Rev* Face Parallel to Planar Face/Datum Plane Perpendicular to Linear Geometry Perpendicular to Face Axis
	Through Point on Curve	Tangent/Normal to Curve/Edge Tangent to Cyl/Con/Rev/Non-Planar Face Parallel to Linear Geometry Parallel to Planar Face/Datum Plane
	Through Linear Geometry (Edge)	Through Point Through Linear Geometry Through Datum Axis Through Face Axis Angle to Planar Face/Datum Plane

	Through Datum Axis	Through Point Through Linear Geometry Through Datum Axis Through Rev Face Axis Angle to Planar Face/Datum Plane
	Thru Cyl/Con/Rev Face Axis	Through Point Through Linear Geometry Perpendicular to Linear Geometry Perpendicular to Datum Axis Through Datum Axis Through Face Axis Angle to Planar Face/Datum Plane
	Angle to Planar Face/Datum Plane	Through Linear Geometry Through Datum Axis Through Face Axis
	Tangent to Cyl/Con/Rev Face	Through Point Through Linear Geometry Through Datum Axis Through Face Axis Tangent to Cyl/Con/Rev Face
Triple Constraint	Through Three Points	

* Cyl = Cylindrical, Con = Conical, Rev = Revolved

** Face Axis means Axis of any Cyl/Con/Rev face

Editing Datum Plane

To edit relative datum plane parameters, use any of the following methods:

- Use **Edit → Feature → Parameters**.
- Select a datum plane on the graphics window, and with the cursor over the selection, click MB3 and choose Edit Parameters.
- Double-click a datum plane.

All methods open the Datum Plane dialog as shown below, where you can edit the parameters of the datum plane. This is the same dialog that you get when you choose the Datum Plane dialog icon .

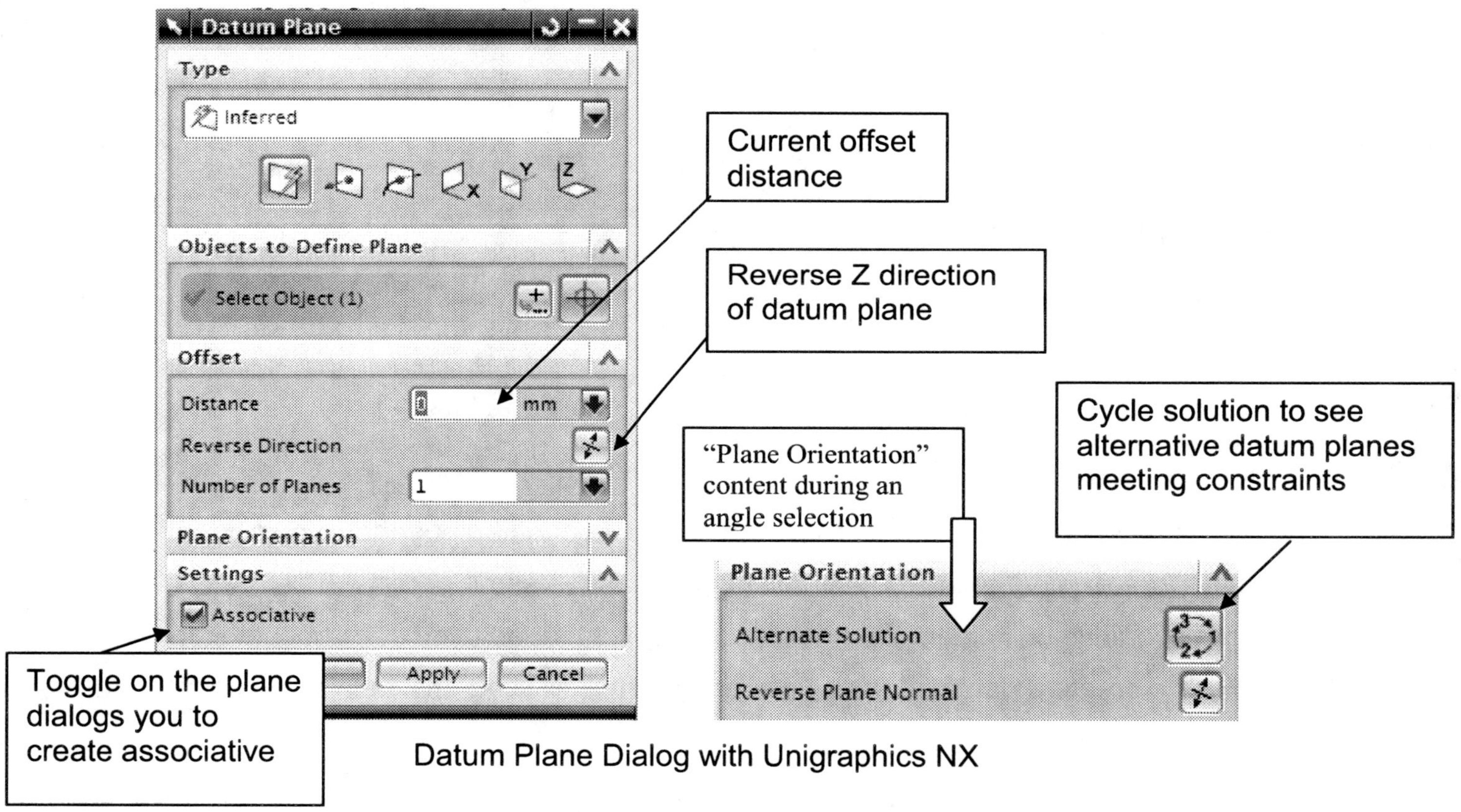

Datum Plane Dialog with Unigraphics NX

Relative datum planes created prior to Unigraphics NX use the old style dialog. When you try to edit these datum planes, the old style dialog appears on the screen as shown below. Dialog options that are not applicable to the selected plane are grayed out. There is also an option to enter the offset directly into a text box instead of using the Datum plane dialog.

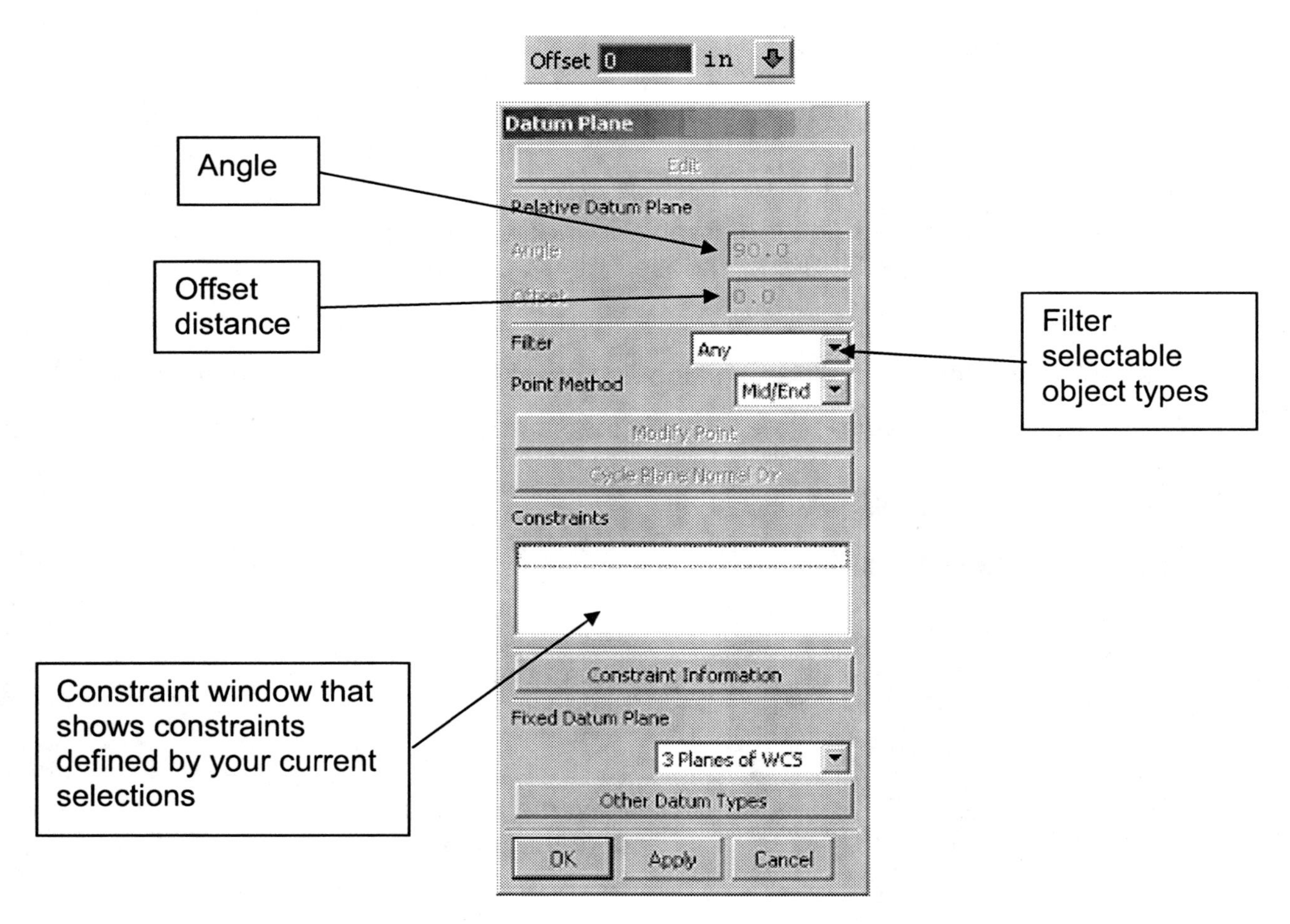

Datum Plane Dialog prior to Unigraphics NX

A dialog displays if you try to create a relative datum plane that is coincident with a pre-existing relative datum plane that has the same parents. The corresponding coincident datum plane is highlighted, and you can choose either "Yes" to create a datum plane coincident with the existing one or "No" to cancel the creation process.

Datum Planes should be placed on different layers and be named to help identify their construction role in the model. According to the default layer standards, layers 61 to 80 are assigned for reference features. It is a good practice to properly change the work layer when you add a reference feature.

Activity 8-1. Creating Relative Datum Planes with a Single Constraint

In Activities 8-1 to 8-3, you will create relative datum planes and datum axes on the pin, and use them to create a hole that is associated, so that updates to the pin automatically result in updates to the hole.

Step 1. **Open** the provided part file **activity8-1.prt** and save it as **xxx_activity8-1.prt** where **xxx** are your name initials.

Step 2. Start the **Modeling** application by choosing **Start→ Modeling**.

Step 3. Change the work layer to 61. Key in **61** in **Work Layer** field at the utility tool bar and hit **Enter**. Alternatively, **Format → Layer Settings** and key in **61** in Work field at the top of the layer setting dialogue and click **OK**.

Step 4. Create a datum plane offset from the bottom face of the head of the pin.

- **4.1.** Click the **General Selection Filters** icon No Selection Filter from the selection tool bar and change the filter to face. If this tool bar is not open, move your cursor to the tool bar area on the near top of the window (right below the UG menu) and click MB3 once. You will see a list of tool bar names appear. Check item "Selection", and then the selection tool bar is displayed.
- **4.2.** Select the bottom face of the pin as shown below. It is highlighted.
- **4.3.** Choose the **Datum Plane** icon or **Insert → Datum/Point → Datum Plane.**

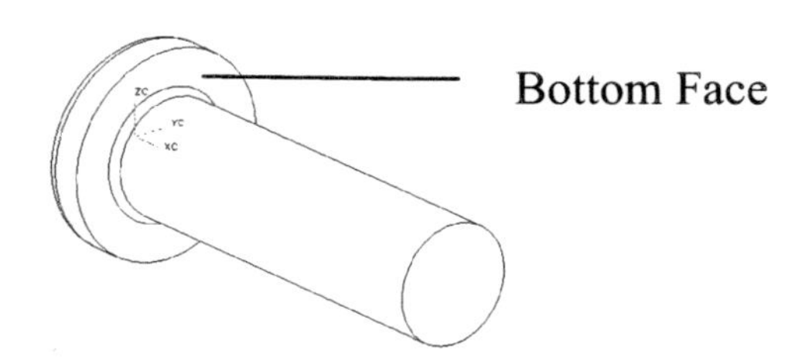

A temporary datum plane is created and a direction vector displays pointing from the selected face and down the shaft of the pin. Also the text box comes up to allow the entry of an offset distance.

4.4. In the **Offset** text box, key in **30.8**. Hit the **Enter** key and a temporary datum plane is created at the offset distance so you can preview what will be created. If this is the correct plane, then click OK to make the plane permanent. If incorrect, key in **-30.8** and hit **Enter**. (Note that after entering the 30.8 offset distance you have the option to choose the Checkmark icon directly and the Datum Plane will be created.)
The first datum plane is created.

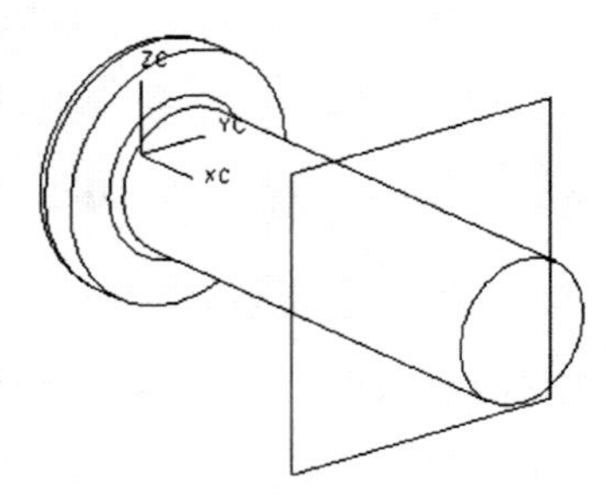

Step 5. Create a datum plane through the axis of the pin. The intersection of this plane with the first datum plane will define the axis for the angle of the hole.

5.1. Choose the **Datum Plane** icon .

5.2. Move your cursor over the cylindrical face of the shaft. The centerline axis of the cylindrical face is highlighted. Select the axis.

5.3. A temporary datum plane is displayed. Click OK to create the datum plane.

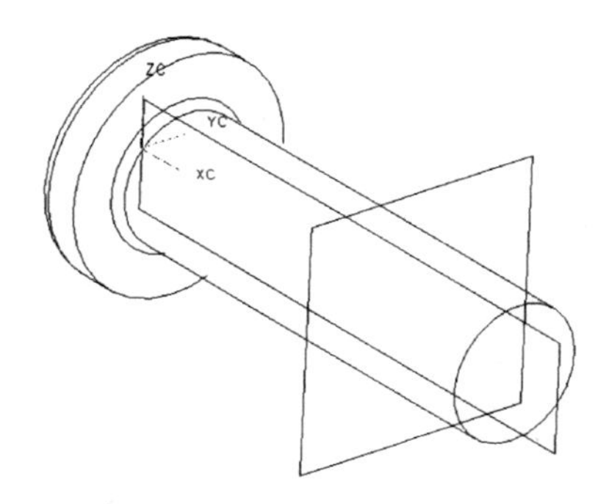

Step 6. If you did not create the datum planes of Steps 4 and 5 according to the reference feature layer standard (layers 61 to 80), you can move them to one of these layers by choosing **Format → Move to Layer** and picking the datum

objects and specifying the destination layer. You can use this command and move any object to any layer. To practice moving objects between layers, move the datum plane from the current layer to layer 71 and move it again to layer 61.

Step 7. **Save and close** this part file.

Activity 8-2. Creating Relative Datum Planes with Dual Constraints

In this activity, you will create relative datum planes using dual constraints.

Step 1. Open xxx_activity8-1.prt that you created in the previous activity, and save it as **xxx_activity8-2.prt** where xxx are your name initials.

Step 2. Change the work layer into **62** (see Step 3 of Activity 8-1 if you are not sure how to do this step).

Step 3. Create a Datum Plane Through an Axis, at an Angle to a Datum Plane. You can create a datum plane at a specified angle from another datum plane, and through the axis of a conical, cylindrical, or revolved feature.

3.1. Choose the Datum Plane icon.

3.2. Change the **Filter** to **Datum** using the selection tool bar as shown:

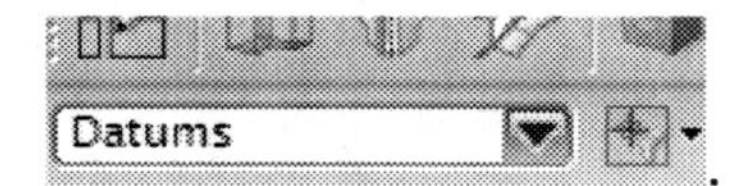

3.3. Select the datum plane that passes through the axis of the pin. A direction vector displays and the datum plane highlights.

3.4. Change the **Filter** to **Any** and move the cursor over the cylindrical face and select **the axis of the face** (read the status line and confirm this selection). A new datum plane is temporarily displayed with the text box defaulted to 90°. This datum plane is perpendicular to the datum plane picked in step 3.3 and goes through the center axis of the cylinder.

3.5. Choose **Apply** icon to ok the creation of the datum plane.

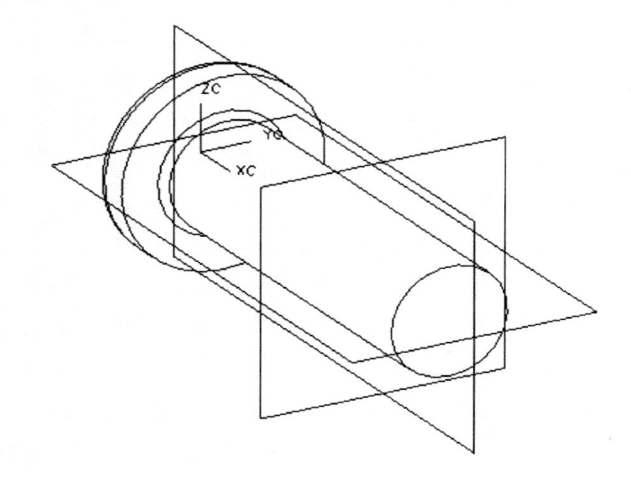

Step 4. Creating a Datum Plane Tangent to a Cylindrical Face.

Since the tangent datum plane could be located anywhere on the cylindrical body, you will specify that the tangent datum plane be parallel to an existing datum plane.

4.1. Choose the **Datum Plane** icon.

4.2. Change the **Filter** to **Face**.

4.3. Select the cylindrical face of the shank.

4.4. Change the **Filter** to **Datum**.

4.5. Select the datum plane on the **XC-YC** plane.

The temporary tangent datum plane is displayed and the direction vector points left. Choose the Cycle Solution icon in the middle of the dialog and you will see an alternative tangent datum plane. Continue to choose the icon until the direction vector point of the tangent datum plane points up. To create a tangent datum plane looking like one in the below figure, continue to click. Choose **OK**. The resulting model looks as below. Note that you can accomplish the same by selecting the datum plane on the **ZC-XC** plane instead of the **XC-YC** plane.

You will use this tangent datum plane to add features to the model.

Step 5. Save and close the part file.

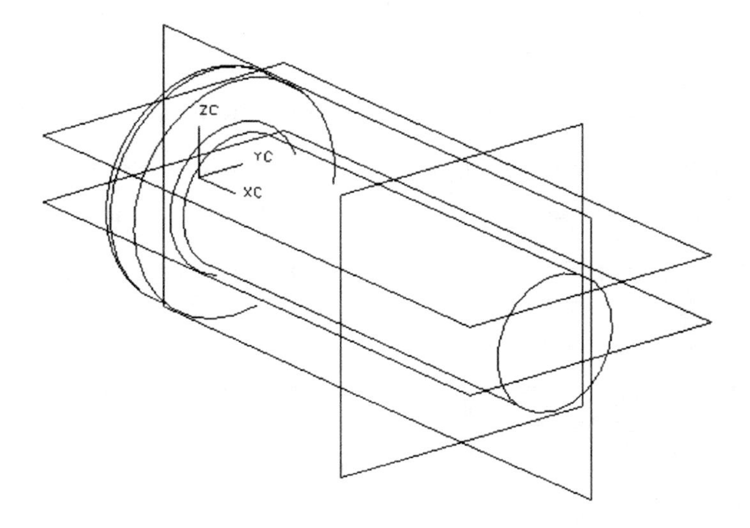

8.2 Datum Axis

Datum Axes are a reference feature, which is an imaginary axis. They can be used to create datum planes, revolved features (see Chapter 9), extruded bodies (see Chapter 9), etc. Datum axes can also be either relative or fixed. Relative datum axes are related to geometric objects (edges, faces, and planes) selected to create them. They are constrained not only by the objects you select, but also by the sequence in which you select the objects. On the other hand, fixed datum axes neither reference nor are constrained by, other geometric objects. This section addresses relative datum axes only.

The user interface and icons for the datum axis is very similar to ones for the datum plane. The Datum Axis option icons are displayed by choosing **Insert → Datum/Point → Datum Axis** or choose the **Datum Axis** icon .

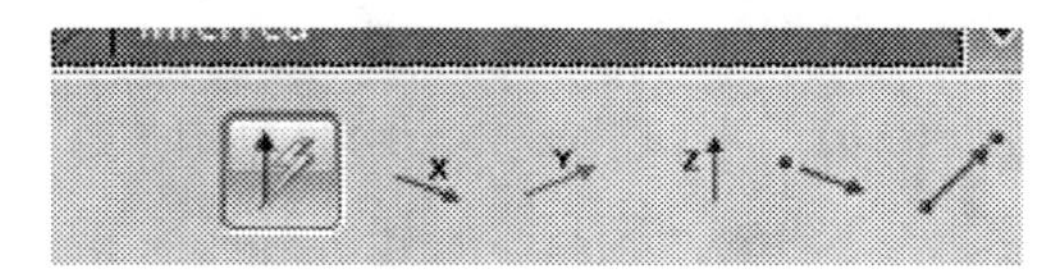

Datum Axis Option Icons

Relative datum axes use the inferred method as a default as well, and are defined and constrained by geometry objects selected. Some of commonly used constraints are discussed below.

Through Edge: This constraint creates a datum axis through a straight edge or the two endpoints of a nonlinear edge. The pick point of the edge determines the positive direction of the datum axis. In the figure below, if the endpoints of the selected edge are V1 and V2, and the pick point is closer to V1, the direction of the axis is from V2 to V1.

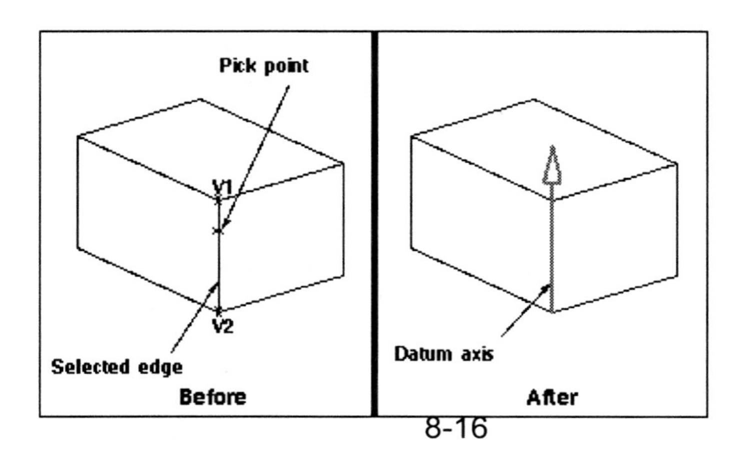

Through Point: Two through point constraints create a datum axis between two points. These points can be edge midpoints or edge endpoints. The direction of the datum axis is from the first pick point to the second pick point. See the figure below.

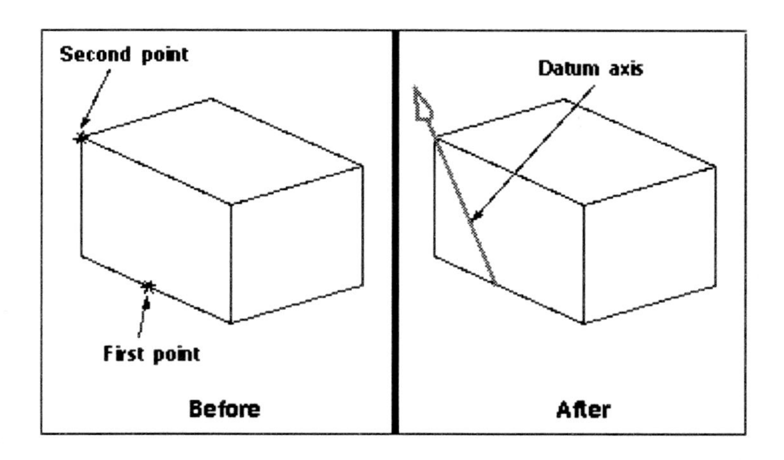

Intersection of Plane: The intersection of two plane constraints creates a datum axis through the straight edge formed by the intersection of two planes. These planes can be either faces or datum planes. The two planes must intersect at some point in space. The direction of the axis follows the righthand rule. That is, place your right hand fingers in the direction of the first face normal and curl your fingers towards the second face normal. The direction in which your thumb is pointing is the direction of the new datum axis (see the figure below).

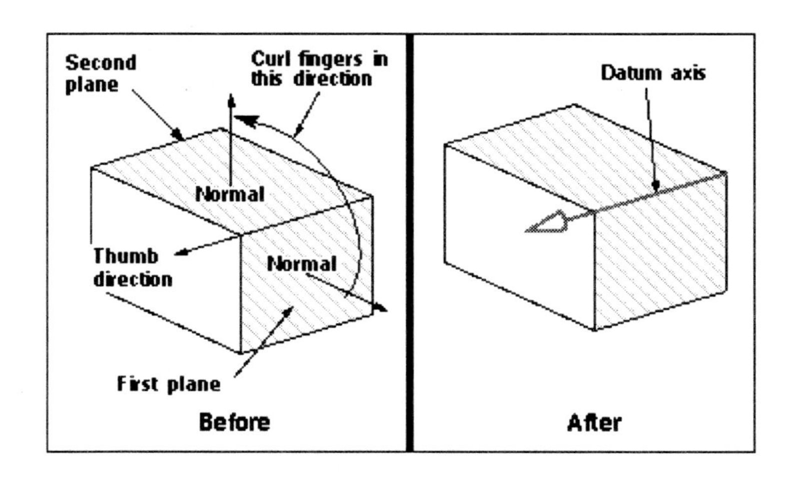

Through Axis of Face: This constraint creates a datum axis through the imaginary axis of any cylinder, cone, or revolved feature. The direction of the datum axis is the same direction as the original creation direction vector for the cylinder, cone, or revolved feature

Activity 8-3. Creating Relative Datum Axes

In this activity, you will create relative datum axes.

Step 1. Open activity8-1.prt and save it as **xxx_activity8-3.prt**.

Step 2. Create three datum planes in layer **61** as shown below. Review Activities 8-1 and 8-2 if you are not sure how to do this step.

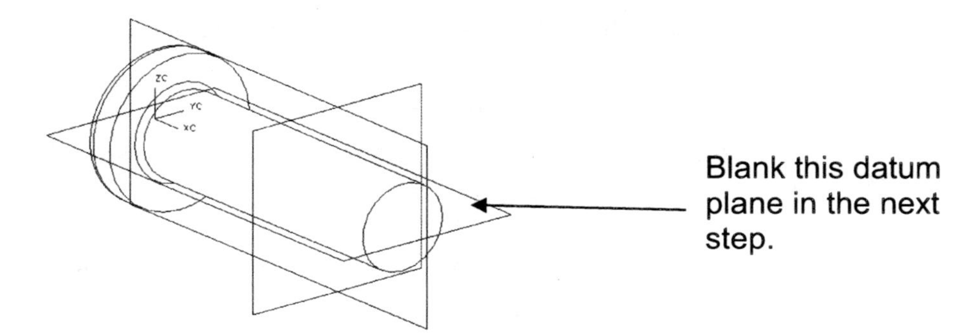

Step 3. **Blank** (temporarily hide from display) the datum plane indicated by the arrow above by choosing **Edit → Show and Hide→ Hide** and picking the datum plane. Alternatively, you can pick it from the graphics window with MB1 (it is highlighted), and click MB3 over it and choose **Hide** on the pop-up menu. The object immediately disappears.

Step 4. Change the work layer to **62**.

Step 5. Create a datum axis through the intersection of two datum planes.

5.1. Choose **Datum Axis** icon .

5.2. Click on the **Filter** options and choose **Datum**.

5.3. Select both of the remaining two datum planes at their edges. A temporary display of the arrow below is shown.

5.4. Choose **OK** to confirm and create the datum axis.

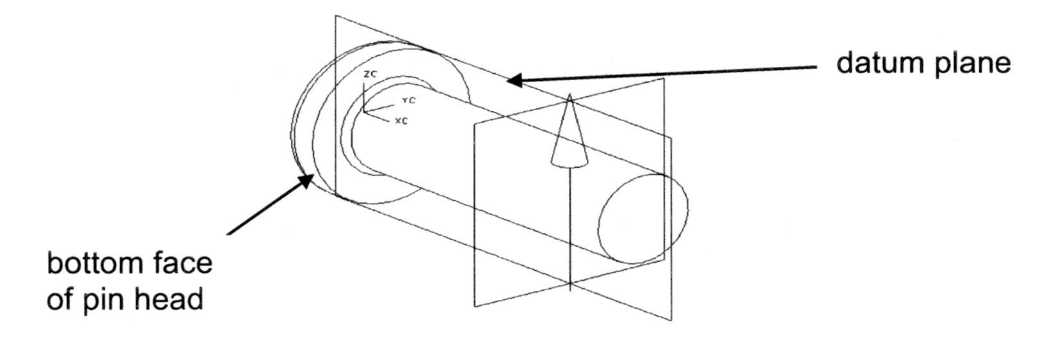

Note that if the datum axis comes in the wrong direction then cancel out and re-select the different side of the geometry until the axis is up as shown. Alternatively, you could create it first and later edit it using the Datum Axis Dialog as shown in Step 8.

Step 6. Create a Datum Axis at the Intersection of a Planar Face and Datum Plane that are indicated by two arrows in the above figure.

- **6.1.** Choose **Datum Axis** icon .
- **6.2.** Change **Filter** option to **Datum** and select the datum plane as shown above.
- **6.3.** Change **Filter** to **FACE** and select the bottom face of the head of the pin as shown above.
- **6.4.** Choose the **Reverse Direction** icon if the arrow is pointing downwards, then choose **OK** to create the datum axis.

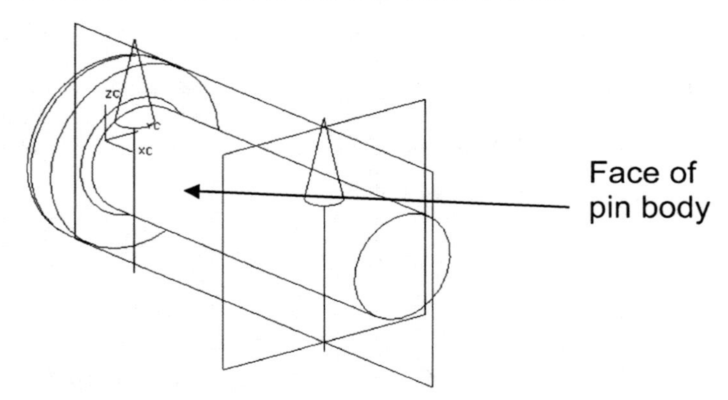

Step 7. Create a Datum Axis Through Axis of Face.

In this step you create a datum axis that goes through the axis of the cylinder face of the pin body as indicated by the arrow above.

- **7.1.** Choose **Datum Axis** icon .
- **7.2.** Change the **Filter** option to **Face** if necessary.
- **7.3.** Select the cylinder face of the pin body.
- **7.4.** Choose **OK** to create the datum axis.

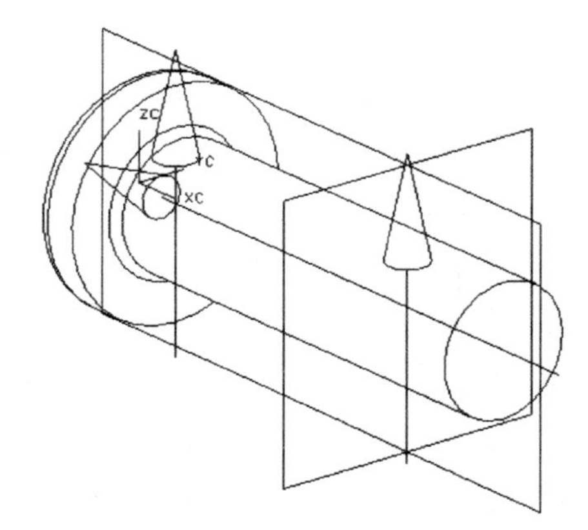

Step 8. Reverse the direction of the datum axis created in the previous step.

8.1. You will open the Datum Axis dialog using one of the following methods.

- **Edit → Feature → Parameters** and select the datum axis created in Step 7 in the list of features or
- from the **Part Navigator** window select the datum axis, use **MB3** and choose **Edit Parameters** off the pop-up menu, or
- select the datum axis on the graphics window, use **MB3** to get the pop-up menu and choose **Edit Parameters** off the menu or
- simply double click the datum axis on the screen.

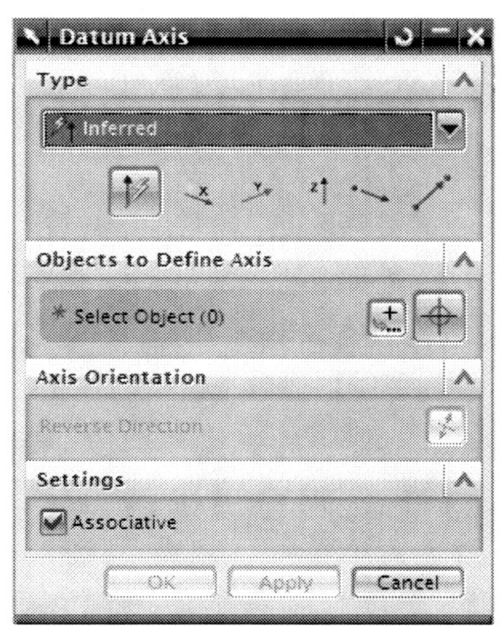

8.2. Choose the **Reverse Direction** icon. The datum axis will reverse direction with a temporary display

8.3. Choose **OK** to confirm the direction and close the dialogue.

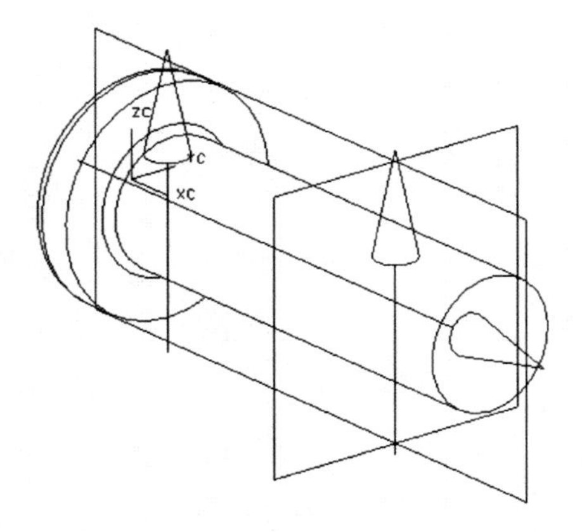

Step 9. **Save and close** this file, which will be used in the following activity

Activity 8-4. Creating an angled hole using reference features

In this activity, you will create two relative datum planes and an angled hole through the pin where the hole is associated to reference features.

Step 1. **Open xxx_activity8-3.prt** that you created in Activity 8-3, and save it as **xxx_activity8-4.prt.**

Step 2. **Delete** the last two datum axes that were created in Steps 5 and 7 of Activity8-3. You do not need them in this activity.

2.1. Choose **Edit** → **Delete** and select the datum axes along the cylinder and the datum axes at the face of the pin head in the graphic window and click the **OK** button. Alternatively, you can use the Part Navigator that was discussed in Chapter 6.

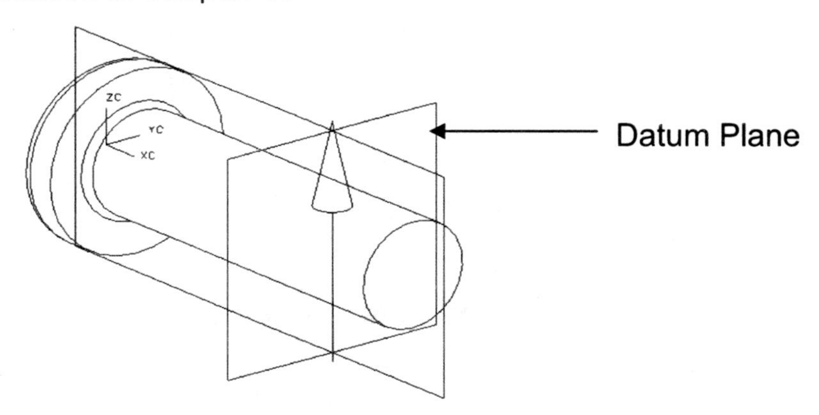

Step 3. Change the work layer into layer **63**.

Step 4. Create a Datum Plane that is angled to the datum plane indicated above.

4.1. Choose **Datum Plane** icon.

4.2. Change the **Filter** to **Datum** and select the datum axis in the graphics area.

4.3. Select the datum plane as shown above.

4.4. The Angle field displays with the positive rotation direction. If the positive rotation direction aligns with the arrow direction in the following figure, key in –45 in the Angle field. Use a positive value (45) if the rotation direction vector is pointing in the opposite direction.

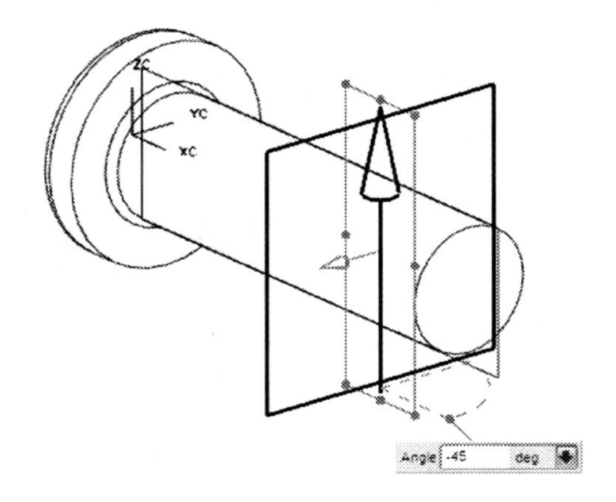

4.5. Choose **OK** to create the datum plane.

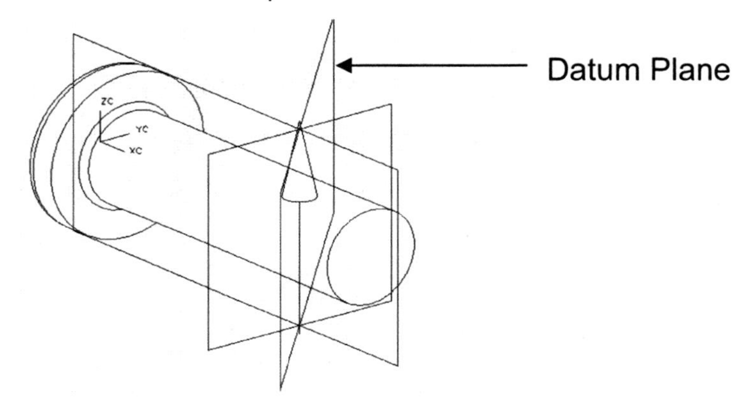

Step 5. Create a Datum Plane offset to the angled datum plane indicated above.

5.1. Choose **Datum Plane** icon.

5.2. Select the angled datum plane as shown above.

5.3. Key in an offset distance of **15** in the **Offset** field, and choose **OK.**

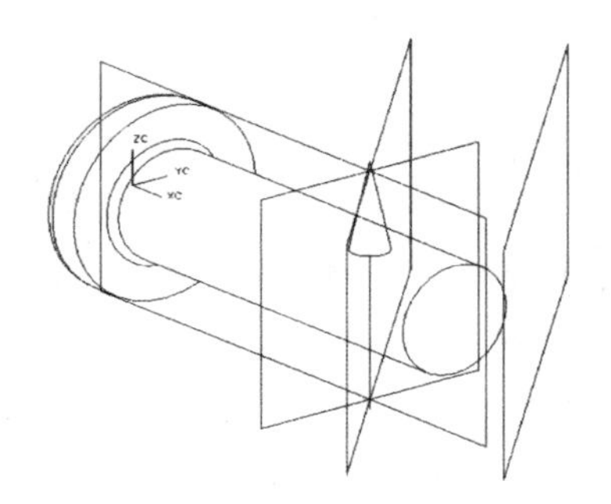

Step 6. Create a through hole on the pin.

6.1 Choose **Insert→ Design Feature→ Hole** or the **Hole** icon. The simple hole is the default.

6.2 Select the offset datum plane created in Step 5 as the placement face. A direction vector displays the direction in which the hole feature will be created.

6.3 If the vector is pointed away from the solid, choose the **Reverse Side** button so that it is pointed toward the solid.

6.4 Select the cylindrical face of the pin. This makes the hole go through the pin.

6.5 Key in **3** for the diameter of the hole and choose **OK**.

The tool solid is initially placed at the center of the datum plane. The **Positioning** dialog is displayed.

Step 7. Position the hole.

You need to position the hole so that it goes through the center of the pin at the location of the datum axis.

7.1. Choose the **Point onto Line** icon.

7.2. Select the datum axis.

7.3. Choose **Edit → Show and Hide → Show All** to bring back the Datum plane which was created in the in **XC – YC plane**.

7.4. Choose the **Point onto Line** icon again and select the datum plane from Step 7.3 in the XC-YC Plane.

7.5. Choose **OK** on the **Positioning** dialog.

The hole moves to align with the datum axis and datum plane. Change your view to **TOP** by choosing MB3 on graphic window → **Replace View → Top**.

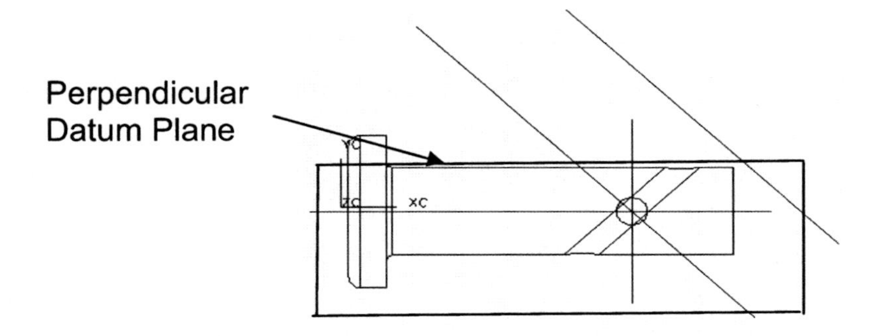

Step 8. Change the angle of the angled datum plane indicated in the above figure. The hole location is controlled by several reference features and affected by their change(s).

8.1. Choose **Edit → Feature → Parameters** and select the angled datum plane in the graphics window as shown below and choose **OK**.

8.2. Key in **–15** in the **Angle** field and choose **OK** twice.

The top view of the new model must look like the following figure. Observe that this edit changed not only the angle of the datum plane but also the angle of the hole that goes through the pin. This is because the placement plane of the hole is the datum plane offset from the angled datum plane.

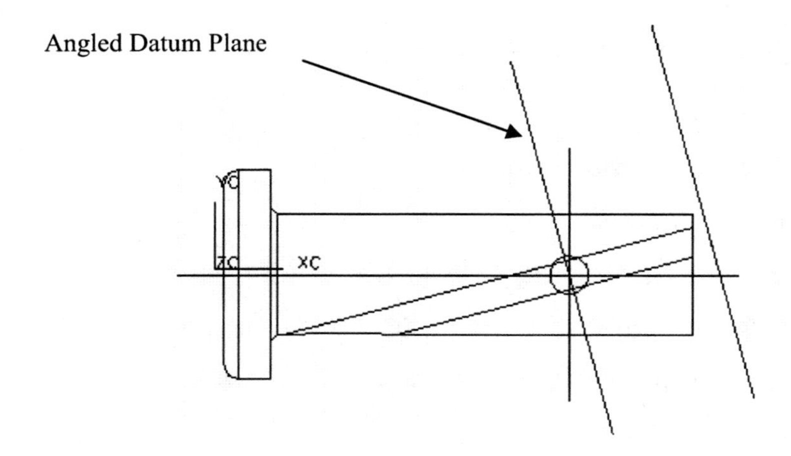

Step 9. **Save and close** this file.

Project 8-1. Spindle

Model the spindle part using the drawings (in inches) below and save it as **xxx_project8-1**.prt in your directory.

Hint: You may have to use trigonometric functions to compute the offset dimension for Chamfer.

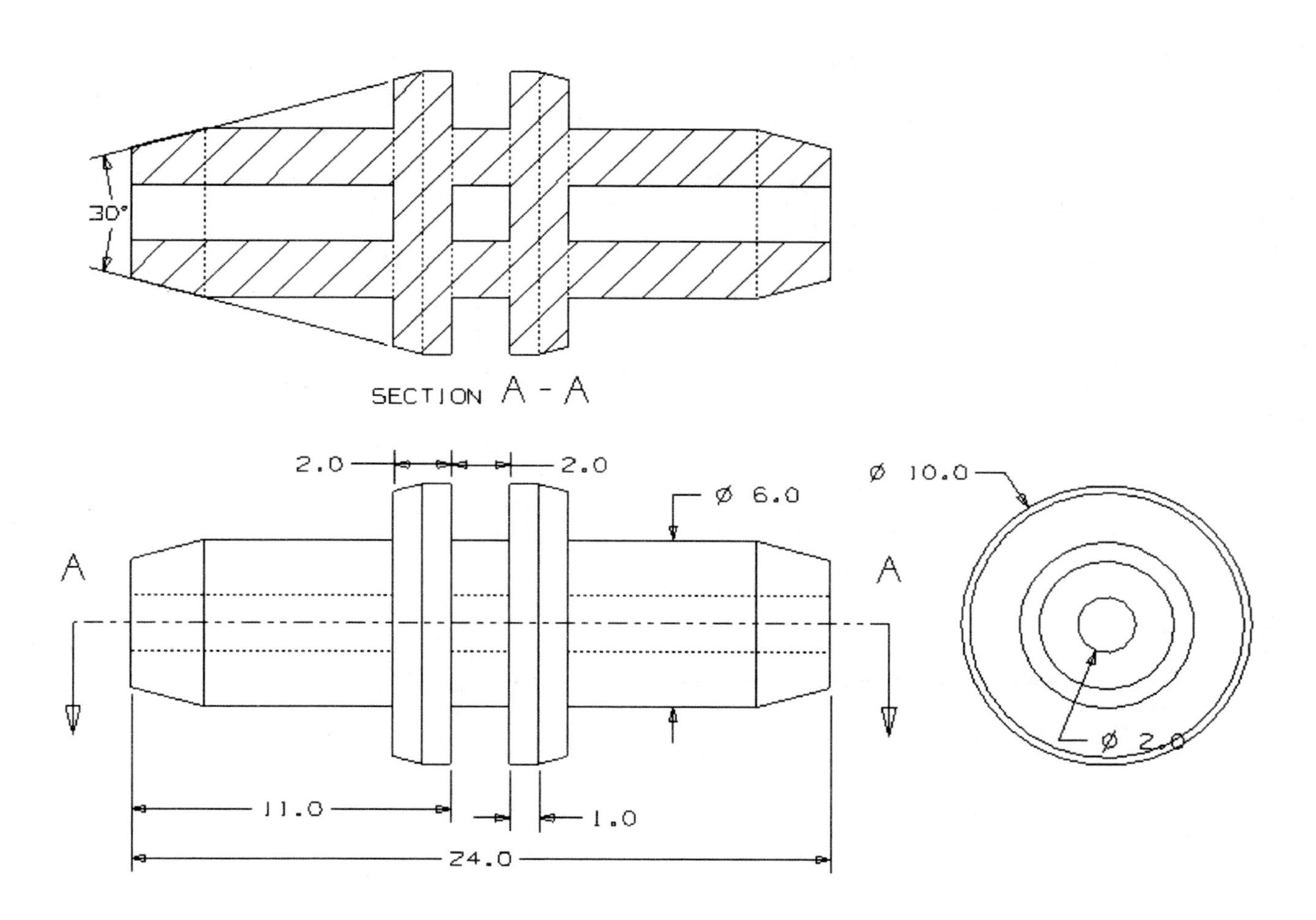

Project 8-2. Spin Block

Model the spin block using the drawings (in millimeters) below and save it as **xxx_project8-2**.prt in your directory. Note that the design intent is to have the hole and the boss always centered in the block even when the block dimensions change.

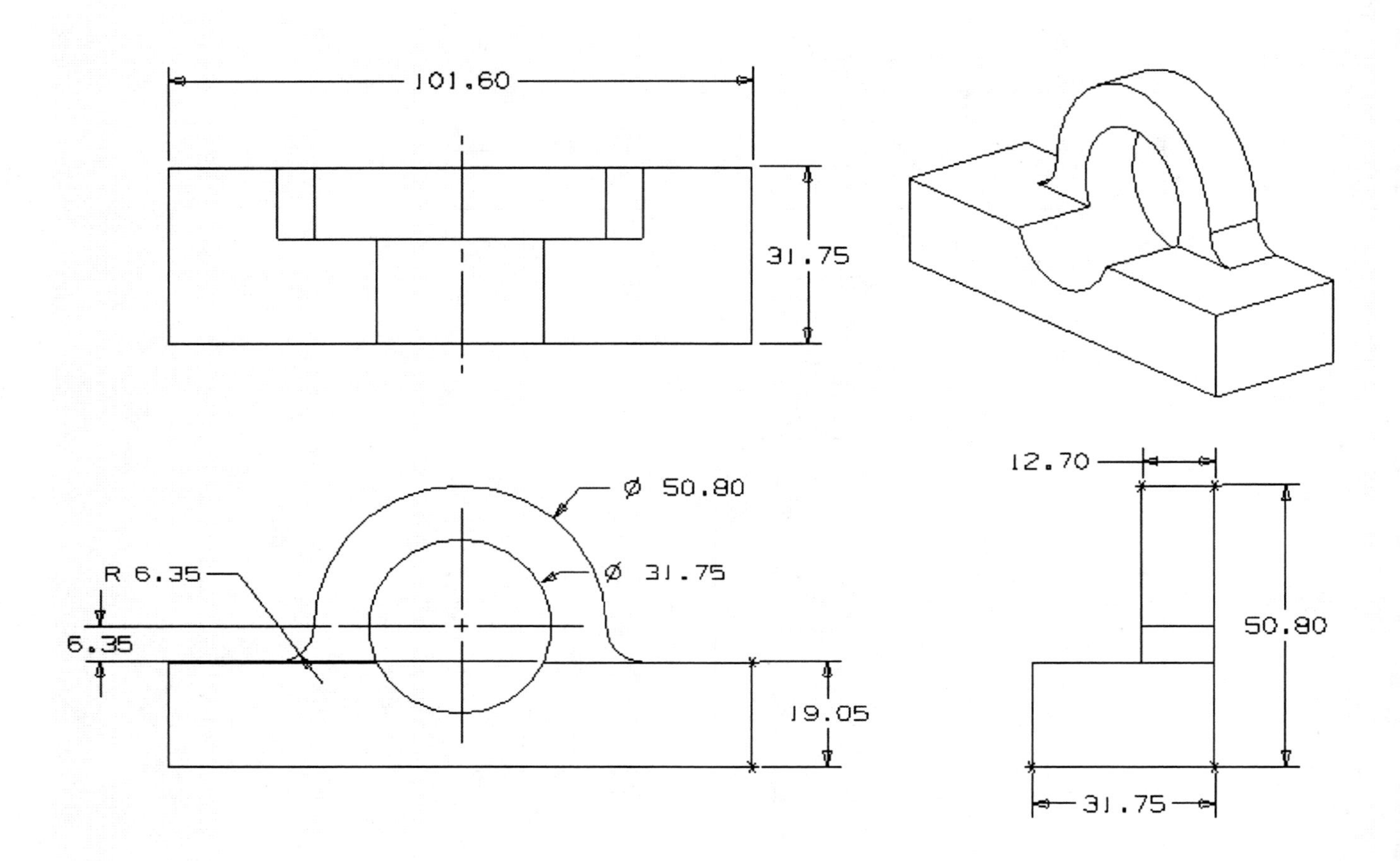

Project 8-3. Retainer

Model the retainer using the drawings (in inches) below and save it as **xxx_project8-3**.prt in your directory.

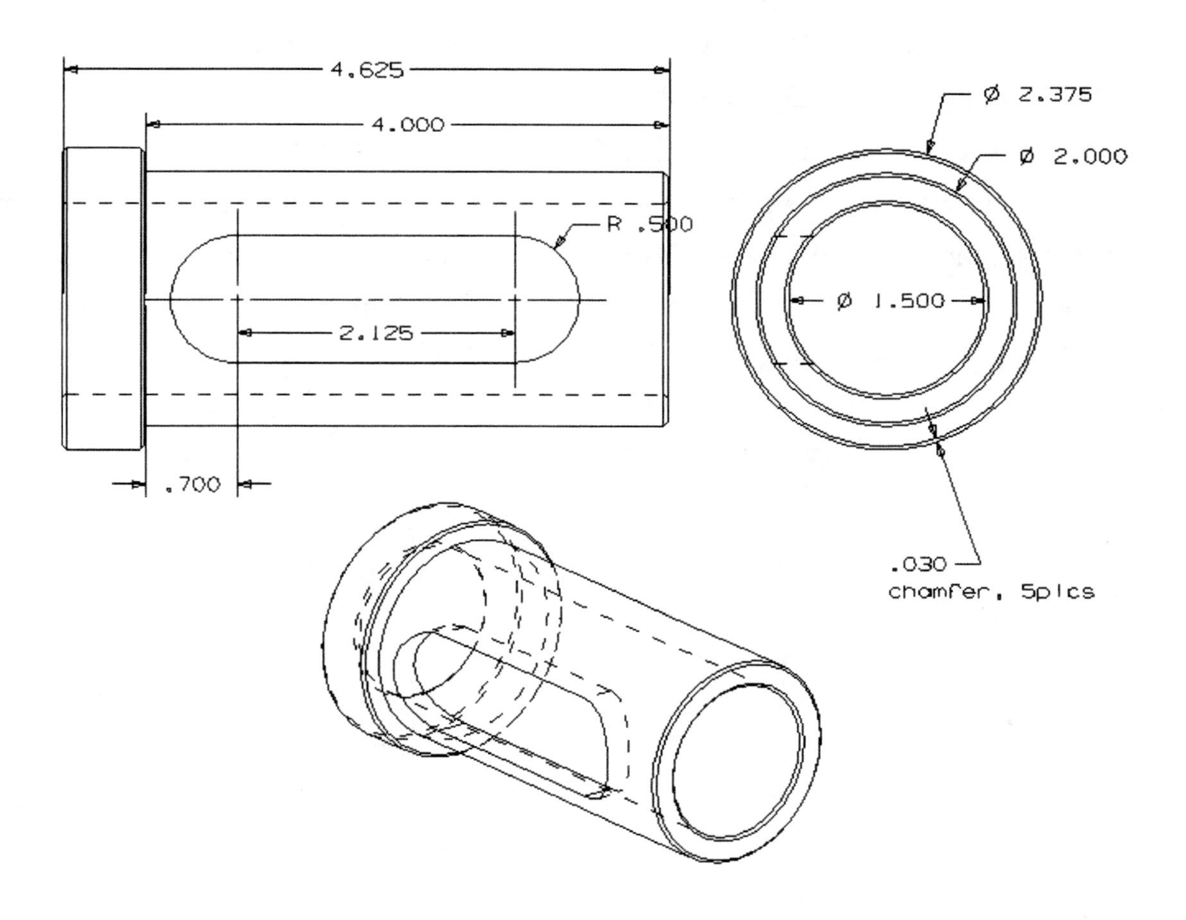

Project 8-4. Block with an Angled Hole

A 1-inch diameter hole is to be created in a block. The origin of the hole is on the top face. The hole is centered in the block along both XC and YC axes. The angle of the hole shall be editable relative to a plane parallel to the bottom face. The block size can change. In the drawing below, the angle is 30° and the block size is (8, 6, 1). After the part is modeled, change the angle or the block size and confirm if the design intent is still met.

Save this project file as **xxx_project8-4**.prt in your directory.

Hint: In this project, one datum axis and four datum planes need to be created to constrain the pivot location of the hole feature. In order to position the hole, use reference features (the datum axis and one datum plane) and the positioning method Point onto Line.

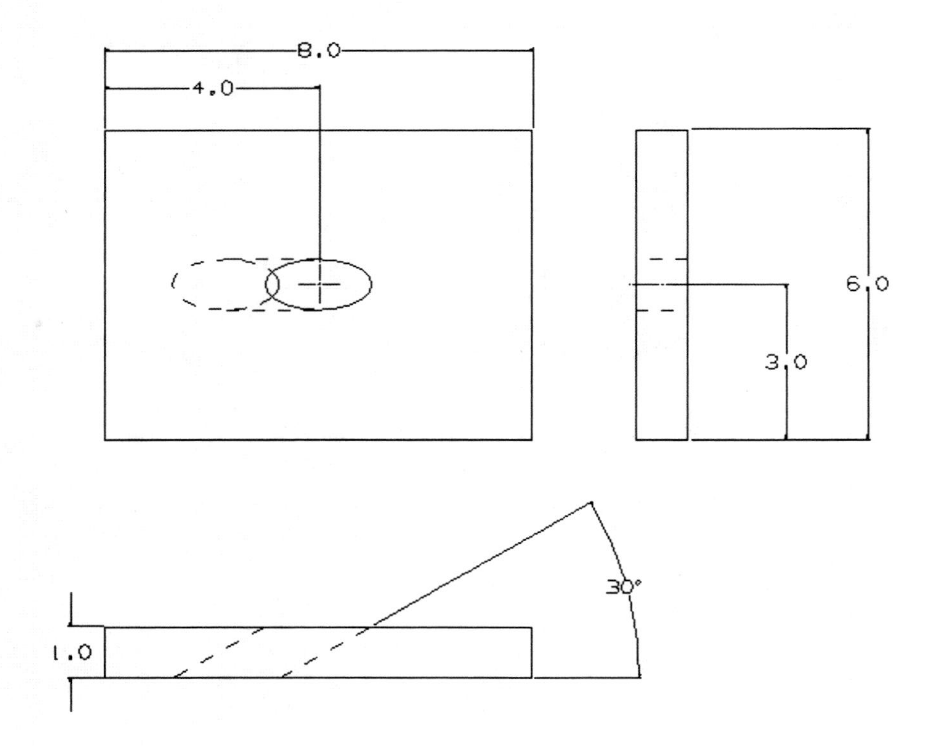

Exercise Problems

8.1 When do you need to use datum planes? Give two examples.

8.2 State two types of datum planes. What is the major difference between the two?

8.3 There are three categories of relative datum planes in term of the number of constraints. Give an example for each.

8.4 The number of constraints used to create relative datum planes is the same as the number of object selections. (True or False).

8.5 Datum Axes can be used to create datum planes. (True of False)

8.6 Datum Planes can be used to create datum axes. (True or False)

8.7 Relative datum planes and axes are associated to objects selected during their creations. What guides designers in identifying this association?

Chapter 9. Boolean Operations and Swept Features

This chapter introduces the Boolean Operations and Swept Features. The Boolean operations are **Unite**, **Subtract**, and **Intersect**. The Swept features are **Extrude**, **Revolve**, **Sweep along Guide**, and **Tube**. This text will cover Extrude and Revolve.

Boolean operations are used to create a single solid body out of two or more solid bodies. When a second solid body is created, Unigraphics knows that one solid already exists and will prompt if the user wants either to have two separate solid bodies or to apply a Boolean operation between the two solid bodies.

Swept features is the general name given to the family of four commands listed above that create a solid by sweeping a set of selected curves, edges, faces, or sheet bodies. Often this newly created solid becomes the second solid and the Boolean operation immediately follows the Swept feature.

This chapter consists of 3 sections. Section 9.1 will discuss the Boolean Operations. Section 9.2 will introduce the Extrude command. Section 9.3 will introduce the Revolve command.

9.1 Boolean Operations

George Boole, an English mathematician, approached logic in a new way by reducing it to a simple algebra, incorporating logic into mathematics. It began the algebra of logic called Boolean algebra published in 1854. Boolean algebra is used in expressing relationships between groups of objects or concepts. Boolean algebra can demonstrate the relationship between groups, indicating what is in each set alone, what is jointly contained in both, and what is contained in neither. We see it applied today in search engines on the Internet using operators AND, OR, and NOT. Therefore, if a search is done on "cats AND dogs", the result will contain information including both cats and dogs, and not each alone.

Likewise, our Boolean operations of Unite, Subtract, and Intersect follow the similar logic. However, we are working with solid bodies so the application is a little different, but conceptually the same.

Another English mathematician, John Venn, published an interpretation of Boole's work and introduced a method of diagramming Boolean algebra. We now call it the Venn diagram. The application of the Venn diagram to our Boolean operations looks like this.

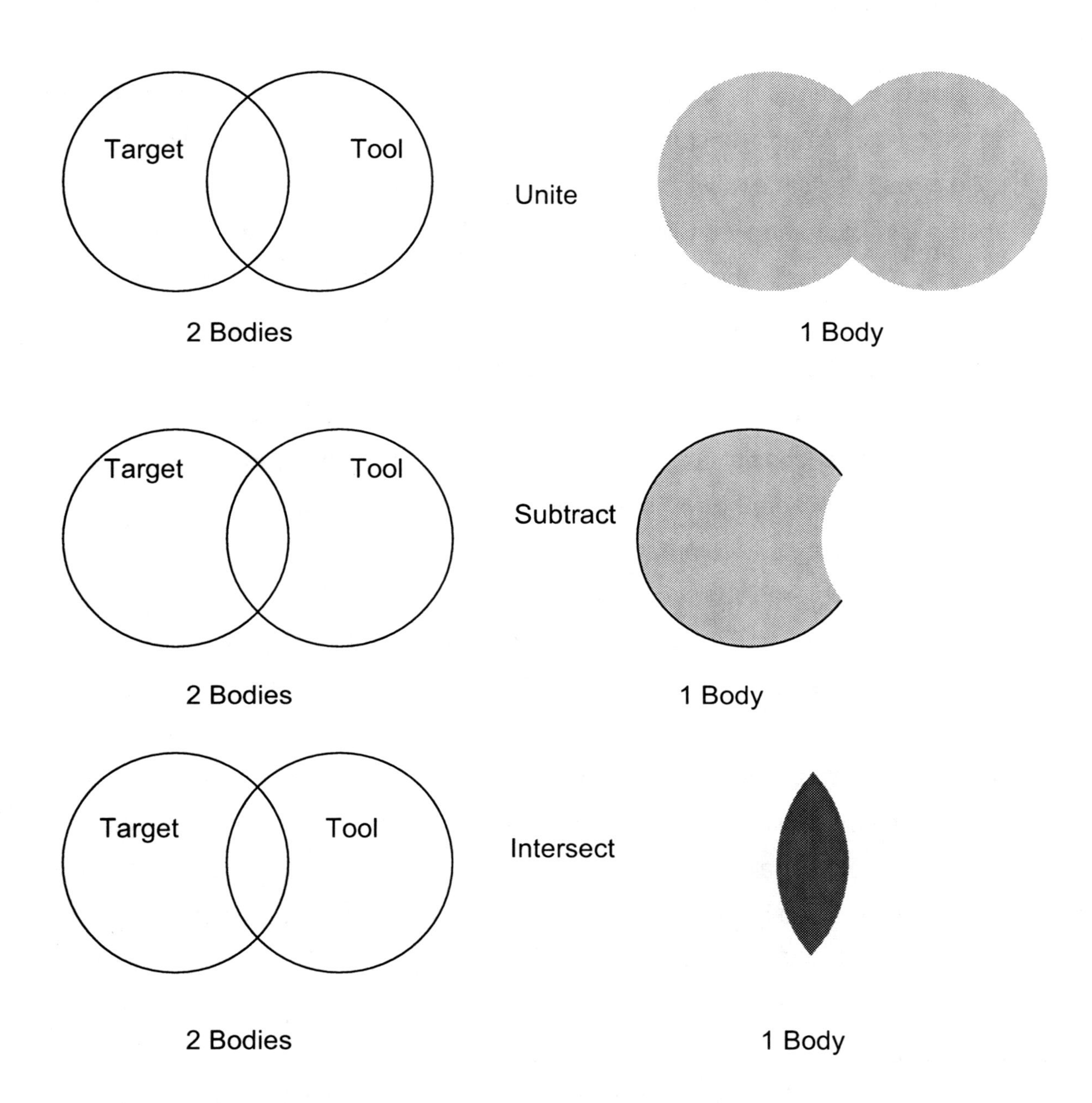

The above figures illustrate that when two solids are combined using a Boolean operator, different results are obtained with Unite, Subtract, and Intersect. One requirement for Boolean operators to work properly is that the two solid bodies must overlap somewhere, i.e., occupy the common space for at least a portion of their bodies. Unigraphics names these two solid bodies as the Target solid and the Tool Solid.

Target Solid

The Target solid is the first solid chosen during the Boolean operation. In the case of a Form Feature (such as Hole, Slot, ...), the Target solid becomes by default the base solid on which the Form Feature is created. The target solid controls properties such as the color, material properties, layer, and any other attributes that may have been assigned. Thus, the resulting body from Boolean operation inherits those properties of the target solid.

Tool Solid

The Tool solid is the secondary solid chosen. It is the solid that is united with, subtracted from, or intersected with, the target solid. Its properties will be lost to the target solid. There may be more than one tool solid as well. When using multiple tool solids care must be taken as to how the operations are carried out because operations might go in a different order and get different results.

Unite

This operator allows two or more solids to be combined into one. The volumes in common with each other unite to become one solid. The portions of the bodies not in the common area remain unchanged. In the Venn diagram above, notice in the Unite example that the overlapping portions of each cylinder were melt together and formed one body. You cannot unite a solid body with a sheet body.

Subtract

This operator allows material to be removed from the target solid by deleting the common volume (overlapping) of the tool solid. The tool solid is then removed completely leaving empty space where the tool solid used to be. In the Venn diagram above, notice in the Subtract example that the overlapping portions of each cylinder was removed from the target solid by the tool solid, thus taking a slice out of the target solid. You can subtract a solid from a solid, a solid from a sheet solid, and a sheet solid from a solid; however, you may not subtract a sheet from a sheet. Furthermore, the overlapping portions of the two solids must overlap some finite amount. Tangency conditions, such as a cylinder tangent to another cylinder, would not subtract. The same rule applies for a cylinder tangent to the planar face of a solid body. In other words, if only point or line contact is common between two solids, then the subtract will not work and you will receive an error message saying “Non-Manifold Condition exists”.

Intersect

This operator unites (or melds) the common volume between the target solid and tool solid(s) and removes (deletes) the rest of the solids that do not overlap each other. It combines the other two operators. In the Venn diagram above, notice the only portion of each solid remaining is the small wedge shape where the two bodies overlap. The remaining solid body will retain the attributes of the target solid. This operator is particularly prone to unexpected results if more than one tool solid is chosen, so it is advisable to perform intersections one at a time. It is permissible to intersect a target sheet and a tool solid body.

Form Features

Form Features are special cases of combining solid bodies where the Boolean is built into the command and the operation done automatically. The Hole feature, Slot feature, Pocket feature, and Groove feature all do a Subtract with the original solid body as the target and with the form feature as the tool. The Boss feature and Pad feature both do a Unite with the original solid body as the target and with the Boss and Pad as the tool. Have you noticed when creating form features that a solid body is displayed after creation and before you position it? That is the tool body that operates on the original solid body.

Creating a Second Solid Body

When a second solid body is created, you make the choice of creating it as a standalone body, or to perform a Boolean on it as the tool body with the existing solid body as the target body. An example of a dialog that comes up looks like this.

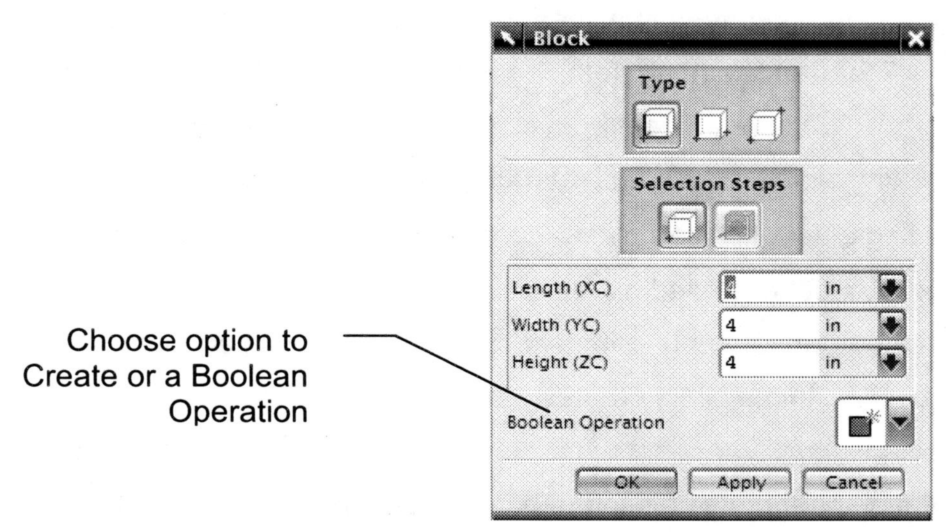

You make this choice before choosing the OK button.

9.2 Extrude

The Extrude option is found under the **Insert** menu bar with the commands of **Insert → Design Feature → Extrude...** or at the **Extrude** icon on the **Form Feature** toolbar. This option allows a solid body to be created by sweeping planar section string geometry in a linear direction over a specified distance.

There is a Preference setting, found in **Preferences → Modeling**, that specifies if a solid body or a sheet body is created when sweeping. With the solid body setting which is a default, the following rules apply.

- Extruding a set of closed planar connected curves creates a solid body.
- Extruding a set of closed planar connected curves with another closed set within the boundary of the first set creates a solid with an interior hole, e.g., circle within a circle creates a pipe.
- Extruding a curve or set of planar connected curves that are not closed creates a sheet body unless the offset option in Extrude is used. The offset option will be explained in Activity 9-3.

When the **Preference** setting is set to create a sheet body, then a sheet body will be created whatever the curve configuration is as listed above. Note that if a closed set of curves is swept, a body will be created that looks like a solid body but will in fact be hollow inside, only a shell with sheets on the outside surface.

General Procedure to Extrude

The general procedure to extrude a set of curves is as follows:

1. Choose the Extrude icon on the Form Feature toolbar.
2. Select the set of curves.
3. Choose the method of Extrusion as Limits and Enter a Start Value and End Value.
4. Verify the direction of the sweep with the displayed arrow. To Reverse the Direction Click the Reverse Direction icon.

   To change to any other direction Click the Direction icon to bring up the Vector Constructor dialog.

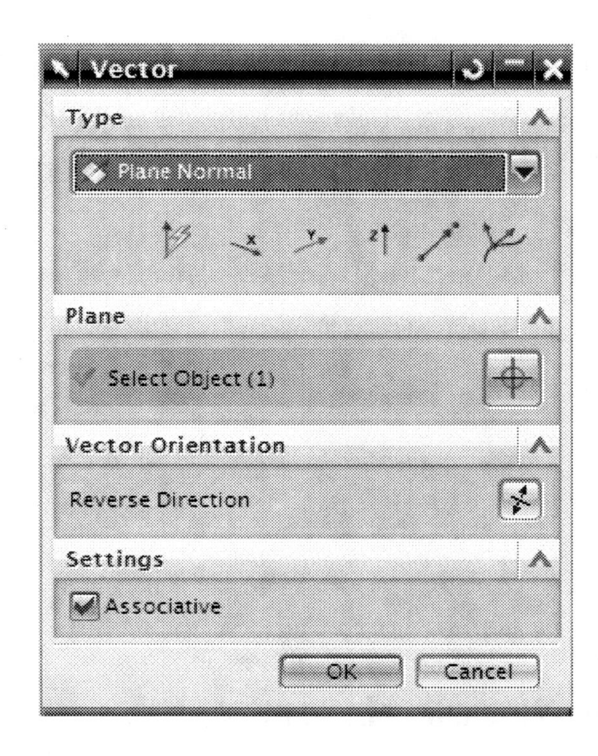

5. To enter the offset parameters Toggle On the Offset button
6. Specify a Boolean operation if a solid already exists and then choose OK.

Activity 9-1. Creating an Extrusion

An extrusion is a feature. It also results in a solid body. Therefore, the Work Layer should be set, color of the body and other attributes that your company wants set.

Step 1. **Open** part file **activity9-1.prt and** save it as **xxx_activity9-1.prt.**

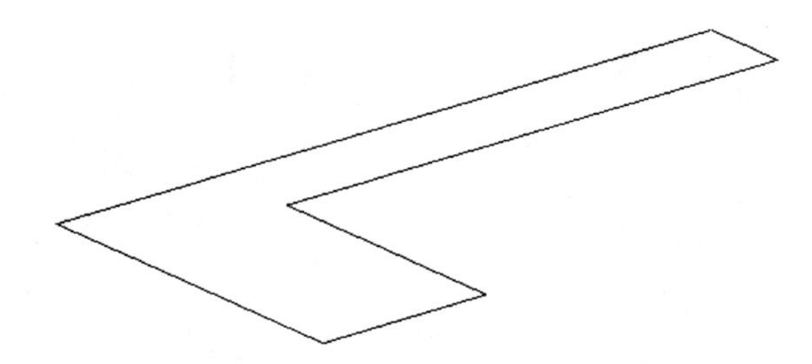

We will extrude the set of curves that are displayed.

Step 2. Open the **Modeling Application**.

Step 3. Change the **Work Layer** to be Layer 1. The solid resulting from extrusion will be placed in this layer.

Step 4. Choosing **Insert → Design Feature → Extrude...** or **Extrude** icon displays the Extrude icon options in the upper left corner of the graphics window.

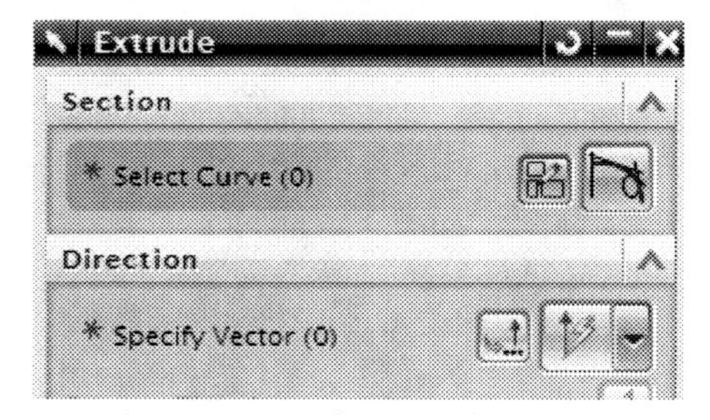

The Icon options use the Inferred method as the default to let you quickly create the Extruded body. The Extrude icon options include the following tools:

Select Section is for selecting existing objects to be extruded. Existing objects could be curves or sketches that will discussed in Chapter 10.

Sketch Section is for creating a sketch "on the fly" before you extrude it. If you use this, the sketch will be "imbedded" in the feature, and cannot be used to create other features. You can, however, extract the sketch from the feature later, so it CAN be used for creation of other features. This is done in the Part Navigator, using MB3→Make Sketch Internal (or Make sketch External).

Inferred Vector is for defining the extrusion direction. (The small arrow at the right of the icon lets you access other vector functions.)

Boolean operations cascade menu – Available operations are Create, Unite, Subtract and Intersect.

Select the Extrude icon to bring up the Extrude Dialog.

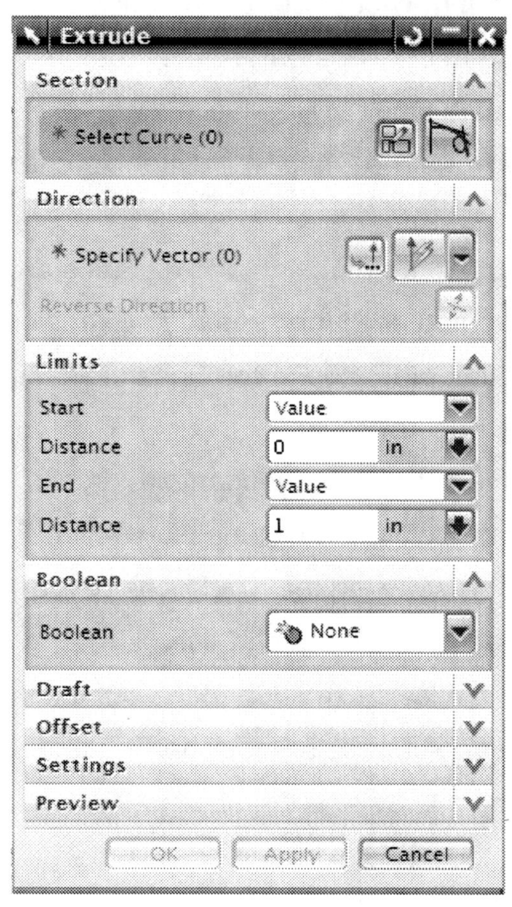

The generator for the sweep may be curves, a solid face, a solid edge, or a sheet body. The **Cue Line** prompts to select the section geometry to extrude. If a sketch is to be the generating curves, you just need to pick one curve in the sketch and the rest will be selected also (How to create a sketch will be discussed in Chapter 10). Furthermore, the resulting solid body is associated to the curves and may be edited by editing the curves.

Step 5. In this activity, the section string consists of 6 lines that form a closed loop. You can pick each of the 6 lines individually. At this point, all 6 lines are highlighted (selected). The Start sub menu under Limits allows six options for extruding that specify the limits and the direction to extrude.

Value —Allows you to specify the direction of the extrude and the length of the extrude by specifying a **Start Value** and **End Value**. The extrude does not have to start at the curves and it may start before the curves. For instance, think of creating the curves in the middle of a part and extruding in both directions. In this case, the Start distance would be negative.

Symmetric Value --Converts the **Start** limit distance to the same value as the **End** limit.

Until Next—With this option the direction and distance are determined by the curves selected and the Face or Plane selected. Extrusion will be normal to the plane of the curves selected and stop at the Face or Plane. The extruded solid will be associated to the Face or Plane.

Until Selected—With this option the direction of extrusion is normal to the plane of the curves selected and be limited by selection of two Faces or Planes. The extruded solid body will be between the two Faces or Planes. The solid body will be associated to both Faces or Planes and curves.

Through all—This option allows extrusion of the curves through multiple solid bodies and then a Boolean subtract is performed. The direction of extrusion is normal to the plane of the curves and the distance is completely through the farthest solid body selected. The extruded solid is the Tool body and the selected multiple bodies are the Target bodies.

Until Extended—This option allows the extruding of curves to a solid body. It is used when the body topology may change and therefore a face would change so the Trim to Face is not an option.

Step 6. We will choose the default option under Limits and Enter the Start and End Value.

The choices here are to accept the default direction (the direction displayed) or to reverse the direction by choosing **Reverse Direction** icon or to re-define the direction using the icon or other options in the pulldown menu.

6.1. Choose the default direction.

Enter parameters for the sweep in the Limits sub menu.

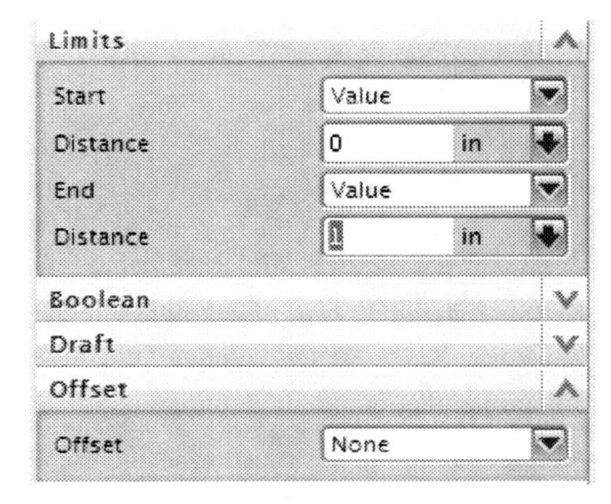

Here is where the parameters for **Start** and **End Distance** are entered. Parameters for **Offset** dimensions and **Taper Angle** may also be entered under their proper option boxes (Offset and Draft respectively).

6.2. Enter the parameters shown in the figure and choose **OK** to accept the values.

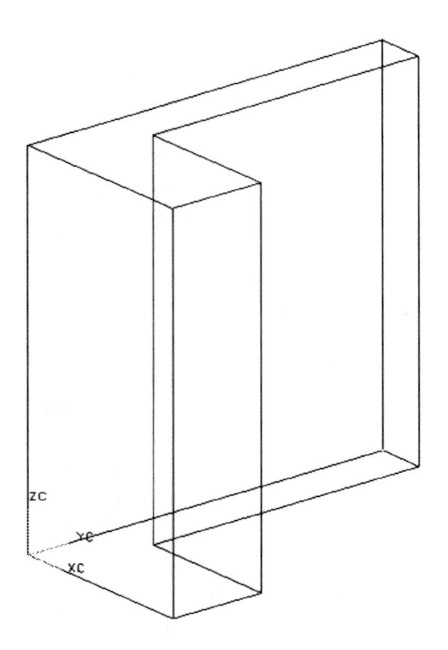

Step 7. The solid body is created immediately. **Save** and **Close** the part file**.**

Activity 9-2. Another Extrude

Step 1. **Open** part file **activity9-2.**prt and save it as **xxx_activity9-2.prt**.

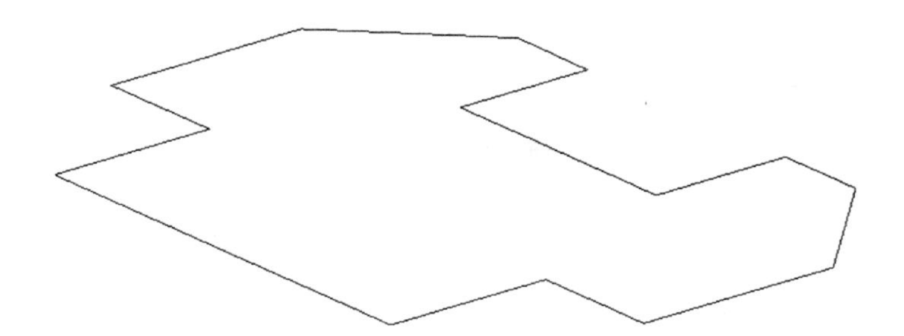

Step 2. Change the **Work Layer** to Layer 11.

Step 3. **Extrude** the curves.

- **3.1.** Choose **Insert → Design Feature → Extrude...** or **Extrude** icon .
- **3.2.** The section string consists of a closed loop of many curves. Instead of picking each of the curves individually, which can be time-consuming, left click and drag the mouse to create a rectangle which includes all the curves. All the curves selected are highlighted and a preview of the extrusion appears on the screen.
- **3.3.** Choose the default option under Limits(Value)
- **3.4.** Ensure the direction vector is pointing up.
- **3.5.** Enter the extrude parameters, **Start Value** 0 and **End Value** 0.75. The **Offsets** and **Taper Angle** are 0.
- **3.6.** Choose **OK**.

Step 4. Create the **Holes** as shown on the drawing below.

Step 5. Save and close the file.

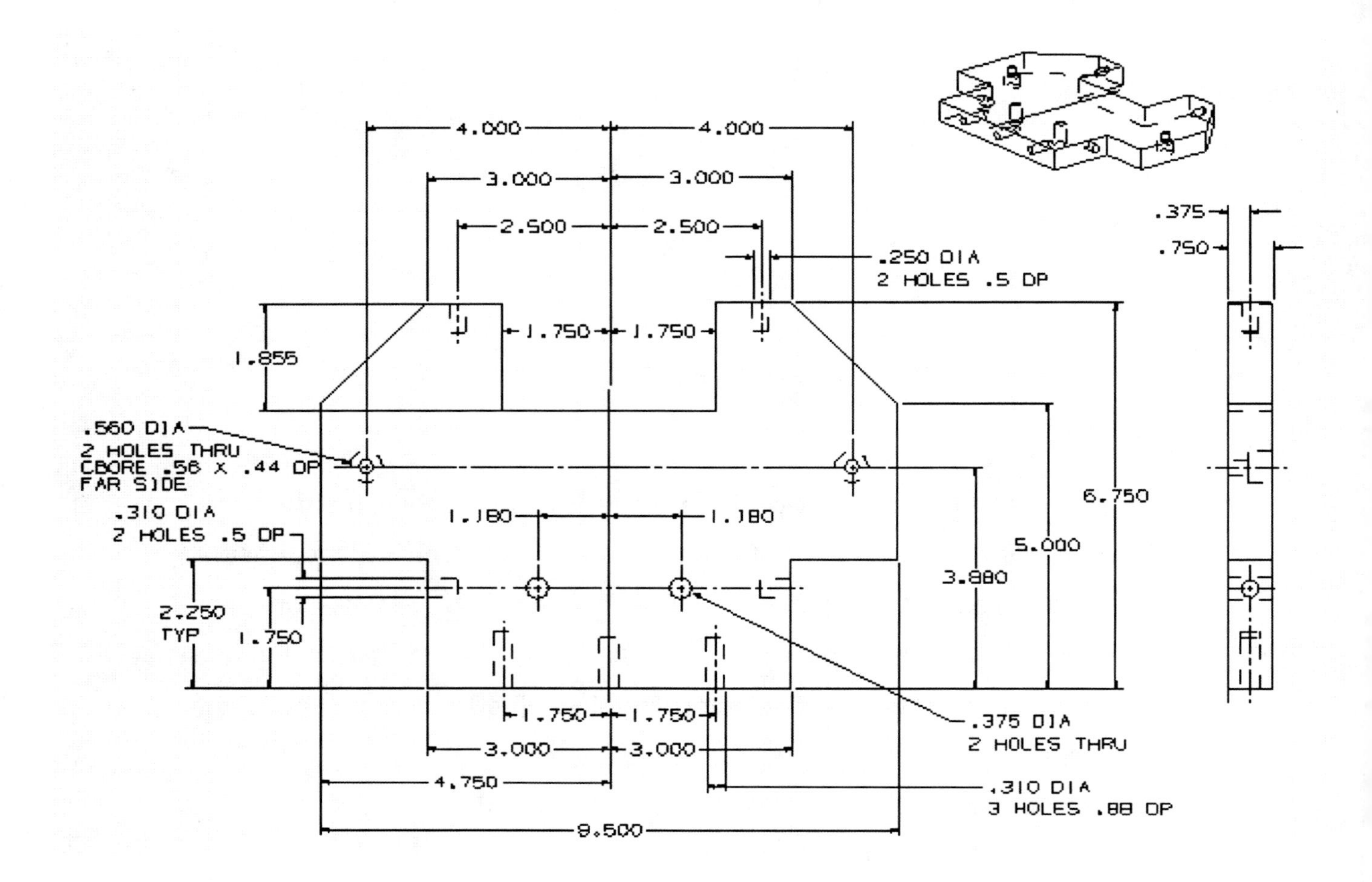

Note that the two counterbore holes have a thru-hole with diameter of 0.25.

Activity 9-3. Extrude with Offset

This activity will demonstrate extrusion of curves with an offset to create an extruded wall of the curves (the offset determines the wall thickness).

Step 1. **Open** part file **activity9-3.prt** and save it as **xxx_activity9-3.prt.**

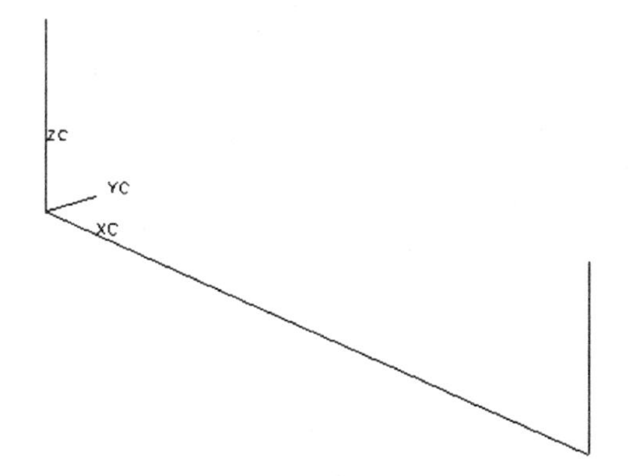

We will extrude these curves to create a length of channel and offset the curves to the outside to create a solid body channel.

Step 2. Choose **Modeling Application** to ensure it is active.

Step 3. Choose **Insert → Design Feature → Extrude...** or **Extrude** icon 

Step 4. Select the three curves to extrude.

Step 5. Ensure the Direction arrow comes up pointing to the back of the screen (in the **+YC** direction).

5.1 Choose the **Reverse Direction** icon if necessary.

Step 6. Enter Start and End Values under Limits and the offset values as shown below.

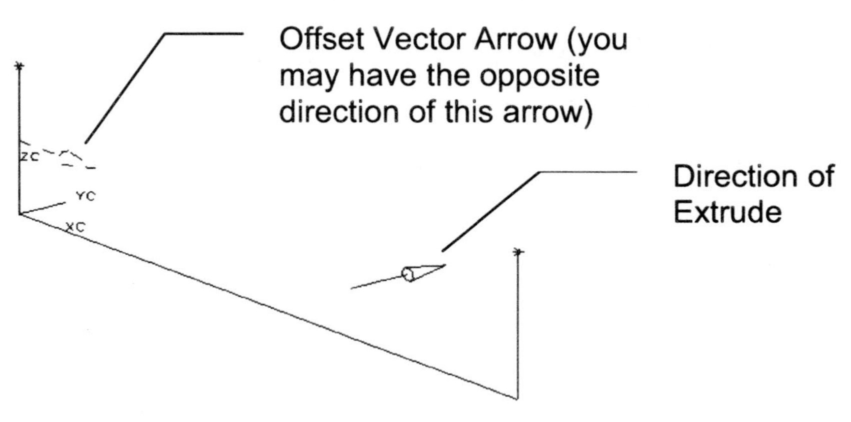

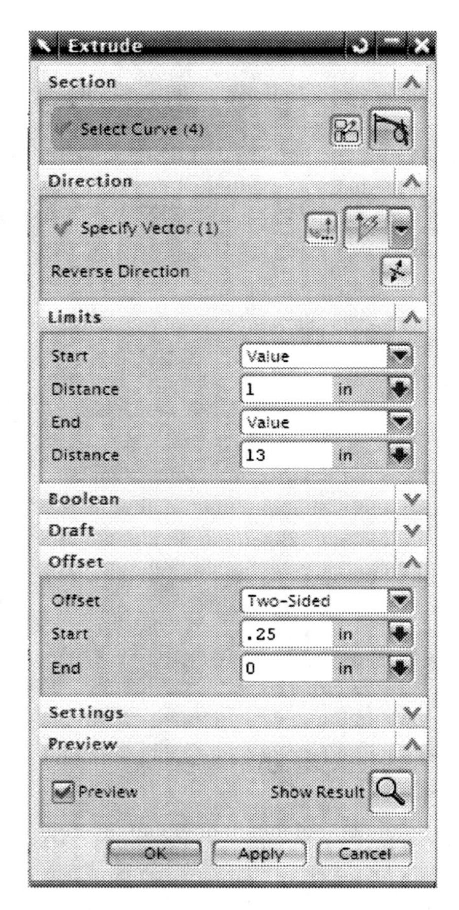

The **Start Value** is 1 and the **End Value** is 13, indicating the extrusion will start 1 inch from the curves and end 13 inches from the curves. Check the Offset field right below the End field if unchecked.

The **First Offset** value is **0.25**, indicating that the offset will be .25 inches from the curves. The **Second Offset** is set to 0. The positive sign indicates the offset will be in the direction of the **Offset Vector Arrow** (dashed arrow in the graphics window). This will create a wall thickness of .25 inch starting with the chosen section string to the offset arrow direction.

Step 7. Choose **OK** to confirm the values and extrude the curves. **Save** and **close** the file.

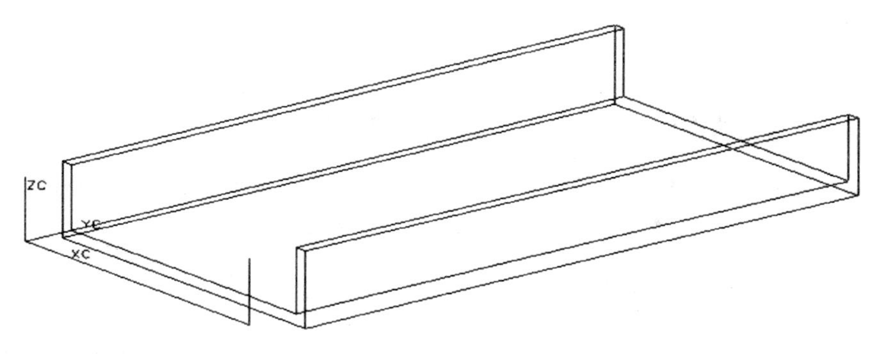

Offset capability gives flexibility to create a range of solids easily that would otherwise require more construction. The difference between two offsets gives the wall thickness of extrusion. Using Offset values has these characteristics:

- If a value is entered only for the First Offset, the value acts like an end value and the offset will extrude from the curves to the offset distance.
- If a value is entered for both the First Offset and the Second Offset, the offsets act like starting and ending distances. The offset extrude will start at the First Offset and end at the Second Offset.

9.3 Revolve

The second type of sweep we will cover is the **Revolve** option. It works conceptually the same as the extrude command except it revolves the objects around an axis and through an angle instead of a direction and distance.

The Revolve option is found under the **Insert** menu bar with the commands of **Insert → Design Feature → Revolve…** or from the **Revolve** icon.

This option allows a solid body to be created by revolving planar, section string geometry around an axis and through an angle. The section geometry may be curves, solid edges, solid faces, and sheet solids and combinations of these types.

Just like the Extrude option, there is a Preference setting, found in **Preferences → Modeling**, which specifies if a body type created when sweeping is a solid body or a sheet body. The default body type is solid. Rules applied for extrusion similarly works for revolving with the following exceptions:

- Revolving open strings of curves will cause the system to automatically cap the end faces to produce a solid body if the rotation is a full 360° and the **Modeling Preferences Body Type** is set to **Solid.**
- The direction of revolve is determined by the Right Hand Rule.

When the **Preference** setting is set to create a sheet body, then a sheet body will be created for all curve configurations. Note that if a closed set of curves is swept, a body will be created that looks like a solid body but will in fact be hollow inside, only a shell with sheets on the outside surface.

General Procedure to Revolve

The general procedure to revolve a set of curves is as follows:

1. Choose the **Revolve** option.
2. Select the set of curves.
3. Choose the method of specifying the rotation and angle.
4. Choose the axis of rotation.
5. Enter the rotation angle and offset parameters and choose **OK**.

Activity 9-4. Revolve to Create Solid

Step 1. **Open** the part file **activity9-4.prt** and save it as **xxx_activity9-4.prt.**

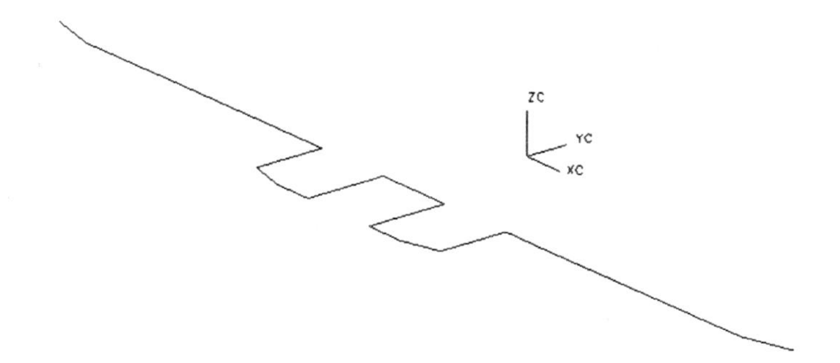

Step 2. Ensure the **Modeling Application** is active.

Step 3. Change the **Work Layer** to Layer 1 if necessary.

Step 4. Choose **Insert → Design Feature → Revolve...** or the icon.

Step 5. Select the string of curves to revolve. This string is a sketch. Notice that you may select any curve and they are all selected. You will learn more about sketch in the following chapter.

Step 6. The methods allow three options of specifying the axis of rotation and the limits to the rotation.

Axis and Angle—This method allows creation of a solid body or sheet solid by selecting a rotation axis and entering the angle of the rotation. The plane of the curves is defined as 0° and rotation occurs around a rotation axis according to the right hand rule. Some caveats:

- Total rotation is determined by entering a **Start Angle** and **End Angle** with the angle not exceeding 360°.
- If the value for the **Start Angle** is greater than the **End Angle**, then the rotation will occur in a negative direction.

Trim to Face—This method allows the creation of a single feature by rotating around an axis from the plane of the curves to the selected trimming face or datum plane on a target solid. You must have a target solid from which you select the trimming face or a datum plane associated with the target solid. To use this option, set the End Angular Limit field to "Until Selected" and select the trimming face or plane.

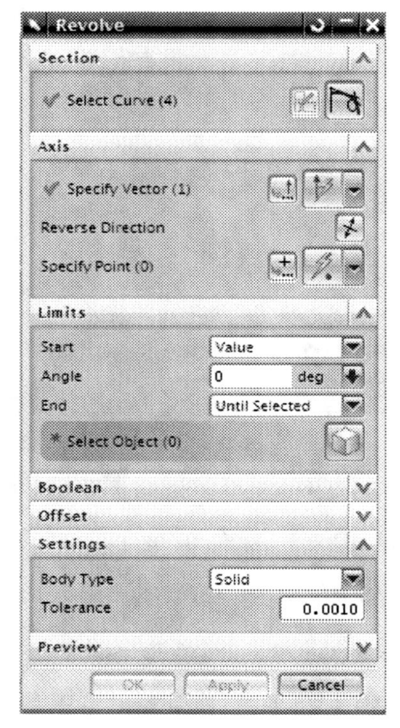

Trim Between Two Faces--This option lets you create a single feature by revolving section curves between two selected trimming faces or datum planes. To use this option, set both the Start and End Angular Limits fields to "Until Selected" and select the trimming faces or planes for both limits.

Step 7. Choose the Inferred vector icon .

The axis of rotation may be an edge, a line or a datum axis. More options are available at the pull-down submenu by clicking the right arrow.

We will use a datum axis.

Step 8. The datum axis is on layer 51 so make layer 51 Selectable.

The axis selection menu will return when you have completed the layer control procedure.

Step 9. Select the datum axis lying along the length of the curve geometry.

A direction vector appears along the length of the datum axis. The right hand rule applies to the direction of rotation. If rotating less than 360°, the direction must be planned and appropriate **Start** and **End Angles** input to rotate in the proper direction. There will be a negative rotation if **Start Angle** is larger than **End Angle**. In our case, we are rotating 360° so we don't need to be concerned about it this time.

Step 10. Enter **Start** and **End Angles** of **0°** and **360°**.

Ensure the **Offset** option is unchecked.

Step 11. Choose **OK**.

The solid body is swept.

Step 12. Change Layer 51 to **Invisible**.

Step 13. Inspect the solid and notice the ends have been closed with a planar face.

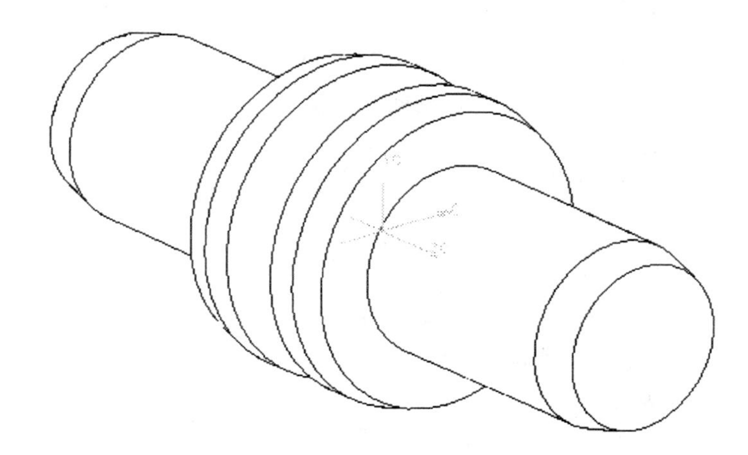

Step 14. Save and close the file.

Activity 9-5. Create the Revolve with Offset

Step 1. **Open** the **activity9-4.prt** to be a new activity and save it as **xxx_activity9-5.prt**.

Step 2. Ensure the **Modeling Application** is active.

Step 3. Change the **Work Layer** to Layer 1 if necessary.

Step 4. Make layer 51 **Selectable**.

Step 5. Choose the **Revolved** icon .

Step 6. Select the string of curves to revolve.

Step 7. Select the inferred vector icon . Choose the **Datum Axis** lying along the direction of the curves.

This time we will change the sweep and do an **Offset** to demonstrate its capability. Put a check mark on the **Offset** field.

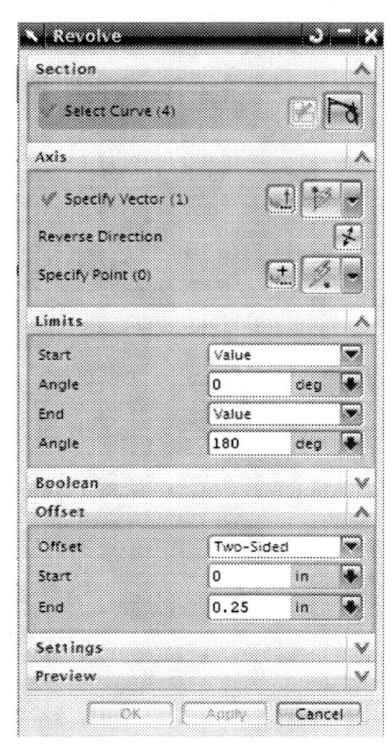

We will do a 180° rotate and 0.25 Offset as shown above. Since the rotation is less than 360° we need to be concerned with the direction of rotation. We want the body of the solid to be under the curves and open to the top.

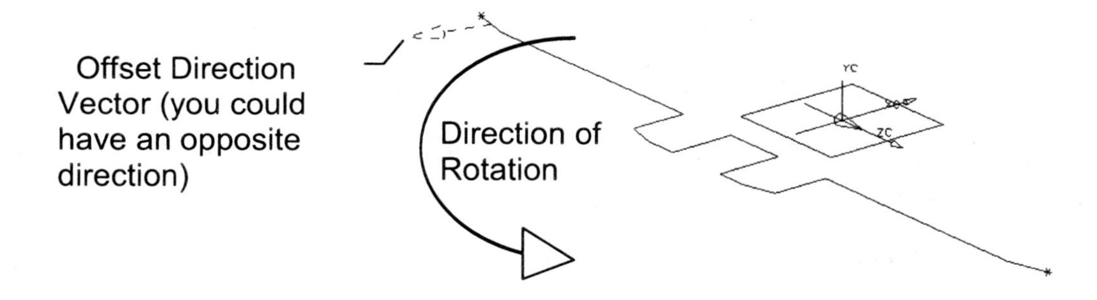

The right hand rule indicates the direction above is the one we want with the angles and the offset specified in the parameter fields. The positive offset value will perform the offset in the direction of the offset direction arrow; a negative offset value will perform the offset in the opposite direction. Therefore, the solid will grow outward from the curves with a .25 wall thickness.

Step 9. Choose **OK** to confirm the values above.

The resultant solid looks like this.

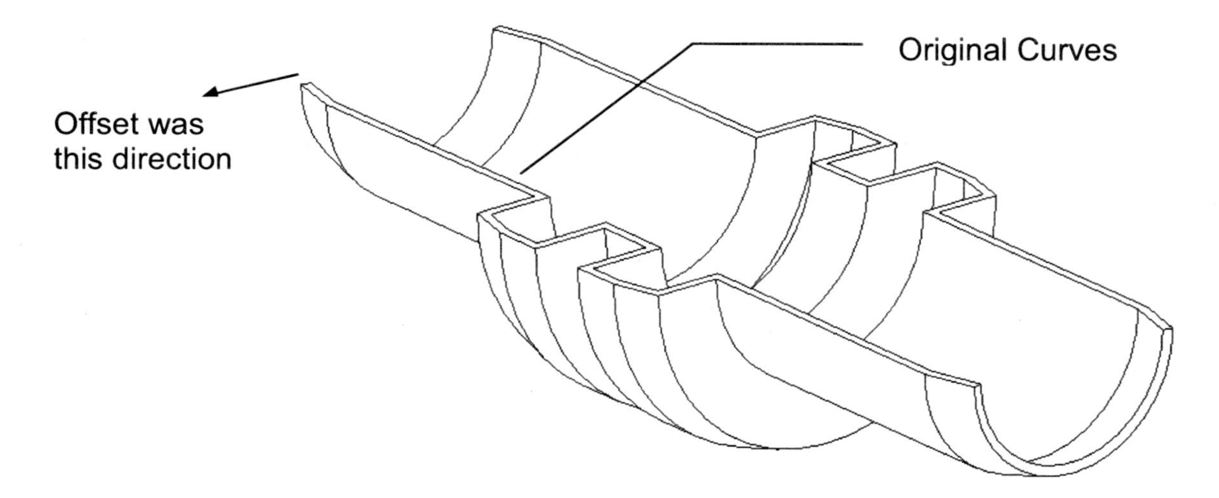

We suggest you to compare this solid with one you created in the previous activity and to see how different offsets and angles can result in different shapes of solids.

Step 10. Change Layer 51 to **Invisible** and save and close the file.

Project 9-1. Retainer 2

The project will practice some revolving techniques with other features such as datum planes and axis, a slot and chamfer. You constructed this same part in Project8-3 of Chapter 8. In Project8-3, you probably used a primitive and other form features without using the revolving feature. This time you are required to build the part using revolve and offset to learn a different way to build the part. The finished part looks like the figure below. The dimensions for the part are the same as those found in Project8-3.

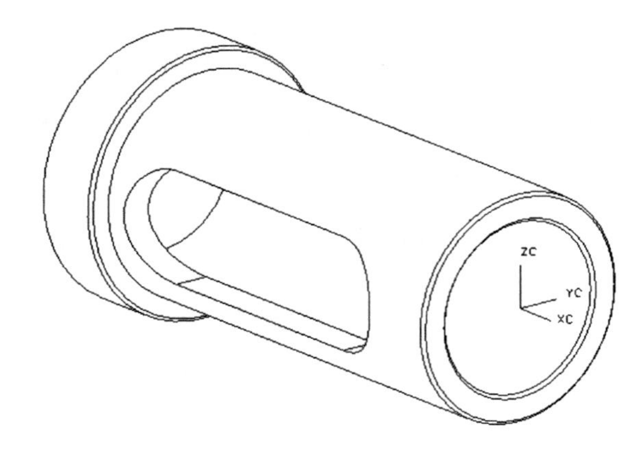

The project steps are to rotate the curves provided for the multi-level cylinders, add the slot and chamfer the ends of the various cylinders. You could apply two revolving features with different section strings and different offset values. Start this project by opening part file project9-1.prt.

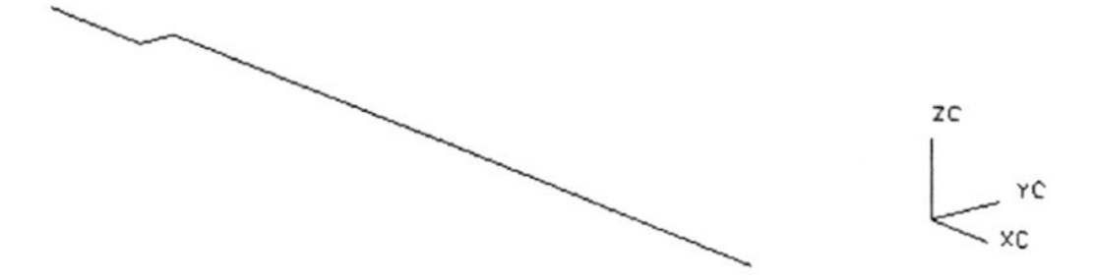

Save the finished part as **xxx_project9-1.prt** where **xxx** is your three letter name initials.

Project 9-2. Angled Bracket

This project will require two extrude operations and an intersect operation along with some simple holes. The finished part looks like below. You will need to sweep and intersect, and interrogate the curves to determine the distance for sweeping by using **Analysis → Distance** (if necessary, refer to Section 6.6 of Chapter 6 for review on how to use **Analysis → Distance**).

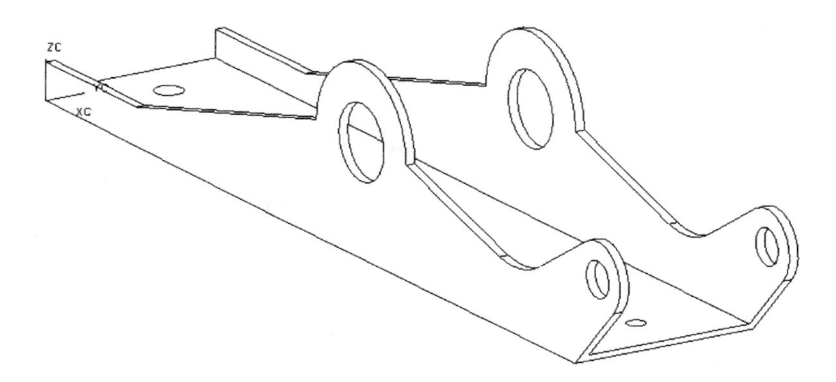

Open part file **project9-2.prt** and the starting geometry looks like this.

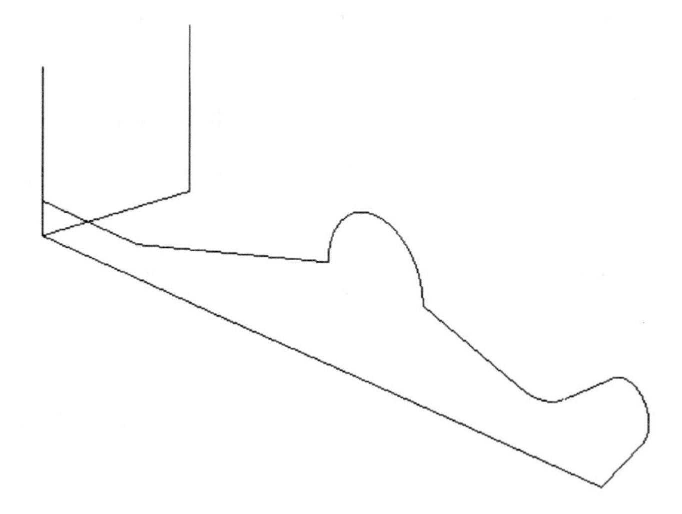

Investigate whether the order of sweeping affects the completed solid body. In general, intersecting three or more solid bodies will give different results depending on the order of intersecting. Then add the holes as shown below.

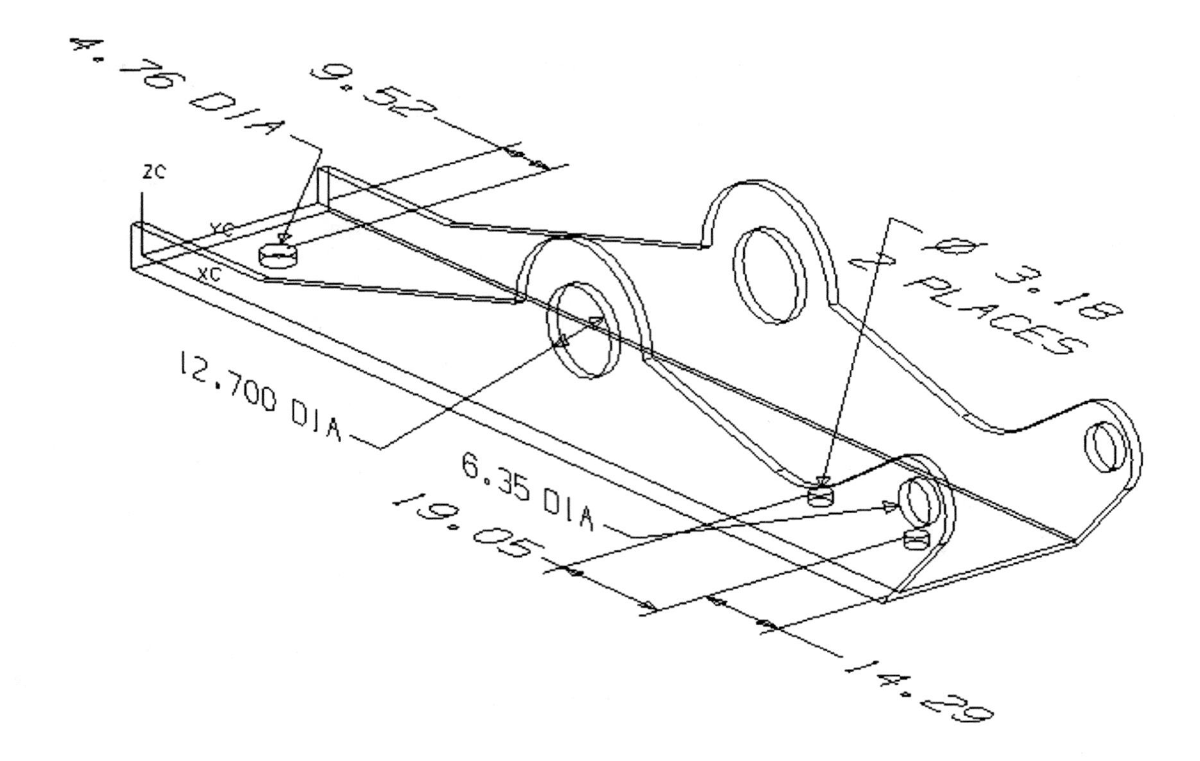

All dimensions above are in millimeters and note that the thickness of the part is 2 mm toward the inside of the profile. Save this file as **xxx_project9-2.prt**.

Optional Challenge: Instead of using **Analysis → Distance**, you could directly measure the distance inside the Extrude feature when you specify the End limit as shown below.

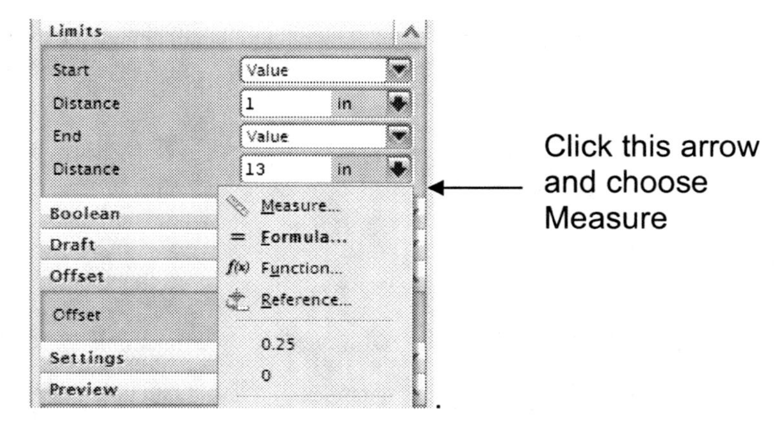

This will create a new measurement expression which will be associative. This means that when you change one of the sketch dimensions to lengthen it, the extrusion length will change accordingly.

Exercise Problems

9.1 What kinds of objects may be selected to sweep?

9.2 What kinds of techniques are available to limit the extrude distance?

9.3 What techniques are available to set the extrude direction?

9.4 What techniques are available to limit revolving?

9.5 What techniques are available to determine the direction of revolving?

Chapter 10. Introduction to Sketching

This chapter introduces sketching, which are special curves. It covers how to create and edit sketches in Unigraphics, and how to use them as a design tool to capture the design intent. The last chapter introduced how to extrude and revolve curves to create solid models, but did not discuss how to create the curves. That chapter is a prelude to this chapter.

So far in this book you have mainly built solid models starting from a fixed set of initial solids called the primitives (base features). You have added to the primitives other features such as holes, pads, pockets, and blends in order to create specific geometries desired.

Sketching is another way to create the initial solid. Extrude and Revolve operations use the sketch as the source of curves and sweep the shape of the sketch into a solid. Sketching can deliver more complex-shaped initial solids than the primitives. Sketching can be also applied to an existing solid that is created from a primitive or a sketch, in order to add intriguing shape to the existing solid. Therefore, sketching can serve as both a base feature and a form feature. Sketching is a feature so that it appears in the feature list of the model navigator.

This chapter consists of 4 sections to cover the topic of sketching. Section 10.1 discusses the concept of sketching of what it is and why we use it. Section 10.2 discusses how to create sketches. Section 10.3 addresses how to put constraints on sketches. Section 10.4 shows how to use sketches for adding non-standard features to an existing solid.

10.1 Concepts of Sketching

A sketch is a collection of two-dimensional curves constituting a profile. A sketch varies from explicit two-dimensional curve geometry because of its ability to parametrically control the profile using a set of rules applied by the user. These rules are called constraints, and constitute the essence of constraint based modeling, which is a very powerful feature of Unigraphics solid modeling.

A sketch may be used anytime to create a solid. There are no specific criteria that suggest when not to use one. Some designers believe that nothing else but sketches should be used to create a solid. These designers create a solid by sweeping curves or sketches. Other designers suggest starting with a primitive and building the solid from there, if the basic solid has a primitive topology. In this case, the sketch is reserved for those shapes that are not one of the primitives in basic shape. Another method is to create explicit curves (non-sketch curves) outside the sketcher and sweep the curves. When changes need to be made to the curves and thus the swept body, the curves are then added to the sketcher and changes are made inside the sketcher.

A sketch is a set of curves joined in a string to form the profile that when swept forms a solid body. The sketch represents the outline of that part. The sketch is created with the intention of sweeping that outline into a solid. The curves are parametrically associated to each other, and they are parametrically associated to the solid that is created when the curves are swept. The curves can be created in the Sketcher and on a plane so they are 2D objects; and they are associated to the plane the sketch is created on. The curves may be created in a free-hand manner with no exact size, or they may be created exactly to size. The free-hand creation may be changed or constrained to exact size at a later time. Sketch curves can be controlled by adding two types of constraints: dimensional and geometric constraints. These constraints define relationships among the curves or between curves and other objects. The constraints may be edited to change sizes if desired.

Thus, the most obvious reason for sketching is when design intent is well known and constraints can be quickly applied to capture the intent. The design intent consists of two items:

- Design Considerations – Geometric requirements on an actual part, including engineering and design rules that determine the detail configuration of the part.
- Potential Areas for Change – Known design changes or iterations, and their effects on the part configuration.

As a general rule, the more design considerations and potential areas for change, the more likely there are benefits from sketching.

The sketch profile outline that is made up of a set of curves is very flexible for creating unusual shapes, but also flexible to allow changes to that profile in the future. The curves are parametric and associative, and they can be easily edited or removed and new curves put in their place. Sketched profiles may be used to create extruded or revolved features or to define section or guide strings for swept features. As the design intent changes, constraints applied to sketched profiles are updated to reflect the intent change and the resulting features and models are immediately and automatically updated, which may lead to a topology or shape quite different from the original one. This is not the case for primitives. For example, a primitive block topology cannot be changed except for its dimensions. However, creating primitives would be easier and simpler.

In summary, the sketch is one of the most useful tools from which to create solid bodies because of flexibility in shapes and the ability to edit and change that shape parametrically and associatively.

10.2 Creating Sketches

The sketch is a feature and is created like other features with a dialog and options. According to the default layer standard, the sketch objects are placed between layers 21 and 40. The sketcher is found off the main menu bar under **Insert**. Choose **Insert → Sketch...** from the menu bar or choose the icon.

With this command, the **Main Menu** changes and four toolbars come up. The main menu changes indicating that a Sketch Task Environment has been created with the new menus used solely for active sketch commands and commands that support sketching (plus, of course, the normal system wide commands).

Task Edit View Insert Format Tools Information Analysis Preferences Help

Note the changes. Some of the menus behind these commands have not changed much because they support the display, etc, i.e., View. Others, like Task are new and support only Sketcher functions.

These four toolbars that come up contain groups of icons to create the sketch, create the curves, constrain the sketch, and perform operations on the sketch. If you cannot see all the toolbars or icons, look for them docked around the periphery of the graphic display window, or choose **Tools → Customize** to access others.

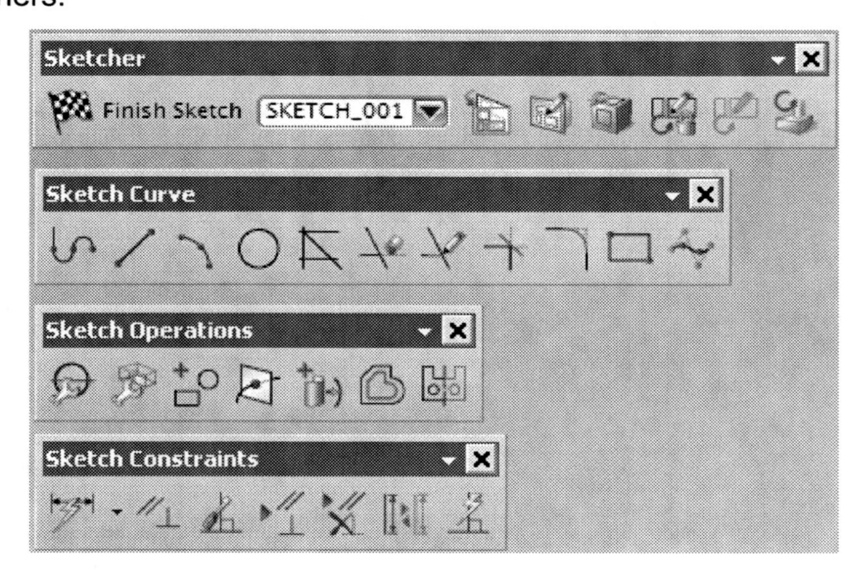

We will give below an overview of these toolbars except the Sketch Constraints toolbar which will be discussed later in Section 10.3.

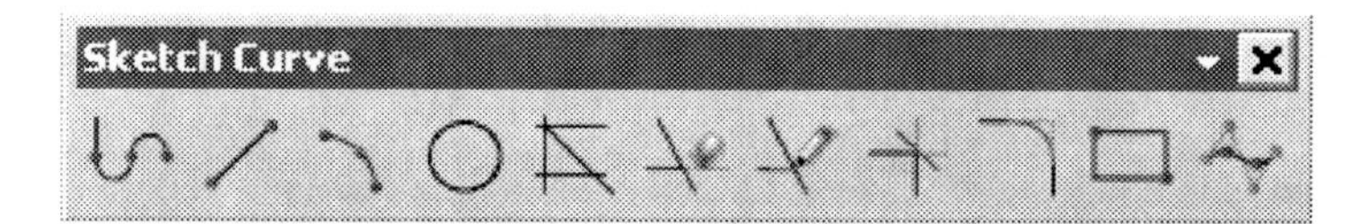

As the name implies, the Sketch Curve toolbar is used to create and edit different types of curves. There are a maximum of 12 options to work with curves.

Profile – to create connected curves in succession without breaks.

This mode of curve creation utilizes the string mode where the system prompts for the start point of the curve, and then it prompts for the end point of the curve. The system is ready for the next curve to be created by selecting only its end point, assuming the start point of the next curve is the end point of the previous curve. This serves two purposes: it is a speedy creation and it ensures continuity of the curves because the end points of two consecutive curves are coincident.

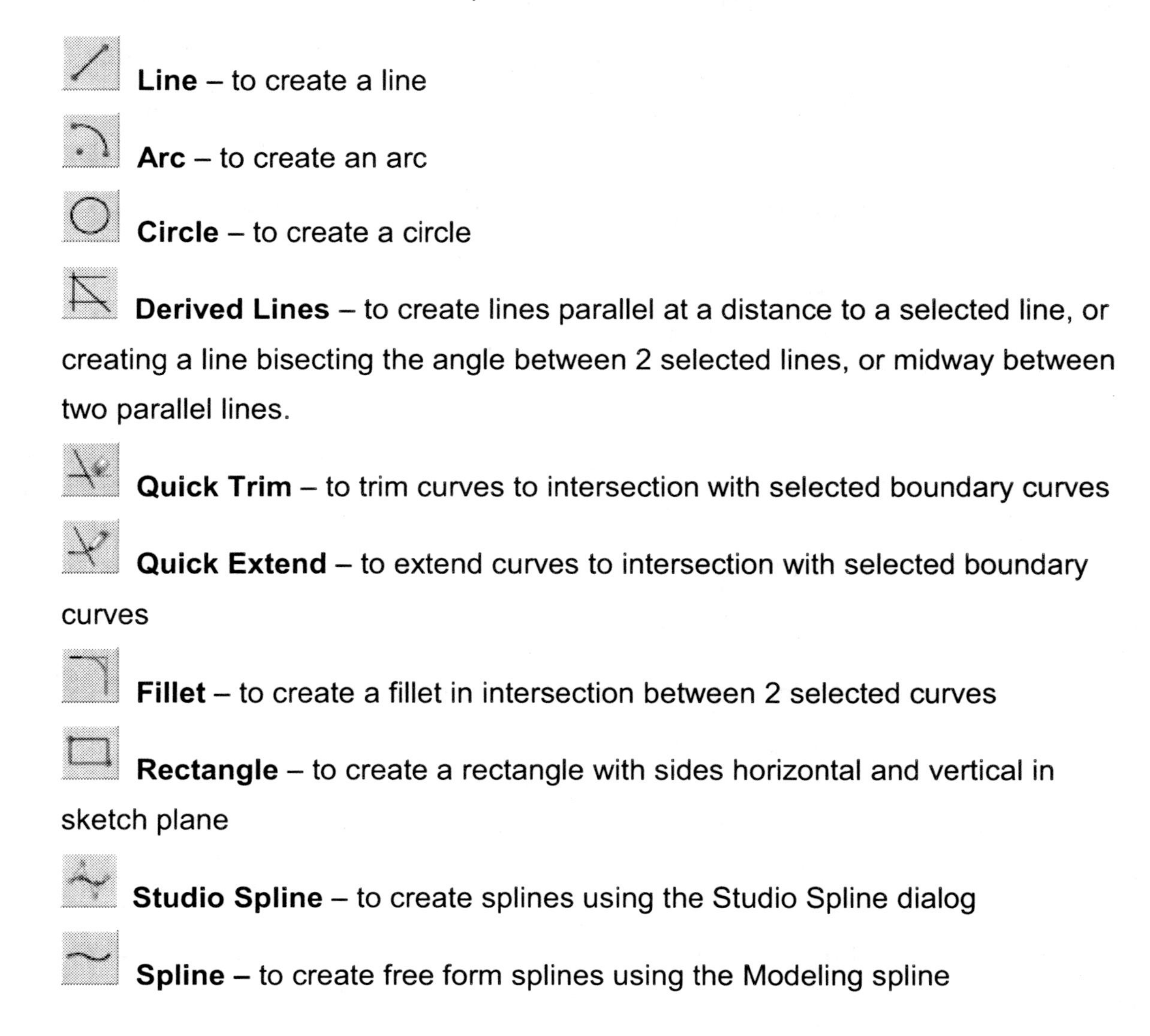

Line – to create a line

Arc – to create an arc

Circle – to create a circle

Derived Lines – to create lines parallel at a distance to a selected line, or creating a line bisecting the angle between 2 selected lines, or midway between two parallel lines.

Quick Trim – to trim curves to intersection with selected boundary curves

Quick Extend – to extend curves to intersection with selected boundary curves

Fillet – to create a fillet in intersection between 2 selected curves

Rectangle – to create a rectangle with sides horizontal and vertical in sketch plane

Studio Spline – to create splines using the Studio Spline dialog

Spline – to create free form splines using the Modeling spline

Text – to create text in the sketch using the text dialog

Point – to create points

Ellipse – to create ellipses or conics

Snap Angle

The Snap Angle is a global preference setting in the Preferences → Sketch dialog that controls how the system automatically infers horizontal and vertical lines as they are created. The default Snap Angle is set to 3° meaning that any line created within ± 3° of true horizontal or vertical, the line will "snap" into true horizontal or vertical. The Snap Angle may be set from 0° to 20°. The 0° setting means that no snap angle will be applied and lines stay at the angle created.

Finish Sketch – to exit the Sketcher

Sketch name – to name the currently active sketch, to rename a sketch and select a sketch to edit.

Orient View to Sketch – to move the current view such that the sketch plane is oriented flat to the screen

Orient View to Model – to return the sketch view to the view the model was in prior to entering the sketcher

Reattach – to move a sketch from its current plane and attach it to a different selected plane

Create Positioning Dimension – to include a pull-down menu for creating, editing, and deleting positioning dimensions

Delay Evaluation – to delay the change update of an active sketch. Normally, sketches are evaluated and updated as changes are made.

Evaluate Sketch – to evaluate and update the changes made to the sketch. This is only available after Delay Evaluation has been used.

Update Model – to update the model by applying changes made in the active sketch. The sketcher mode is still active for further edits.

Mirror – to mirror sketch geometry through any existing line in your sketch

Offset Projected Curves – to associatively offset curves that you have extracted with the Add Extracted Curves to Sketch option below

Edit Curve – to edit sketch curves

Edit Defining String – to add or remove curves from a sweep or guide string

Add Existing Curves – to add to a sketch most curves created outside the sketcher, i.e., ones already created by explicit curve creation methods

Project – to add external, extracted curves projected to the sketch

Note that in this chapter we do not intend to cover all the above tool bar options, but we will focus on a selected subset of options that are commonly used in an introductory level of sketching practice.

General Procedure to Create a Sketch

The following steps are the general procedure to create a sketch, add curves, and constrain them to have the sketch ready to sweep it to a solid.

1. Choose **Insert → Sketch**.
2. Change the name of the sketch if so desired.
3. Specify the sketch plane on which sketch curves are to be created. The sketch plane is defaulted to the XC-YC plane of WCS. A datum plane and two datum axes are displayed in temporary. If this plane is acceptable, you may proceed, but if not then use the following Sketcher Plane toolbar 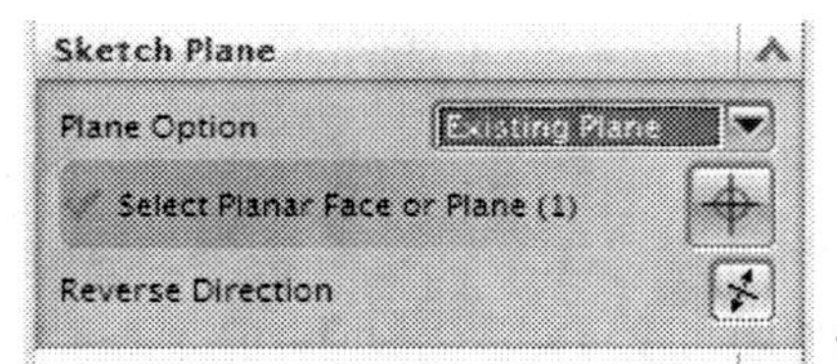

to define another plane. When the desired plane is selected, choose OK on the toolbar to activate the sketch. The selected plane will move to flat on the screen.
4. Create sketch curves by selecting proper curve creation methods on the Sketch curve toolbar.
5. If desired, constrain the curves using the Sketch Constraints toolbar.
6. If desired, position the sketch using the Sketch Constraints toolbar or the Sketcher toolbar.
7. Exit the sketcher by choosing the **Finish Sketch** icon.
8. If desired, replace view to a Trimetric and sweep the sketch into a solid body.

Activity 10-1. Creating a Free-Hand Sketch

Step 1. Create a new part file (unit: inches) and name it **xxx_activity10-1.prt.**

Step 2. Choose **Start → Modeling** to ensure it is active.

Step 3. Change the Layer Setting **Work Layer** to layer **21**.

Step 4. Choose **Sketch** icon .

Step 5. Notice the default **XC-YC** as the sketch plane. A datum plane and two datum axes are displayed.

Step 6. However, in this activity we want to use the **YC-ZC** plane for the sketch plane. Use the **Sketch Plane** toolbar and select the **YC-ZC** plane for the sketch. The datum plane and datum axes move to be in the **YC-ZC** plane.

Step 7. Rename the current sketch. In the sketch name window SKETCH_00C on the **Sketcher** toolbar, type in the name **s21_profile** and hit the Enter key.

Step 8. Choose OK OK to activate and create the sketch.

Notice right away the view moves and places **YC-ZC** plane flat on the screen

Step 9. We do not want the datum plane and axes to be displayed so move them to Layer **61**.

- **9.1.** Choose **Edit → Object Display**.
- **9.2.** Select the Datum Plane and both Datum Axes.
- **9.3.** Choose the green **OK** to execute the selection.
- **9.4.** In the Edit Object Display dialog enter **61** in the **Layer** field and choose **OK**. Those objects are moved and made invisible.

Step 10. We are ready to create curves, so choose the **Profile** icon from the **Sketch curve** toolbar.

This helpful small toolbar comes up to aid us in selecting the kind of line or arc we want to create.

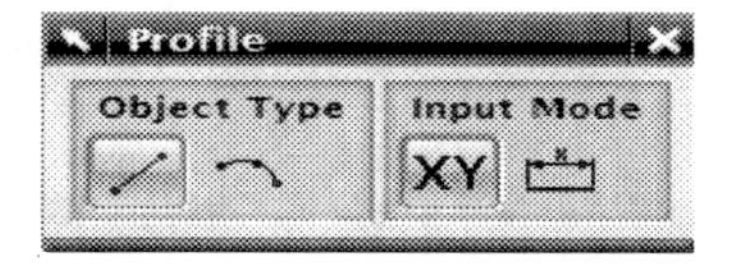

Step 11. Ensure that the **Line** icon is active.

The toolbar will not change. It means that a line will be created. When an arc, a circle, or a fillet is needed, then select those icons.

Step 12. You will construct the profile as shown below.

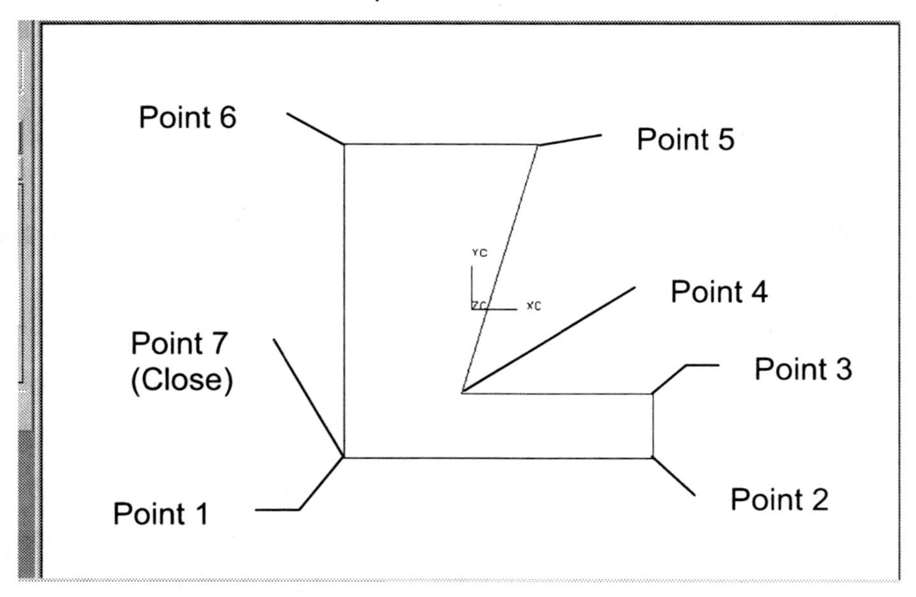

Do not worry about the exact size and location on the screen. Concentrate on approximating this shape. We will adjust the exact size and location later. The point number indicates the order in which you will select the points.

Notice the **Cue Line** prompts first to select the start point. The text field moves with the cursor and gives the coordinates of the location of the cursor. You don't need to pay too much attention to these coordinates for this purpose.

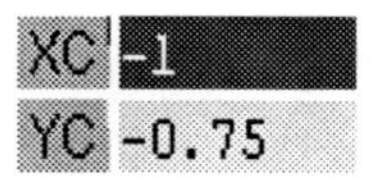

Once the start point has been selected, the **Cue Line** prompts for the line end point each time for the next line, assuming the start point for each line is the end of the previous line.

The next point will create a horizontal line.

12.1. Note that if lines are constructed within the Snap Angle (3°) they will snap to horizontal or vertical. Also note that when a line is horizontal or vertical the system will create a dashed line and vector arrow indicating that it is horizontal or vertical as shown below.

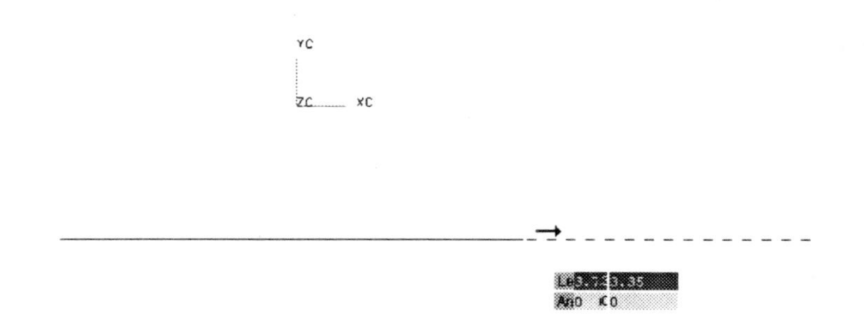

Note that the text field has changed to reflect the defining of the length and angle of the line to be created. The line rubber-bands on the cursor. Choose the end point by clicking **MB1** once any place along the line. The first curve that has just been created is a horizontal line.

12.2. Next we will create a vertical line as shown below.

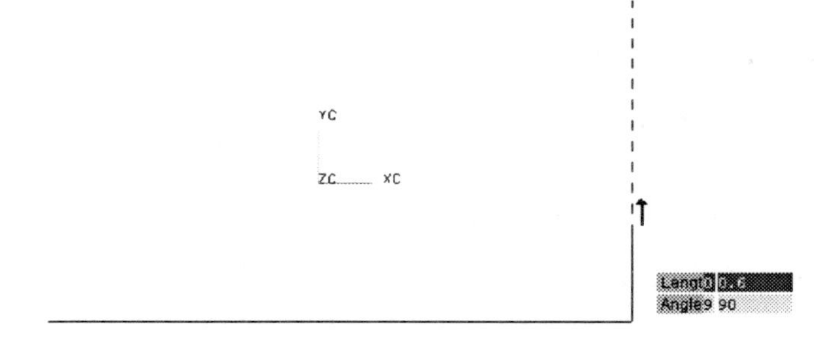

The end point of the horizontal line becomes the start point of the vertical line connected to it. Notice when the line is vertical, the dashed line and arrow appear to make us aware the line is vertical.

12.3. Continue around the profile selecting each point approximately as shown and not paying too much attention to the text field.

12.4. We want Point 7 to be the same as Point 1 and the line from Point 6 to Point 7 to be vertical. Notice as you drag the cursor over to Point 6 from Point 5 that when you make the line horizontal the dashed line shoots out from it. As you drag the Point 6 to the left to line it up with Point 7 vertically, be very careful because as you get lined up with Point 7 a dashed vertical line will appear indicating vertical alignment between Point 6 and Point 7. Then press **MB1** and place the Point 6. See the figure below.

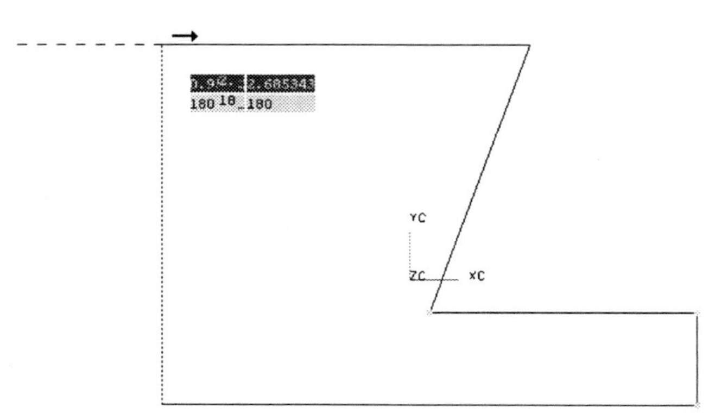

12.5. After Point 6 is selected, the top horizontal line is created. Then slide down toward Point 7 and select it to create a vertical line and complete the profile.

12.6. After all the points are selected, use **MB2** or select the Profile icon, to break the string mode.

Step 13. Add a fillet into the corner of the profile.

13.1. Choose the **Fillet** icon**.**

13.2. The **Fillet** text box comes up.

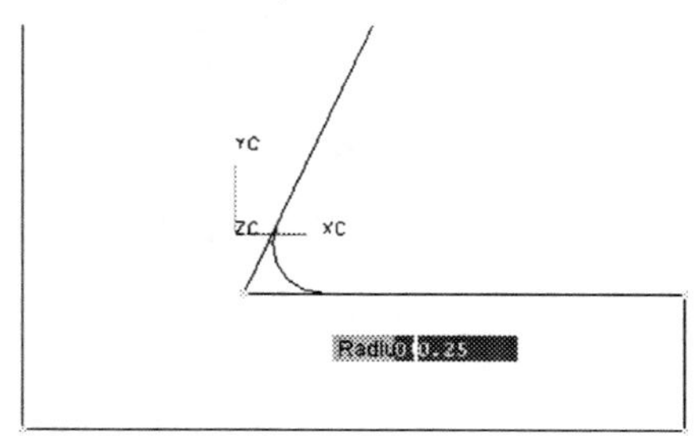

13.3. Enter .25 inches in the **Radius** text box.

13.4. Locate the cursor over Point 4 such that the cursor circle encompasses both lines at the intersection and select at point 4.

The fillet arc is immediately created. The profile now looks like the figure below.

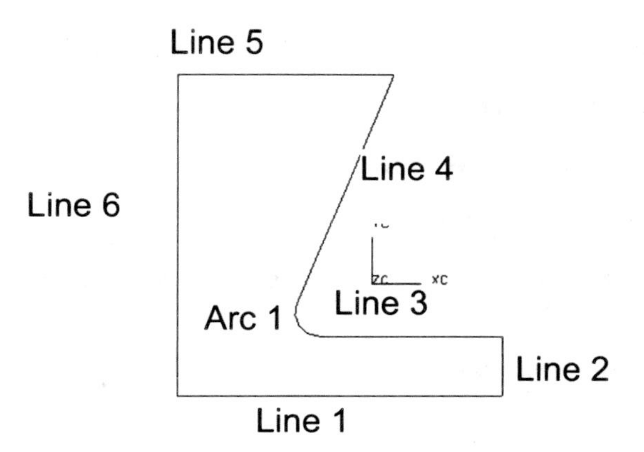

The goal of the construction is that lines L1, L3 and L5 are horizontal and lines L2 and L6 are vertical.

If your profile looks more like this, then that's ok too.

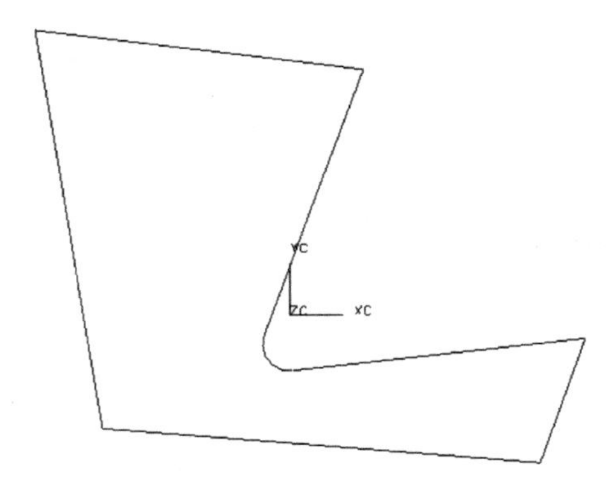

We will show you how to fix this later in this chapter.

Step 14. Exit the sketch mode by selecting the **Finish Sketch** icon .

Step 15. Choose **File → Save**.

Activity 10-2. Sweep the Profile into a Solid Body

The purpose of this activity is to illustrate how a solid is created from the sketch. It will use the **Extrude** command, covered in Chapter 9, and just the necessary steps will be shown assuming you know how to extrude.

Step 1. Open the previous file **xxx_activity10-1.prt** if not already open, and save it as **xxx_activity10-2.prt**.

Step 2. Replace the current view with the Trimetric, TFR-TRI.

Step 3. Change the **Work Layer** to layer 1.

Step 4. Choose **Insert → Design Feature → Extrude…** from main menu bar.

Step 5. Click the Extrude dialog button from the options that are displayed on the upper left corner of the screen

Step 6. Select one of the sketch curves. Notice all of the curves are highlighted.

Step 7. Enter a Start Value of 1 and End Value of 5 under **Limits** and accept default direction.

Step 8. Choose **OK** in the Extrude dialog.

Step 9. Use **MB3** and choose **Fit** (or use the **Fit** icon).

You see the solid has been created. **Cancel** the dialog.

Step 10. Choose **File → Close → Save and Close**.

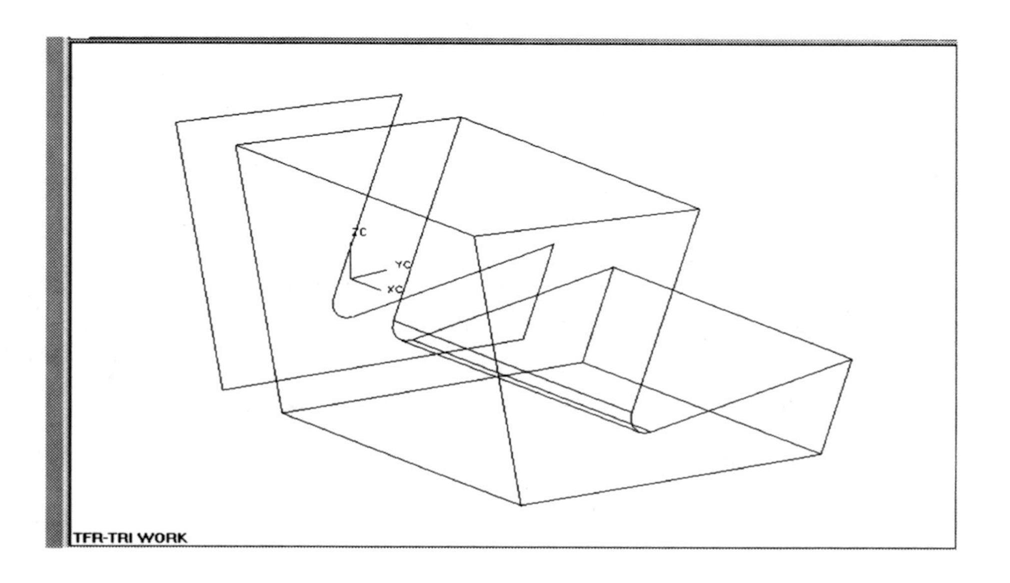

Activity 10-3. Creating Curves in Profile mode

In this activity we will create a sketch profile that include arcs and circles. We will also constrain it and sweep it. Below is the profile we will create.

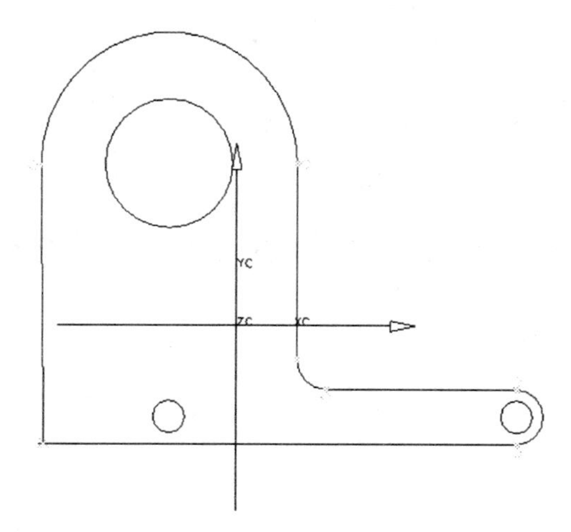

Step 1. Create a new part file (inches) and name it **xxx_activity10-3.prt**.

Step 2. Start the **Modeling** application. Change the view to the TFR-TRI view.

Step 3. Change the Layer Setting **Work Layer** to layer **21**.

Step 4. Choose **Sketch** icon [icon].

Step 5. Notice the default **XC-YC** as the sketch plane. A datum plane and 2 datum axes are displayed to aid in construction.

Step 6. However, we want to use the **Sketch Plane** toolbar and select the **YC-ZC** plane for the sketch plane. The datum plane and datum axes move to be in the **YC-ZC** plane.

Step 7. In the sketch name window on the toolbar, type in the name **s21_profile** and hit the **Enter** key.

Step 8. Choose OK [OK] to activate and create the sketch.

Notice right away the view moves and places **YC-ZC** plane flat on the screen

Step 9. We do not want the datum plane to be displayed so move it to Layer **61**. Use **Object Display** as done in the previous activity. Make layer 61 selectable.

Step 10. We are ready to create curves, so choose the **Profile** icon [icon] from the **Sketch curve** toolbar if not active. This helpful small toolbar comes up to aid us in selecting the kind of line or arc we want to create. The line option is highlighted.

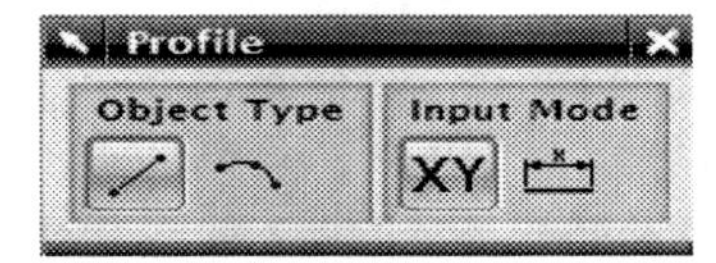

Step 11. Create the lower horizontal line stopping it at a length of 7.5 or so, approximately where it is shown below.

Step 12. Select the arc icon from the small toolbar in the corner.

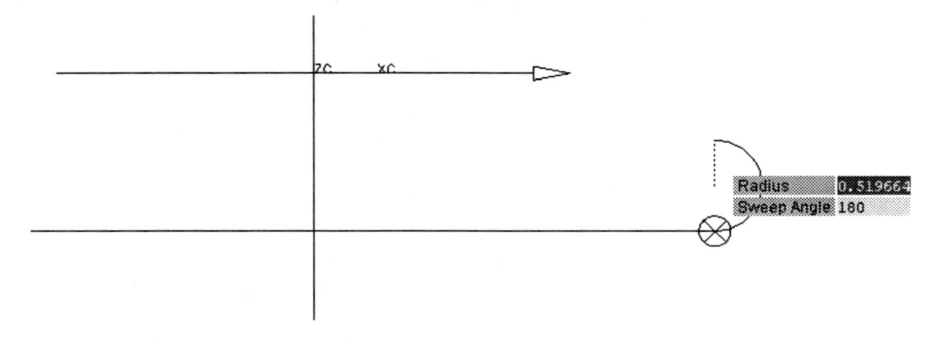

Notice a couple of things that happen when you move the mouse to select the arc:

1. After selecting the end point for the line and moving the mouse toward the arc icon, the line is dragged along with the mouse up toward the icon.
2. After the arc icon is selected and you move back down toward the end of the line, an arc radius is dragged on the arc so it starts out rather large and decreases in radius as you approach the line end point. You can make the arc whatever size you want by dragging it around.

Step 13. Drag the arc until it looks like above with the vertical dashed line down to the connecting point between the line and the arc. At this moment, type in the radius of .5 and then press **MB1** to place the arc. Be careful not to move your mouse when try to enter .5. Otherwise, you may lose the .5 dimension and the marker of the vertical line. You may have to repeat that step.

Step 14. After placing the arc end with **MB1** the default curve is the line, therefore drag out a line to the left, keeping it horizontal utilizing the horizontal dashed guide line. Stop the line at a length of 3.0 inches and press **MB1**.

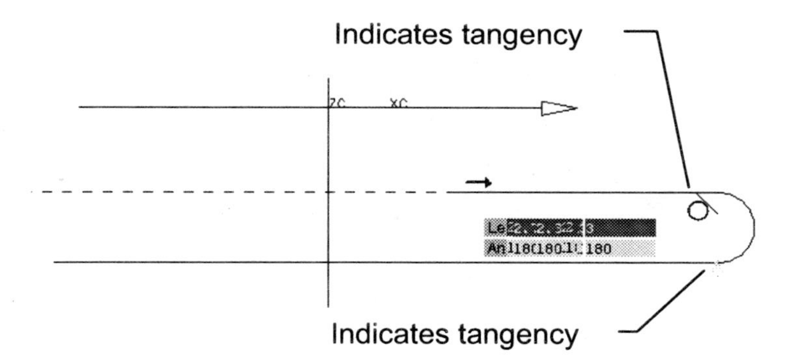

Step 15. We want to add an arc again, so choose the arc icon from the small toolbar. Drag the arc to near a .5 radius and drag until the dashed line appear indicating tangency. Type in the radius of .5 and choose MB1.

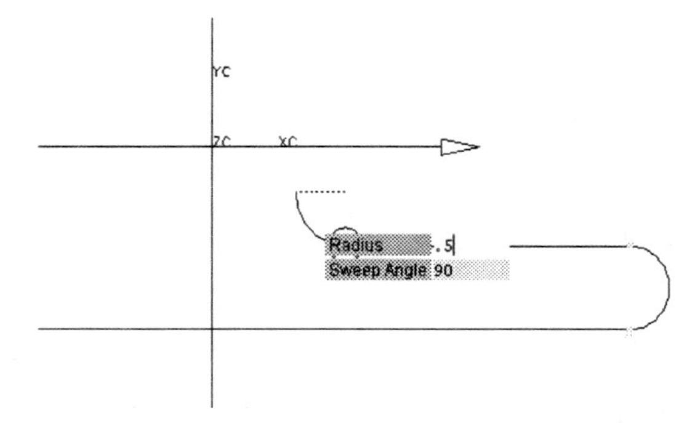

Step 16. The curve creation mode switches to a line again and we want a vertical line 3.0 inches so drag the line until that is obtained and hit **MB1**.

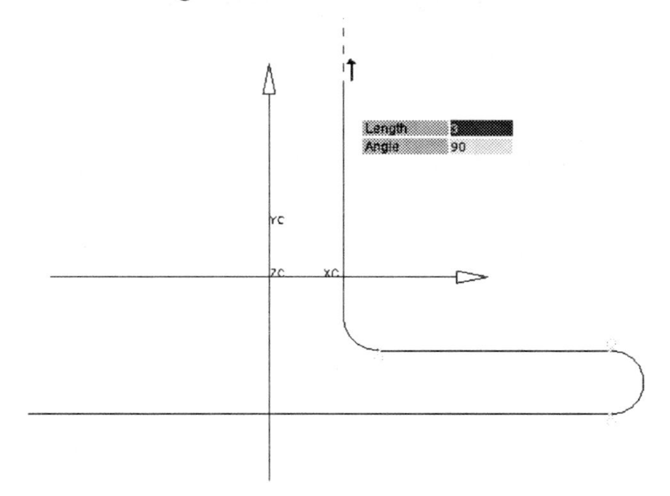

Step 17. We want an arc so select the arc icon, drag the arc in a bigger circle now, line up the dashed lines with the original point and with the arc end point to ensure tangency and press **MB1**. This step looks like this.

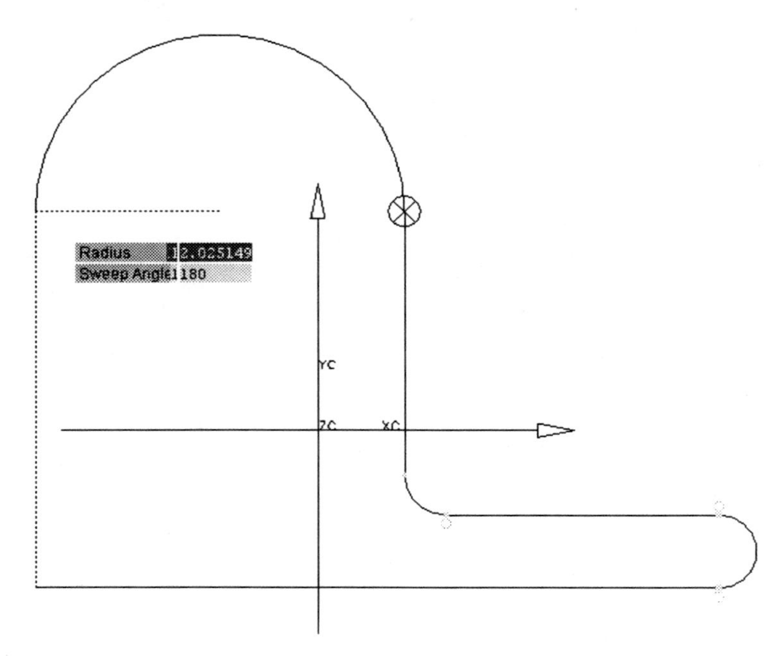

The vertical dashed line indicates the point is vertical above the end point of the start line. The horizontal line indicates the point is horizontal with the other end of the arc. The sweep angle of 180° also indicates a half rotation of the arc. Use **MB1** and select that point. The radius is an odd number and we'll fix that later.

Step 18. Lastly, drag the line down and select the start end point of the first line.

Step 19. Choose the Profile icon or **MB2** to break the string mode.

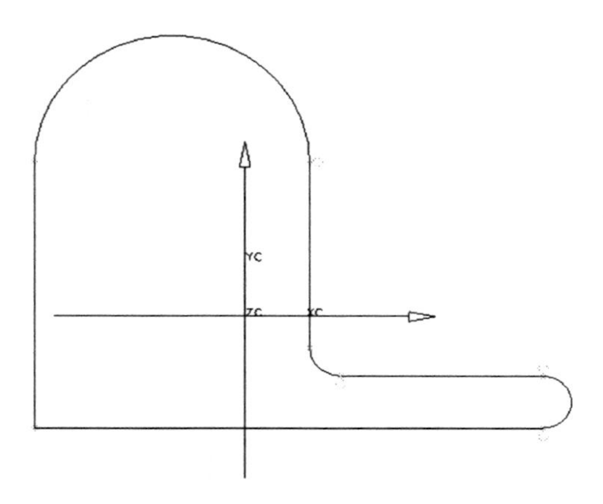

The profile is complete. Now we will add the holes.

Two of the holes will be located at the arc centers of the end arcs of the profile. The third hole will be lined up with the other two.

Step 20. Select the Circle icon from the Sketch Curve toolbar.

Step 21. Move the mouse over the small arc on the end of the profile.

The Cue Line prompts to select the center point of the circle. As you move the mouse around the arc different points are highlighted that may be selected: the end points, midpoint, or arc center point etc., depending on options you have turned on in the snap point toolbar (see page 3-10 for detail). The Status Line identifies various points highlighted. Turn on arc center if inactive.

Step 22. Click **MB1** once at the moment when the Status Line changes to the Arc Center and the center point is highlighted on the graphic screen.

Step 23. In the text box, type in the desired diameter of .5 inches and press the **Enter** key.

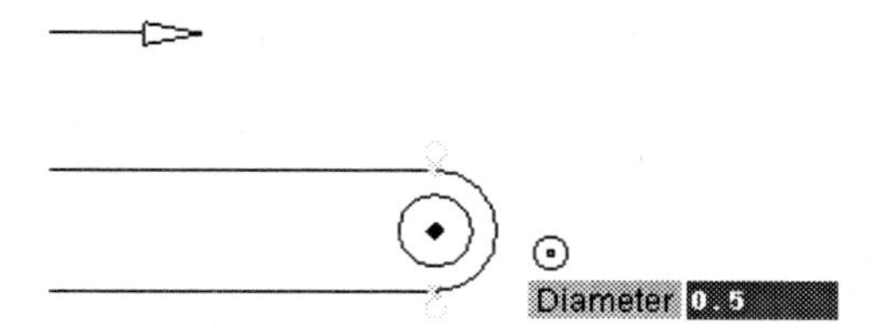

The circle is created immediately and the system is ready to create another circle and the Cue Line is prompting to select a center point for the circle.

Step 24. We want a 2.00 diameter circle concentric with the top arc. Move the cursor up to the arc of the circle and move around until the center point of the circle is highlighted and the Status Line indicates the center point.

At this moment, click **MB1**. The center point highlight looks like below.

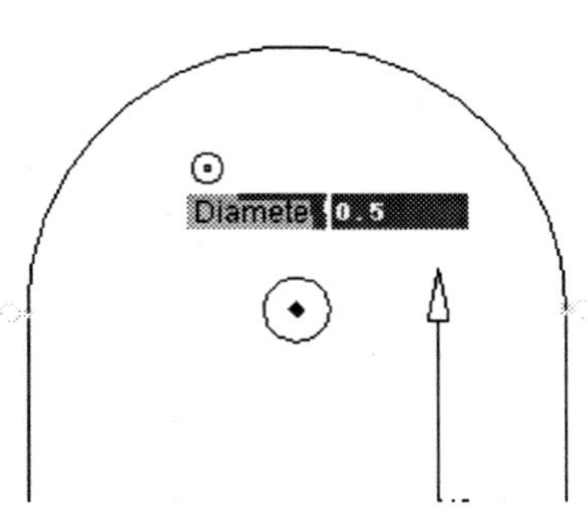

Step 25. Enter a diameter so type in the 2.0 diameter and press the **Enter** key. The hole diameter changes.

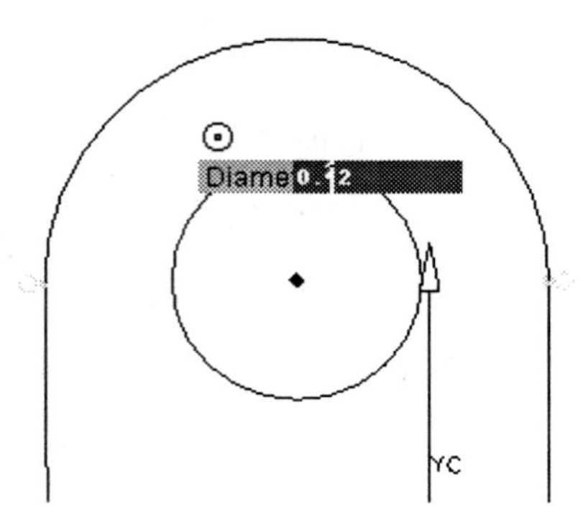

Step 26. Ensure that the 2.0 diameter circle is concentric with the top arc (you might have to drag your cursor around slightly) as shown above and press **MB1** to place the circle.

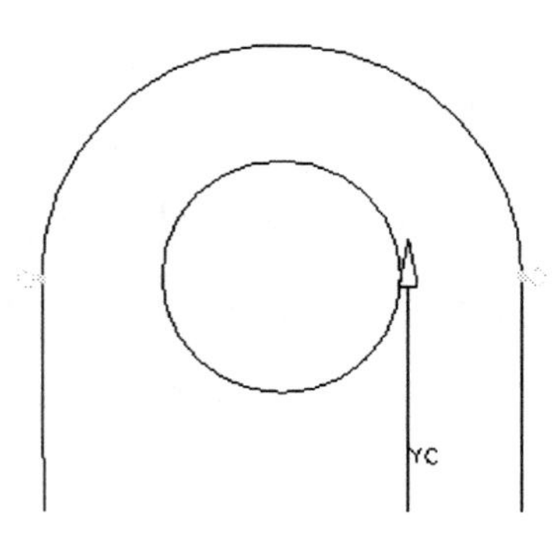

The system is still ready to create another circle and the Cue Line is prompting to select the center point. This time we want a .5 diameter circle lined up with the other 2 circles and in the bottom corner of the profile.

Step 27. Create the circle by moving the cursor down until the horizontal and vertical dashed lines are created to indicate orthogonal alignment with the other 2 circle centers. Press **MB1** to locate the center. A diameter box will appear.

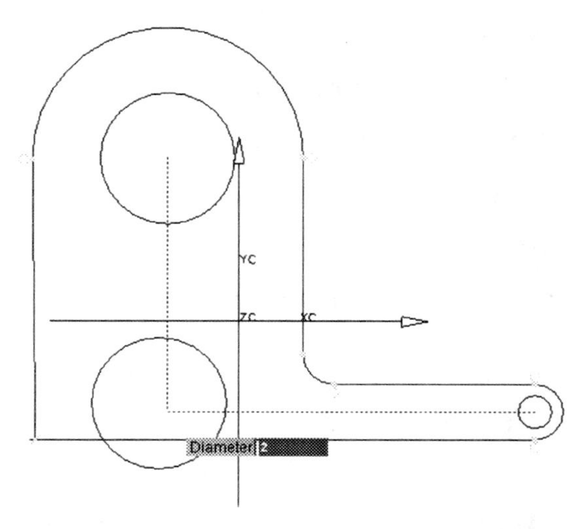

Step 28. The text box is prompting to enter a diameter, so type in .5, hit the **Enter** key or **Tab** key. The diameter changes to a .5 circle.

Step 29. Ensure the horizontal and vertical dashed lines are still visible and press **MB1** to place the circle.

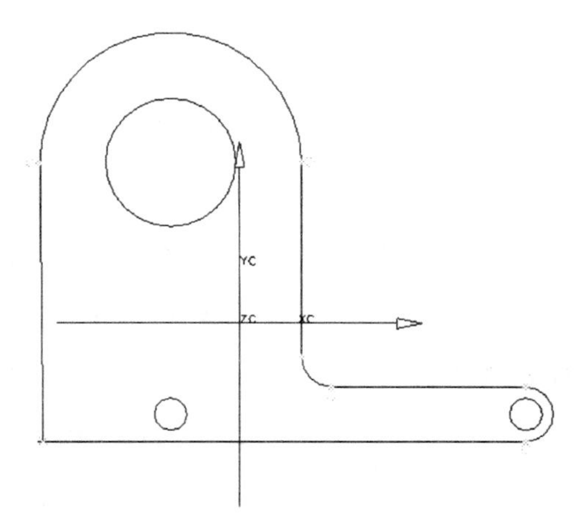

Step 30. Exit the sketcher by choosing Finish Sketch icon and Choose **File → Close → Save and Close**. We'll constrain this sketch later.

10.3 Sketch Constraining

As was seen with Activities 10-1 and 10-2, sketches do not have to be constrained for Unigraphics to function and to build parts. However, Unigraphics provides the capability to sketch with constraints in order to capture the design intent and to easily make design changes as the design intent changes over time. Thus, sketches may contain any continuum from no constraints at all, through just a few constraints, to being completely constrained. With usage of sketch constraints, the designer can capture the design rules, relationships between geometry objects, and guidelines specified for part design.

Dimensional and Geometric Constraints

There are two types of sketch constraints: Dimensional constraints and Geometric constraints. Dimensional constraints are very much like dimensions on a drawing done in drafting. They define the physical size and orientation of curves by a dimension. The dimension can be easily edited.

There are 9 types of **Dimensional** constraints.

Horizontal—the horizontal distance between two points parallel to the horizontal direction defined on the sketch plane.

Vertical—the vertical distance between two points perpendicular to the horizontal direction defined on the sketch plane.

Parallel—the shortest distance between two points.

Perpendicular—the perpendicular distance from a selected line to a point.

Angular—the angle between two selected lines measured counter-clockwise from the ends of the lines on the end nearest where the lines were selected.

Radius—the radius for an arc or circle.

Diameter—the diameter for an arc or circle.

Inferred—This infers what type of dimension to apply, depending on objects selected, selection order, and dimension placement location.

Perimeter — The perimeter for an arc, circle or a polygon.

Geometric constraints define the geometric characteristic of a curve or define the relationship between 2 or more objects. They may be added or removed from curves to make changes.

There are 22 **Geometric** constraints. These are listed alphabetically, not in order of the most widely used.

Associative Trim—A constraint on a spline that has been trimmed using the associative output option from the Edit Curve-> Trim Curve dialog.

Associative Offset--A constraint on an extracted curve that has been offset using the Offset Extracted Curves option.

Coincident—Defines two or more points as located on top of each other (occupying the same point in space).

Collinear—Defines that two or more lines all lie in the same straight line.

Concentric—Defines two or more arcs or circles as having the same center.

Constant Angle—Defines a line having a fixed angle.

Constant Length—Defines a line having a fixed length.

Equal Length—Defines two or more lines having the same length.

Equal Radius—Defines two or more arcs or circles having the same radius.

Fixed—Fixes the location of points or objects in space. We suggest to avoid using this constraint since it gives the least control/flexibility to the sketch objects fixed.

Horizontal—Defines a line to be fixed horizontal.

Midpoint—Defines the selected point on a curve to be coincident with the midpoint of another curve.

Mirror—Defines two objects as mirror images of each other.

Parallel—Defines two or more lines as parallel to each other.

Perpendicular—Defines two lines as being perpendicular to each other.

Point on Curve—Defines the location of a point on a curve as lying on another curve.

Point on String—Defines that a point on a curve is located lying on a selected string of curves.

Slope of Curve—Defines that a spline selected at one of its defining points is tangent to another curve at its selected point.

Scale, Non-Uniform—Defines that when a spline is changed, it will only scale in the horizontal direction and stay the same in the vertical direction.

Scale, Uniform—Defines that a spline will scale proportionately in both the horizontal and vertical directions when the horizontal length changes.

Tangent—Defines two curves are tangent to each other, i.e., arc and line, or two arcs.

Vertical—Defines a line to be fixed vertical.

Overview of Sketch Constraints Toolbar

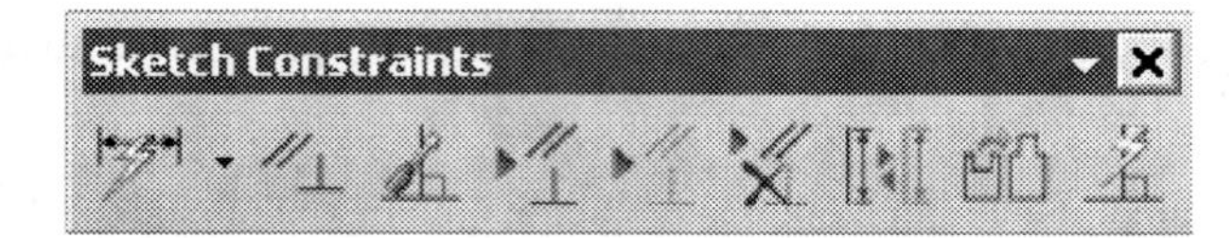

Inferred... – to create or edit dimensional constraints

Constraints – to create geometric constraints

Automatic Constraints – to let the system create many constraints at once automatically

Show All Constraints – to display all geometric constraints on the model

Show No Constraints – to remove the display of geometric constraints

Show/Remove Constraints – to list all geometric constraints in a dialog and allow them to be removed

Animated Dimensions – to dynamically display the effect of varying a given dimension over a specified range

Convert To/From Reference – to convert curves or sketch dimensions between active and reference

Alternate Solution – to change from one solution to another where more than one solution is possible when sketch constraints are applied

Infer Constraint Settings – to control which constraint types to infer during curve creation

In Activity 10-1, we created a sketch similar to the one below.

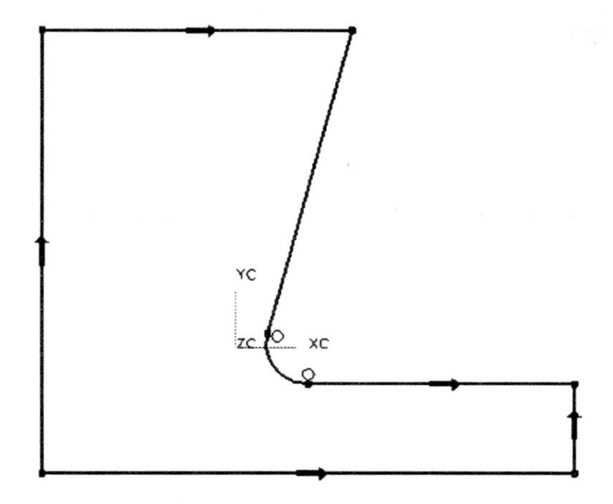

We can see by inspection on this figure that the system has already created some geometric constraints:

- The arrows indicate the line is horizontal and the vertical arrows indicate the line is vertical.
- The small circles indicate tangency. In this case, the arc is tangent to the line at each end.
- The solid squares indicate the points at the ends of the curves are coincident—occupying the same point in space.

The system has inferring ability. It observes the way the curve objects are created, and creates geometric constraints accordingly and implicitly. In the following example, we will explicitly create sketch constraints.

Activity 10-4. Application of Constraints to a Profile

Step 1. Open **activity10-4.prt** and save it as **xxx_activity10-4.prt**. This file contains a sketch that might look slightly different from what you created in the earlier activity.

Step 2. Ensure that **Modeling Application** is still active.

Step 3. Change the **Work Layer** to that of the sketch, layer 21, and change the solid body layer to **Invisible**.

Step 4. Activate the sketch.

4.1. Choose **Insert → Sketch** or the **Sketch** icon.

4.2. Select the sketch name **s21_profile** from the Name window.

The view flips around to show the sketch full, flat on the screen.

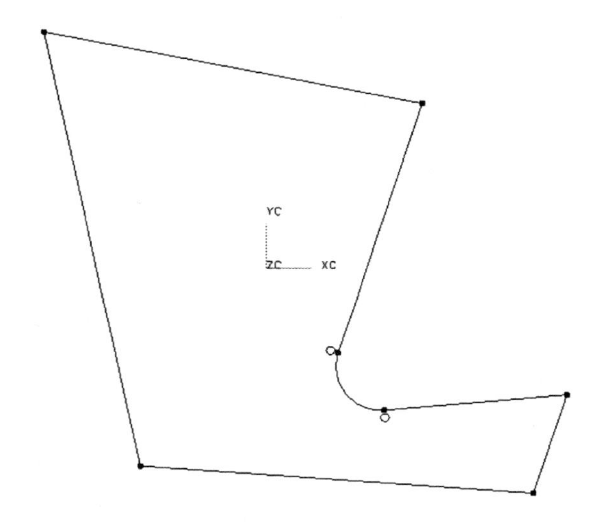

We will use this profile to demonstrate sketch constraints.

Step 5. Interrogate the model and determine if any constraints already exist that were applied by Unigraphics as a result of our geometry construction.

5.1 Notice by inspection that the squares are at each point indicating the end points of the curves are coincident.

5.2 Also notice that the arc is tangent at both of its ends with the lines.

5.3 Choose the **Show/Remove Constraints** icon.

5.4 Toggle the **All in Active Sketch** button to **ON**.

5.5 Change the **Show Constraints** pull-down option to **Both**.

5.6 The current status of all **Geometric** constraints is displayed in the window. In this case the tangency of the fillet to the adjoining lines and the coincident end points are in the model so far.

5.7 The constraints in the list may be selected, or the **UP** and **DOWN** arrow buttons may be used to select each constraint, and the affected geometry will be highlighted and named to graphically show which objects are affected.

5.8 **Cancel** this dialog.

Step 6. View the sketch for constraining.

6.1. Choose the **Constraint** icon on the **Sketch Constraints** dialog. Note that 3 significant actions occur: the graphic area of the sketch displays arrows, the **Cue Line** is prompting to select objects to constrain, and the cursor icon changes to include the symbols for the Geometric constraints.

Let's discuss the arrows in the graphic window first. These are Degree of Freedom arrows and represent the directions that each point is free to move because its location is not defined. Note the arrows are parallel to horizontal and vertical. Because this is 2D by definition, and defined to be on a plane, there is no freedom to move in a Z direction. When working in a sketch, remember that the conventional way of thinking about a curve of being locked in space is no longer valid; everything is free to move until defined by a constraint.

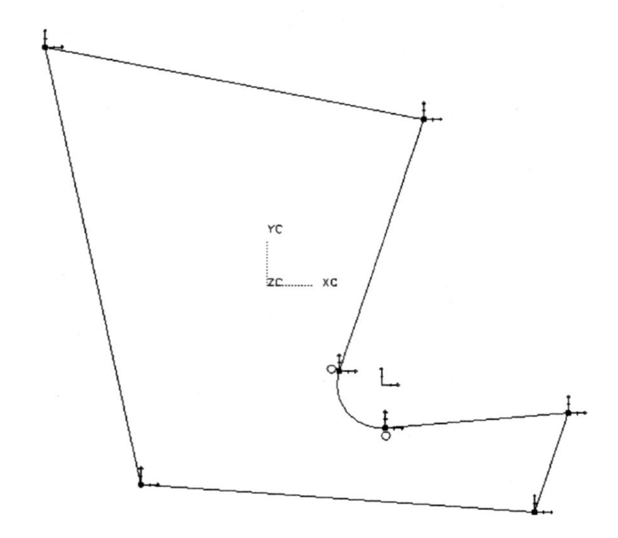

The degree of freedom arrows indicate the direction the point is free to move. Technically this means that Unigraphics is unable to define the exact location of the points. As the constraints are added, and Unigraphics is able to locate a point exactly, the arrows will disappear one by one. When a sketch is fully constrained, there will be no degree of freedom arrow. These arrows appear at the defining points on a curve. These defining points include end points of lines, center points of arcs, and defining points of a spline. In the sketch above there are degree of freedom arrows at the end points of each line, at the center of the radius, and also the end points of the arc of the fillet.

We know that some lines should be horizontal and some lines should be vertical so we will start constraining with these curves.

Step 7. Use Geometric constraints on the horizontal and vertical lines.

The **Cue Line** is prompting to select curves to create constraints. Select a line in the middle, do not select on the end points. You want to select the line object and not the points. If end points are selected, different constraint options will be available.

The below figure is the desired profile after we add horizontal and vertical constraints. Note the labels of the lines for reference.

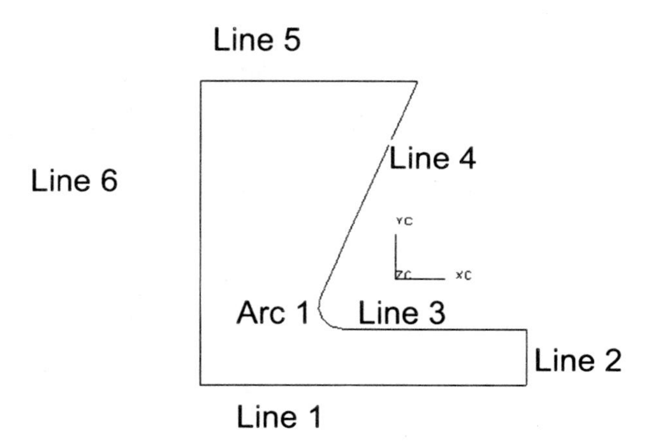

7.1. Select the lines L1, L3, and L5 to be horizontal.

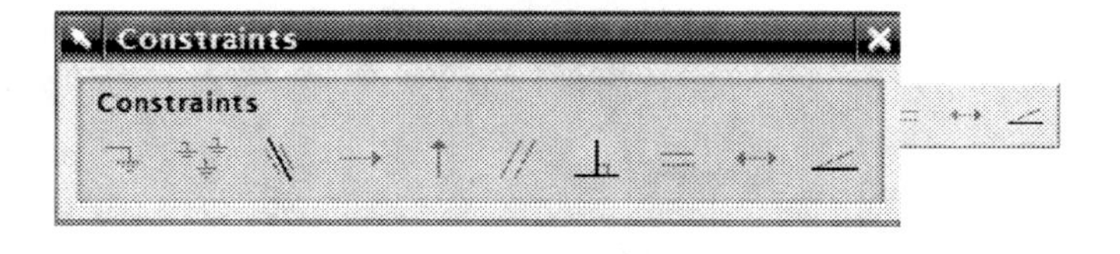

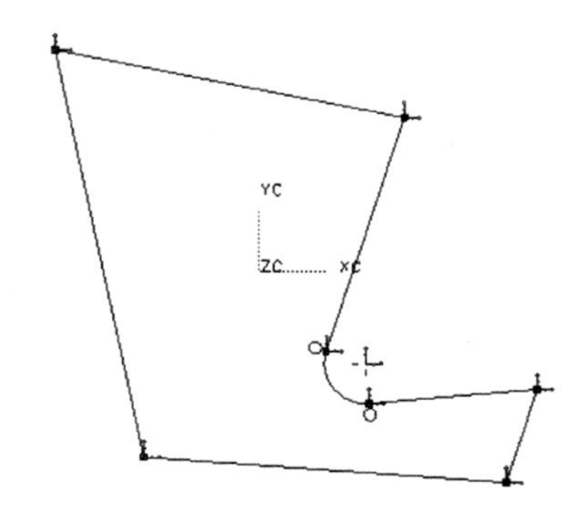

The icons of Geometric constraints appear at the upper left corner of the graphics window. They represent the types of constraints that can be applied to the specific selection of curves we made. The list will vary depending on which and how many curves were selected. And the list will update each time a curve is selected, so it is dynamic and interactive.

7.2. In our case we want the lines to be horizontal, so choose the **Horizontal** icon .

7.3. Immediately those 3 lines are modified to be horizontal and labeled with the horizontal icon of the right-hand arrow. If you do not see the horizontal icon, choose the Show All Constraints icon .

7.4. Select the vertical lines L2 and L6 and remember to select in the middle of the lines.

7.5. Choose the **Vertical** icon from the toolbar. Again, the lines snap to vertical.

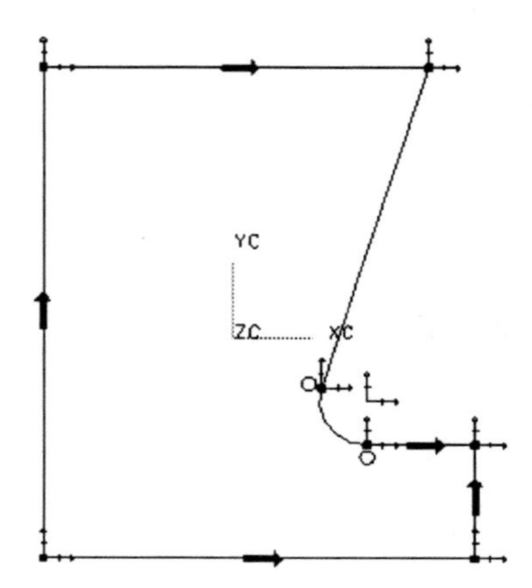

Our profile now looks like this. Yours may be different and that's ok.

Step 8. Interrogate the sketch again for constraints by repeating Step 5 and you will see that the **Horizontal** and **Vertical** constraints have been added.

We will now use the **Dimensional** constraints. We will first add all the constraints that are necessary and then we will edit the dimensions later to obtain the desired shape.

Step 9. Add Dimensional constraints.

9.1. Choose the **Infer Dimensions** icon on the **Sketch Constraints** toolbar. The Dimensions icon appears in the upper left corner. Select the icon and the **Dimensions** dialog comes up.

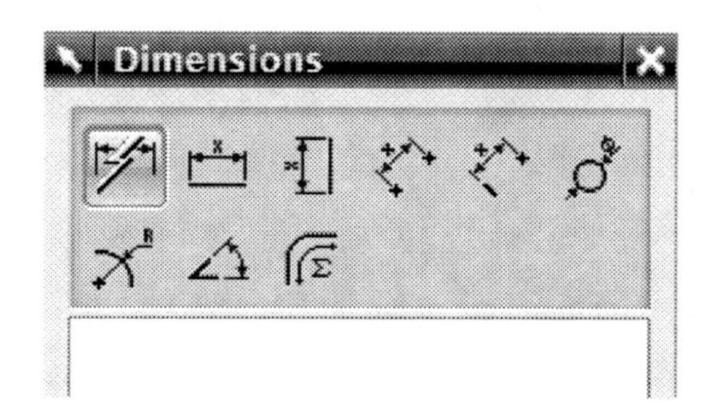

9.2. Choose the **Infer** dimension type (the upper left icon) if necessary. Notice the degree of freedom arrows will disappear individually as the dimensions are placed.

9.3. Add 3 horizontal or vertical dimensions for the bottom line, the left side vertical line, and the right end line as follows. You may need to zoom down the size of the model to make room for dimensions.

- Select a line or arc away from an endpoint.
- Wait a second and a dimension will appear.
- The dimension is on the cursor so drag it to a convenient location and press **MB1** to place the dimension.

(If it is difficult to select or get the correct dimension type, then switch from the Infer dimension type to the specific dimension type and try again.)

9.4. Choose the **Radius** constraint dimension from the dialog, select the arc and drag the dimension to a convenient location.

9.5. Choose the **Angular** constraint. Select the two lines L3 and L4. Drag and place the angle dimension.

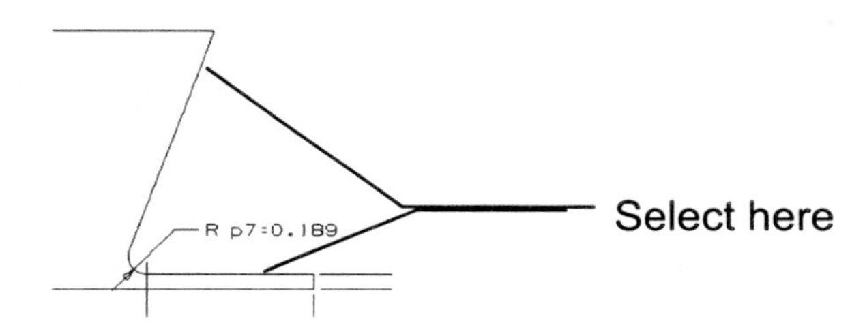

9.6. Specify the horizontal distance from the left corner point to the center of the radius. You may use the **Inferred** type or the **Horizontal** type. The technique to make the selections is listed here:

- 2 points will have to be selected. Select the left endpoint of the corner of the profile.
- Now we want the arc center so place your cursor over the arc but don't select it.
- Notice the arc highlights, as well as its end points and arc center.
- Without selecting anything, slide your cursor over to the arc center
- Select the arc center. If the selection box comes up, choose the number that corresponds to the arc (it won't say center—just arc).
- The dimension appears so drag it to a convenient location.

The dimensionally constrained profile looks like this now. And each dimension appears in the dimension dialog window.

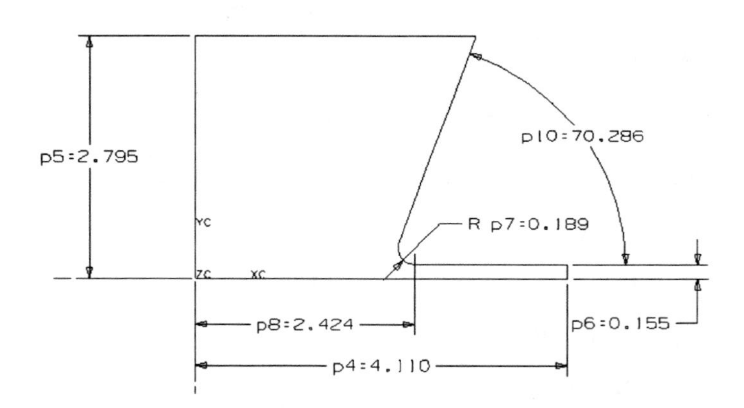

The Status Line says that the sketch needs 2 constraints to fully constrain it. Next we will position the sketch relative to the WCS origin. We use the **Datum Axes** for this.

Step 10. Constrain the location of the sketch in space.

10.1. Change layer 61 to **Selectable** so the **Datum Plane** and **Axes** are visible. Choose **Format → Layer Settings.**

10.2. Choose the **Constraint** icon on the **Sketch Constraints** dialog.

10.3. Select the lower left corner of the sketch by selecting an end point of one of the lines.

10.4. Select the vertical datum axis. It may be selected anywhere.

10.5. The **Constraint** toolbar in the upper left corner shows only one option, the **Point on Curve** constraint. Choose it. The sketch jumps horizontally to make the end point coincident with the vertical datum axis.

10.6. Now let's do the other direction. Select the same lower corner end point on the profile and select the horizontal datum axis. Choose the Point on Curve constraint on the toolbar. The sketch jumps vertically to make the end point coincident with the horizontal datum axis.

Now the Status Line says that the sketch is fully constrained. No more degree of freedom appears on the screen.

Step 11. Change layer 61 to **Invisible** so the Datum's will be out of the way again.

Step 12. Edit the dimensions to get the desired shape.

Each of the dimensions in the dimension dialog may be edited by selecting it in the dimension window or in the graphics area, and typing in a new dimension in the **Current Expression** window.

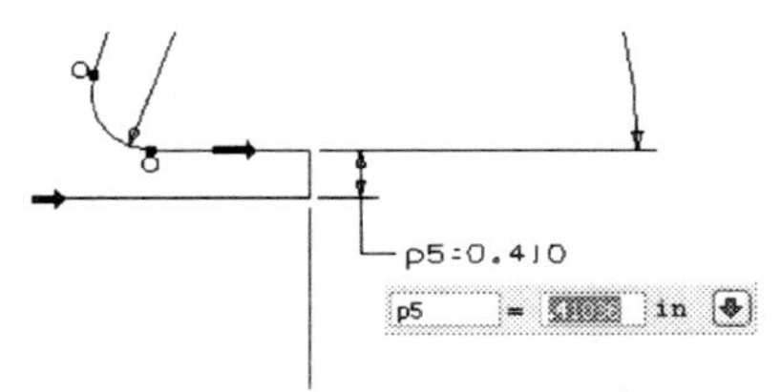

The value is highlighted and a new value may be typed in. Hit the **Enter** key when finished typing each new value and the new number will take affect and change the sketch. Observe the sketch to ensure it is changing in the manner you are expecting it to. Note also that the profile is fixed in the lower left corner which is coincident with the WCS origin. Dimensions changed will be applied away from that, or vertically upward from the bottom line and horizontally to the right from the left line.

12.1. Choose the **Infer Dimensions** icon on the **Sketch Constraints** toolbar.

12.2. On the **Dimensions** dialog, select each of the dimensions and enter a new value to approximate the desired profile. You can undo the last change by pressing two keys, Ctrl and Z simultaneously. You can undo last several changes by repeatedly pressing those keys.

12.3. You may experiment with values until you arrive at what is pleasing to you.

Please note that some dimension values will not work. The values that are entered must still have physical reality behind them and make sense physically. Ours looks like the one below when completed.

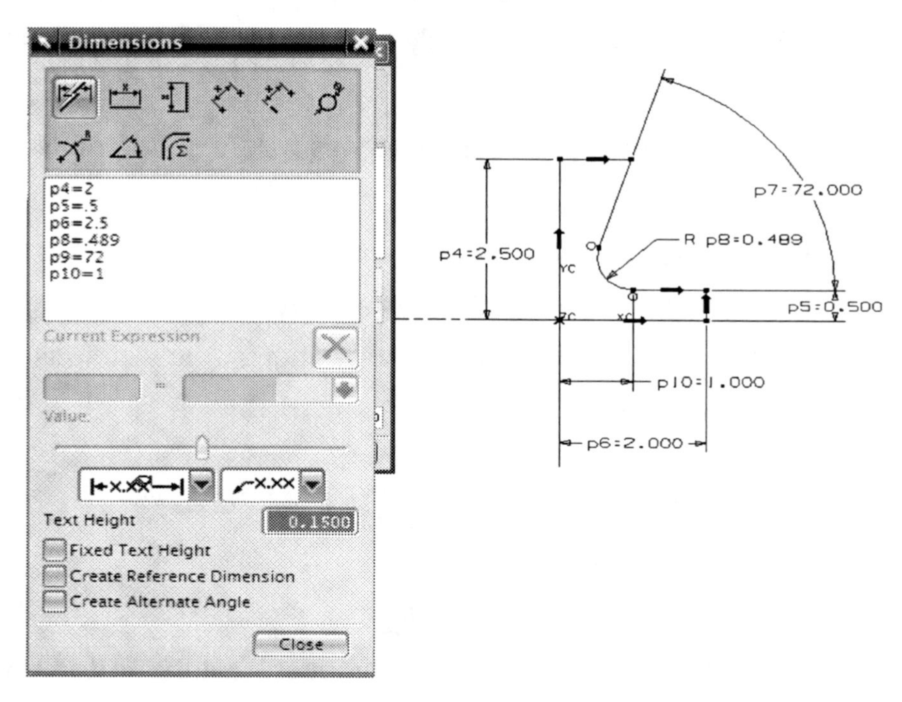

Continue to the next activity to update the solid body using this new profile.

Activity 10-5. Update the solid body

Step 1. **Cancel** out of the **Dimensions** dialog.

Step 2. You may want to **Replace** the view to the TFR-TRI so you can see the change happening.

Step 3. Change the layers to make the body **Selectable**, layer 1.

Step 4. Choose the **Update Model** icon from the **Sketcher** toolbar, or choose the **Finish Sketch** icon from the toolbar, or simply save the file. Observe as the solid body quickly updates to the new shape of the profile.

Step 5. Exit Sketcher by choosing the **Finish Sketch** icon if you have not done so in Step 4.

Step 6. File → Close→ Save and Close.

Activity 10-6. Constraining a Sketch

We will constrain the sketch from Activity 10-3 in this activity.

Step 1. Open **xxx_activity10-3.prt** and save it as **xxx_activity10-6.prt**. Start **Modeling**.

Step 2. Open the sketch **s21_profile**.

Step 3. Choose the **Show All Constraints** icon .

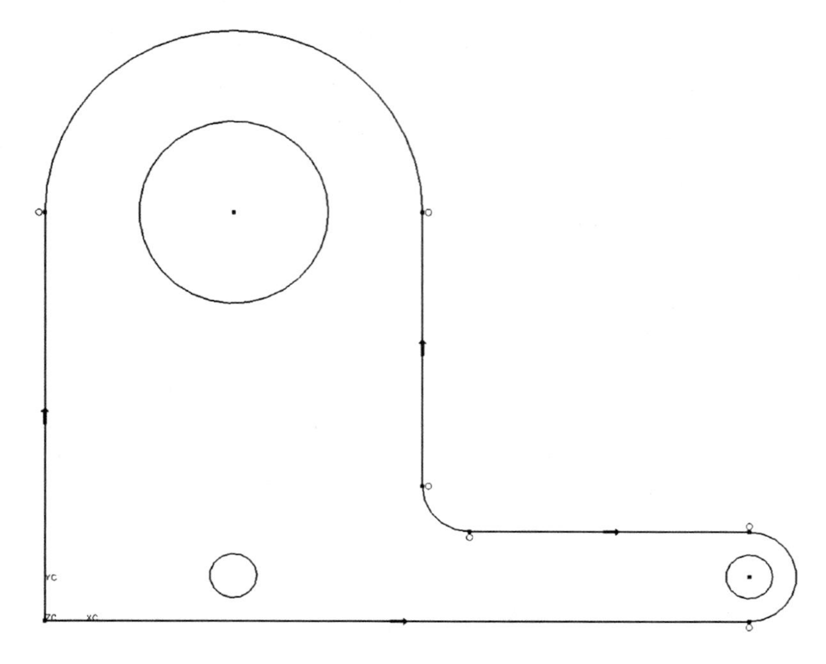

We can see by inspection that all the lines are either horizontal or vertical as they should be. We see that all of the arcs are tangent to their adjacent lines and all the connections between curves are coincident.

Step 4. Choose the **Show/Remove Constraints** icon . Change the options in the dialog as follows:

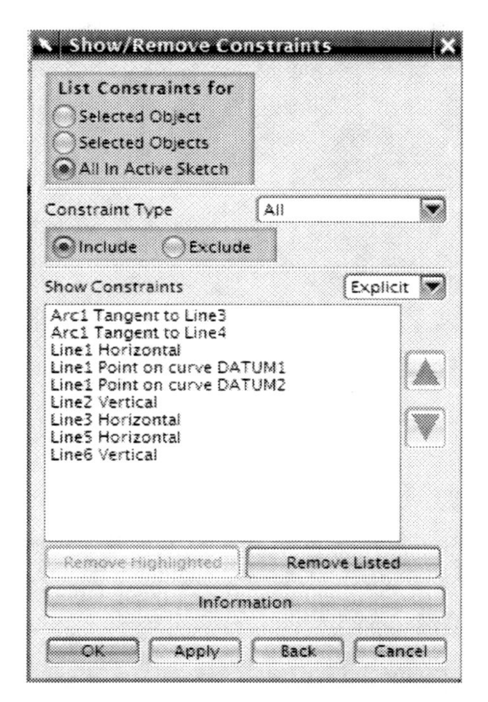

We notice on this list that there are several tangent, horizontal, and vertical constraints. For example, Arc2 is tangent to Line2 and Line 3, Arc 3 is tangent to Line3 and Line4; Lines 1 and 2 are horizontal while Lines 3 and 4 are vertical. We need to add some more Geometric constraints to the sketch.

Step 5. Add the geometric constraints to correct the profile.

5.1. Choose the **Constraints** icon.

5.2. Select Arc3 and the large circle. The menu bar appears, choose the Concentric icon. The circle moves to be concentric and the Concentric symbol appears.

5.3. Select Arc1 and its small circle in it. On the menu bar choose the **Concentric** icon and the circle moves to be concentric and the **Concentric** symbol appears.

The sketch with the new constraints now looks like this.

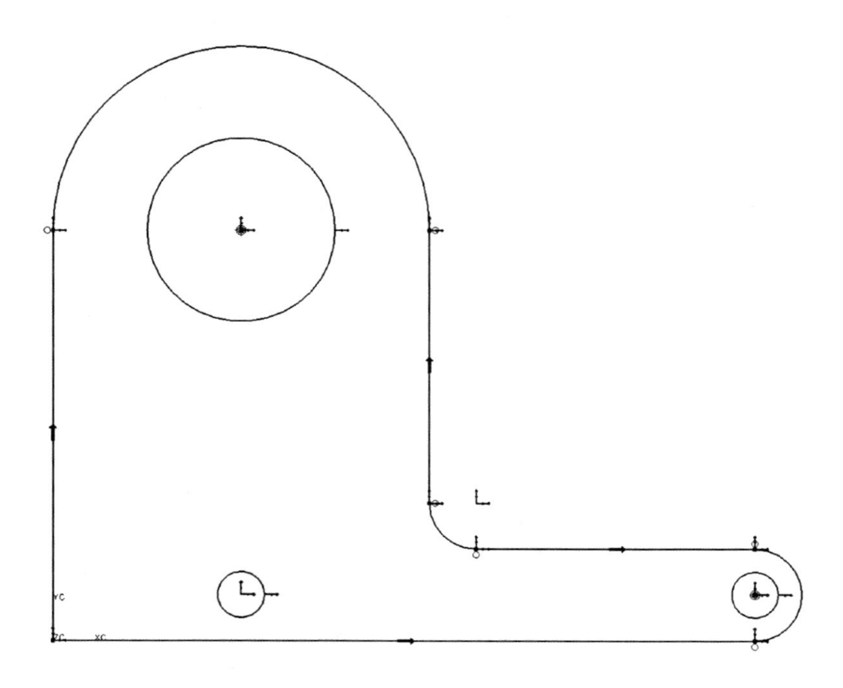

Next we are going to add some dimensional constraints to get the sizes we want, particularly for the large arc that ended up an odd diameter, and to lock down some geometry.

Step 6. Choose the Inferred… icon and then Sketch Dimensions Dialog icon from the top left of the screen.

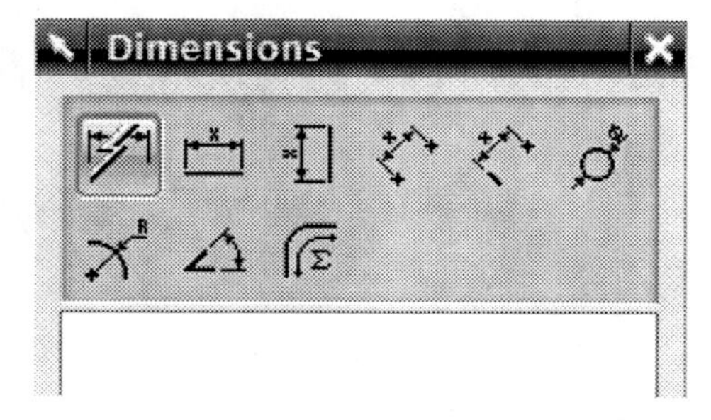

6.1. Choose the **Radius** icon from the dialog and select the large arc. Place the dimension in a convenient place above the arc.

6.2. Immediately, while the dimension is still selected, change its radius to 2.0 by typing in 2 and hitting **Enter**.

6.3. Choose the **Diameter** icon and choose the small hole from the left corner and place the dimension. Enter .5 if the value is not .5

6.4. Choose the **Perpendicular** icon to locate the hole from the left corner. Also put another dimensional constraint of radius=.5 on the small arc as shown below.

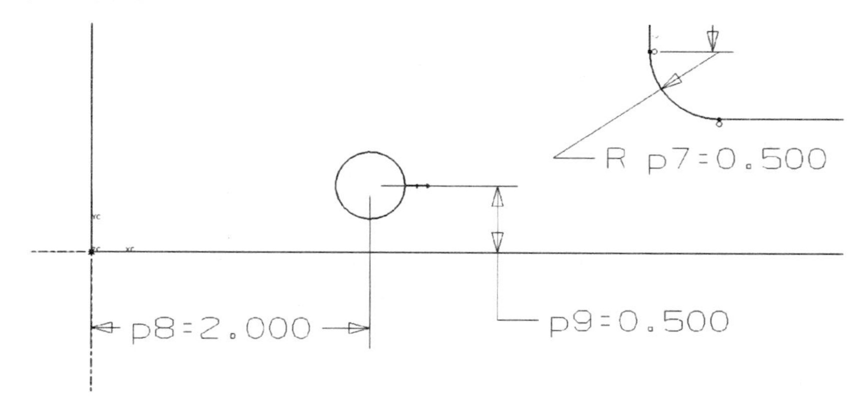

6.5. Choose the **Vertical** icon and mark the dimensions of the two vertical lines, 4.5 and 3, respectively.

Step 7. We want to position the sketch such that the lower left corner is located at the intersection of the two datum axes, which is the WCS origin.

7.1. Choose the **Constraints** icon to create the geometric constraints.

7.2. Select the **end point** of Line1 (horizontal line), and then select the horizontal datum axes.

7.3. The menu bar appears with only one choice that is the **Point on Curve** icon. Choose the icon and the sketch moves to align with the datum axis.

7.4. Select the **end point** of Line4, and then select the vertical datum axis.

7.5. The menu bar appears again with only one choice the **Point on Curve** icon, so select it and the sketch moves to align it again.

The sketch now looks like this. You may have to drag some dimensions and place them in new locations as the sketch moved to a new location. In order to do this, you need to choose the **Dimensions** icon from the **Sketch Constraints** toolbar.

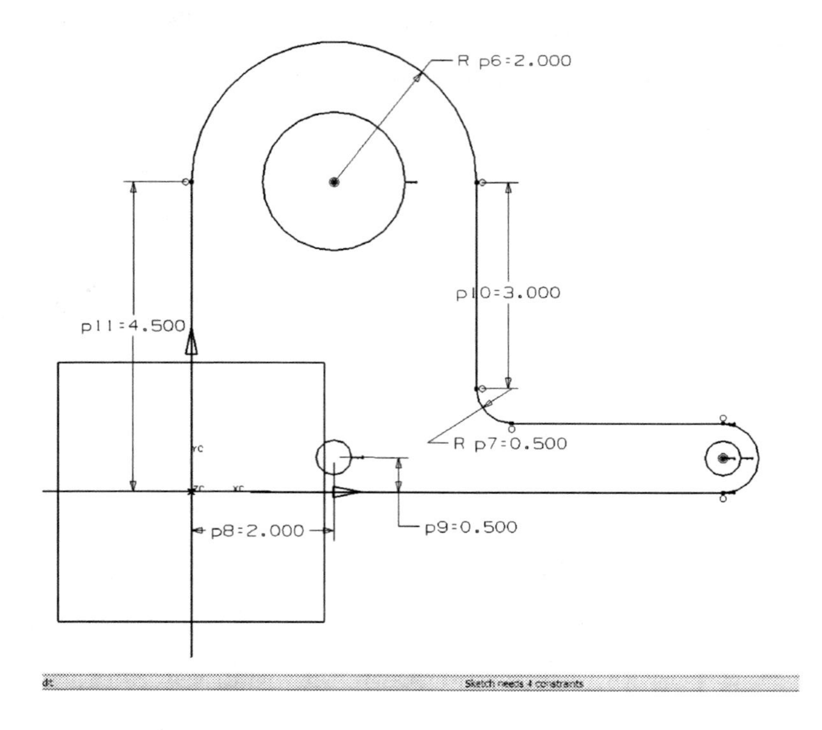

The sketch is positioned now where we want it and the dimensions are correct. We don't need additional constraints to define the sketch or to sweep the part. Remember a sketch does not have to be fully constrained to sweep it and create the solid. We only constrained the sketch to define shape we wanted. Notice that the **DOF** arrows are still present because the sketch is not fully constrained. In fact, the status line on the bottom right side indicates that the sketch needs still four constraints to be fully constrained.

Step 8. It is a common practice to **Extrude** or **Revolve** the sketch to create the solid. We suggest you to do so.

Step 9. File → Close→ Save and Close.

This completes the activity.

10.4 Using Sketches to Create Form Features

Sketches may be used in many ways to enhance the design and carry out the design intent of the designer. One of its more flexible usages is to put sketches on the planar face of a solid body or to create the sketch on a datum plane away from the solid body and extrude the sketch into the solid body to create a feature.

An example is a figure like the one below.

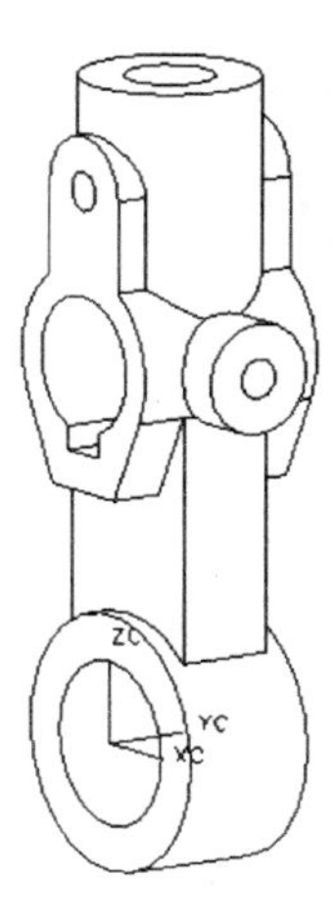

In the following activity we will create this part demonstrating this usage of sketching.

Activity 10-7. Creating the Rotator Arm

The dimensions and views for the part are shown below. The completed solid model for the part is shown above. Note that the part is in mm.

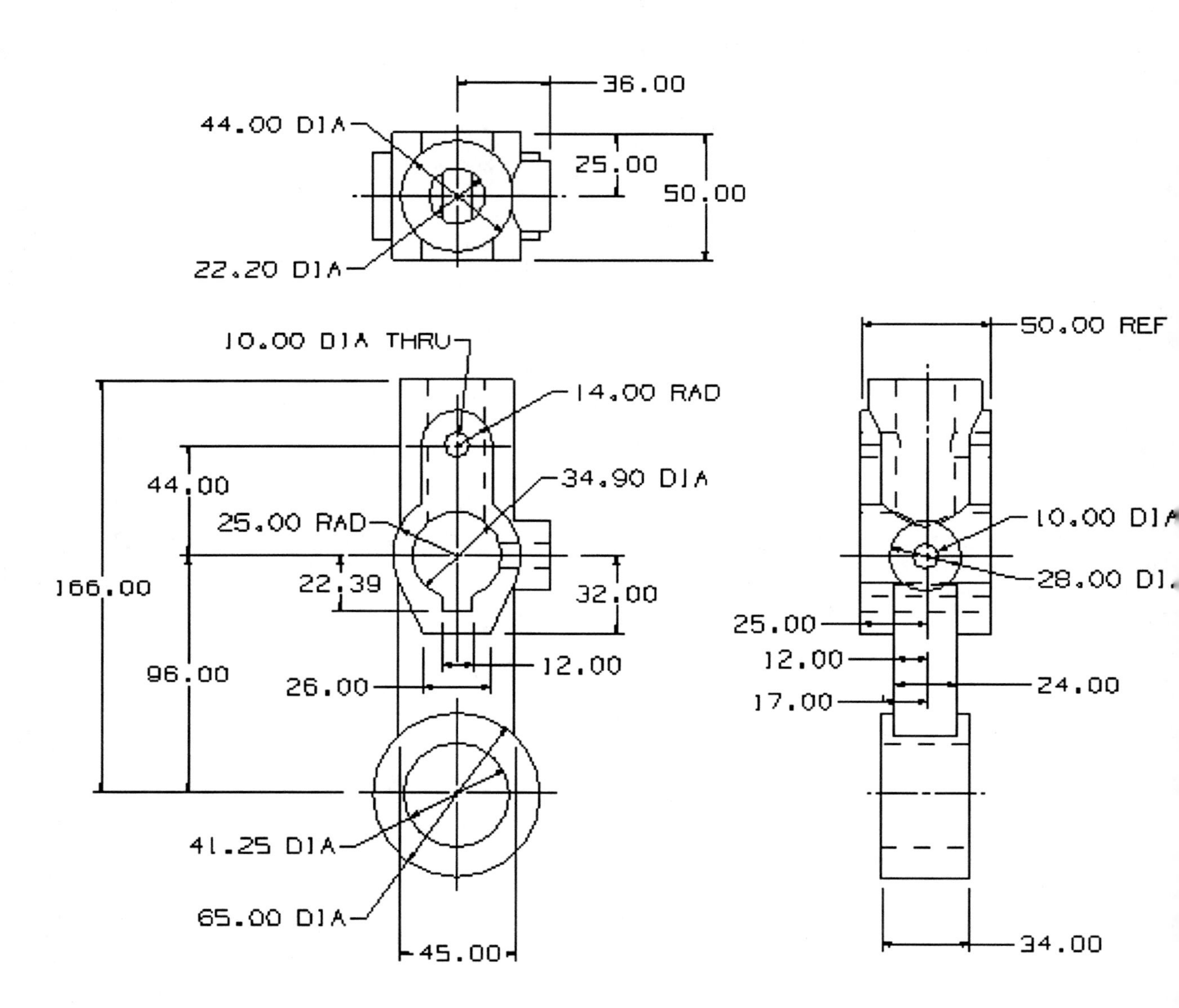

Step 1. Open a **New** part file in mm and name it **xxx_activity10-7.prt**.

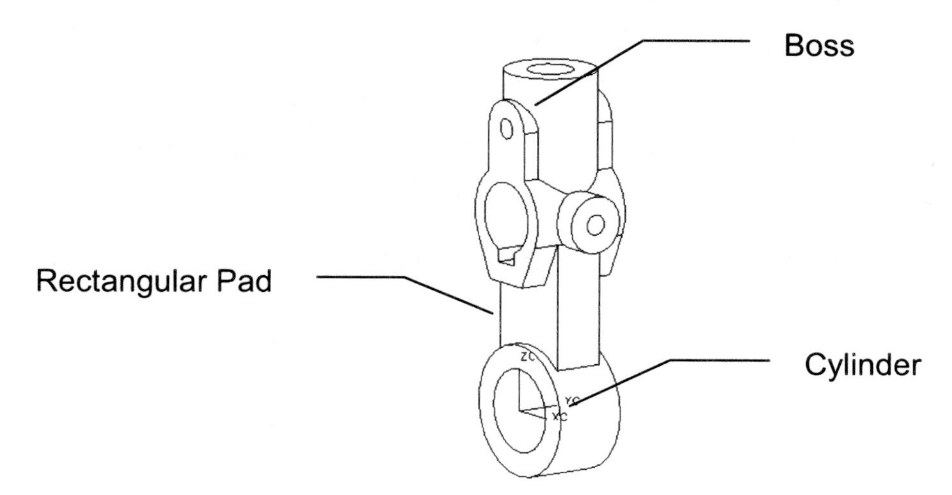

Step 2. Construct the body of the part using **Cylinder**, **Pad**, and **Boss** with the dimensions given.
Hint: Create the cylinder first. Then you will need some datum planes to create the pad.

After creation of the cylinder, pad and boss, the model looks like this and we are ready to add the sketch feature.

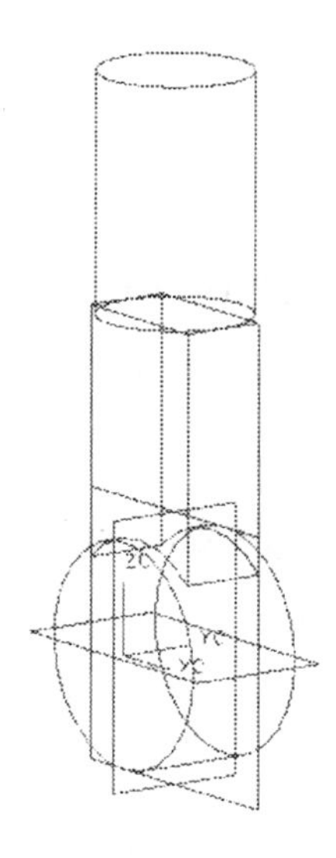

The sketch must be placed on a planar object and we have a couple of choices on how to proceed.

- One option is to construct the sketch on the existing datum plane running through the middle of the part. Since the sketch is in the middle of the part, the **Extrude** is defined to be at a **Start** of –XX to **End** of +XX.
- The other option is to create a new datum plane parallel to the same existing datum plane but **Offset a** distance away. The **Extrude** would then happen from **Start** of 0 to **End** of XX.

Let's choose the first option and build the sketch on the datum plane in the middle of the part.

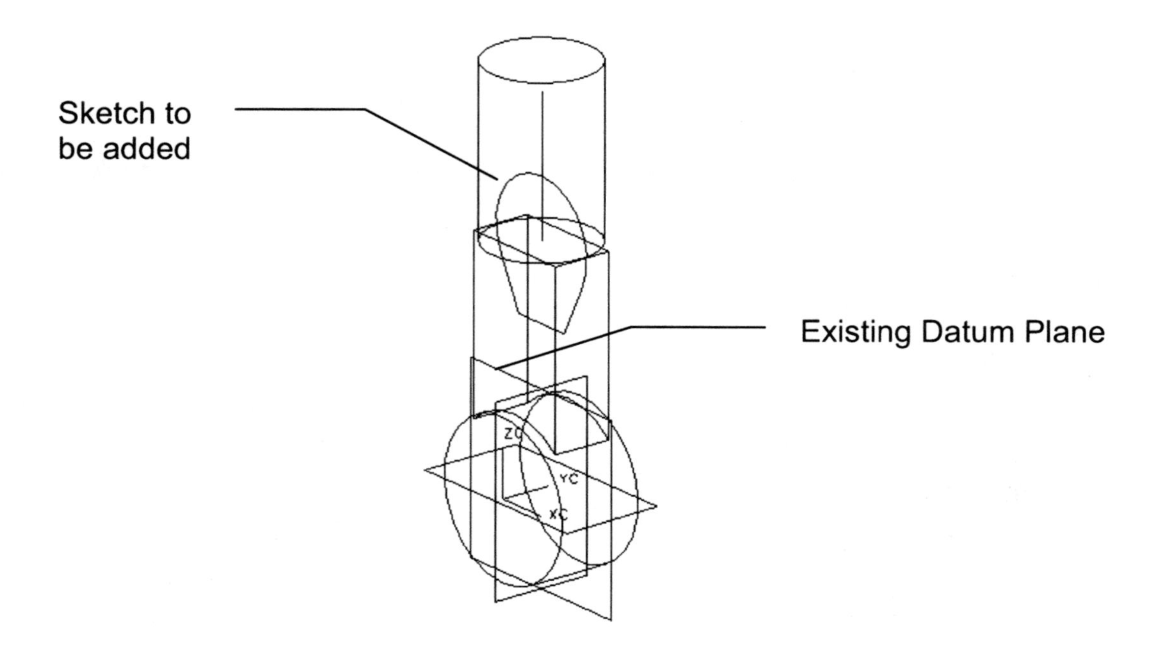

We now are ready to create the sketch on this datum plane. We give a brief note here about the feature we are adding. It does have a keyway in it so it is worth thinking ahead and determining how we are going to create that keyway. It is true that our sketch will contain the outside profile of the feature, but it could also contain the inside profile of the keyway and hole through the part. However, general recommendations are that sketches only profile one feature at a time. Therefore, that means creating a sketch for the outside profile and then creating a separate sketch for the inside profile. But notice, using a Hole Feature and a Slot Feature could also create the inside profile in this case. Both options will work but we will use the Hole/Slot option since it is easier to create the first time.

Step 3. Create the sketch for the outside profile.

3.1. Change the **Work Layer** to 22.

3.2. Choose the **Sketch** icon.

3.3. Type in the name for the sketch as **s22_feature** on the **Sketcher** toolbar and hit the **Enter** key.

3.4. The **Cue Line** prompts to select the sketch plane. Select the newly created datum plane labeled in the figure above. Choose the **Checkmark** to ok the selection.

The view should jump to place the datum plane flat on the screen. The system uses the default horizontal direction as to the right and places the WCS in that orientation.

Adjust the geometry so the place where the sketch is to be located is in the middle of the screen. You may need to **Zoom Out** and **Pan** a bit as well.

3.5. Visually determine approximately where the sketch would be located relative to the solid body.

3.6. Choose the **Profile** icon from the **Sketch Curve** toolbar if inactive.

3.7. Ensure the **Line** icon is selected from the toolbar in the upper left corner.

3.8. Create the profile shown as close to shape as possible. Let's walk through it.

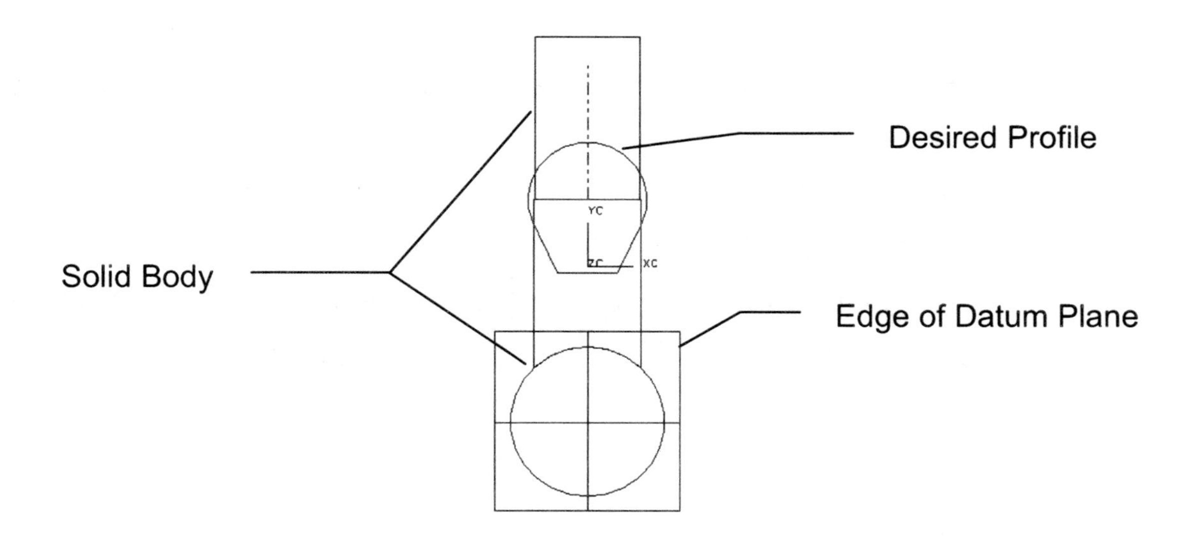

The lines are a little difficult to separate on this picture but will be easier in Unigraphics because the colors will be different.

3.9. We are going to show the figure without the solid present because in black and white it is hard to distinguish, however, you probably want to keep the solid visible so you can relatively place the sketch in the correct location.

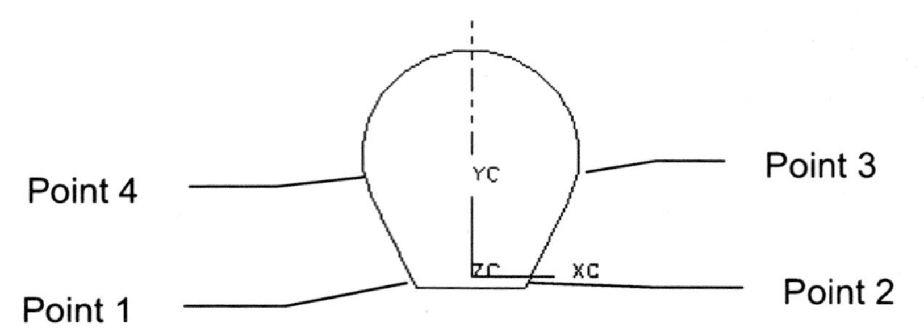

The following steps will use this figure and points to guide our path.

3.10. With the **Line** icon chosen from the toolbar in the upper left corner, select **Point 1** approximately where shown.

3.11. Select the end point approximately at **Point 2** keeping the line between Point 1 and Point 2 horizontal. Note the horizontal dash line.

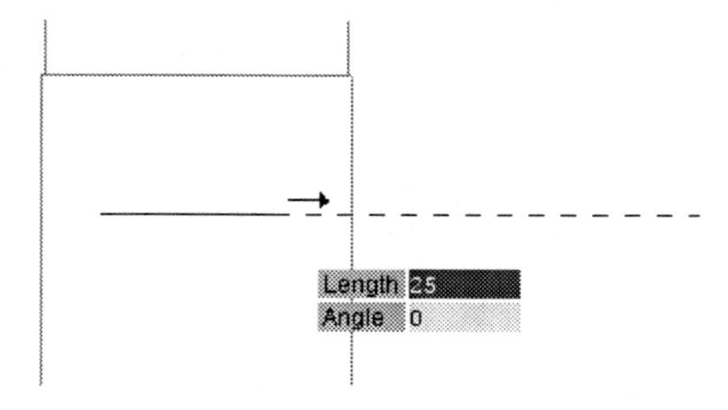

3.12. Select the end point approximately at **Point** 3 insuring the line is at an angle as shown.

3.13. An arc is needed so choose the **Arc** icon from the toolbar in the upper left corner. The **Cue Line** prompts for an end point, so select approximately where **Point** 4 is located. The arc is shown and you can drag around the **Point** 4 and vary the radius somewhat as well.

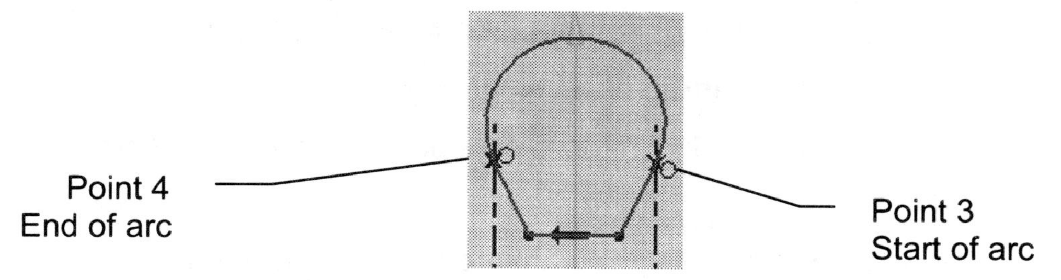

This method of placing an arc asks for the start point (in this case the end of the previous line), the end point of the arc.

3.14. The system automatically assumes the next curve will be a line so just select **Point 5** to be coincident with **Point 1**.

3.15. Cancel the construction mode by de-selecting the Profile icon.

The curve creation is now complete and some constraints need to be checked. We know that we want a line to be horizontal and end points should be coincident. This is evident from the figure with the arrow and squares at the curve endpoints. Also you can see that the arc and one of the lines are tangent. The additional constraints we need should be added now.

Step 4. Apply tangent constraints where needed. If you have already two tangent constraints, skip this step.

4.1. Choose the **Constraints** icon to add Geometric constraints.

4.2. Select arc 4 – 3 and line 4 – 1 and choose the **Tangent** icon from the small toolbar in the upper left corner. Repeat this for the other side.

Step 5. Apply **Dimensional** Constraints. Select the **Infer Dimensions** icon.

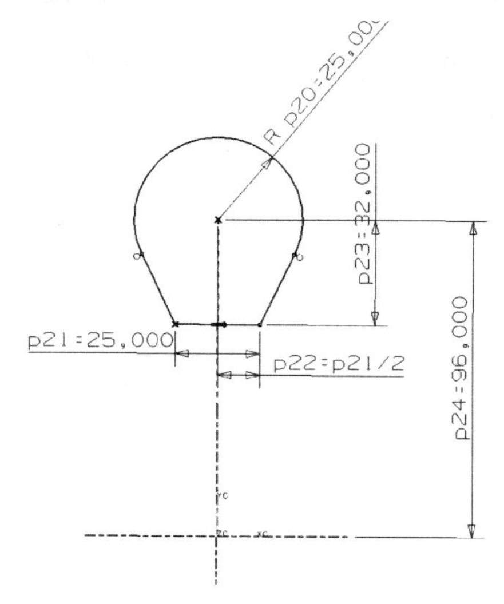

5.1. With the **Inferred** option selected, add the **Horizontal** dimension for the bottom line and the **Radius** dimension.

- Select the bottom line in the middle and drag the dimension down a bit.
- Select the arc and drag the dimension to a convenient spot above and to the right.

Most of these dimension constraints are straightforward, like dimensioning a part in drafting. However, there is some technique involved you may not be familiar with. The next steps will walk through some of those techniques.

5.2. Add the **Vertical** dimension from the arc center to the bottom line of the sketch. To select the center of the arc, put your cursor over the arc but don't select it. Notice the arc highlights and notice points associated to the arc highlight as well: the two end points of the arc and the center point. Without selecting anything yet, slide the cursor over the center point and select the center point. Watch for visual verification in the correct point selected because the asterisk marker will appear at the point actually selected. Then select the line on the bottom and drag the dimension to the right of the figure.

Another different dimension is the relational dimension at **p103 = p22/2**. This type of expression is used to ensure that the arcs stay centered over the flat bottom part of the sketch. Please review Chapter 6 to get a fuller description of **Expressions**. If the dimension letters are too small or large, change accordingly the value of "Text Height" field at the bottom of the **Dimensions** dialogue.

5.3. The relational dimension is created by selecting the **Horizontal** dimension type, then selecting the arc center and selecting the end point of the line 1-2.

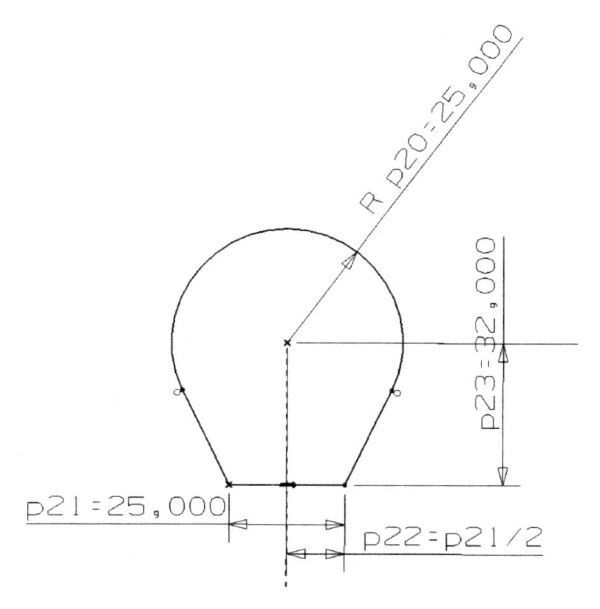

5.4. With the expression highlighted in the Value window, type in **p22/2** and hit **Enter** (your p# may differ and use it in place of p22).

5.5. Edit the dimensions to the correct values as shown above to comply with the drawing dimensions shown at the beginning of this activity.

When the dimensional and geometric constraints are complete, the positioning constraints need to be added to locate the feature correctly with the solid body. The sketch will be located using the arc centers and edges of the datum planes with dimensional constraints.

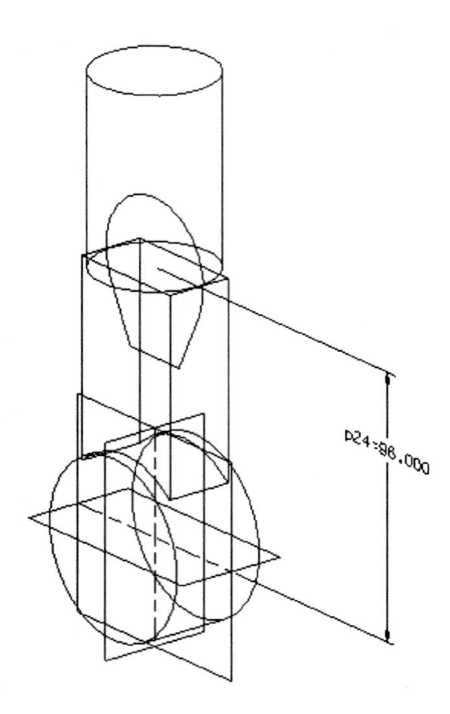

The figure above represents the sketch with the sketch dimensions blanked out for clarity. The positioning dimension is shown as applied.

Step 6. Apply positional constraints using the same **Dimensions** dialog.

6.1. Choose the **Inferred** dimension icon.

6.2. To create the constraint corresponding to the 96 mm dimension, select the arc center of the sketch arc and choose the datum plane on the bottom as shown above. Then place the dimension. This locates the sketch vertically from the arc center of the cylinder at the bottom of the solid body.

6.3. To create the constraint to center the sketch in the body, choose the **Constraints** icon and select the edge of the vertical datum plane that we see as an edge and the center point of the arc in the sketch, and select the **Point on Curve** icon in the toolbar in the upper left corner. That constraint locates the sketch in the center of the body

The sketch should be fully constrained now.

Step 7. Extrude the sketch.

7.1. Choose the **Finish Sketch** icon and the graphic will change to Trimetric.

7.2. Choose the **Extrude** icon.

7.3. Choose the Extrude Dialog icon from the upper left corner to display the Extrude dialog.

7.4. The **Cue Line** prompts to select the curves to extrude. Select one of the curves in the sketch and notice that all the curves are highlighted and selected. This is because the sketch curves form a feature.

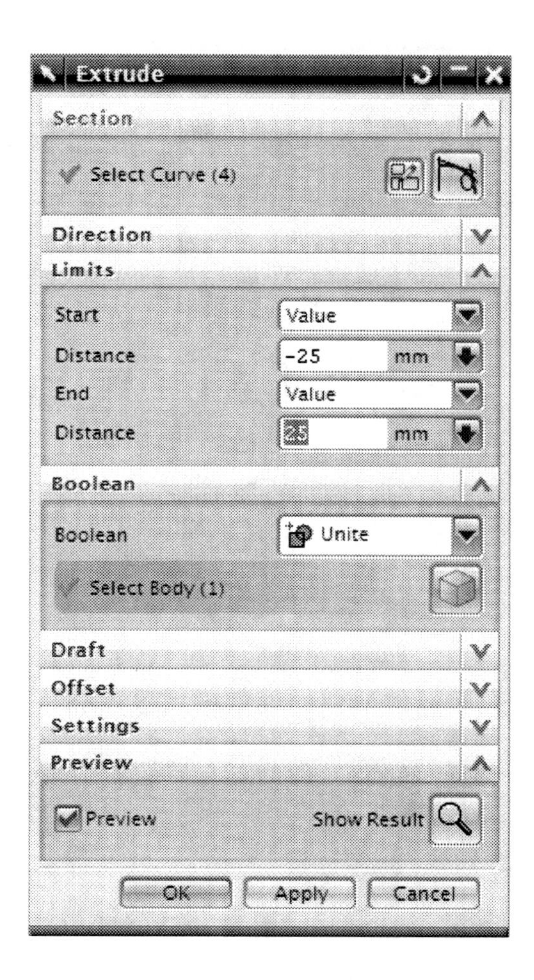

7.5. Enter -25 and 25 for the **Start** and **End** values in the **Limits** menu respectively, and change the **Boolean** to **Unite**. Choose **OK** in the Extrude Dialog.

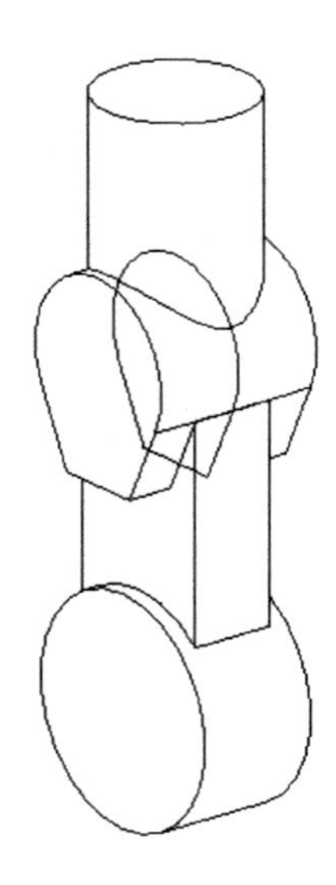

The solid body looks like this after these operations.

We now need to create the extrusion extending on top of the extrusion we just did. This portion looks like the figure below.

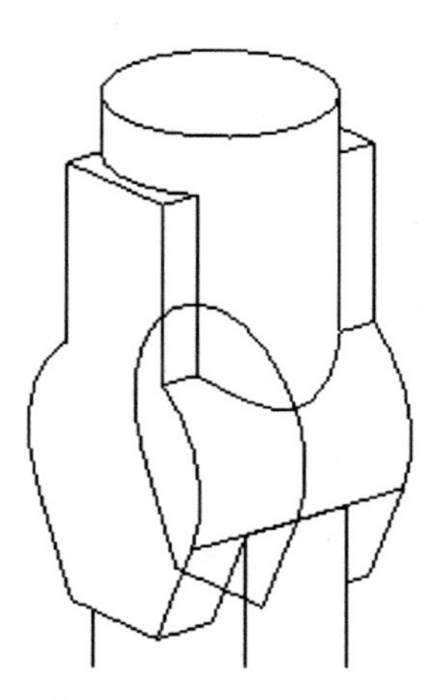

We will accomplish this portion of the body by extruding a line in both directions, with the offset. We will create the line in the same sketch.

Step 8. Create the vertical line.

8.1 Change the work layer to 23.

8.2 Re-enter the sketcher choosing the **Sketch** icon and create a new sketch by typing in the name of the sketch **s23_line** and choosing for the sketch plane the same vertical datum plane that was used in the previous sketch.

8.3 Choose the **Line** icon from the **Sketch curve** toolbar.

8.4 Construct a line from the arc center of the previous sketch vertically upward past the arc and observe two things:

a. observe the text box and select a point when the line is 58 mm long and

b. insure that the line is vertical by observing the vertical dashed line and the vertical arrow after the line is created.

8.5 Add a **Vertical** dimension to the line, and insure the line to be 58 mm long (44 mm + 14 mm). Edit it if it is not.

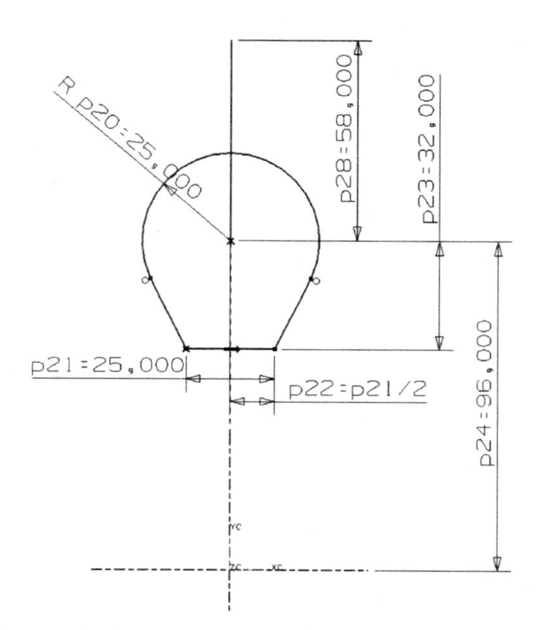

Step 9. Extrude the line to create the upper portion of the flat area of the feature.

9.1. Choose the **Finish Sketch** icon to exit out of the sketch menu and the graphics will change to the Trimetric view.

9.2. Choose the **Extrude** icon followed by the Extrude Dialog icon and select the vertical line we just created.

9.3. Enter the values in Limits menu as shown in the dialog below.

9.4. Toggle the offset menu on and enter a start and end value for the offset.

9.5. Change the direction of the sweep to be in the +YC direction if necessary and toggle Enable preview on.

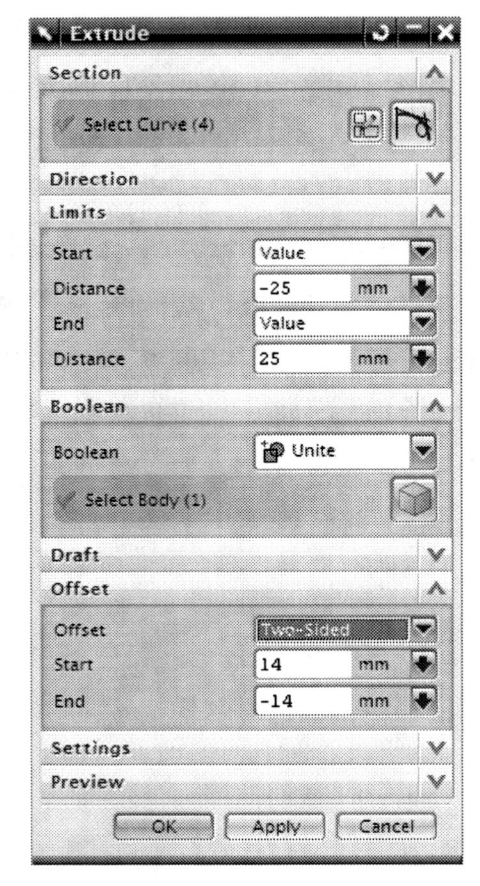

9.6. Change the **Boolean** to **Unite**.

9.7. Choose **OK** to complete the sweep.

Step 10. Blend the ends of the sweep to create the round at the top. Choose four edges and apply Edge Blend with radius = 14 mm.

Step 11. Change the layers to make the sketches invisible and make layer 1 the work layer.

Step 12. Create the **Boss** on the side of the body.

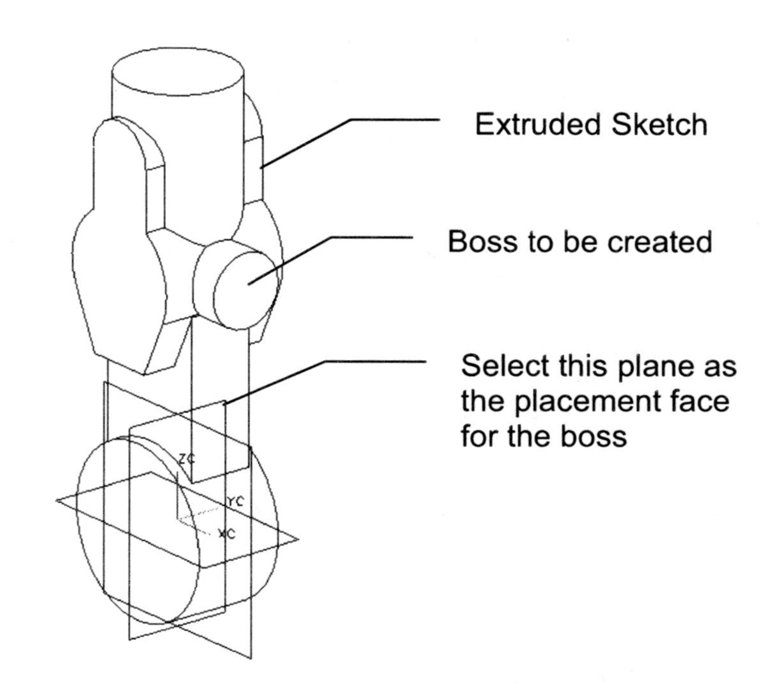

12.1. Choose the **Boss** icon.

12.2. Select the planar placement face to be the datum plane indicated above.

12.3. Ensure the direction arrow is pointed to the right.

12.4. Enter the parameters for the boss.

12.5. Use the positioning dimensions to locate the boss in the proper location relative to the solid body using the dimensions on the drawing. You could use a combination of Point to Line and Perpendicular positioning methods. You may need to use a negative value.

Step 13. Add five holes to the solid body as indicated on the drawing. Choose the **Hole** icon and create and place the holes per the drawing.

Step 14. Create and place the keyway in the body.

We chose that the **Rectangular Slot** (alternatively, Rectangular Pocket works) would be used to create the keyway because it allows the **Thru** faces to be selected and associated to the faces in case of future change. A datum plane will need to be created for the planar placement face for the slot.

14.1. Change the work layer to 64 and choose the **Datum Plane** icon.

14.2. Several ways may be used to create the plane, but the easiest and quickest is to offset it from the plane at the bottom that is already created and in the correct orientation.

14.3. Select the bottom datum plane and enter the offset distance of 96 mm.

14.4. Choose the **Slot** icon, the Thru option, the **Rectangular** option and select the datum plane as the placement plane. Enter the rest of the information for the slot and create the slot. To position the slot, you may need to use two Line onto Line methods using different datum planes and center lines of the slot.

Step 15. File → Close→ Save and Close.

This completes the construction of the rotator arm.

10.5 Create a Sketch-Based Hole 

In Section 5.3, we discussed how to create a form-feature based hole. In this section, we introduce a different method of creating a hole, which is sketch-based. When you choose Insert → Design Feature → Hole, you will be prompted with the following dialog:

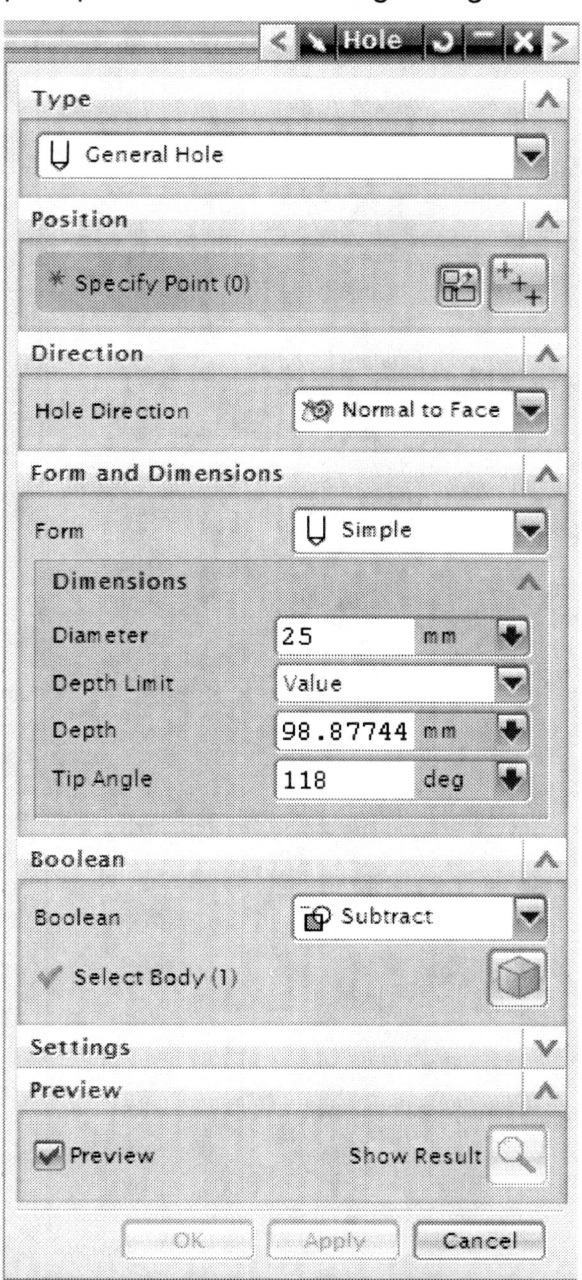

Hole type: Simple, Counterbored, and Countersunk

Hole Dimensional Parameters

Creating a hole such that its center is coincident to the center of a circular edge:

When the above Hole dialog is presented, pre-select the circular edge as shown below by hovering your cursor over the edge. When the status line says "Arc Center – Edge …", select the edge. Now the hole center is defined as the circular edge center. In the dialog, you can enter the hole parameters such as Diameter and Depth.

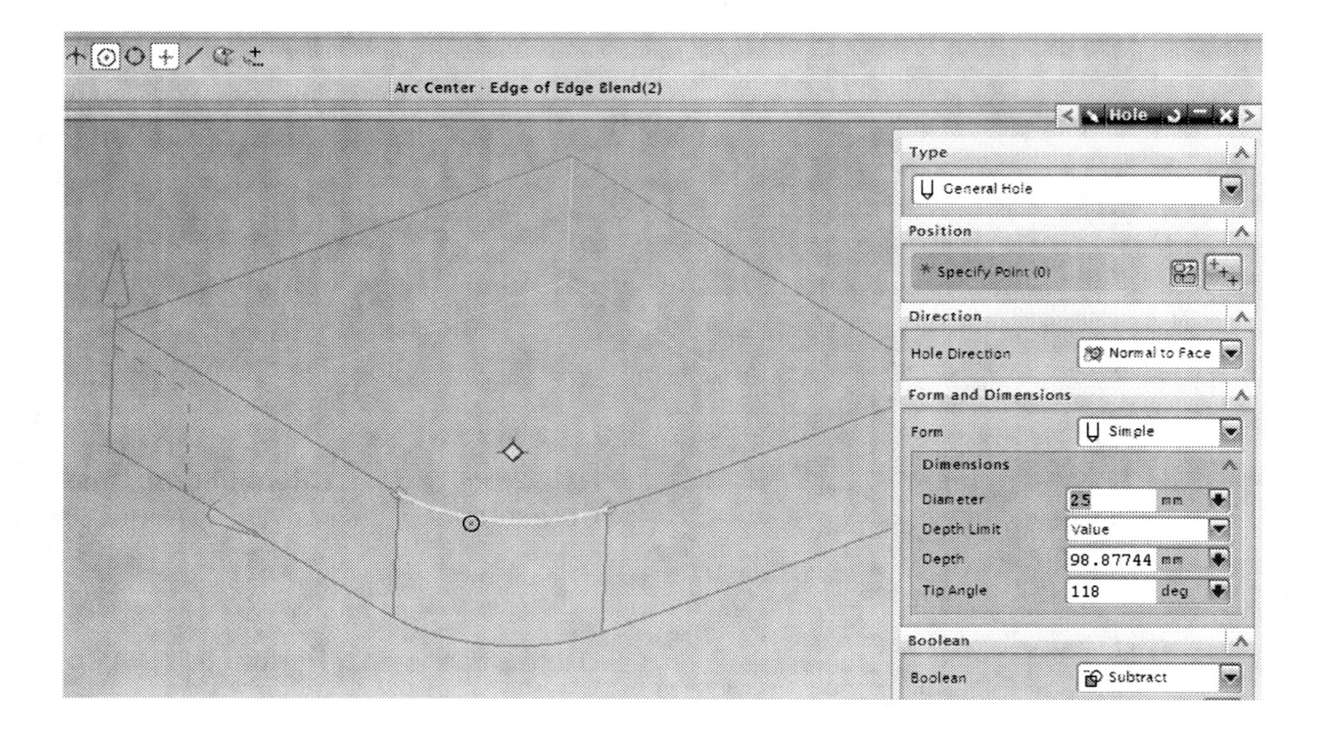

If you want to make a thru hole, select the down arrow next to Depth Limit, choose the option Through Body as shown below.

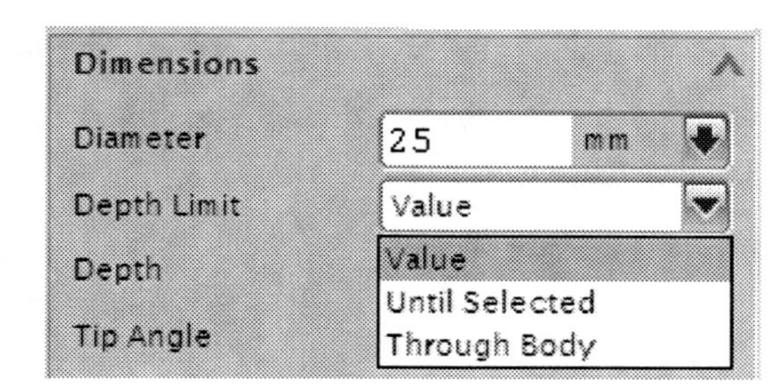

Make it sure that the Boolean setting is set to Subtract (Default) and choose OK.

Creating a hole such that its center is perpendicular to linear edges or other objects:

In the above Hole dialog, select a top face of the block as a hole placement face. Now you will be in the Sketch creation mode and be presented with the following dialog. The green square dot is the current hole center location, which is the screen location of the top face you selected.

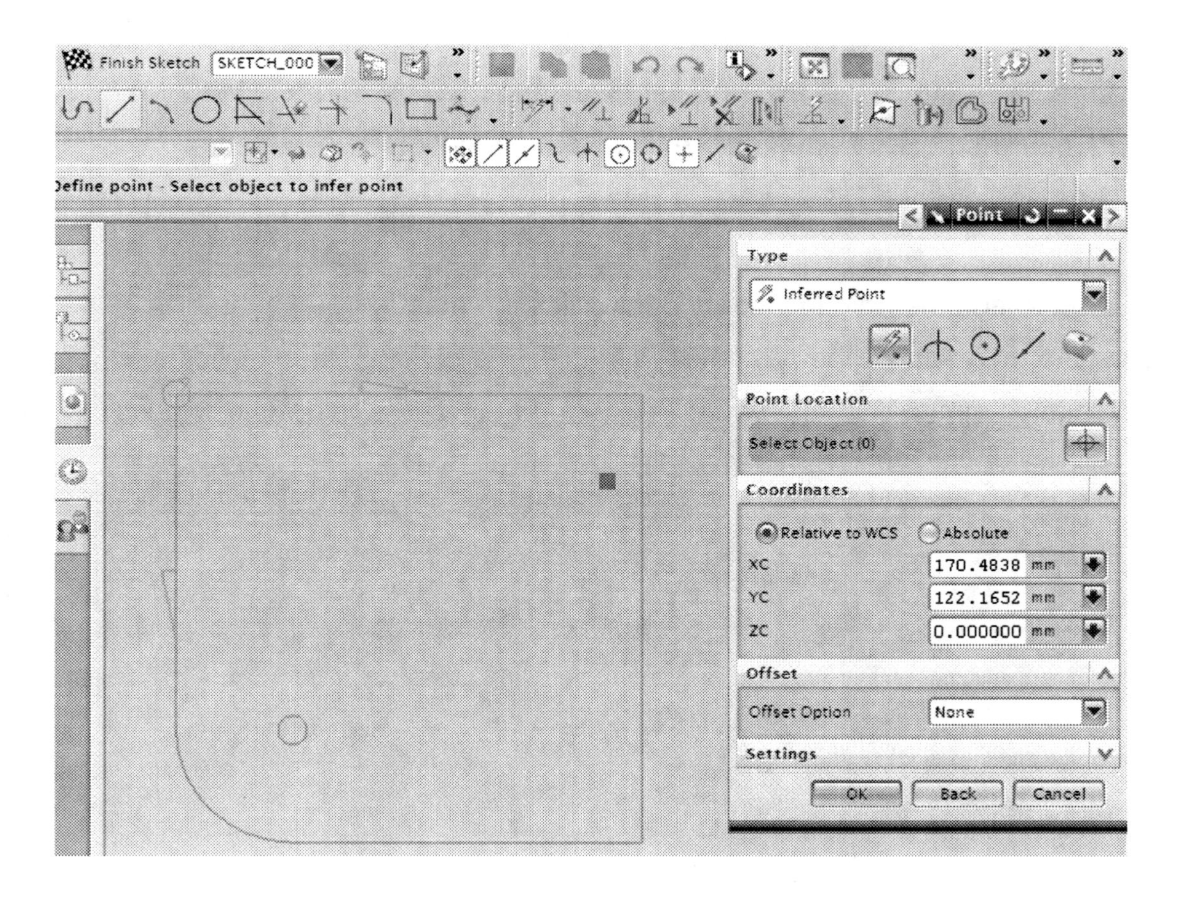

Choose OK until the above Point dialog is closed. The green dot is replaced with a sketch point (green + symbol). As is, the sketch point is unconstrained. You could use Sketch Dimensions discussed in Section 10.3 in order to position the sketch point (hole center) at the desired exact location. The below figure shows two dimensional constraints applied to the sketch point, making it fully constrained.

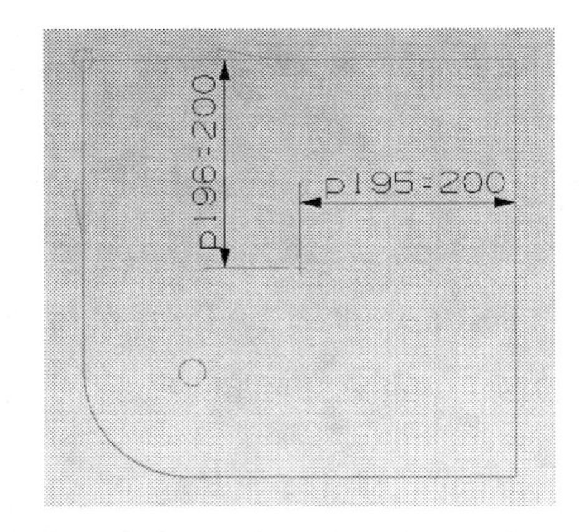

Choosing Finish Sketch will exit from the sketch mode and return to the Hole dialog. Enter the hole parameters, set the Boolean option to Subtract if necessary, and choose OK.

If you want to edit the hole location, double click the hole feature in the graphics window and then you will be prompted with the Hole dialog. Double click a specific sketch dimension you want to edit and enter a new value and choose OK.

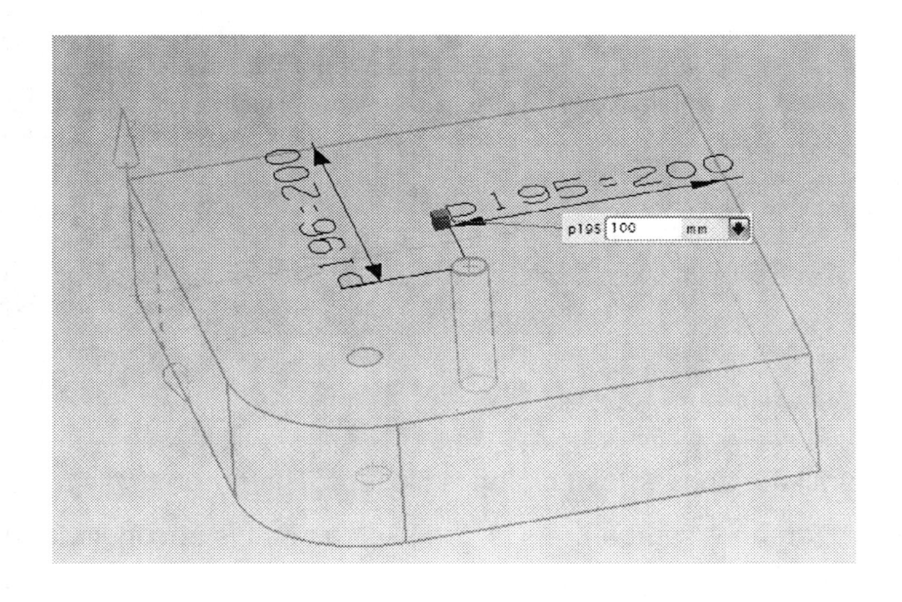

Project 10-1. Fixture

It is a relatively simple design that could be built with primitives, but your assignment is to build it with a sketch. Primitives are not allowed. The part is in mm. Save this file as **xxx_project10-1.prt**.

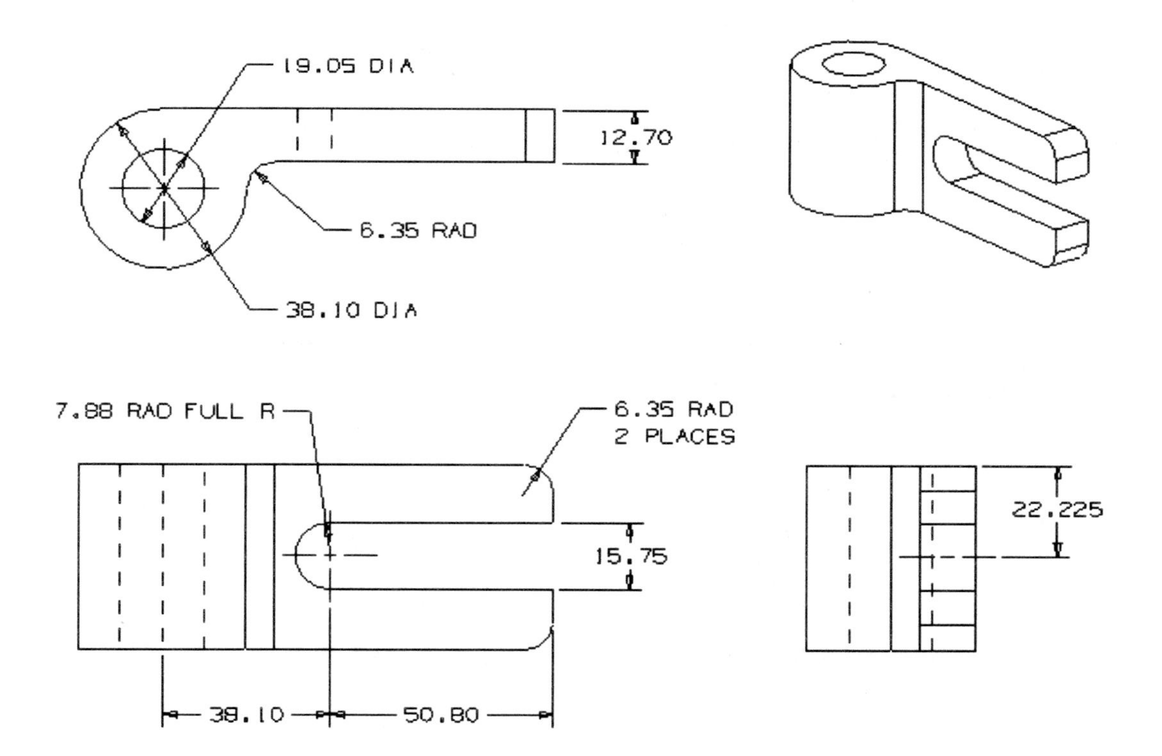

Project 10-2. Adaptor Block

Model this part using sketches and form features only—no primitives. The part is in mm. Save this file as **xxx_project10-2.prt**.

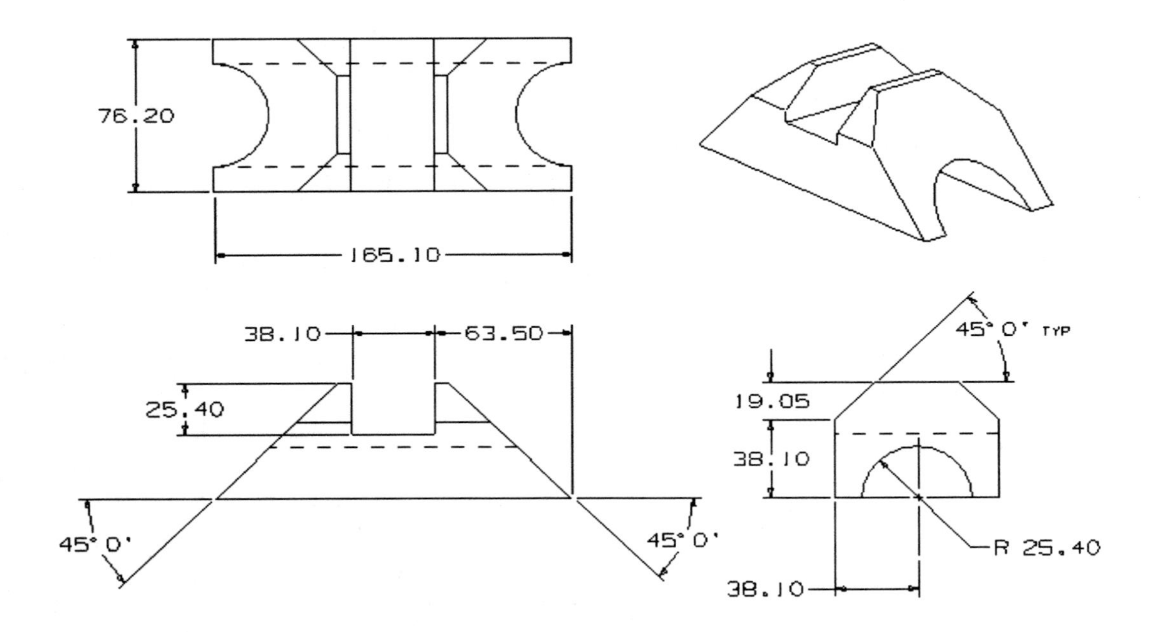

You can use either one or two sketches with some other form features.

Project 10-3. Bracket Using Sketches

Create a model of the part using the drawings and dimensions (in inches) given below. Name your part as **xxx_project10-3.prt** where xxx is three-letter initials of your name. This project is the same as Project 5-2. In Project 5-2, you created this part using a primitive and form features such as pre-NX5 holes, pads and bosses. Now in Project 10-3, you are required to create this part using only sketches, extrusions or sketch-based holes discussed in Section 10.5.

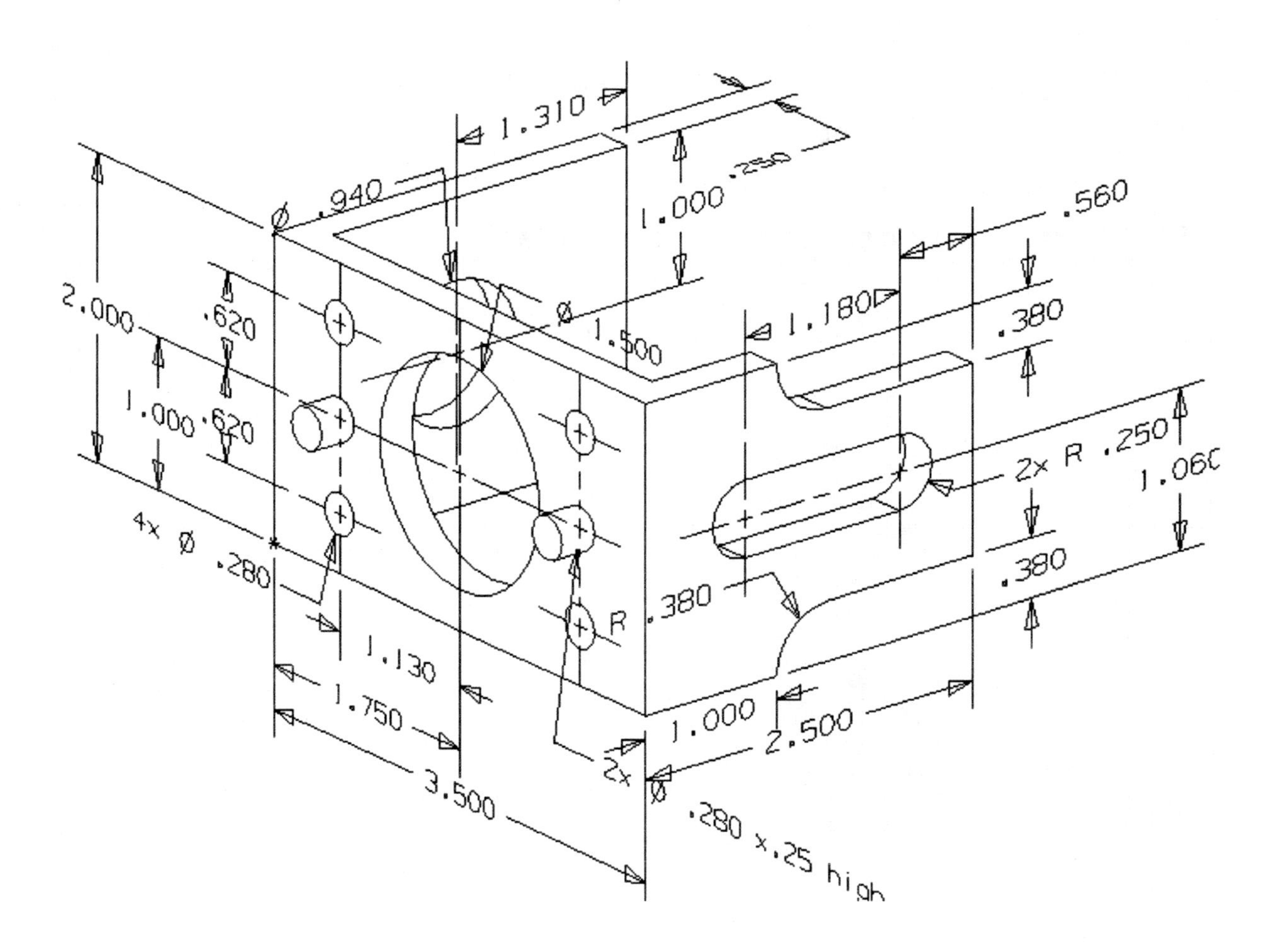

Exercise Problems

10.1 What is a sketch?

10.2 Why would you use a sketch?

10.3 Are sketches the fastest way to model a part?

10.4 What functionality is used in conjunction with a sketch, but after the sketch?

10.5 What are the basic building blocks of a sketch?

10.6 Which kinds of objects are used to build sketches?

10.7 What is the purpose of constraining a sketch?

10.8 Do sketches have to be constrained?

10.9 What are the types of constraints?

10.10 Are constraints able to be changed or deleted?

Chapter 11. Instance, Trim Body, and Thread Features

This chapter introduces feature operations that are useful to create complex geometries. These feature operations include Instance, Mirror Body & Feature, Trim Body, and Thread. With Instance and Mirror Body & Feature, you can quickly create copies (instances) of desired existing feature(s), obviating repeatedly creating each of them. With Trim Body, you can create a complex solid body by cutting away a portion of the solid determined by the shape of a face or plane. Thread facilitates creation of any object with threads such as bolts, nuts, etc. Sections 11.1 to 11.4 discuss these four features in order.

11.1 Instance

An *instance* is a shape-linked feature, similar to a copy. You can create one or more instances of a feature, or group of features. Since all instances of a feature are associated, you can edit the parameters of the feature and those changes are reflected in every instance of the feature. Thus, this feature allows you to quickly create a number of similar features and add them to the model in one step and to edit all instanced features in one step.

As the instances are created, the Boolean operation is defined by the features that was selected for instancing. For example, if you select a boss and a hole, the instance of the boss is added (united) and the instance of the hole subtracted, from the solid body to which they are attached.

Five Instance Options:

There are three instance options for this feature. Choose **Insert →**

Associative Copy → Instance or the **Instance Feature** icon to see the dialog.

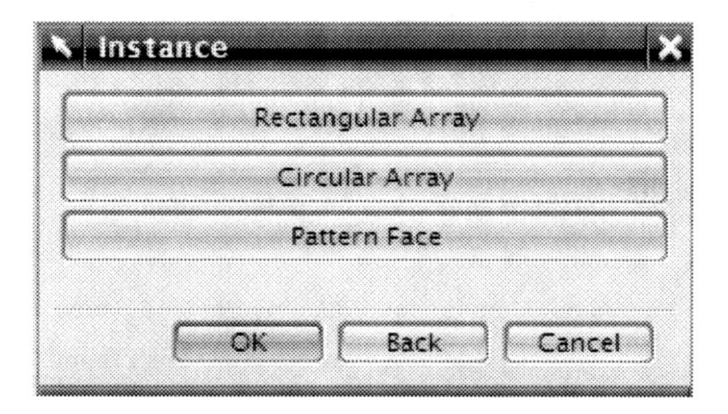

Rectangular Array	To create a linear array and rectangular array of instances from one or more selected features.
Circular Array	To create a circular array of instances from one or more selected features.
Pattern Face	Opens the Pattern Face dialog under Direct Modeling, to let you make copies of a face set. (This is out of scope of class.)

Instance Array Types

You can create three types of rectangular and circular instance arrays:

General	Creates an instance array from existing features and validates all geometry. An instance of a General array is allowed to cross an edge of the face. Also, instances in a General array can cross over from one face to another. This is the default type.
Simple	Similar to a General instance array, but it speeds up the instance array creation by eliminating excessive data validation and optimizing operations.
Identical	The fastest way to create an instance array; it does the least amount of validation, then copies and translates all the faces and edges of the master feature. Each instance is an *exact* copy of the original. No checking is done for invalid geometry. You can use this method when you have a great many instances, and you are sure they are all exactly the same.

The figure below illustrates a case where a General array has been created where one instance crosses the edge of the body. The Simple and Identical arrays look the same, but the Identical array is created faster, since the least amount of validation is done.

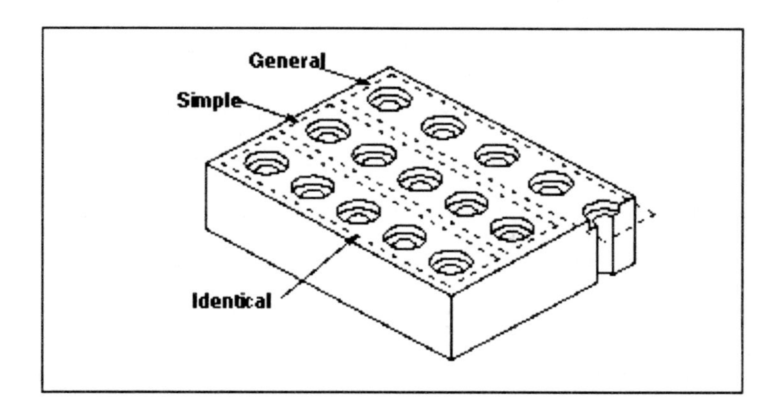

The next figure shows a part with a master feature. It also illustrates a valid Simple array, and shows how an Identical array could not be created in this situation.

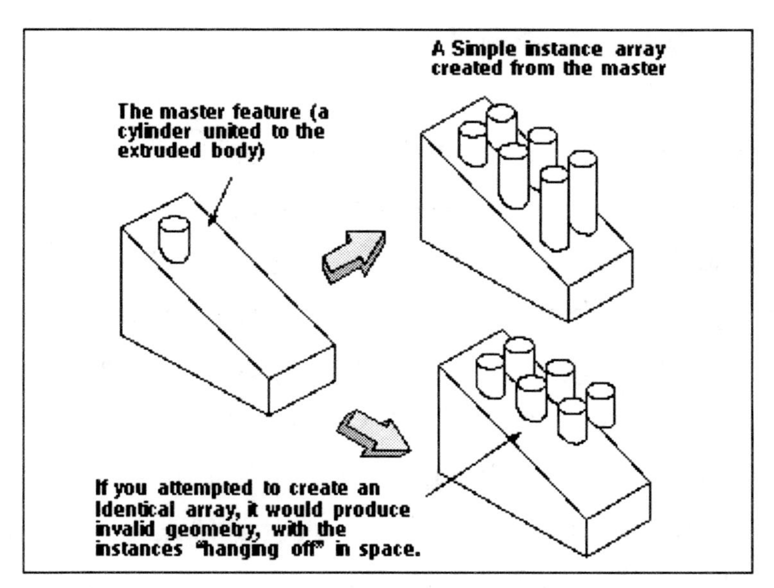

General Guidelines

- If you create a blend or chamfer on an instance after the instance set is created, the **Blend All Instances** (in the blend creation dialogue) or **Chamfer all instances** options allow you to add the blend or chamfer to all the *current* instances in that set. These options have no effect on instances created after the blend or chamfer.

- You cannot instance the following objects: hollows, blends, chamfers, trimmed features, datums, instance sets, taper features, free form features, offset sheets, and trimmed sheet bodies,
- The feature you select to instance is replaced by a member of the instance array. For example, if you select a hole and specify a rectangular array of 3 x 4, the Model Navigator replaces it with one array feature and 12 instance features, one of which is the original selected hole.
- You can create an instance array from an existing feature or an existing instance.
- Features that are instanced must reside within the target body. In the event you are instancing the target body itself, the created instances must intersect as each is created. You cannot instance a feature in relation to another target solid.
- Before creating the instance array, the system temporarily displays the position of the instances based on the parameters you input. (If you selected multiple features to instance, only one of them is displayed, to save time.) At this point, you can accept these results and create the instances by choosing Yes, or re-specify the positioning before the system adds the feature to the model by choosing No.
- You can edit an instance using the **Edit →Feature → Edit Parameters** option. To edit the size parameters of a set of feature instances (height, depth, width, etc.), choose the **Feature Dialog** option. To edit the parameters of an instance array (number of instances, distance, angle, reference point, etc.) choose the **Instance Array Dialog** option.
- If the master feature is not fully positioned, you may use the **Edit → Feature → Move** option to move instances. This move applies to the entire instance array.
- You can delete the entire array by choosing the array feature (not an instance feature) in the Model Navigator with MB3 and choosing the Delete option in the submenu.

11.1.1 Rectangular Array

This option lets you create a linear array of instances from one or more selected features. Rectangular instance arrays can be either two-dimensional in XC and YC (several rows of features) or one-dimensional in XC or YC (one row of features). These instance arrays are generated parallel to the XC and/or YC axes of WCS based on the number and offset distance you enter. You can change the orientation of the WCS (the XC and YC directions) using **WCS** → **Origin**, **Rotate**, or **Orient** as discussed in Chapter 3.

After you select the desired features to instance a rectangular array, the following parameter dialogue appears:

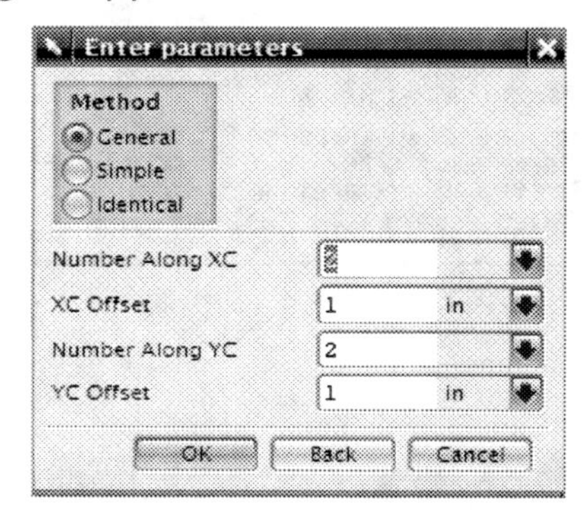

The field Number Along XC defines the total number of instances to be generated parallel to the XC axis of the WCS. This number includes the existing feature you are instancing. XC Offset defines the spacing for the instances along the XC axis. This spacing is measured from a point on one instance to the same point on the next instance along the XC axis. Negative values position the instances in a negative direction along the axis. Number Along YC works the same as Number Along XC (except YC in place of XC). This is true for YC Offset as well. As an example, to create a one-dimensional array in the XC direction, set the XC value to the number required and set the YC value to one.

The down arrow is called the DesignLogic Parameter Entry Options button for whenever you're creating a feature and want to enter a formula, reference an existing value or derive a value from a measurement, instead of cutting and pasting values into the parameter entry field.

The figure below shows a 3 by 4 rectangular array of 12 instances. Each instance is spaced .75 in the XC direction and 1.0 in the YC direction.

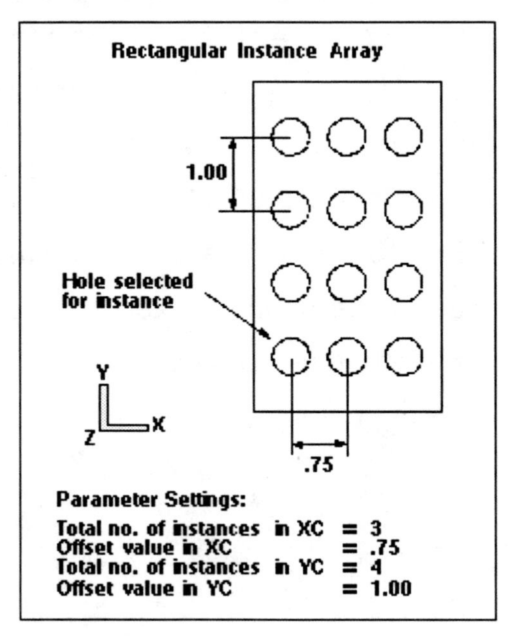

Activity 11-1. Creating a rectangular array of instances

In this activity, you will create a row of five hole instances 7 inches apart between any two adjacent holes.

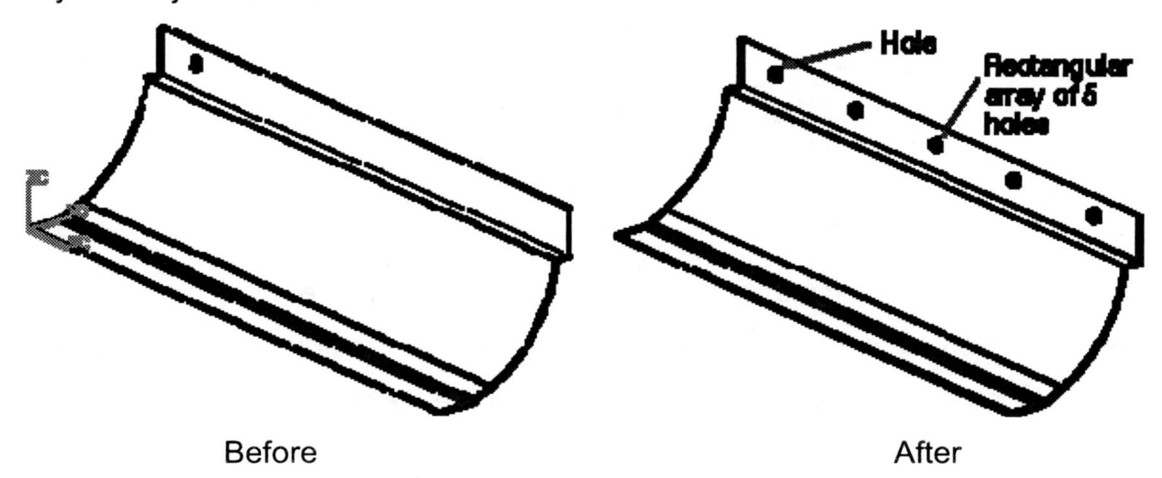

Before After

Step 1. Open activity11-1.prt and save it as **xxx_ activity11-1.prt**.

Step 2. At the **Modeling** application, choose **Insert → Associative Copy → Instance** or the **Instance** icon  .

Step 3. Choose the **Rectangular Array** option.

Step 4. Choose **SIMPLE_HOLE(2)** listed in the dialog as the feature to instance and then **OK**.

Step 5. The instance parameter dialogue appears. Key in the following values:

Number Along XC = 5

XC Offset = 7

Number Along YC = 1

YC Offset = 0

Choose **OK**. A temporary display of the elongated hole locations are displayed in the graphics area.

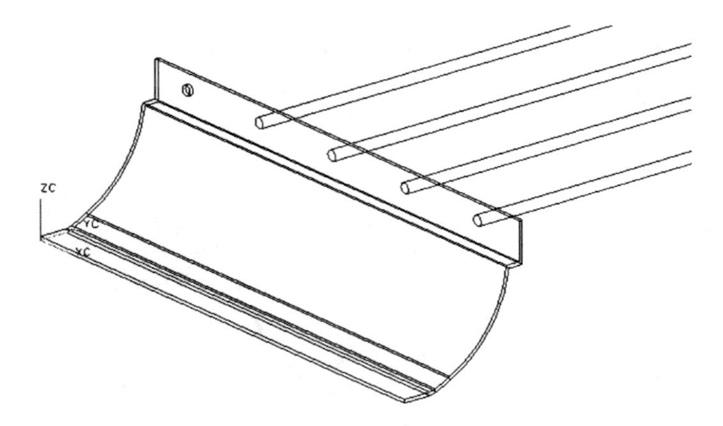

Step 6. Verify the hole instances are created. Because these holes are located where you need them, you can complete the instance array by choosing **Yes** on the **Create Instances** dialog. **Cancel** the **Instance** dialog.

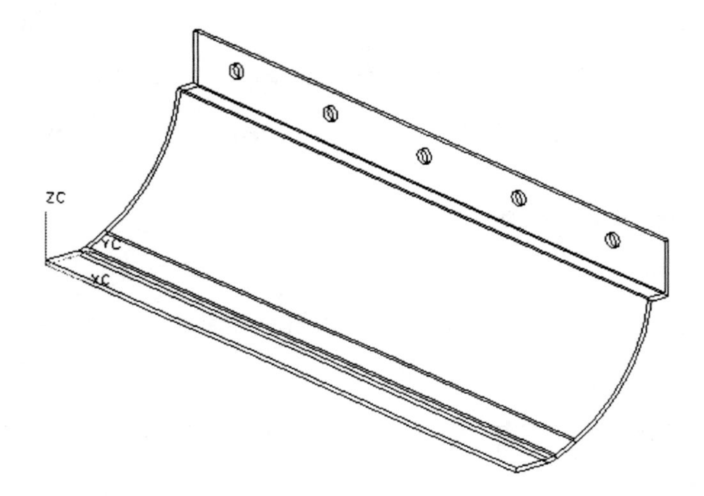

Step 7. **Save and close** this file.

11.1.2 Circular Array

This option lets you create a circular array of instances from one or more selected features. The procedure to create a circular array is as follows:

1. Select the features you want to instance.
2. In the Enter Parameters dialog, specify the array method (**General**, **Simple**, or **Identical**), the total **Number** of instances (including the original feature) and the **Angle** between instances. Then, choose **OK**.

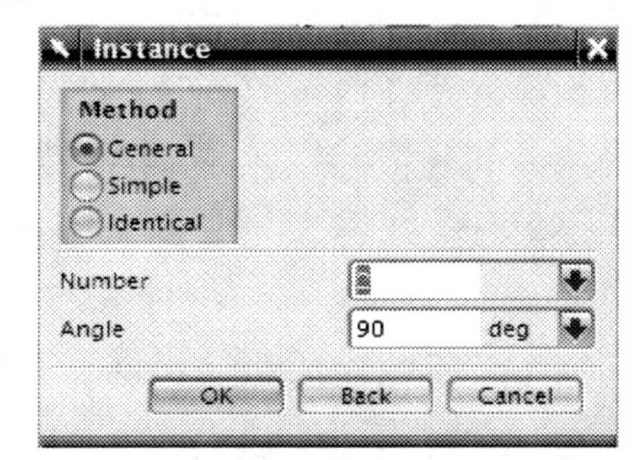

3. Choose **Point_Direction** or **Datum Axis** to establish the rotation axis.

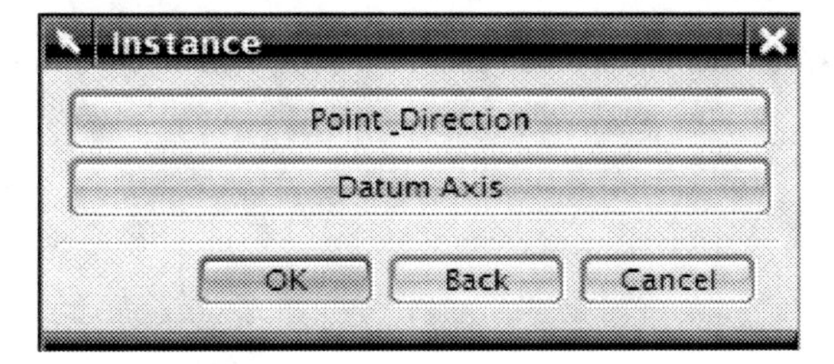

If you choose **Point_Direction**, use the *Vector Constructor* to establish a direction and the *Point Constructor* to establish a reference point. If you define the axis using the *Vector Constructor*, you can change it to a datum axis later using **Edit → Feature → Edit Parameters** and selecting the instance. An example for this option is depicted below.

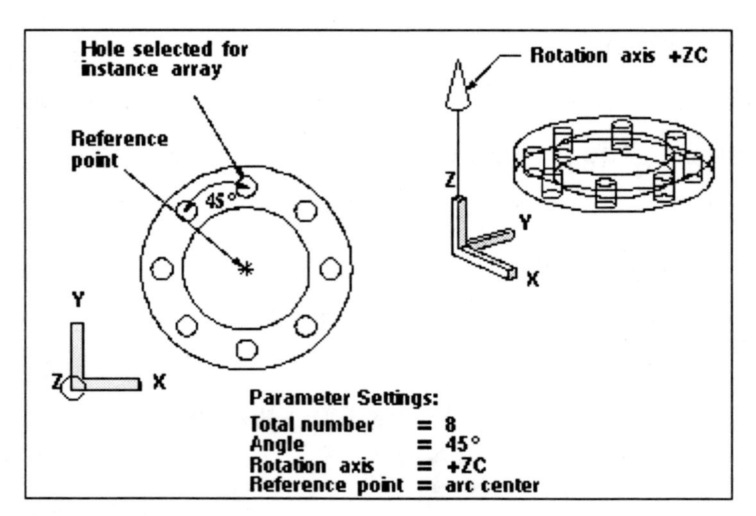

If you choose **Datum Axis**, select a datum axis. The radius of the array is calculated as the distance from the datum axis to the local origin of the first feature you selected. The rotation axis of the array will be associative to the geometry used to define the datum axis.

4. A highlighted representation of the array is displayed. Choose **Yes** to create the instance array, or **No** to return to the Enter Parameters dialog.

Activity 11-2. Create a circular array of instances

In this activity, you will create instances of some features using a circular array to quickly model a desired part as shown below.

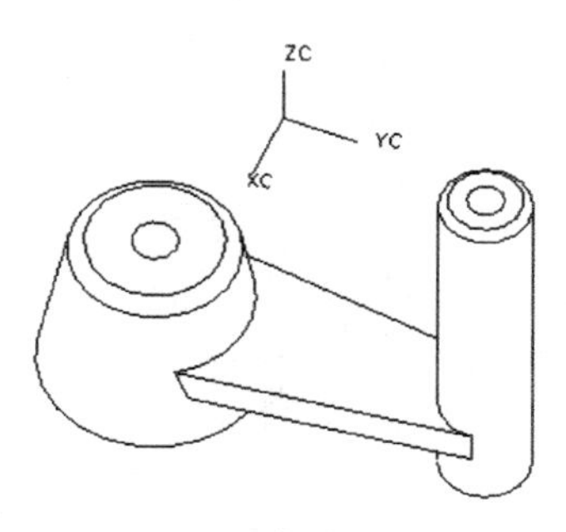

Before

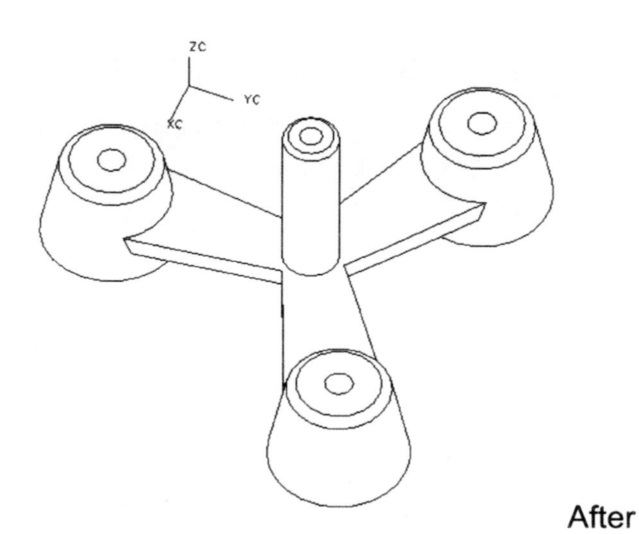

After

Step 1. **Open activity11-2.prt**, and save it as **xxx_ activity11-2.prt**.

Step 2. Create a datum axis that will be used as the rotation axis for the circular array.

2.1. Change the work layer to **62**.

2.2. Choose the **Datum Axis** icon, select the long cylinder face on the graphics window, and choose the **OK** icon. The datum axis is created as shown below.

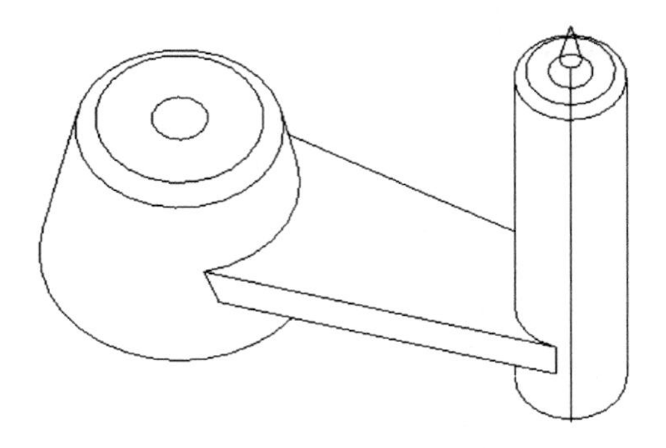

Step 3. Change the work layer to **1** and choose **Insert → Associative Copy → Instance**.

Step 4. Choose the **Circular Array** option. The following **Instance** dialogue appears, displaying features for which you can create instances.

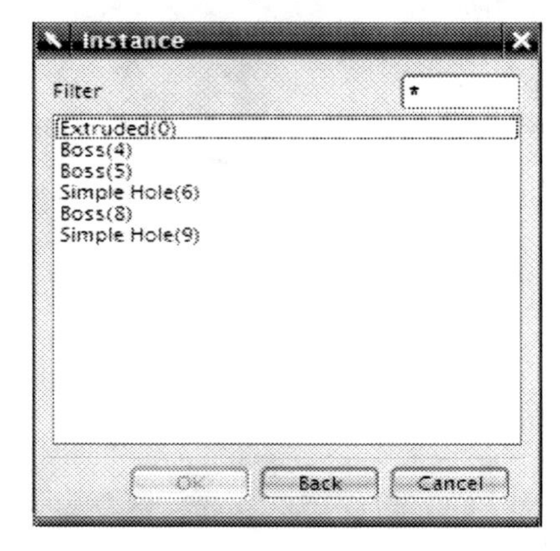

Note that the Chamfer feature is not listed in this dialogue since it is one of the features that cannot be instanced.

Step 5. Select three features that you want to instance.

5.1. Select the EXTRUDED(0), BOSS(8), and SIMPLE_HOLE(9).

To select multiple features, press the Ctrl key and select features that you want to add to instance. Notice that as you add them, they are highlighted in screen in red.

5.2. Choose **OK**.

Step 6. Enter the following instance parameter values:

Number to be **3** and **Angle** to be **360/3**.

Choose **OK**.

Step 7. The cue line prompts you to select a rotation axis. Choose the **Datum Axis** button.

Step 8. Select the datum axis (one created in Step 2) from the graphics window.

Step 9. A highlighted representation of the array is displayed as shown below.

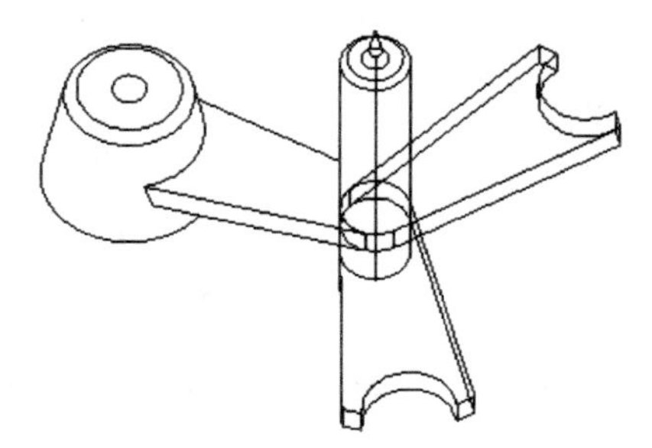

9.1. Choose **Yes** to create the instance array.

9.2. Choose **Cancel** to close the **Instance** dialogue.

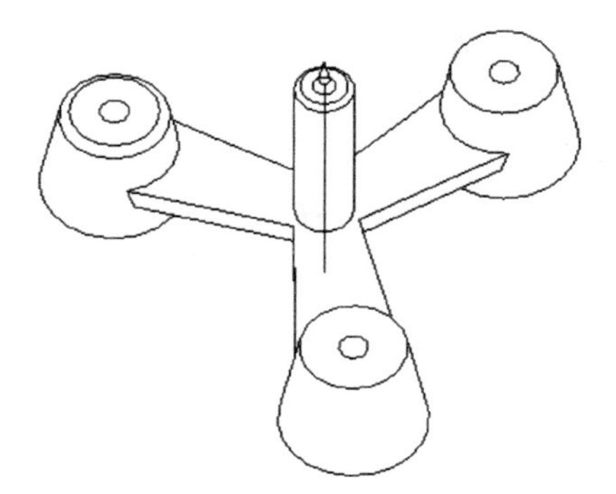

Step 10. Create a single-offset chamfer with offset distance=3 to the circle edges of the two instanced bosses (refer to Section 7.2 for detail on Chamfer). If you select just one circle edge (either one), put a check mark on the field "Chamfer All Instances". Choose **Apply**.

You have completed the part and choose **Cancel** to close the Chamfer dialog.

Step 11. Now edit this instance feature to make **6** instances instead of **3**.

11.1. Open the **Part Navigator (PN)** window by clicking the icon once. Or you can click the PN icon twice and anchor on the left as follows.

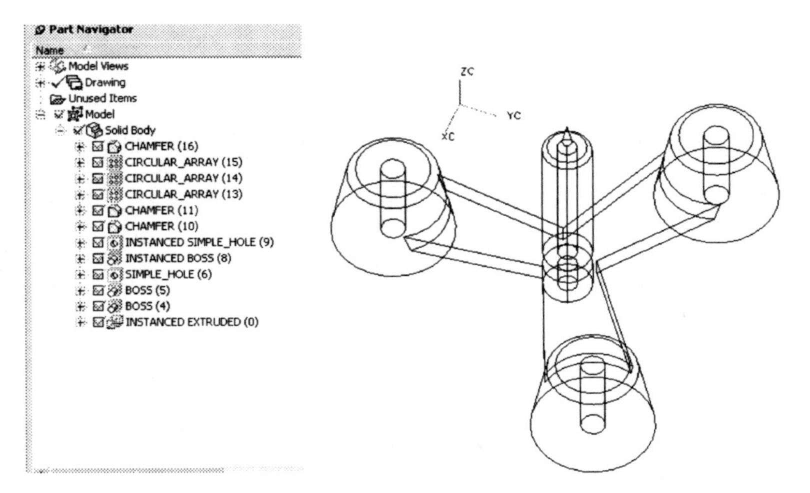

11.2. Move your cursor over the feature **INSTANCED BOSS 8** in the **PN** window and click the right mouse button (MB3) once. The following submenu appears.

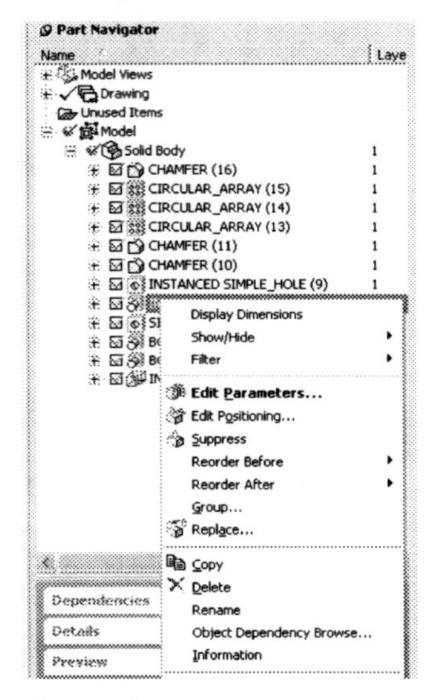

11.3. Move your cursor over **Edit Parameters** and click the left mouse button (MB1) once. The following dialog appears.

11.4. Choose the **Instance Array Dialog** button.

Now the circular array dimensions are displayed. However, the Angle is not available to edit.

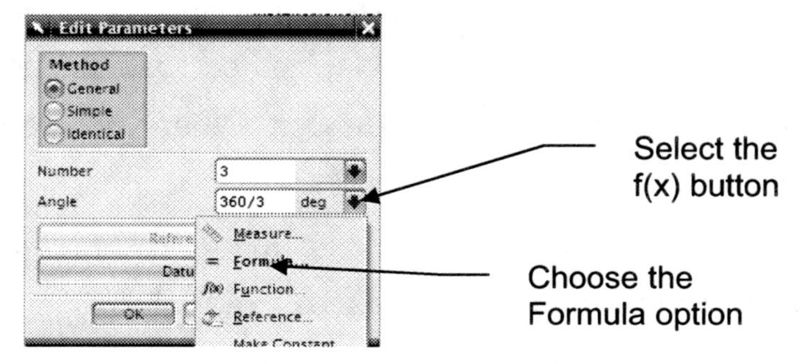

11.5. Enter **Number** to be **6.**

11.6. There is a formula in the number field so the formula must be edited. Select the f(x) button as shown, then the option **Formula** and the Expressions window comes up to edit the expression formula.

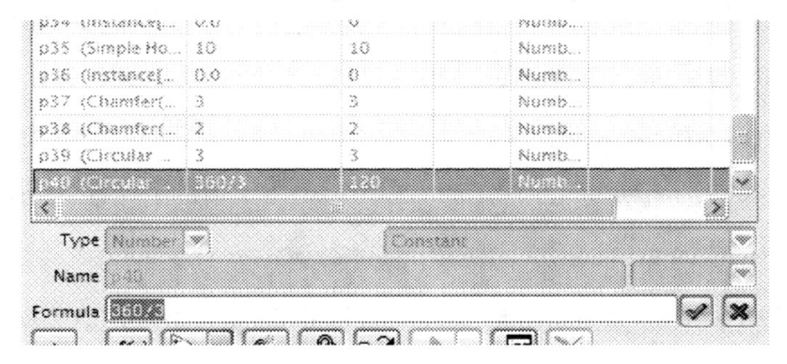

11.7. Correct the Formula to be 360/6 and choose **OK** three times.

The part is completed.

Step 12. **Save and close** your part file.

11.2 Mirror Body and Mirror Feature

Mirror Body (**Insert → Associative Copy→ Mirror Body**) lets you mirror an entire body about a datum plane. You can use this, for example, to form the other hand of a left-hand or right-hand part. When you use this option, the system creates a feature whose name is Mirror.

The following statements describe the Mirror feature and its relationship to the original body and the datum plane:

- When you mirror a body, the Mirror feature is associative to the original body - it has no editable parameters of its own. If you change the parameters of a feature in the original body, causing the original body to change, those changes are reflected in the mirrored body.
- If you edit the parameters of the associated datum plane, the mirrored body changes accordingly.
- If you delete the original body or datum plane, the mirrored body is also deleted. If you move the original body, the mirrored body also moves.
- You can add features to the mirrored body. However, be sure that you do not reorder these features, such that they would occur before the Mirror feature.
- You can combine the original and mirrored bodies, using the **Unite** option, to create a symmetrical model, as shown in the figure below.

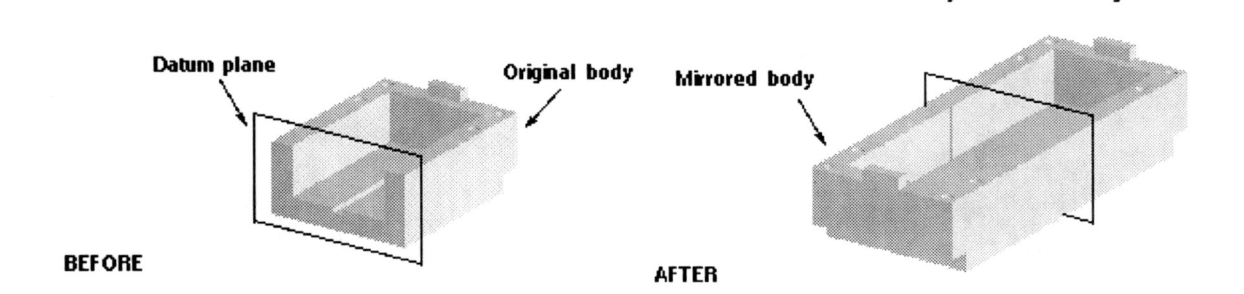

Activity 11-3. Creating a mirror body

In this activity, you will use the Mirror Body option to create the left half of this part. The results are two solid bodies. Because these two solids will have coincidental faces at the datum plane, they can be united.

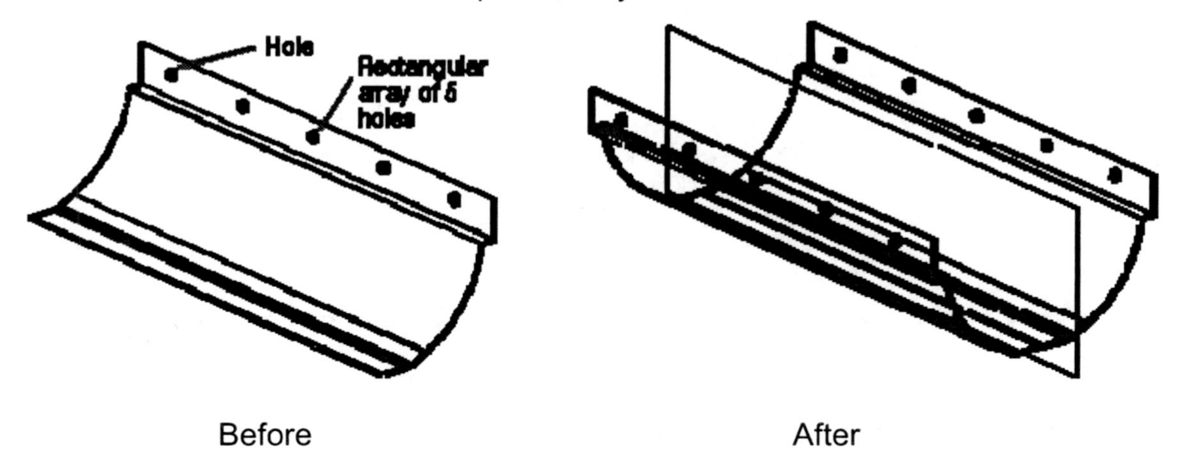

Before After

Step 1. Open xxx_activity11-1.prt that you created in the last activity, and save it as **xxx_ activity11-3.prt**.

Step 2. At the **Modeling** application, change the work layer to **61** and create a datum plane coincident on the face depicted above (i.e., offset value 0).

Step 3. Change the work layer to **1** and choose **Insert → Associative Copy → Mirror Body** or the **Mirror Body** icon.

Step 4. The **Selection** bar displays to help in your selection. Select the green solid body.

Step 5. For **Mirror Plane**  select the datum plane and choose OK. The mirrored body is created as shown above. Cancel the Mirror Body dialog.

Step 6. Unite the two bodies.

6.1. Choose **Insert → Combine Bodies → Unite.**

6.2. Select the solid on the right side for the target body.

6.3. Select the solid on the left side for the tool body and choose **OK**.

Step 7. Delete the **Rectangular Array** feature by opening the Part Navigator and choosing this array feature with **MB3** and choosing **Delete** in the submenu. To

open Part Navigator, select the icon from the resource bar on the right side (refer to Section 6.2 if you want detail on the Part Navigator). Observe the change in the mirrored body as shown below.

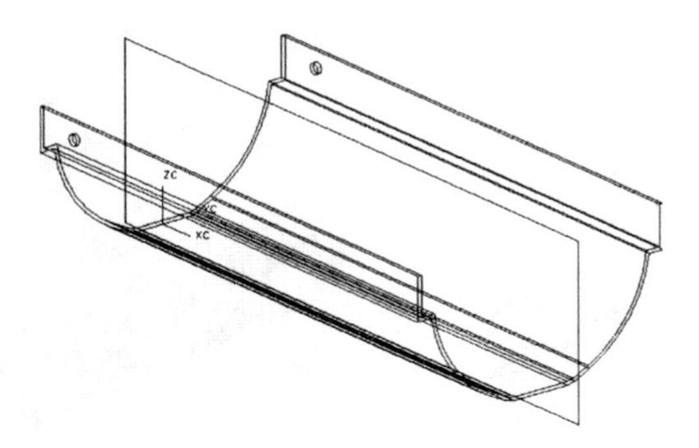

Step 8. Change the value of the expression **p10** from 10 to 30. In this part, the extrusion was created from a sketch and sketch dimension **p10** controls the height of the vertical face.

8.1. Choose **Tools → Expression** and change the Listed Expressions filter in the dialog to All.

8.2. Highlight the expression **p10** in the list of expressions in the dialogue.

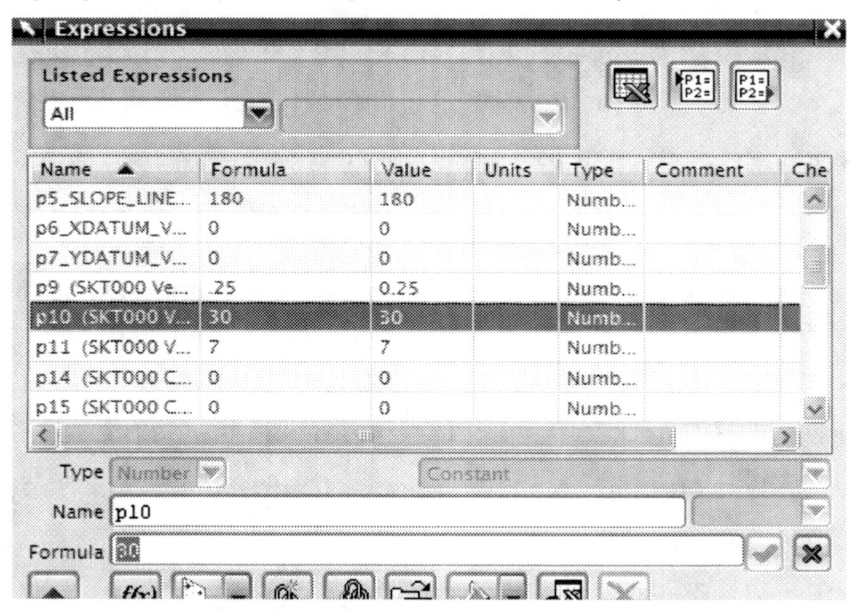

8.3. In the Formula value field beneath the list of expressions, change it to **30** and choose **OK**.

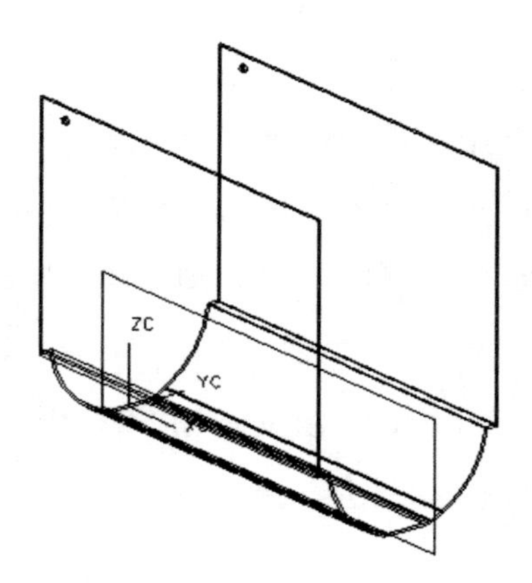

The two previous steps illustrates that the mirrored body is associative to the original body. If you delete the original body or datum plane associated with a mirrored body, the system deletes the mirrored body.

Step 9. **Save and close** this file.

Mirror Feature (**Insert → Associative Copy→ Feature**) lets you create symmetrical models by mirroring *selected features* through a datum plane or planar face. To create a simple mirrored body you would normally use the **Mirror Body** option. **Mirror Feature**, however, lets you mirror features within a body. Output from this option is a feature named MIRROR_SET. During edit of a MIRROR_SET feature, you can redefine the mirror plane and add/remove features to/from it.

The Mirror Feature dialog looks like this:

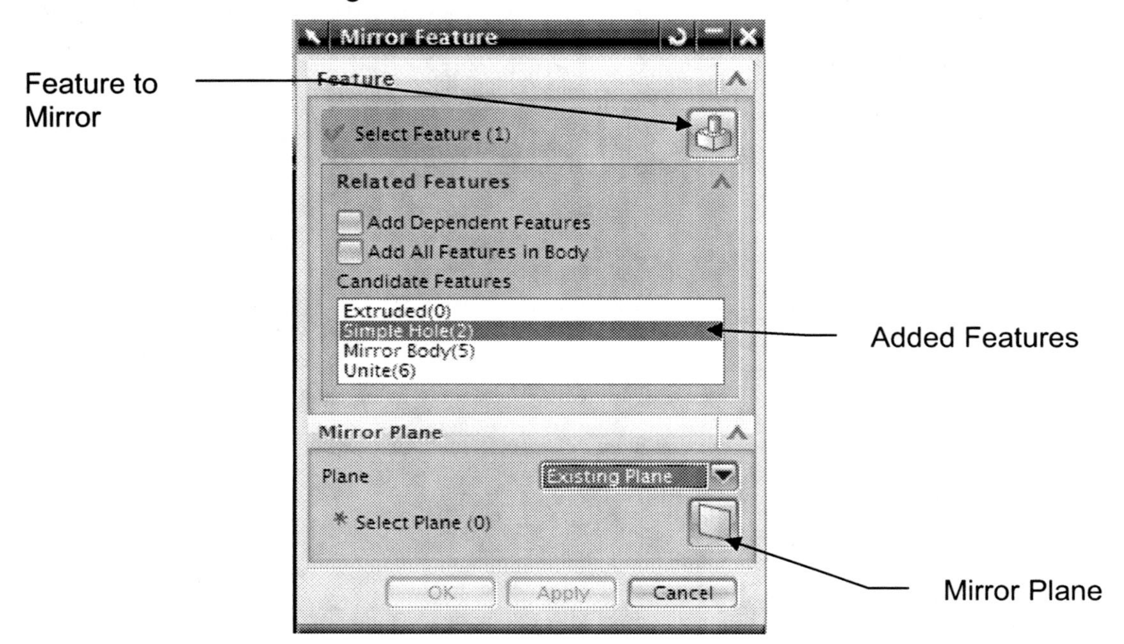

Mirror Feature Procedure

1. From the Mirror Feature dialog, click the **Feature to Mirror** selection step option (see the above figure) if necessary.
2. If desired, turn on the **Add Dependencies** option, to include feature dependencies of the selected features that are to be mirrored.
3. If desired, turn on the **All in Body** option, to mirror all features present in the body.
4. Highlight the features in the **Candidate Features** listing that you wish to Mirror, and then click to add them to the listing. You can make multiple selections by highlighting multiple items in the **Candidate Features** listing

and holding the **Control** key while clicking MB1. You can remove features in the **Candidate Features** listing by holding the **Control** key again while clicking MB1.

5. Click the **Mirror Plane** selection step option (see the above figure). Move the cursor to the graphics window and select the datum plane or planar face to be used to reflect the feature during the mirror operation.
6. Click **OK** or **Apply**. The features in the **Candidate Features** listing are mirrored across the mirror plane. The result is a separate body composed of a single MIRROR_SET feature. If desired, you can use **Unite** to join the new body with the other solid body in the part file.

11-4. Creating a mirror feature

In this activity, you will use the Mirror Feature option to create the left half of this part with selected features.

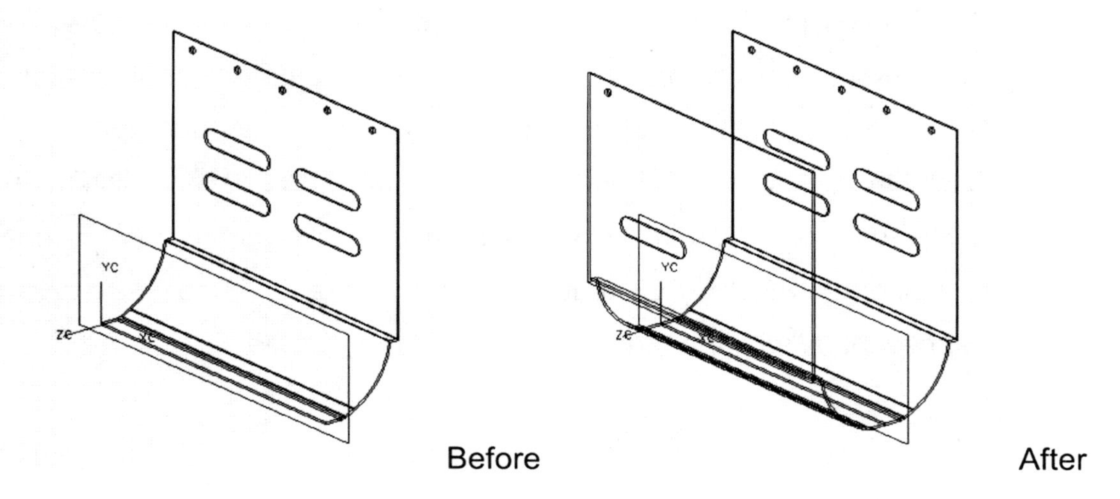

Before After

Step 1. **Open activity11-4.prt** and save it as **xxx_ activity11-4.prt**.

Step 2. After choosing the **Modeling** application,

choose **Insert → Associative Copy → Mirror Feature**.

Step 3. Turn off the **Add Dependencies** and **All in Body** options if necessary.

Step 4. Select the features to mirror.

4.1. Highlight the following three features from the **Candidate Features** listing:

EXTRUDED(0),

INSTANCE[0](1)/SIMPLE_HOLE(1),

INSTANCE[0](6)/RECTANGULAR_SLOT(6).

When you highlight more than one feature, hold down the **Ctrl** key and click features to mirror. To deselect a highlighted feature, do the same.

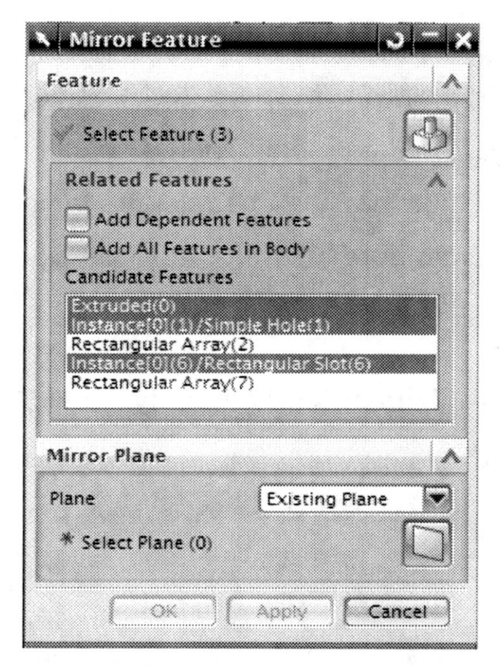

4.2. Choose the **Mirror Plane** icon on the **Selection Steps** window near the top of the dialog.

4.3. Select the datum plane.

4.4. Choose **OK**.

The mirrored features are created as shown above. Cancel the **Instance** dialog.

Step 6. Unite the two bodies.

6.1. Choose **Insert → Combine Bodies → Unite.**

6.2. Select the solid on the right side for the target body.

6.3. Select the solid on the left side for the tool body.

6.4. Choose **OK**.

Step 7. **Save and close** this file.

11.3 Trim Body

This feature operation trims one or more target bodies using a face, datum plane or other geometry. You select which portion of the bodies you want to keep, and the trimmed bodies take the shape of the trimming geometry.

We give the general procedure below for the Trim Body operation. Detail descriptions follow the steps.

1. Choose **Insert → Trim** or the Trim Body icon . The following dialog appears.

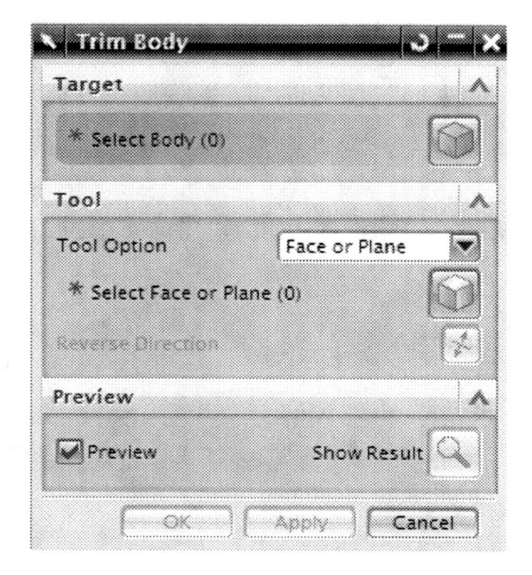

2. Select at least one target body and choose the tool icon in the Selection Steps.
3. Select a face or datum plane, or define other geometry to trim the target bodies.
4. A vector is displayed. The portion of the target bodies in the direction of the vector will be removed. Choose OK to accept the direction of the vector or reverse it by choosing the reverse trimming direction icon .

For step 3, you can select from the graphics window the existing face or datum plane as the trimming object. One important point to remember is the associativity issue. If you intend to maintain associativity of the trim to the solid body, you must use for the trimming object relative datum planes or features created through the use of the Geometry Linker (will be discussed in Assembly Modeling in Chapter 12).

The Trim Body operation can fail under certain conditions. When you trim a body using a face, the face must be large enough to cut through the body completely. If it does not, the following error message is displayed: Non-Manifold Solid. Also the trim operation may fail if:

- A face of the trimming sheet is tangent to the face of the solid body.
- The selected trimming geometries are faces from different solid bodies.
- The selected trimming geometries are disjoint (i.e., unconnected) faces from a solid body.

For Step 4, the direction of the normal vector determines which portion of target bodies is kept. The vector points away from the body portion that will be kept. This is illustrated by the below figures in which the trimming object is defined by a cylinder face. You can change the direction of the vector by choosing the reverse trimming direction icon 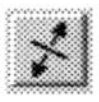 in the Trim Body dialog.

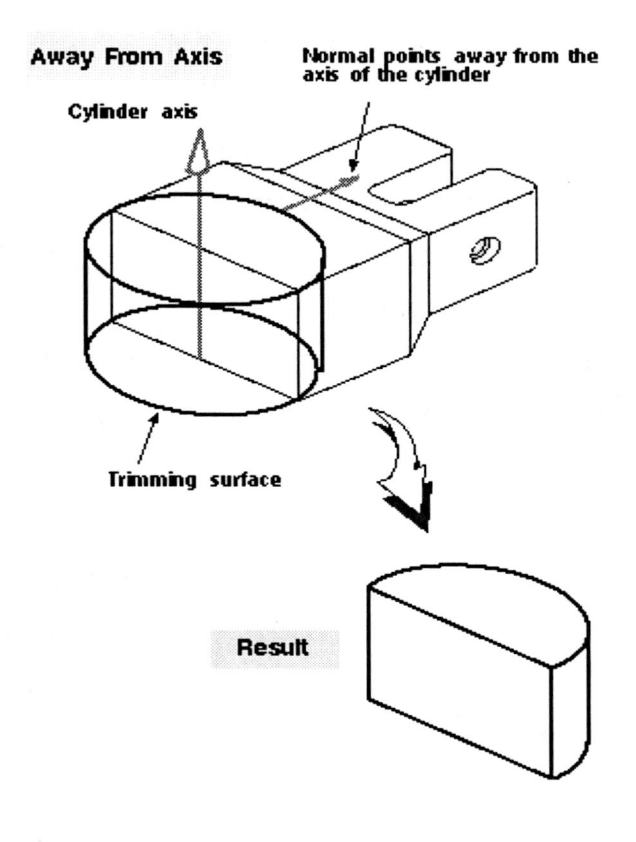
Away From Axis
Normal points away from the axis of the cylinder
Cylinder axis
Trimming surface
Result

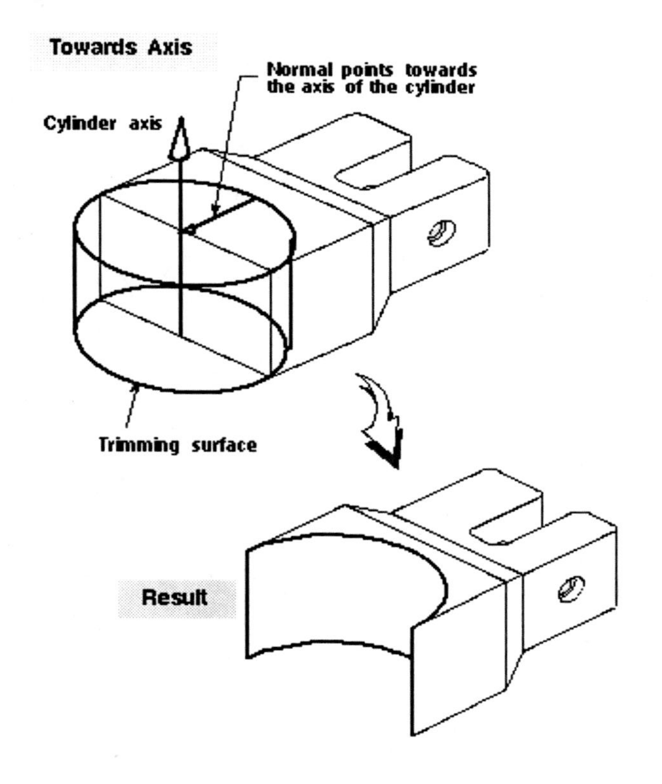
Towards Axis
Normal points towards the axis of the cylinder
Cylinder axis
Trimming surface
Result

Activity 11-5. Trimming a Solid Body

This part, shown below, is a pin with an angled thru hole. There are four datum planes and one datum axis that are used to create the hole. You are going to trim this pin with the angled datum plane.

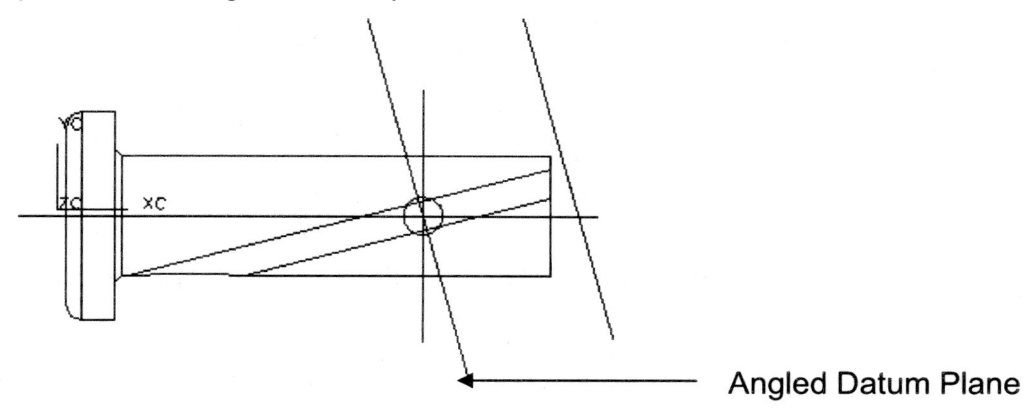

*Note:** You need to be in the **Role** Advanced with full menus.

Step 1. **Open activity11-5.prt** and save it as **xxx_activity11-5.prt**.

Step 2. At the **Modeling** application,

choose **Insert** → **Trim** → **Trim Body** or the **Trim Body** icon .

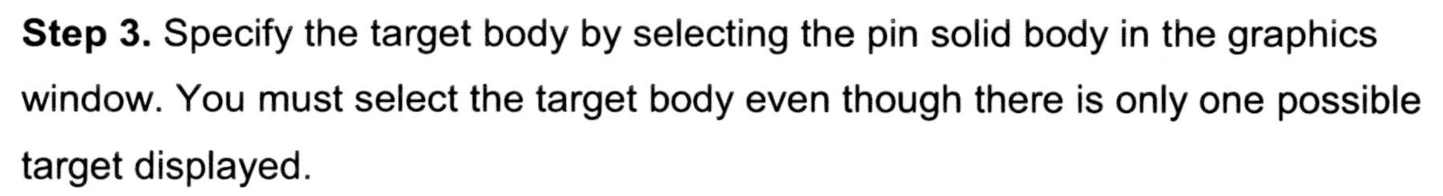

Step 3. Specify the target body by selecting the pin solid body in the graphics window. You must select the target body even though there is only one possible target displayed.

Step 4. Choose the tool body icon in the Selection Steps. Specify the trim body by selecting the angled datum plane that was indicated above. The trim direction vector appears.

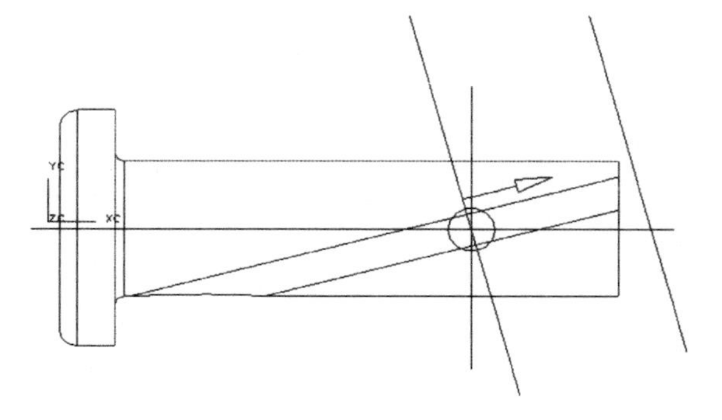

Step 5. Choose **OK** to accept the default trim direction. The resulting model looks like the below figure.

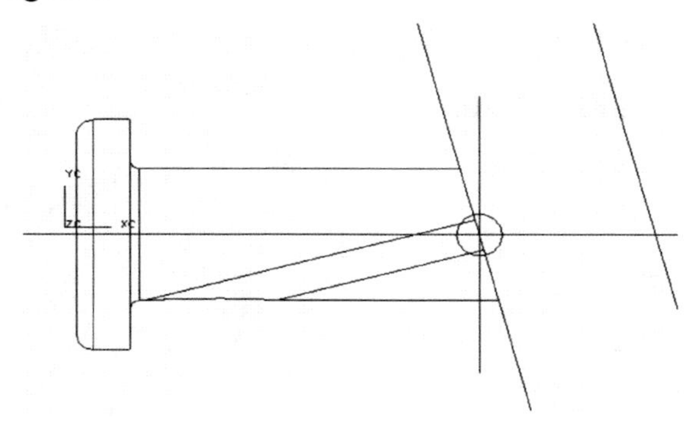

Step 6. Edit the trim body (the angled datum plane) by changing its angle from –15 to –45.

6.1. Choose **Edit → Feature → Edit Parameters** and select in the feature list **Datum_Plane (8)**, which is the trim body. (Alternatively, you can select it in the graphic screen). Choose **OK.**

6.2. The **Datum Plane** dialogue appears. Key in **–45** in the **Angle** field.

6.3. Choose **OK** twice. The trim body has been updated, which affects the trimmed solid pin because of the associativity between the angled datum plane for the trim body and the pin for the target body.

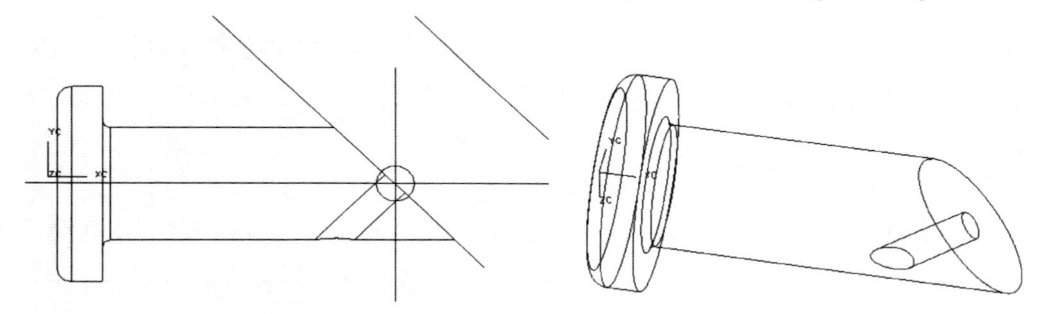

Step 7. **Save and close** the file.

11.4 Thread

This feature operation creates threads on features with cylindrical faces. These features include holes, cylinders, bosses, and subtractive or additive sweeps of circular curves. Some key parameters of threads are shown in the following figure.

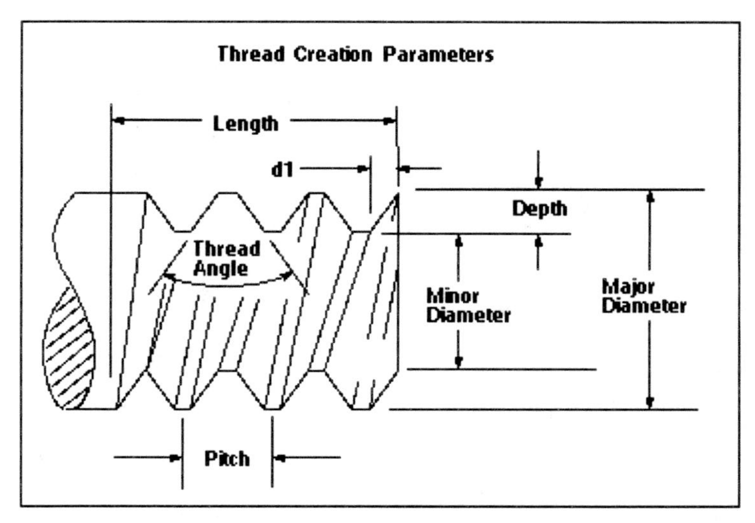

Threads can be right-handed or left-handed. A right-handed thread winds in a clockwise direction while a left-handed thread winds in a counterclockwise direction. This is illustrated by the following figure.

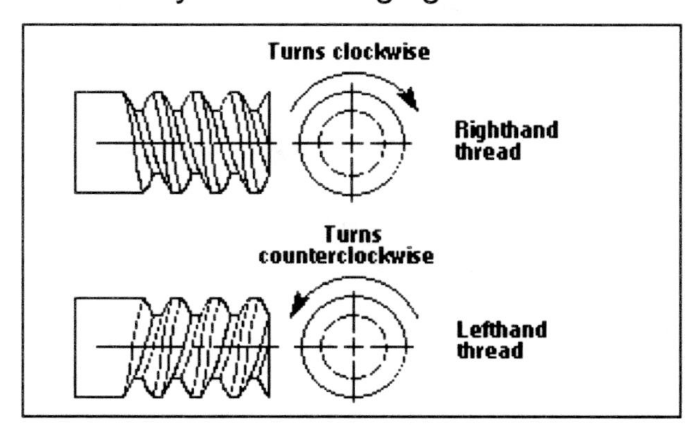

This feature operation can create two different types of threads: symbolic and detailed. Symbolic threads are displayed as a dashed circle on the face or faces that will be threaded. Symbolic threads get their values from external look up tables that are provided in the **thd_english.dat** and **thd_metric.dat** files. The files are from one industry standard: Machinery's Handbook, 25th edition, 1996, published by Industrial Press Inc. There are other industry standards[1] and you can modify these tables as required. Symbolic threads cannot be copied or instanced once created, but you can create multiple and instanced copies at creation time. See the figure below for examples of symbolic threads.

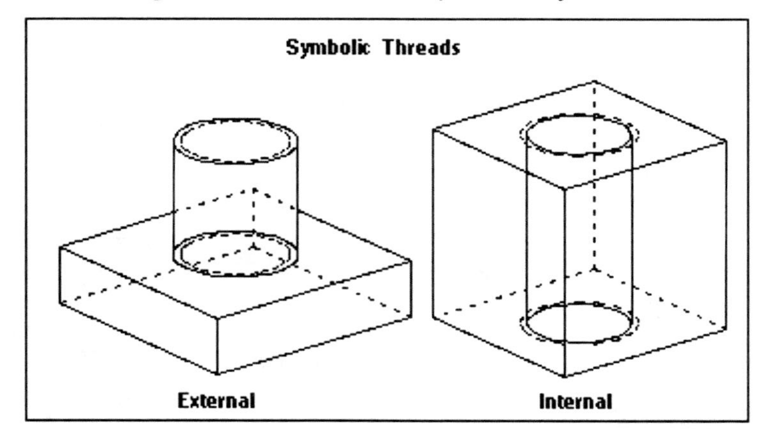

Detailed threads look more realistic but also take considerably longer to create and update because of their complex geometry and display. Detailed threads use embedded tables for the default parameters. Detailed threads must be created one at a time, but can be copied or instanced after creation. Detailed threads are fully associative; if the feature is modified, its thread updates accordingly. This is not true for symbolic threads. See the figure below for an example.

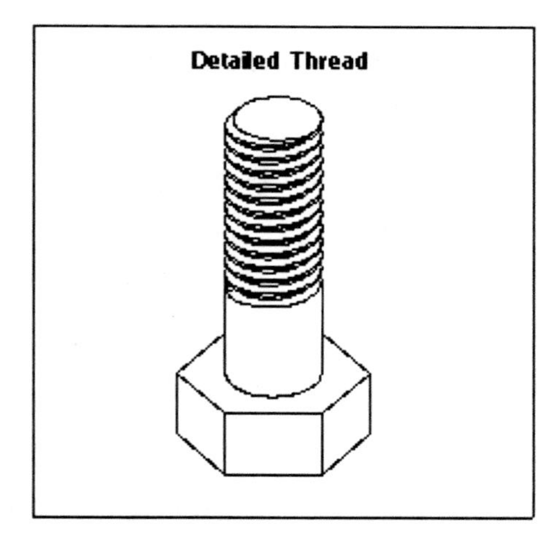

There are certain guidelines to follow for creation of threads.

---

1 Engineering Design Communication by S.D. Lockhart and C.M. Johnson, 2000, Prentice Hall, Inc.

- You can only create threads on previously unthreaded cylindrical faces.
- If the cylindrical face has only one available planar end, it will be used as the start of the thread.
- If a cylindrical face has two planar (flat) ends that could be the start of the thread, your pick location will indicate the starting location and direction of the thread as illustrated below.

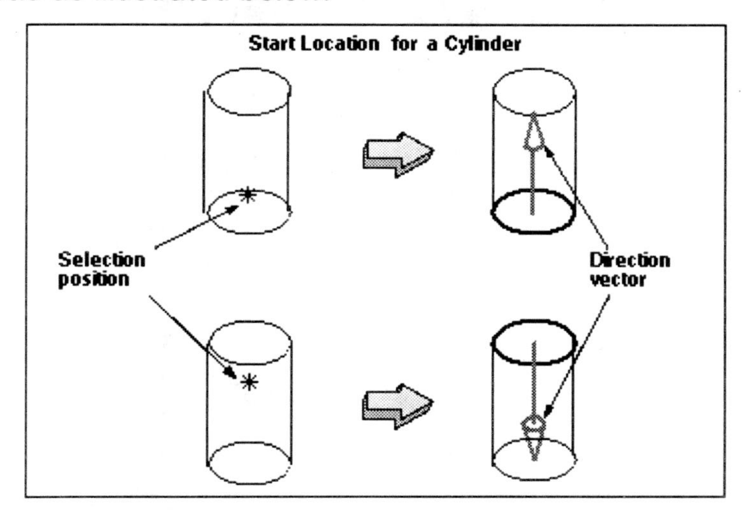

- If the selected cylinder has been united to a larger body, a non-united end is the start for the thread (see the figure below). If the other body is smaller than the selected cylinder, the end closest to the selection position will be the start face.

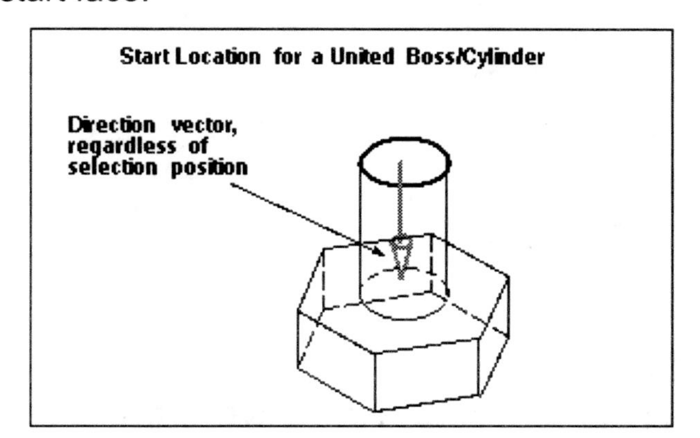

- The selected cylinder must have planar ends. If the system cannot find a planar end to be used as the start face of the thread, you must select a start face that is planar (for example, a datum plane).

Procedure to create threads

At the Modeling application, choose **Insert → Design Feature → Thread** or the **Thread** icon . The thread creation dialogue appears. Different options are available depending on whether you choose a Symbolic or Detailed thread.

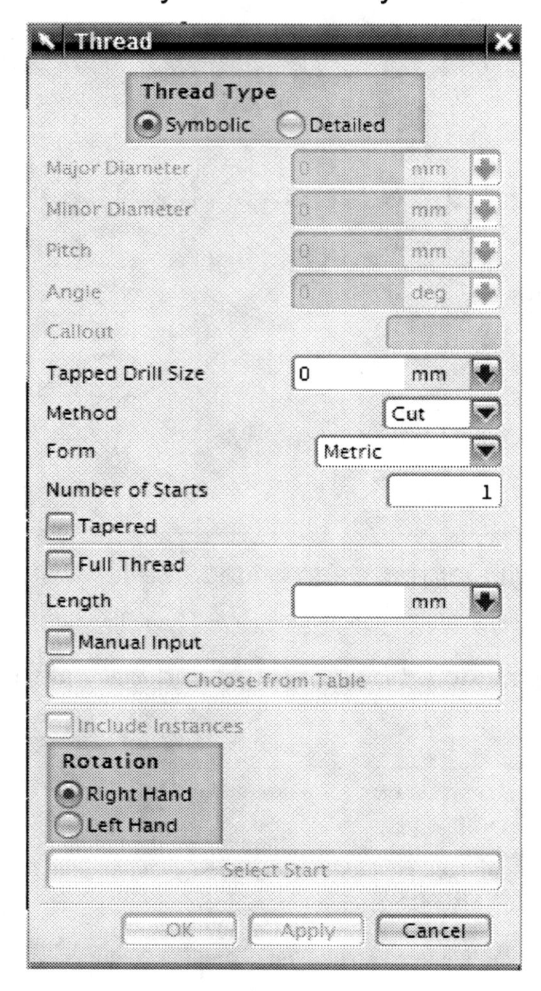

To create symbolic threads:

1. Choose **Symbolic** at the top of the dialogue.
2. Choose the thread manufacturing **Method**. Four manufacturing methods are available. They are Cut, Rolled, Ground, and Milled.
3. Choose the **Form** for the thread. This determines which lookup table is used to obtain the parameter defaults. Examples of Form options include Unified, Metric, and Acme.

4. Select one or more cylindrical placement faces. Based on Form and on the first selected face, default values for **Major Diameter**, **Minor Diameter**, **Pitch**, **Angle**, **Length**, and **Shaft Size** (for external threads) or **Tapped Drill Size** (for internal threads) appear. Most of the parameters that were grayed out are now active. If you do not want default values to be supplied from a lookup table, choose **Manual Input** and enter the parameters. **Number of Starts** specifies whether this is a single thread or multiple threads.
5. Modify the parameters as desired. (Some parameters, such as Callout, cannot be modified directly.)
6. **Callout** references the thread table entry that provides the default values. **Choose from Table** lets you choose a different entry (and, therefore, a different set of default values). If you edit the parameters manually, **Callout** grays out to indicate that the current parameters do not come directly from a table.
7. Choose **Tapered** if you want the thread to be tapered.
8. If you want the thread to update when the cylinder changes, choose **Full Thread**. (**Length** grays out.)
9. If a selected face belongs to an instance array, you can apply the thread to the other instances by choosing **Include Instances**.
10. Decide how you want the **rotation** of the thread to be, either a **Right Hand** or a **Left Hand** thread.
11. Choose **Select Start** if you want to specify a new **starting location** for the thread, and select a planar face on a solid body, or a datum plane.
12. Choose **OK or Apply**. With **Apply** you can continue to create more threads.

To create detailed threads:

1. Choose **Detailed** for the Thread Type. The following shortened dialogue appears.

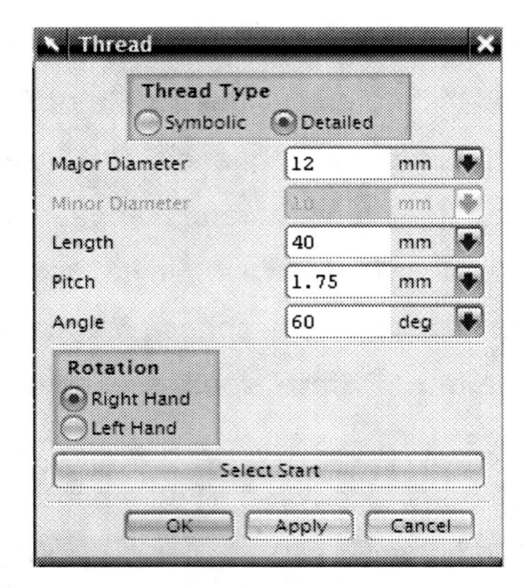

2. Select a cylindrical placement face. Default parameter values appear, based on the diameter of the face.
3. Modify the parameters if required.
4. Choose **OK or Apply**.

Activity 11-6. Creating a thread

In this activity, you will practice creating two threads.

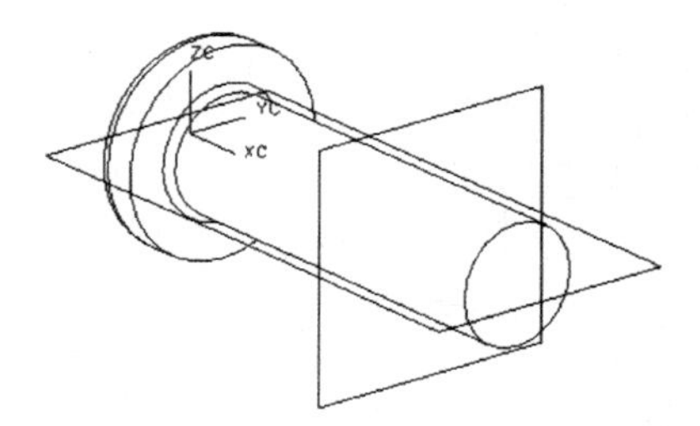

Step 1. **Open activity11-6.prt** and save it as **xxx_activity11-6.prt**.

Step 2. After choosing the **Modeling** application,

choose **Insert → Design Feature → Thread** or the **Thread** icon . The **Thread** dialogue opens.

Step 3. Choose **Detailed** for the thread type.

Step 4. Select the cylindrical face of the pin shaft. Default values of thread parameters appear with the thread direction vector.

Choose **OK** to create a thread on the entire pin shaft.

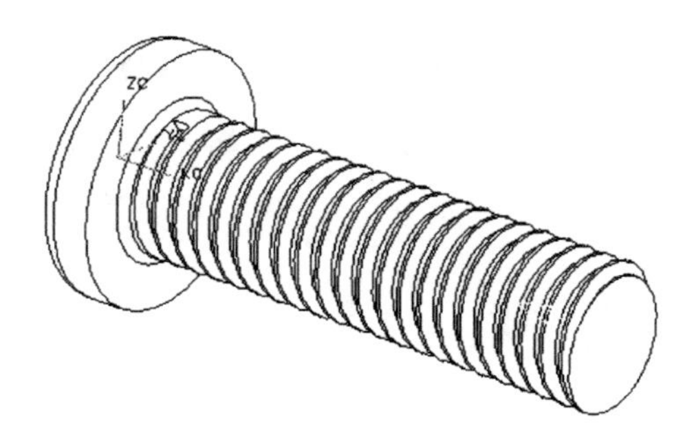

Step 5. Let's **Undo** Step 4 by pressing two keys Ctrl + Z at the same time.

Step 6. Choose the **Thread** icon . Select the same cylindrical face of the pin shaft. Default parameter values appear.

Step 7. Specify a new start face that will be used to create a thread on the partial cylindrical face of the pin shaft.

7.1. Choose button "**Select Start**" at the bottom of the thread dialogue.

7.2. Select the datum plane indicated below.

7.3. Choose **OK** to accept the default thread axis direction as shown below.

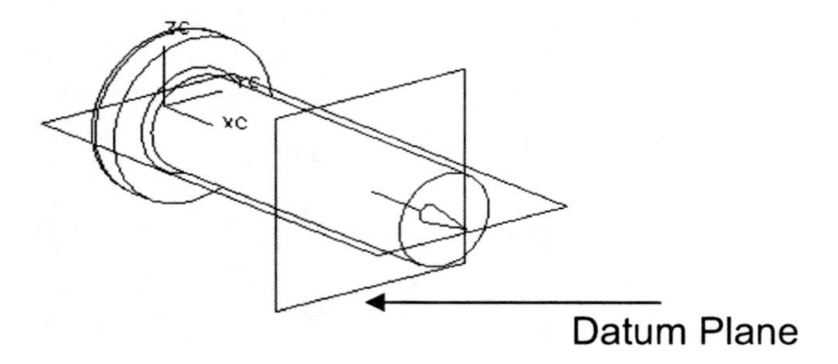

7.4. Choose **OK** to create the thread on the **Thread** dialog.

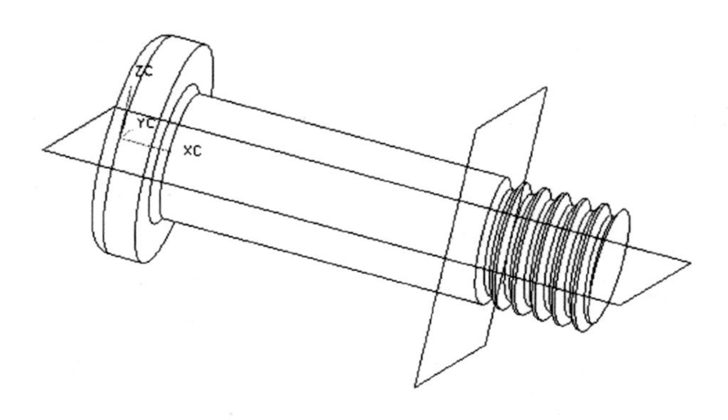

Step 8. Edit the datum plane used in Step 7 to change the offset distance, and observe how the thread feature changes. For example, enter 15 for the new offset distance.

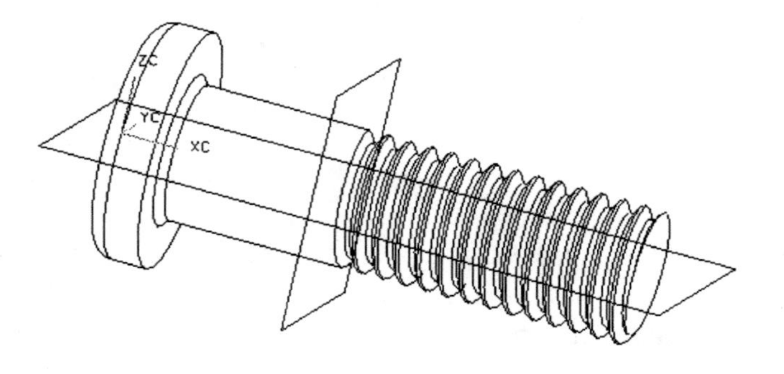

Step 9. **Save and close** the file.

Project 11-1. Ratchet

Model the ratchet part using the drawings (in inches) below and save it as **xxx_project11-1.prt** in your directory.

Hint: You may start with a cylinder. In order to create a gear tooth, you can create a sketch, extrude it, and subtract it from the cylinder. And then you can instance it to create all the gear teeth.

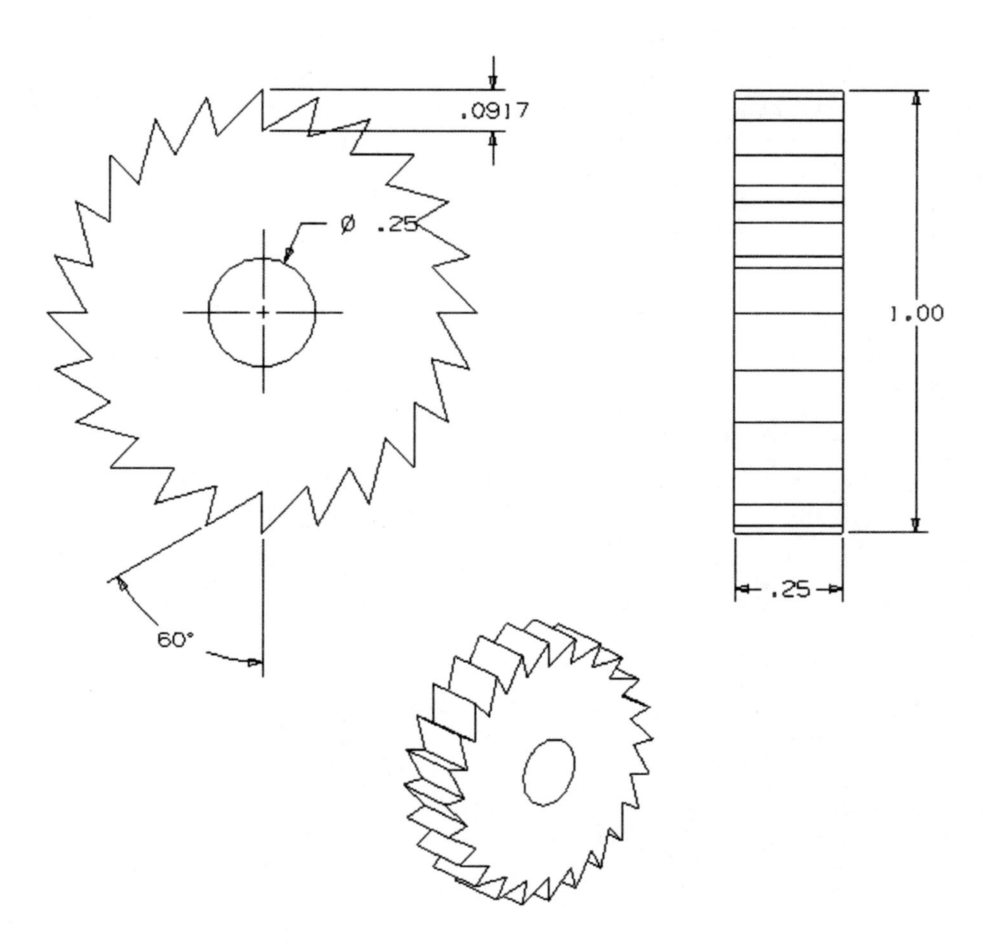

Project 11-2. Protector

Model the protector part using the drawings (in inches) below and save it as **xxx_project11-2.prt** in your directory. Note that all six holes are simple thru holes.

Hint: You may find it useful to use features like rectangular array and mirror body.

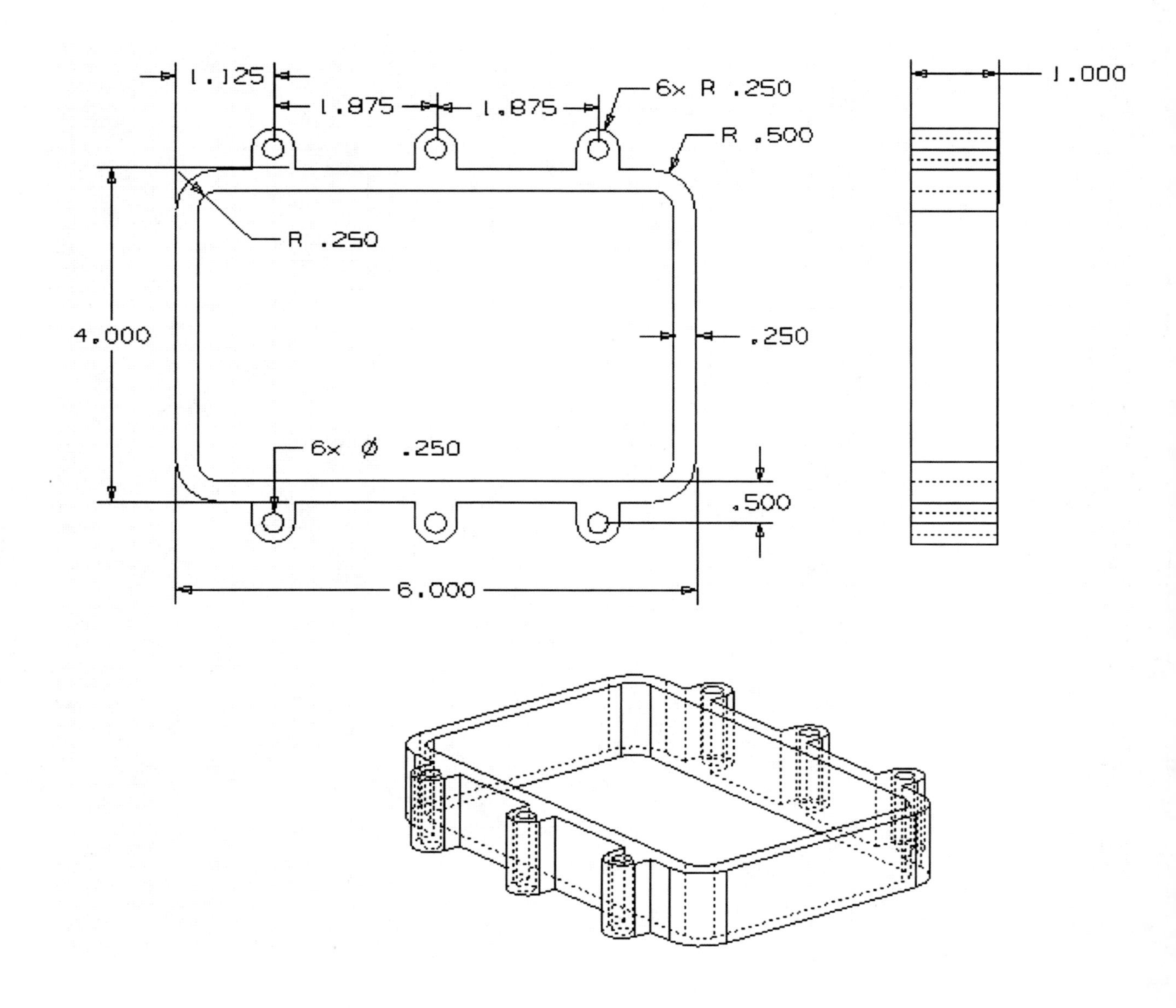

Project 11-3. Shelf Stud

Model the shelf stud part using the drawings (in inches) below and save it as **xxx_project11-3.prt** in your directory.

Hint: You may find it useful to use features like trim body and circular instance.

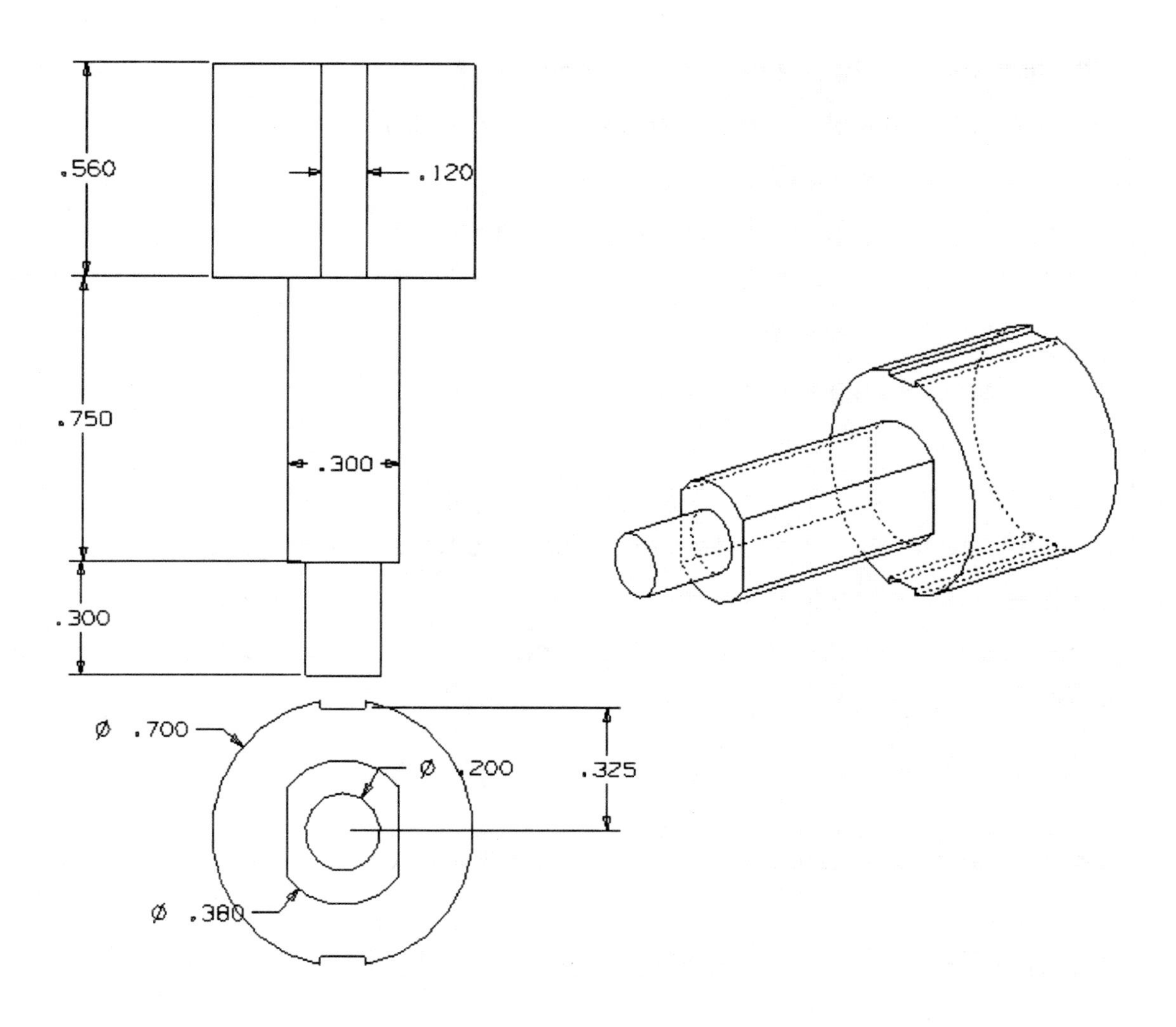

Project 11-4. Flat Screw

Model the flat screw part shown below. The flat screw consists of two parts, head and shaft. The screw head has cross-shaped pockets for a tool. As you model it, capture the following design intent into your model:

Top diameter of screw head = base diameter of screw head + .75

Shaft diameter = base diameter of screw head

Shaft height (including the non-threaded angled tip)

= 5 times height of screw head

The pockets are centered on the screw head and its dimensions are

Length = base diameter of screw head

Width = length / 5

Depth = height of screw head / 2

The initial dimensions of the screw are given as shown in the drawings (in inches) below. As for the parameter values for the detailed thread, use the default values provided by the system. After you finish modeling the part, change the values of the key design variables such as screw head diameter (its current value shown as 2) and observe the model update to confirm if the design intent is met.

Save the part file as **xxx_project11-4.prt** in your directory.

Hint: Create expressions with meaningful names to capture the design intent. You may need to create a datum plane to indicate the start face for the thread.

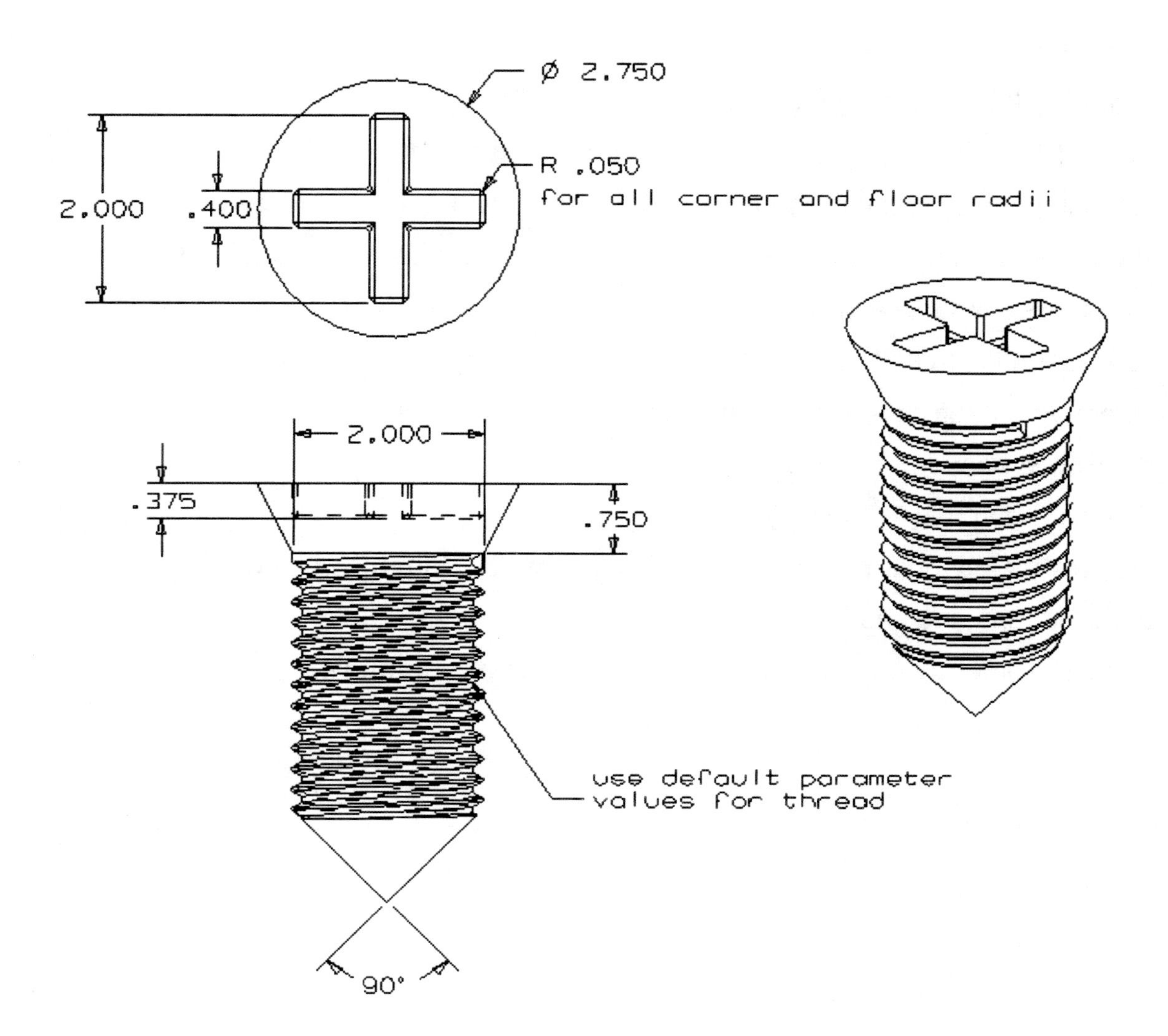
Ø 2.750
2.000
.400
R .050
for all corner and floor radii
2.000
.375
.750
use default parameter
values for thread
90°

Exercise Problems

11.1 What are the advantages of using Instance?

11.2 Name two options of Instance that were discussed in this chapter.

11.3 What is the major difference between Mirror Body and Mirror Feature?

11.4 Rectangular Array Instance can create instances of a feature in XC and YC directions. What coordinate system is used for this purpose?

11.5 In creating Circular Array Instance, there are two options Point and Direction or Datum Axis to establish the rotation axis. What is a major difference between the two?

11.6 When do you use Trim Body?

11.7 What is the advantage of using a datum plane as a trimming object over other options such as Define Plane?

11.8 Describe differences between symbolic and detailed threads.

Chapter 12. Introduction to Assembly Modeling

In this chapter, we will present an introduction to the Unigraphics assembly modeling. Assembly modeling has significant importance not only to design but also to other disciplines as it allows different departments to work concurrently for product development processes. The fundamentals of assembly modeling included in this chapter are discussed in the seven sections. Sections 12.1 and 12.2 provide an overview and definitions of the assembly modeling and terms. Section 12.3 teaches how to build the assembly using the Bottom Up method. Section 12.4 introduces the Assembly Navigator Tool. Section 12.5 teaches how to build an assembly using the Top Down method, along with how to design in context. Also introduced in Section 12.5 are Interpart Modeling and the WAVE Geometry Linker to associate a part in the assembly to another part in the assembly using geometry. Section 12.6 refines the assembly process by using reference sets to filter out objects that are not wanted in the assembly. Section 12.7 introduces Mating Conditions tools to allow associating parts in the assembly in a way to insure that parts that mate together stay together.

12.1 What is Unigraphics Assemblies?

We use many different assembly products in our daily lives. These include bicycles, electric razors, automobiles, telephones, and computers, etc. All these products have in common that they contain individual parts that are made to fit together and work together.

A Unigraphics assembly is a part file that contains the individual piece parts added to the part file in such a way that the parts are virtual in the assembly and linked to the original part. All parts are selectable and able to be used in the design process for information and mating to insure a perfect fit as intended by the designers. We will elaborate this concept in this chapter.

Unigraphics assembly modeling is the tool that allows and facilitates the collaboration among designers, draftsmen, engineering analysts, manufacturing people, and others, to insure their assembly works together. This enables individuals in different disciplines to work concurrently, resulting in faster delivery to market for products. This concept is related to master modeling that will be discussed in detail in Chapter 13.

12.2 Unigraphics Assemblies Definitions and Terminology

In this section, we describe some commonly used terms in Unigraphics assembly modeling.

Unigraphics Assembly

This is a Unigraphics part file containing virtual copies of piece parts and sub-assemblies represented by pointers to the original parts. These pointers are called component objects.

Component Objects

A component object is the pointer residing in the assembly part file that links to the original part file. The pointer contains the information of the part file name, location in the assembly space and its orientation in assembly space.

Component Part

A component part, also called piece part, is the name given to the original digital model part file created by a designer. This is to distinguish the component object that resides in the assembly part file from the original part that resides in a file outside the assembly. When the Master Model method is implemented (Chapter 13), the component part will also be called the Master Model part.

Sub-Assembly

This is a unique Unigraphics part file in that it is a component object because it resides in an assembly part file, but it is also a part file itself because it contains component objects that reside in it. A sub-assembly resides both in an assembly part file and outside the assembly part file.

The following diagram illustrates the concepts of an assembly part file, a sub-assembly part, and five component parts. Note that five component parts are represented outside the sub-assembly and assembly part files, corresponding to the leaves of the hierarchy tree. The component objects reside in the sub-assembly and assembly part files. The assembly part file corresponds to the root of the tree.

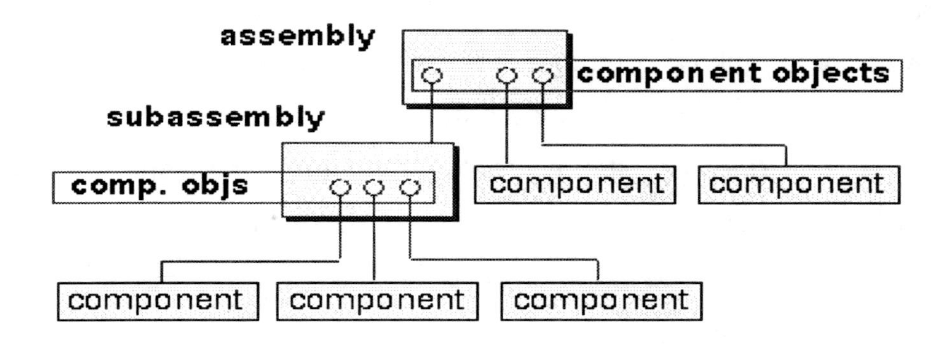

Top-Down Modeling

The assembly part file is created first and then component objects and component parts are created in the assembly part file. This type of modeling is useful in new design.

Bottom-Up Modeling

The components parts are created first in the traditional way and then added to the assembly part file. This technique is particularly useful when parts already exist from the previous designs and are being re-used.

Work Part

The work part is the component part where all creating and editing of geometry is stored. The header of the graphics window displays the part file name that is the work part.

Displayed Part

The displayed part is the part(s) that is seen (visible) in the graphics window. In the case of a component part file, the displayed part and the work part will be the same part. In the case of an assembly part file, the displayed part and the work part may or may not be the same. The assembly part file when opened typically becomes both the displayed part and the work part. When top-down modeling is used to create a component part, the work part is changed to the component part so that all creation is stored in the component part. In general, whenever creating or editing is done on a component in the assembly, the work part must be changed to the component being worked on.

Design in Context

The ability to make a component of an assembly the work part while leaving the assembly itself as the displayed part allows the assembly to be designed in context. All new geometry that is created is added to the work part. Edits can be made to the features and expressions that reside within the work part while all the assembly is displayed in the graphic window.

Mating Condition

Mating conditions are a set of constraints between component objects that describe how the objects fit together (i.e., mate together). Collectively, they describe a position of an object in space relative to another object as defined by the designer.

Reference Set

A reference set is a filter created and stored in a part file that determines which geometry is displayed in a higher-level assembly part file. A common use is to create a filter such that only the solid body is added to the assembly part file and all sketches and datum remain solely in the component part file. However, reference sets may be created in any part file.

Bottom-Up Design

A Bottom-Up assembly modeling approach starts by identifying the lowest level component parts that will make up your assembly, and by creating component parts and subassemblies as you move up the assembly level hierarchy.

The component parts are designed and modeled separately from an assembly level part file. In the below figure, we give an example of Bottom-Up assembly modeling. The support base and the support angle parts are created first separately and then added to the wheel support assembly file as an component object. Notice that the wheel support assembly has two support angle parts, thus, the assembly file has three component objects. The second support angle has to be repositioned when it is added. Later components to be added are bolts, nuts and the wheel mechanism.

support base—component part **support angle—component part**

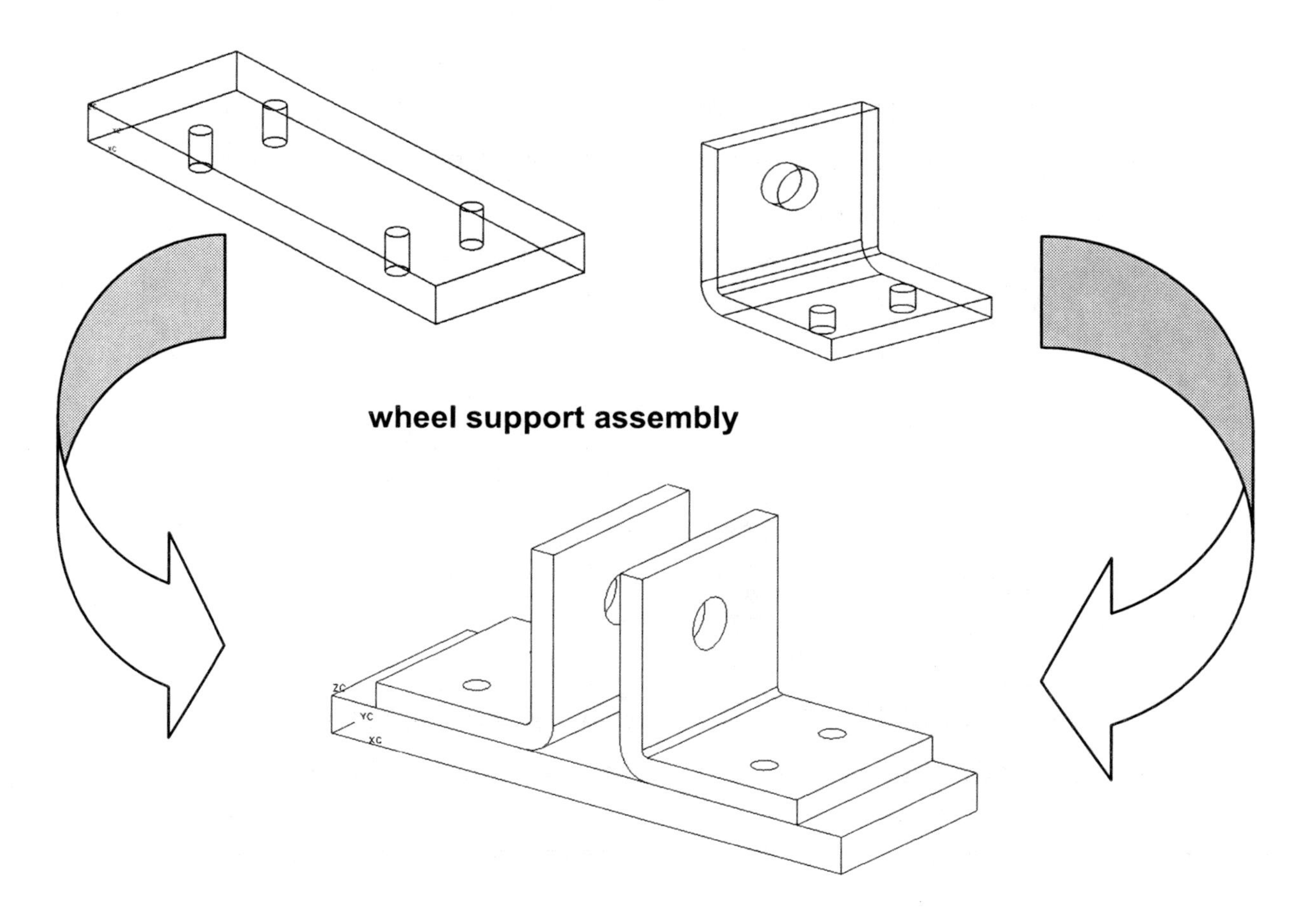

Bottom-Up Modeling--Add a Component

In order to use many assembly functions, first enable the assembly application by choosing **Start → Assemblies** if inactive. in order to add components to an assembly use the Assembly pull-down menu and choose Components, which then brings up the cascade menu as shown below. This menu can be used to create, edit, and manipulate the assembly structure as well.

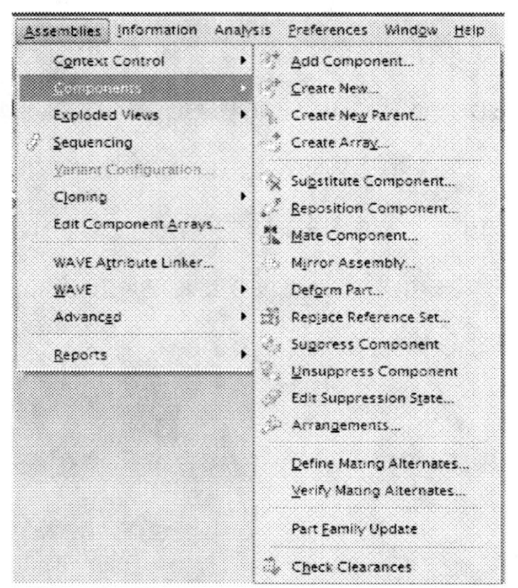

To create an assembly by the bottom-up method, choose the Add Component command from cascade menu. This starts the dialog to select the component part to be added to the assembly.

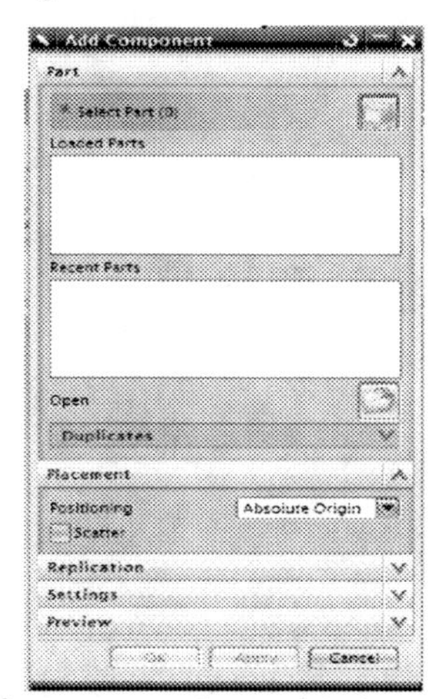

The Add Component dialog allows you to select a part that has already been loaded as shown by the parts listed in the open window; or, you may browse for a part in a directory by selecting the Open command [icon]. With click of OK after selecting a part, you will be prompted with Add Component dialog to allow the setting of several parameters about the component object being created.

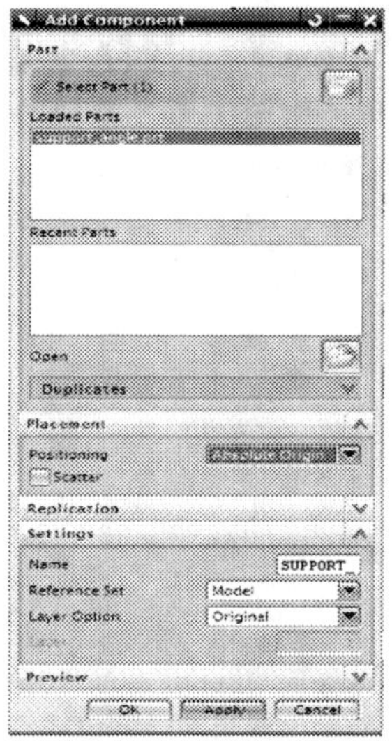

- The component object may be named. It defaults to the component part name.
- The reference set may be chosen to bring to the assembly. We'll discuss this option later.
- Multiple copies of the part may be added by using an array or chain selecting thus speeding the placement of many components.
- The layer of the assembly part file on which the component object resides may be chosen: Work Layer (current work layer of the assembly part file), Original layer (the same layers used in the component part file), or a layer to be specified by a user.
- How to position the component object in the assembly may be chosen among four available options:
    - Absolute Origin – place the component at the Absolute Coordinate System 0,0,0
    - Select Origin - places the component at a selected point.
    - Reposition – place and orient the component in one prompting sequence
    - Mate – use Mating Conditions to locate the component being added relative to other components already existing in the assembly

When we select “Reposition” among the four Positioning options and choose OK, the system displays the Point Constructor dialog and asks for a location point to be selected in the graphics window (assembly space), where the part is initially placed. Then the Reposition Component dialog comes up as follows.

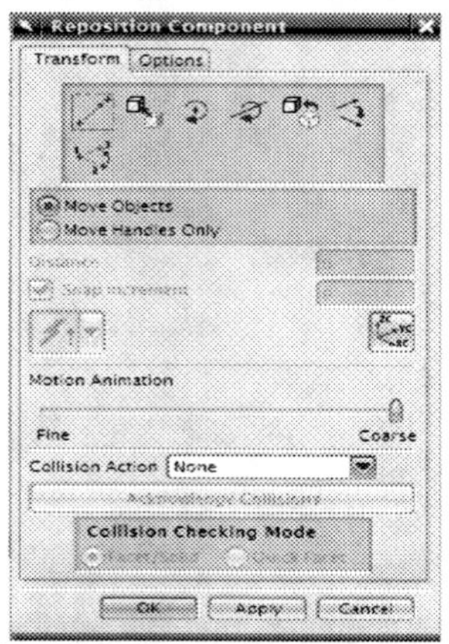

Choose the reposition method and orient the part exactly where you want it using one or a combination of the seven Transformation options that are available in the dialog.

Once the part is placed exactly where you want it you end the repositioning with OK and UG prompts to Add another component object of the same component part. If done, then choose Cancel.

Repositioning a Component

There are seven repositioning methods on the dialog. One will be used later in the lesson and discussed in detail. Here is a brief description of each.

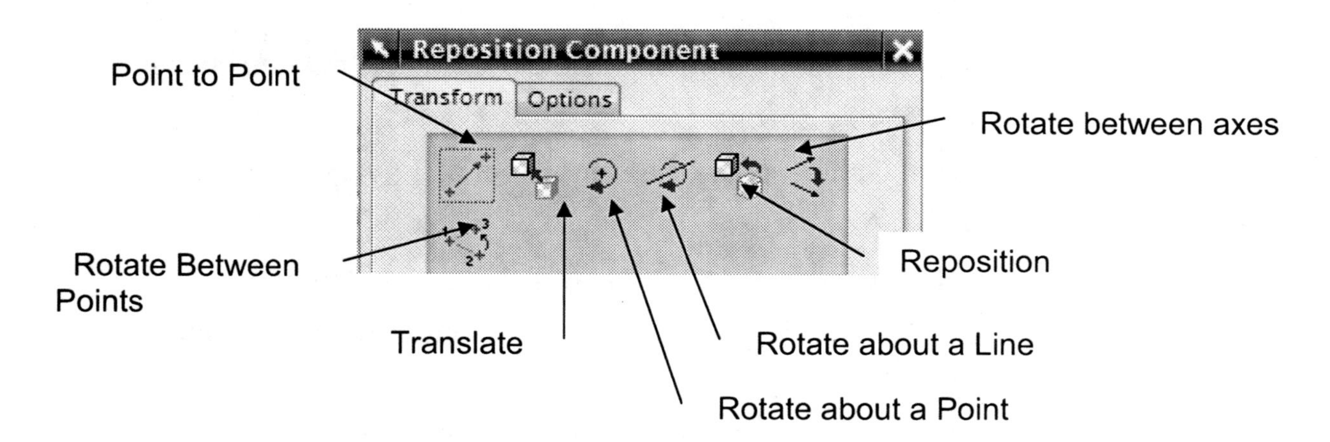

Point to Point—Allows the movement of the component by defining and selecting a reference point to move From and a destination point to move To. The orientation of the component part stays the same.

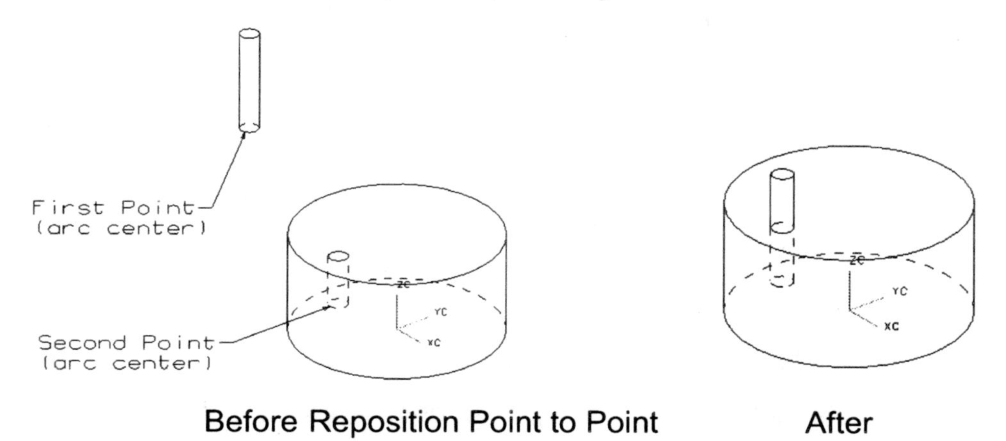

Before Reposition Point to Point After

Translate—Allows movement of the component from one point to another by entering Delta distances from its current location in the DX, DY, and/or DZ direction. The orientation of the component part stays the same. The dialog and the Delta directions X, Y, and Z are shown below.

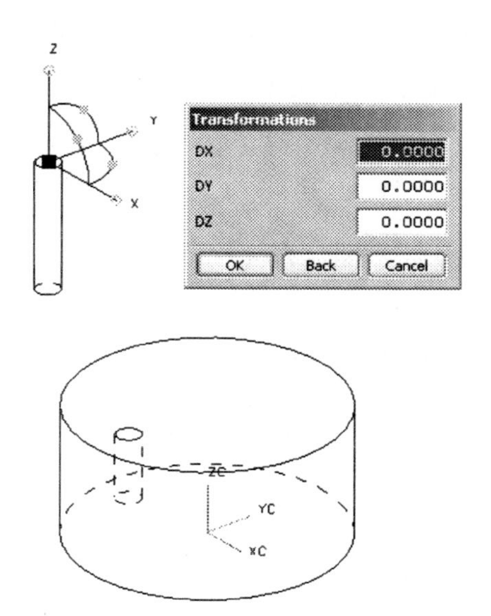

Rotate About a Point—Allows rotation about a single point in the XC-YC plane. The ZC axis is parallel to the axis of rotation. The direction of rotation follows the right-hand rule—counterclockwise. For this method, you need to select a rotation point and to provide the rotation angle.

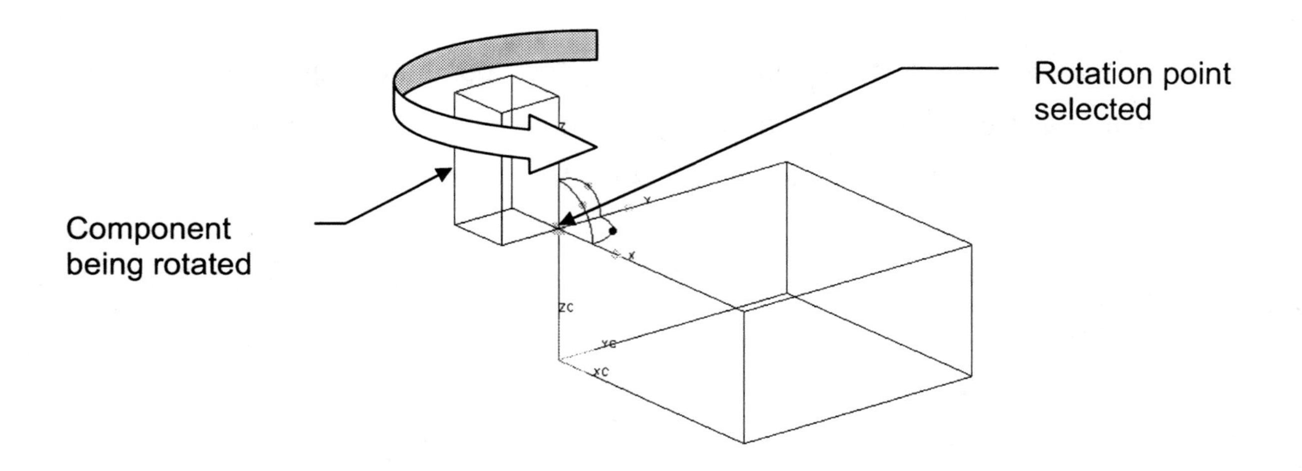

Rotate About a Line—Allows rotation about a line that you select. Line may be in any orientation in 3D space. The line may be chosen from an existing line, edge, two points, or a point and vector. The direction of rotation follows the right-hand rule about the rotation axis. The positive direction of the rotation axis is determined by which side of the line is selected, that is, it points to the nearer end point from where you pick in the line). The same result as the above example in "Rotate About a Point" can be obtained by selecting the edge location shown below. For this method, thus, you need to select a rotation point and a rotation line and to provide the rotation angle.

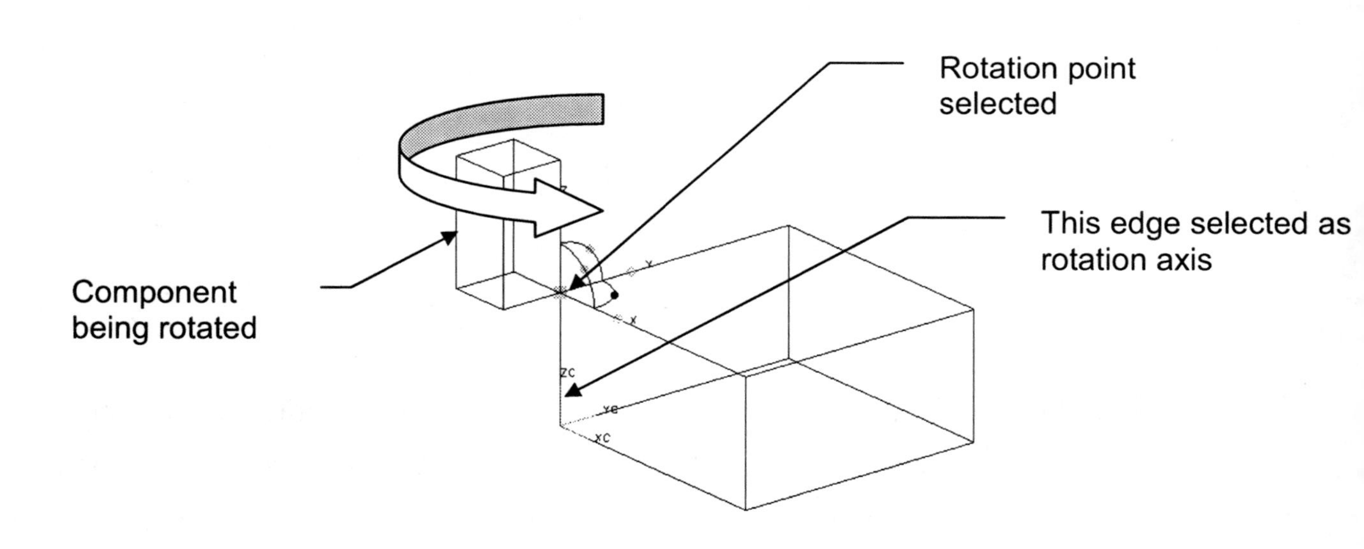

Reposition—Allows free relocation of the part from its current orientation to a destination orientation with complete freedom of orientation and location.

- To accomplish this you select a ***from*** reference coordinate system for the component in the current location.
- This action establishes a ***from*** X, Y, Z temporary csys about the component.
- Then you select a ***to*** temporary coordinate system for the position you would like the component to move to.
- This action establishes a temporary X, Y, Z csys for the component to move ***to***.
- When the OK is given to move the component, the ***from*** X will be coincident with the ***to*** X, the ***from*** Y will be coincident with the ***to*** Y, and the ***from*** Z will be coincident with the ***to*** Z.

Rotate Between Axes—Allows rotation about a selected point with the angle of rotation defined by selecting two lines. These two intersecting lines define the rotation plane. The rotation will take place around the rotation point by the angle from the first line (reference axis) to the second line (destination axis) along the rotation plane. We will give an example for this in the following activity.

Rotate Between Points—Allows rotation between selected points. This is similar to Rotate Between Axes and the difference is that the points selected establish the two lines instead of picking lines directly.

Activity 12-1. Creating an Assembly Bottom-Up

This activity will help to practice building an assembly bottom-up and positioning components in the assembly. You will complete the assembly from the previous picture

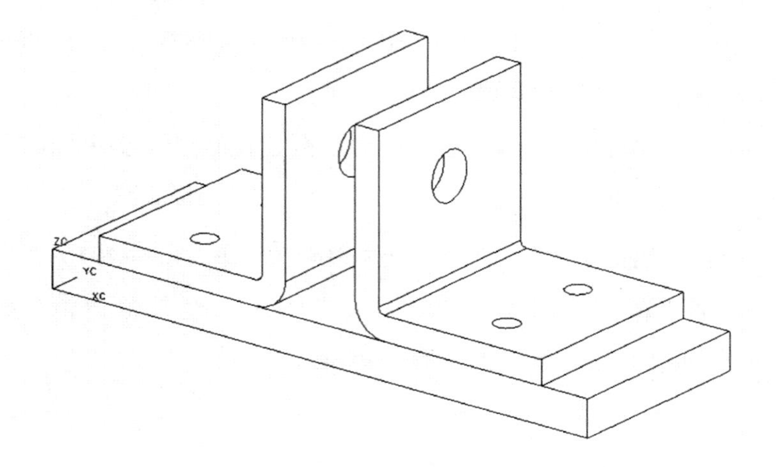

by adding the following parts .

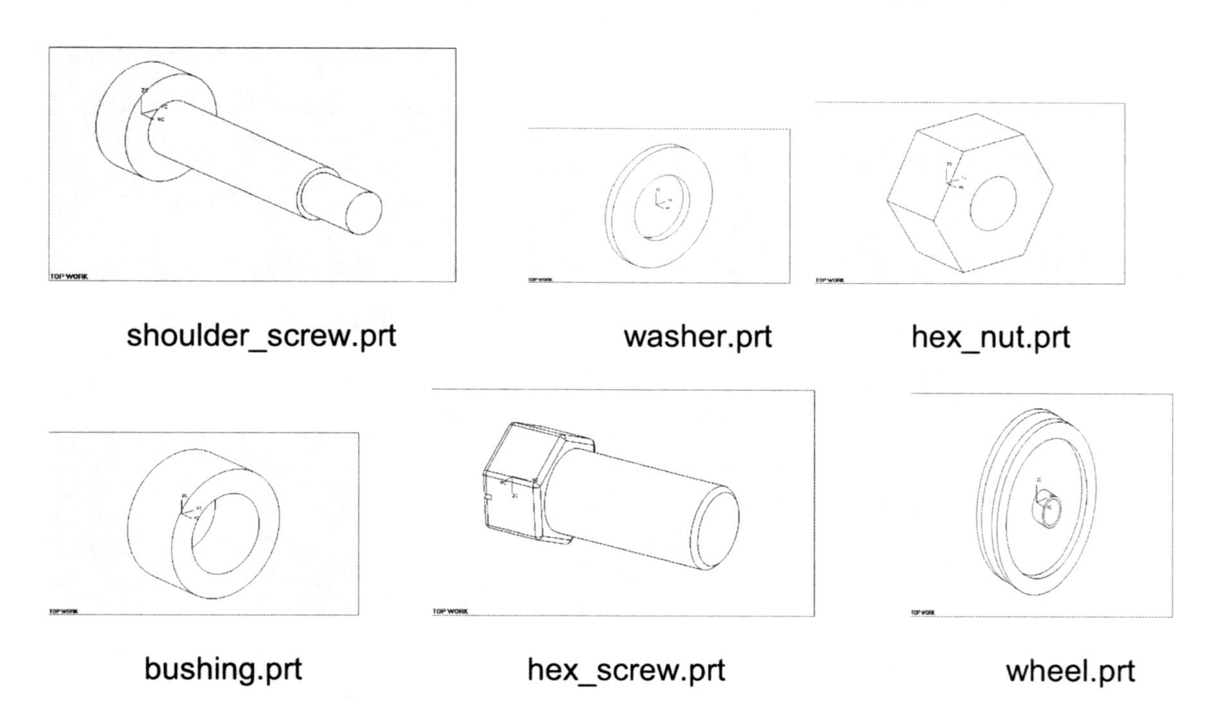

shoulder_screw.prt washer.prt hex_nut.prt

bushing.prt hex_screw.prt wheel.prt

Step 1. To create the assembly bottom-up design, open the assembly **wheel_support_assm.prt** under directory Chapter12-partfiles\Support and save it as **xxx_wheel_support_assm.prt.**

Start the **Modeling** Application and then start the **Assemblies** Application.

Step 2. Add the 4 hex_screws to the assembly.

2.1. Choose **Assemblies → Components → Add Component**. The dialog comes up.

2.2. Select the **Open** file button.

2.3. Choose **hex_screw.prt** from the **Part Name** window. It appears in the Loaded Parts window.

2.4. Move down the dialog and change the **Positioning** to **Reposition** and then **Layer Option** to **Work**.

2.5. Choose **OK**. The **Point** dialog comes up, asking for a point to locate the placement of the screw.

2.6. Choose a point in space close to the assembly.

The screw is drawn on the screen. Note it is oriented wrong to go into the holes in the angle and base. We must reposition the screw so it is oriented vertically. The **Reposition Component** dialog now comes up ready for you to choose a reposition method to locate the screw into the bolt hole in the angle. We can achieve this by aligning the CSYS of the hex head base circle edge with the CSYS of one of hole edges in the support angle (see the figure below).

2.7. Choose the fifth option **Reposition** . Next the **CSYS** constructor dialog appears. Unigraphics prompts first to select the object to define the "from csys".

2.8. Select the Type pull down and choose the option **CSYS of Object** icon for the "from" CSYS.

2.9. Choose the circle edge at the base of the hex head (see the figure below) as the **"from CSYS"** and choose **OK**. A coordinate system appears at the location.

2.10. The Cue Line is prompting now to "Select objects to infer CSYS". It is actually wanting you to choose the method of selecting the "to CSYS". Use the option as before **CSYS of Object** for the "to CSYS" and choose one of the hole edges in the support angle as shown in the below figure. Choose **OK**.

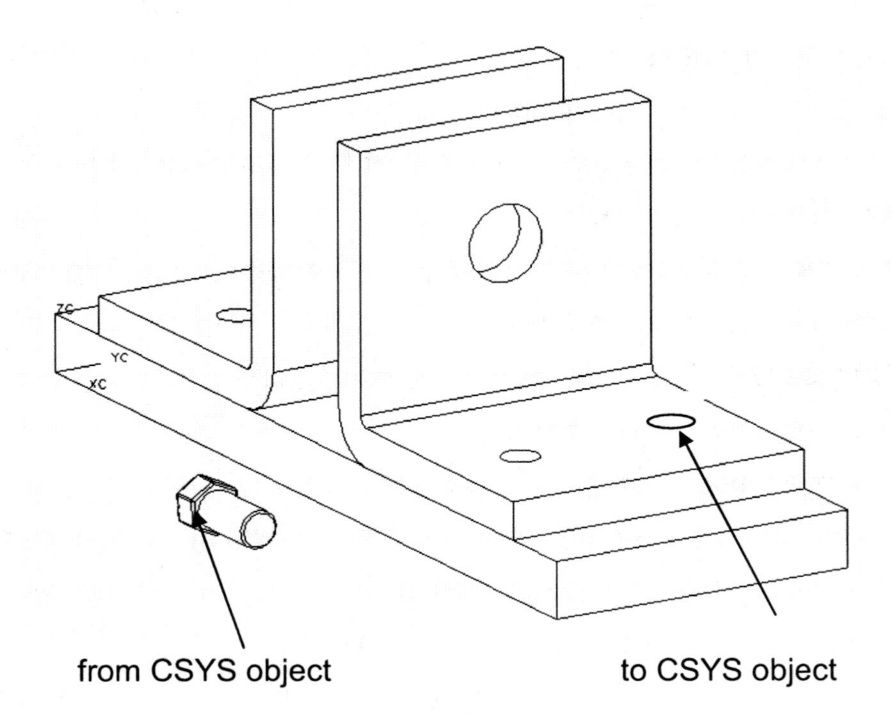

Now the screw moves to the hole in the support angle. Choose **OK** to confirm this reposition move in Reposition Component dialog. Note that if you do not confirm with OK, you will lose the reposition move, although the graphics window reflects the reposition move already.

We will add one more screw to the assembly using a different Reposition method.

2.11. Add another **hex_screw** component part to the assembly, following Steps 2.1 through 2.6.

We are going to reposition the screw using a combination of two reposition methods: **Point to Point** to locate it at the hole, which will be followed by **Rotate Between Axes** to bring it upright and down into the hole.

2.12. On the **Reposition** dialog, choose the **Point to Point** icon

2.13. Zoom in on the screw. On the **Point** constructor dialog choose the **Arc/Ellipse/Sphere Center** toggle to insure selecting the arc center.

2.14. Select the arc at the base of the screw shaft under the bolt head.

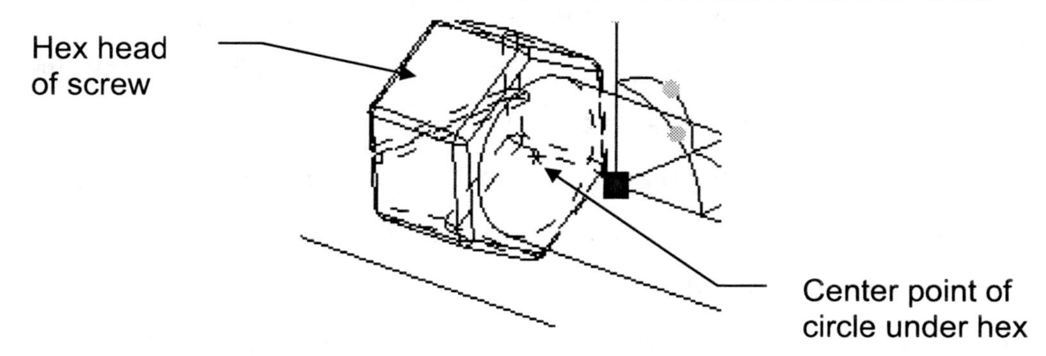

2.15. Unigraphics is prompting for the second point to be selected. Insure the Arc center is still chosen, and select the hole edge in the support angle next to the screw already placed.

2.16. As soon as the selection is made the screw moves to the selected location in a temporary display.

2.17. Inspect the location and if correct then choose **Apply** on the dialog to make the change permanent. The part now looks like this.

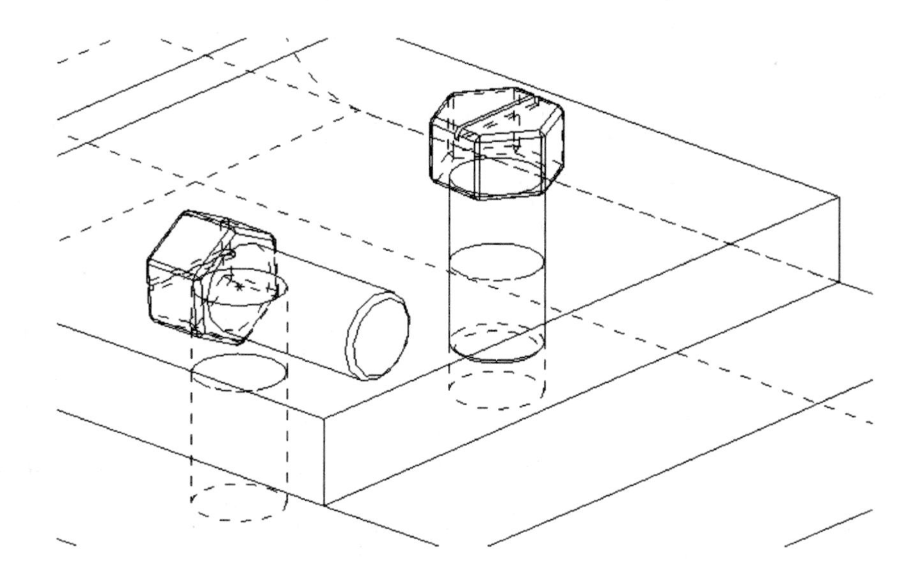

2.18. Now we need to rotate the part into position. In its current orientation the screw needs to rotate in the XC-ZC plane. In other words, the tip of the screw needs to rotate clockwise 90°. We should look for two axes that would fill the 90° requirement and if we select them in the correct order we should get the direction correct as well.

2.19. Choose the **Rotate between Axes** icon from the **Reposition** dialog.

2.20. **Point** constructor dialog appears, asking to select the reference point which is the rotation point. Choose the **Arc/Ellipse/Sphere Center** toggle in the dialog and select the arc at the base of the screw shaft under the bolt head.

2.21. We need to select 2 lines for our axes that lie in a vertical plane. And they must be chosen on the correct end to get the vector in the correct direction so the rotate happens in the correct direction.

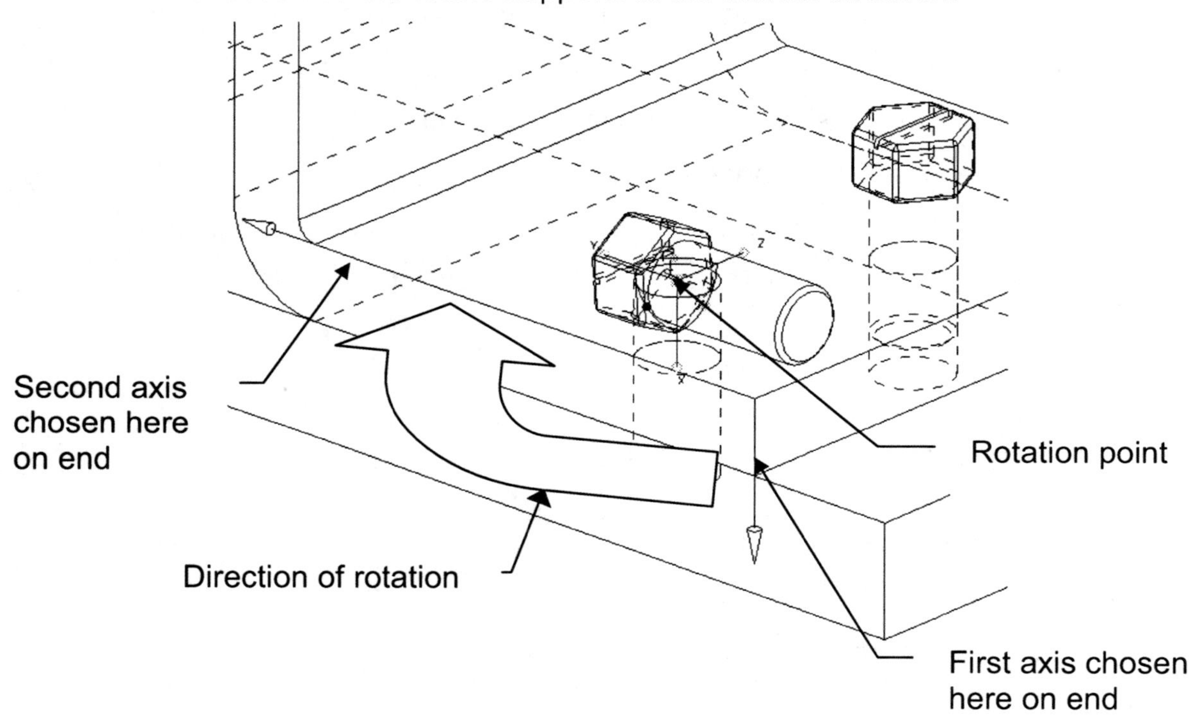

2.22. Select the vertical line indicated above toward its bottom end without selecting the end point. An arrow appears. Choose **OK** to confirm the vector.

2.23. Select the horizontal line indicated above toward its left end without selecting the end point. An arrow appears. Choose **OK** to confirm the vector.

We have established the point of rotation; the two lines have established the plane of rotation and the direction of rotation.

2.24. The **Reposition** dialog appears. Choose **OK** on the **Reposition** dialog and the screw is rotated into proper position.

2.25. Add the two remaining hex_screws. You might try different methods. For example, **Rotation About a Line** followed by **Point to Point** will work as well.

Step 3. **Save** the file. If you continue to do Project 12-1, you do not need to close this file. Otherwise, close it.

Project 12-1. Bottom-Up Assembly Modeling

Open **xxx_wheel_support_assm.prt** that you created and saved in Activity 12-1, if it is not open. Using the similar procedure of adding a component to an assembly part file as in Activity 12.1, finish the assembly that looks like below. To do so, you need to add two bushings, two washers, one hex_nut, one shoulder_screw, and one wheel. Refer to p.12-10 for the drawings of these parts. Save your file as **xxx_project12-1.prt**.
Hint: In order to position the wheel correctly, i.e., center it between the two walls of the support angle part, you may have to use **Analysis → Measure Distance**.

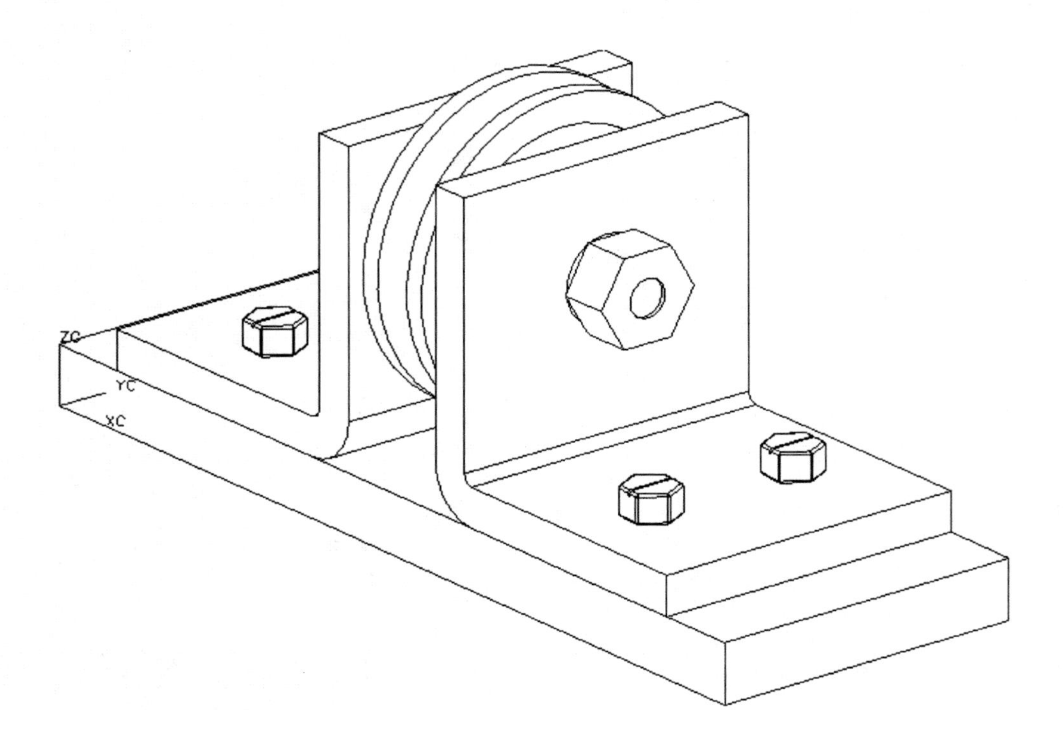

12.4 The Assembly Navigator Tool (ANT)

The Assembly Navigator Tool provides a graphical display of the assembly structure (hierarchy) of the components within an assembly, and a quick and easy method to manipulate the components in an assembly.

The Assembly Navigator window for the completed wheel support assembly is shown below.

The Assembly Navigator window above shows the current assembly as the work part with the hierarchy of component objects shown indented in a list below it—the assembly tree. There is an item in the list for each component, including multiple copies of the same component, displayed as a node in the assembly tree structure.

To open the Assembly Navigator Tool (ANT), choose the Assembly Navigator icon on the resource bar on the side of the graphic window.

The ANT is useful as a shortcut for many component operations without going through the menu structure. With ANT, you can also verify the assembly structure and the presence of components.

By selecting one of the components, several options are available. First, the component selected will be highlighted in the graphics window for quick identification of the component visually. In addition, choosing a component in ANT with MB3 will bring up the pop-up menu with additional options and operations that may be done. We will just highlight some options here. The pop-up menu looks like this.

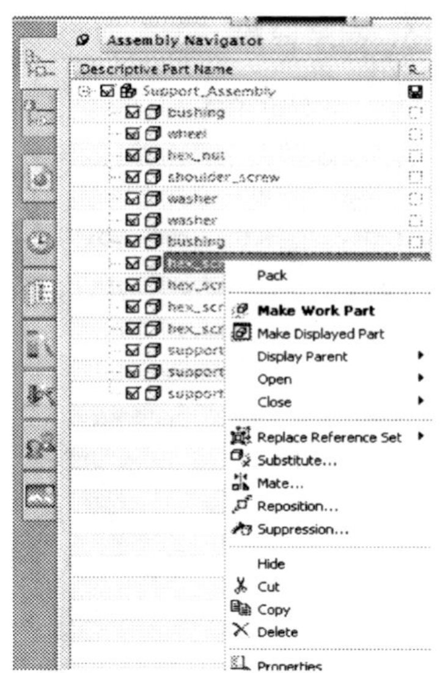

Note that some of the operations available on the pop-up, which are also available elsewhere in the menus.

Make Work Part—Changes the component to the Work part so creating and designing of additional geometry can be stored in the component part. The other parts in the assembly will still be visible and selectable, but grayed out as in the background. This is the configuration for how Design in Context would be done.

Make Displayed Part—Changes the component to the Work Part and the Displayed part. It is the same as working in the component part file with no surrounding parts.

Display Parent—Changes the displayed part to a part file that is the parent assembly/subassembly of the component. You can have the parent assembly/subassembly displayed if the Make Displayed Part was invoked on a component. This just restores the display only and the component part is still the work part.

Replace Reference Set—Changes the reference set displayed in the assembly. We will discuss reference sets in a later section.

Open/Close—Allows the opening or closing of part files, same as the File → Open/Close menu does.

Properties—This option supplies other information on the selected component such as non-graphical attributes and physical attributes.

12.5 Top-Down Design

In Top-Down design, you create an assembly at the top level of the hierarchy and move down the hierarchy, creating subassemblies and component parts. Thus, Top-Down assembly modeling presupposes that you will be "designing in context", creating new component parts relative to other components, i.e., building parts around other parts.

An “assembly” part file is just another Unigraphics part file. One main difference between regular part files and assembly part files are that component objects reside in assembly part files. But Unigraphics does not treat an assembly file as a different kind of part file but as just a regular part file. So creating and editing done in the assembly part file may be done as in any part file. The Work Part and Displayed Part become even more important when creating and editing geometry, and creating component objects because all creating still goes into the Work Part whether the Work Part is the assembly or a component part.

There are two basic methods used to create a component object top-down.

Method One:

Create component objects first (Work Part is the assembly).

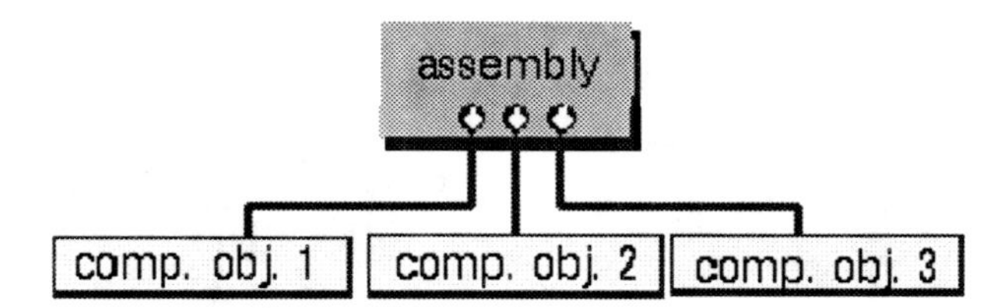

Make a component object the Work Part.

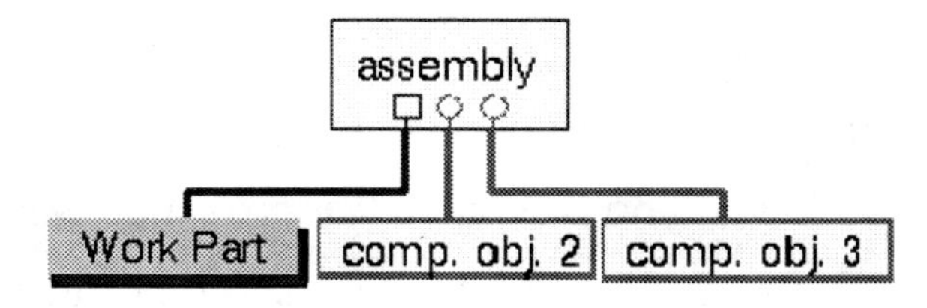

Create geometry in the component part that is now the work part.

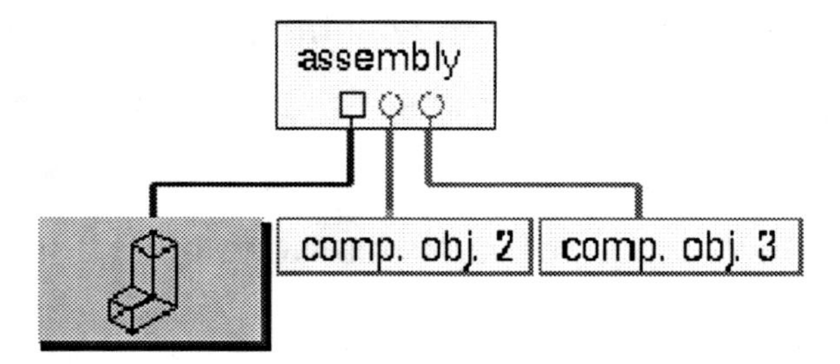

Method Two

Create geometry in the assembly part (Work Part is the assembly).

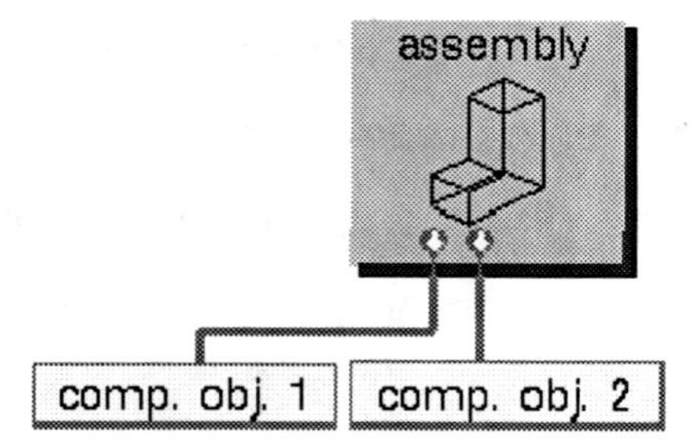

Create a new component and add the geometry to it.

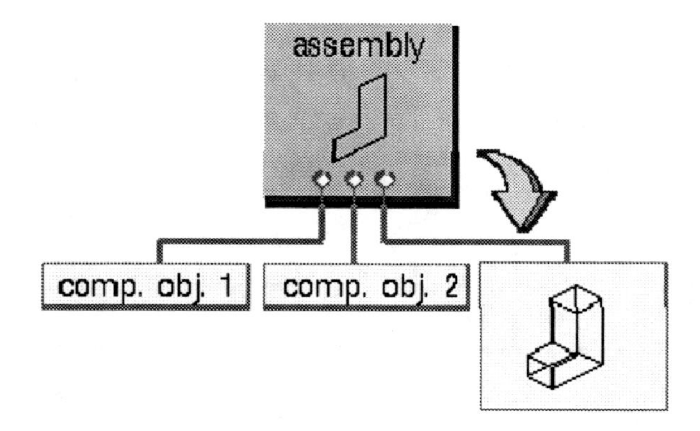

Note that for both methods described above, both geometry and component object must be created. Both methods may be used in mix depending on the design needs, what information you have available for the design criteria and design intent, and other constraints you may have.

In practice, the Top-Down and Bottom-Up approaches will both be used, as you often use existing parts and create new parts in order to meet your design needs. For example, standard bolts and nuts may be used in many assemblies.

Creating a New Component

The steps to create a new component will be shown here and then an activity to practice the steps will follow. We will create the component object and then add geometry to it by working in the context of the assembly and using surrounding parts.

Select the pull down dialog **Assemblies → Components → Create New**

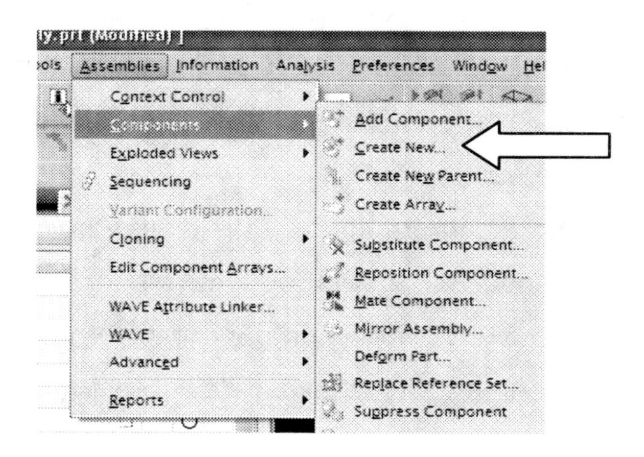

Choosing the **Create New** button creates a new component. Make sure that the current Work Part is the assembly part file.

The class selection bar comes up to allow the selection of geometry to be added to the new component part when created (Method Two). If Method One is being used and geometry will be created later in the component part with the component part as the work part, then accept the class selection dialog without picking any geometry by choosing **OK** or the OK .

The **Select Part Name** dialog comes up to name the component part file, so type in a name.

After the component part file name has been defined, you may define information related to the new component object in the **Create New Component** dialog.

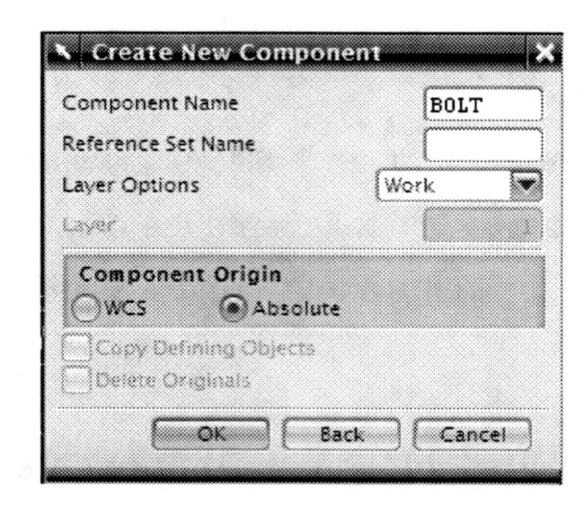

Similar to the Bottom-Up method, you have the opportunity to specify some parameters of the component object: component name, a reference set, what layer the geometry will be created on, and the orientation and location in space of the component part geometry.

Enter the appropriate information in this dialog, or leave the default values, and when complete choose **OK**. If Method One is used, then the normal procedure would be to make the new component the work part and start constructing geometry. An activity follows below to practice these concepts.

Activity 12-2. Creating an Assembly Top-Down

In this activity, you will add an existing component to the assembly. You will then create a new component in the assembly and add geometry to it by working in context of the assembly.

Note: In order to complete this section, the interpart modeling switch variable should be set as **Allow Interpart Modeling**. This setting is located within Unigraphics under **File → Utilities → Customer Defaults → Assemblies → General** and then select the Interpart Modeling tab. If this is not set properly, you cannot access the Geometry Linker. Consult with your instructor or system administrator if you have a problem with this.

Step 1. Create a new file **xxx_fix_assm.prt** (units: inches) and save it in your specified directory. This empty part will be used as the assembly. Several components will be added to this assembly.

Step 2. Add the part **fix_baseplate.prt** from the fixture subdirectory to the origin of the assembly. Replace your view with TFR-TRI if necessary.

Hint: Choose **Assemblies → Components → Add Component** with use of Absolute for Positioning.

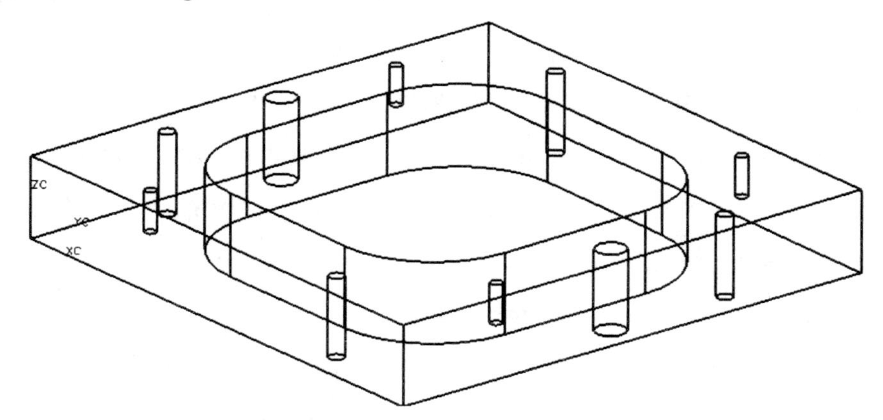

The first method to create components in an assembly is to create the empty component part then add the geometry to it.

Step 3. Choose **Assemblies → Components → Create New**.

Next the selection bar displays in the upper left corner. This is where you are prompted to select the geometry you want to add to the new component.

Step 4. Ok by selecting the **Checkmark** to create the component without adding any geometry to it.

The **Create New Componnet** dialog (which is also the template dialog for Modeling) prompts you to:

- Specify the directory path to the save the component,
- enter **xxx_fix_locpin** as the new component part name,
- choose the template name you usually choose for Modeling,
- and Choose **OK**.

Step 5. The **Create New Component** dialog comes up to name the component, specify a reference set and a layer option to create the part on.

Choose Work layer and choose **OK**.

Creating Linked Components

The next step is to create the geometry in the new component. There are two ways you can do this: if you want unlinked geometry, you can simply model new geometry in the new component. On the other hand, if you would like to take advantage of the component linked to geometry in some other component, you can link geometry from one component to the other.

There are complications to creating geometry within the assembly. That is, not all creation methods are available directly. You can create a primitive and create the size of the primitive by selecting points on other parts within the assembly; however, you cannot extrude a solid by picking edges, curves or sketches existing on other component parts.

The implications of this are that for unlinked geometry you create the component completely within itself without referencing other geometry. The component is the work part and all objects and features are within it. If you want to create geometry that references geometry on other parts (i.e., an outline as a sketch of another part, a hole of another part as the diameter of a new hole of the current work part, or a hole of another part as the positioning location for a new hole) then you must link the new geometry of the current work part to the existing geometry of other parts that you want to reference. The **WAVE Geometry Linker** is the tool that is used to create the link between the existing geometry and the new geometry in the work part that is to be created.

The link that is created using the Geometry Linker is a *feature* of the work part, thus appears in the part navigator window. The linked geometry is associative to the parent geometry from which they were selected so that when the parent geometry is changed, the linked geometry will be updated accordingly. In turn, the solid that was made from the linked geometry will be also updated.

Interpart Modeling

Interpart Modeling is another term given to describe design in context in an assembly when modeling in a work part and selecting geometry in other parts to use as reference for creating and positioning new geometry.

Different types of objects may be selected to use as linked geometry: points, curves, sketches, datums, faces, regions of faces, bodies, and mirrored bodies. We will only include a few of these for illustration.

There are some considerations and caveats to consider when choosing to use the Geometry Linker:

- Consider the permanence of the geometry chosen, like selecting a single curve versus selecting a whole sketch. A single curve might be deleted and break the link, but a curve in a sketch could be deleted, replace with another curve and the link and updating would still be intact.
- The Geometry Linker should only be used when the parts have a strong physical constraint relative to an adjoining part and the parts will be used in the same assembly. This means that you really need parts to be related such that if a parent geometry changes, the children geometry need to update.
- The Geometry Linker should not be used just because you can. The link creates another level of complication and could have downstream impacts.

We will now go back to our activity and create the geometry for the locator pin using the Geometry Linker. Be sure to pay attention to the steps and the reason for each step as the pin and subsequent parts are created.

Activity 12-2. (continued)

In the previous step we created the component **xxx_fix_locpin** with no geometry in it. Now with Linking discussed we will create geometry in **xxx_fix_locpin** using a hole from the baseplate as reference. Here are the steps to do it.

Step 6. In the **ANT** window, move your cursor over **xxx_fix_locpin** from the list click it with **MB3** and choose **Make Work Part**. This makes **xxx_fix_locpin** the work part. All geometry to be created will be saved in this work part. Confirm this by reading the banner line of UG window above the menu line, which shows:

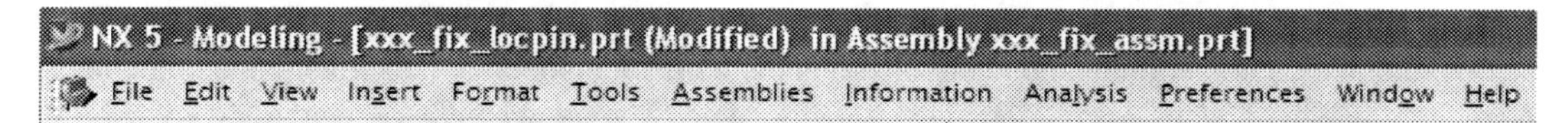

Also notice that the currently displayed part, the baseplate geometry in the assembly file, turned gray since this is not a work part any longer.

Below is a picture of our goal in designing the pin.

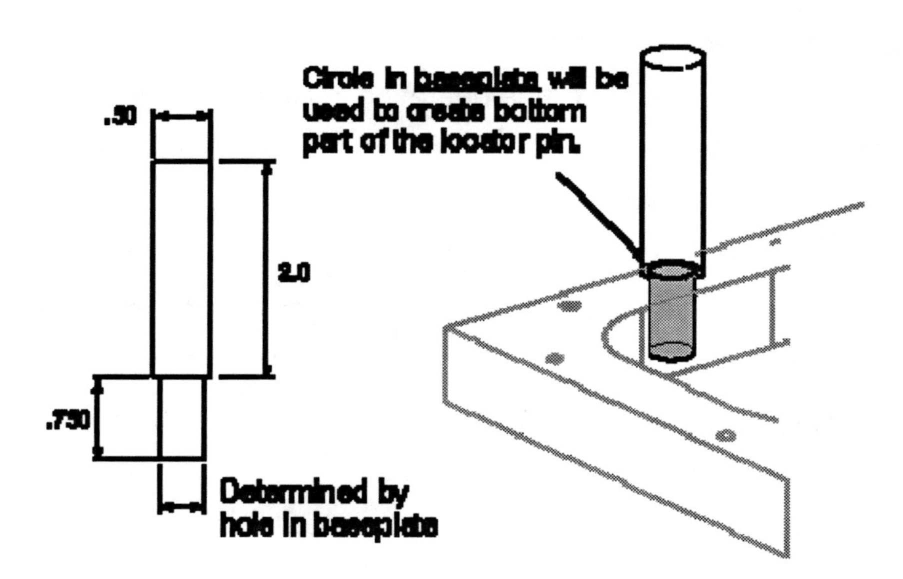

The first step is to create the link with the geometry that will be needed.

Step 7. Choose **Insert → Associative Copy → WAVE Geometry Linker** The Geometry Linker dialog appears with the Type pull down of objects to be selected at the top of the dialog.

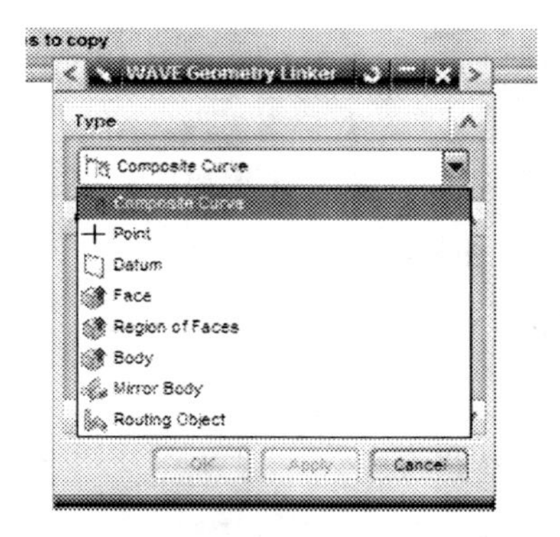

We will use an arc in the baseplate to create the link; therefore the Composite Curve option is used to create our pin as shown above.

Step 8. Choose the **Composite Curve** in the above dialog pull down menu.
Step 9. Select the top edge of the hole highlighted in the baseplate picture above. Choose **OK**. A circle was created superimposed on the curve selected.
Step 10. Construct the lower part of the locator pin (smaller diameter).

We will use the **Extrude** command to create a cylinder down into the hole using the arc we just created. See Extrude operation in Chapter 9 for details. The arc was created in the Work part and the creation below of the locpin will be done in the work part and component part.

Step 11. Choose the arc as the generator curve, the length of the extrude is .75 inches and down into the hole.
Step 12. We have just completed the bottom portion of the pin. Now we will create a top portion with a **Boss**.

The boss is placed on top of the cylinder just created, is .5 diameter and 2 inches long. It is centered using Point to Point option.

The locator pin is created and its lower diameter is linked to the hole in the baseplate.

Step 13. The entire assembly must be saved to save the creation in the assembly and to save the component part file in the directory.

13.1. Change the work part to **xxx_fix_assm.prt** by clicking **xxx_fix_assm** with **MB3** in the ANT window and choosing **Make Work Part**.

13.2. File → Save. You do not need to close this file since you will continue to use it in the following activity.

That completes this activity.

Activity 12-3. Checking Linkage between components

In this activity, we will check the linkage between the two components added in the assembly file, **xxx_fix_assm.prt**. We will do so by changing the diameter of the hole in the baseplate and observing its effect on the locator pin.

Step 1. Change the baseplate to the Work Part.

Step 2. Choose **Edit → Feature → Edit Parameters…** from the main menu.

or

Choose the **Edit Feature Parameters** icon .

The **Edit Parameters** dialog comes up with the list of features in the work part.

Step 3. Choose **Simple Hole(1)** from the list or select it from the graphics area, then **OK** to accept.

Step 4. In the **Edit Parameters** dialog, select the **Feature Dialog** option.

Step 5. The new **Edit Parameters** dialog shows the current hole diameter of 0.375. Change it into 0.1, and then choose **OK** three times to change the part with the new diameter.

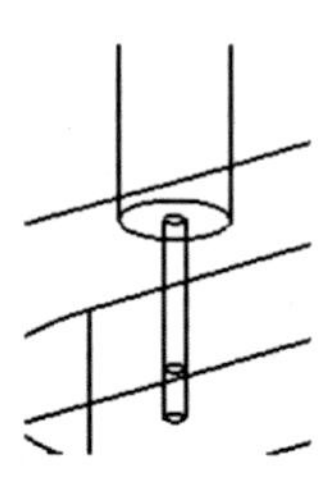

Notice that both the hole of the baseplate and the lower cylinder of the locator pin are modified. This is because of the linkage between the baseplate hole and the locator pin via **WAVE Geometry Linker**.

Step 6. Undo the previous step by either pressing both keys **Ctrl-Z** simultaneously or choosing **Edit →Undo List → Edit Feature Parameters** in order to change the diameter of the hole back to the original.

Notice now that the hole in the baseplate and the diameter of the pin went back to their original design. If you are not sure, then use **Information→ Object** and pick the both objects and compare their diameter values.

Project 12-2. Add Another Locator Pin

Add the second locator pin as shown in the below figure. You can do this by taking two approaches and we suggest you try both. The first approach is to add it using the already existing part, **xxx_fix_locpin.prt.** The second approach is to create a new component **xxx_fix_locpin2.prt** and add necessary geometry to it using the WAVE Geometry Linker. Name the resulting two assembly files **xxx_project12-2_assm.prt** and **xxx_project12-2_assm2.prt**.

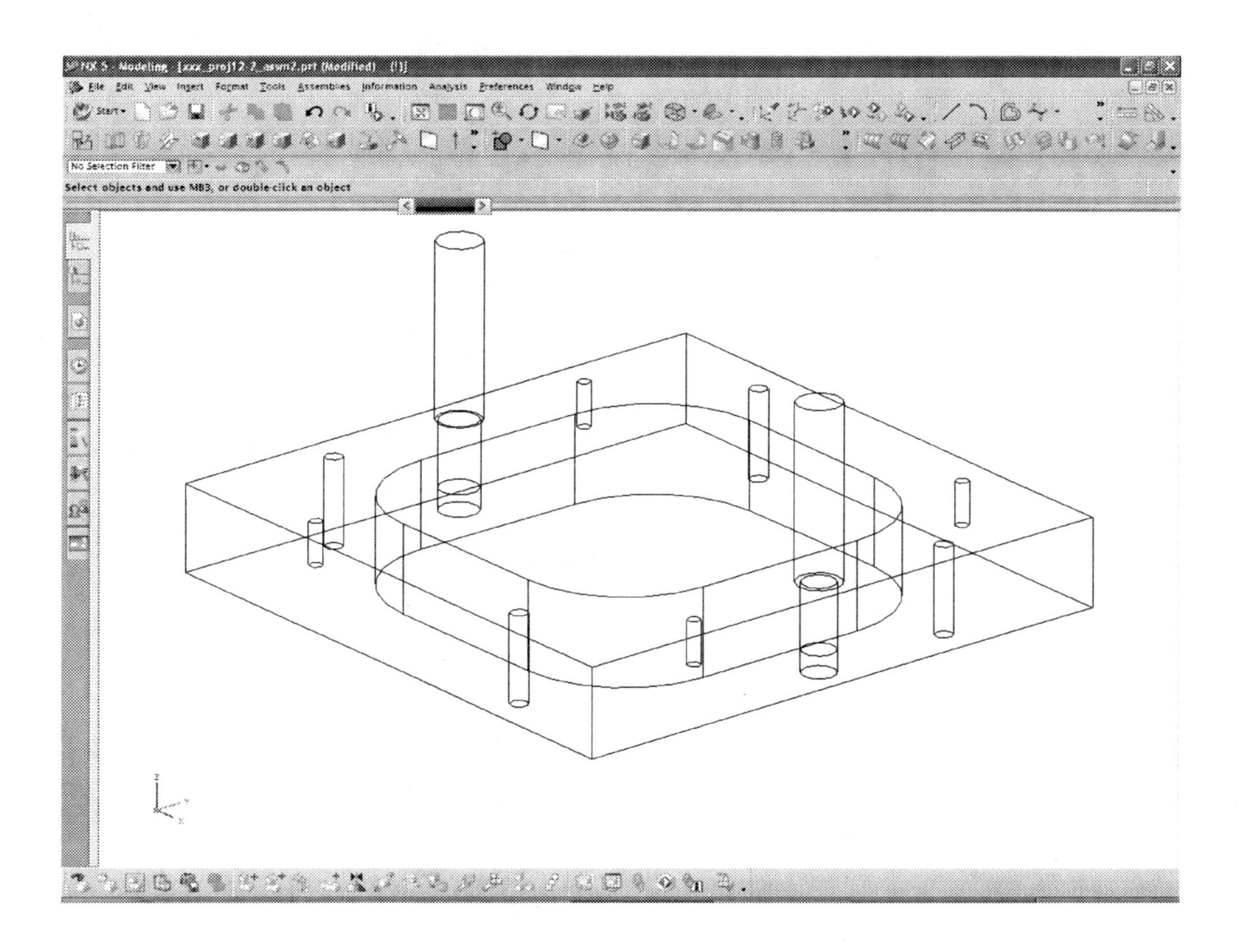

12.6 Reference Sets

A reference set is a tool used to control which objects of a component part are loaded into an assembly. The reference set is a named group of objects and features contained in the component part file that will be loaded into the assembly part file. Objects not in the reference set are not loaded to the assembly.

A reference set is a method to filter out features and objects that you do not desire to display in the assembly. These may include the datum planes, datum axes, sketches, and other construction geometry that may have been created first in order to help create a solid body representing the part. We will use the wheel support assembly for illustration.

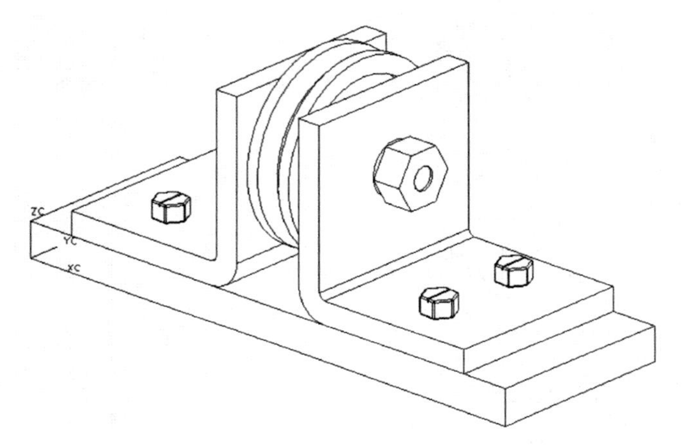

(a) Only solid object displayed

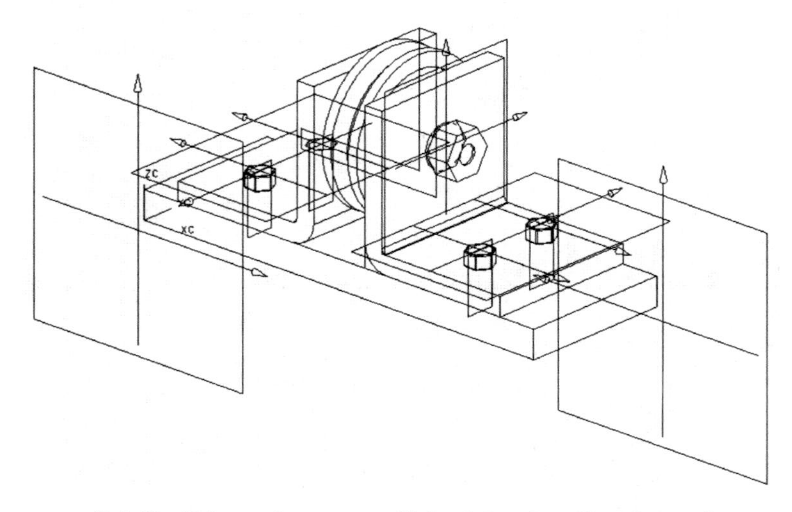

(b) Solid and non-solid objects displayed

The above figure (b) is cluttered with many non-solid objects displayed that are not needed in the assembly. We could manipulate the layer settings (**Format → Layer Settings**) to make all the non-solid objects invisible in the assembly and turn it into figure (a) above. However, on a large assembly with numerous objects of many different types, this might be very difficult to do. The assembly part file would be much larger as well. Changing layers would require finding all non-solid objects and making their layers invisible so that only the solid is visible. The reference set is an easier and more flexible method than layer manipulation in controlling visibility of objects of various components in the assembly. A reference set is created in the component part file and contains only the geometry that is desired in the assembly file.

Operations of Reference Sets

Choose **Format → Reference Sets**... will bring up the **Reference Sets** dialog.

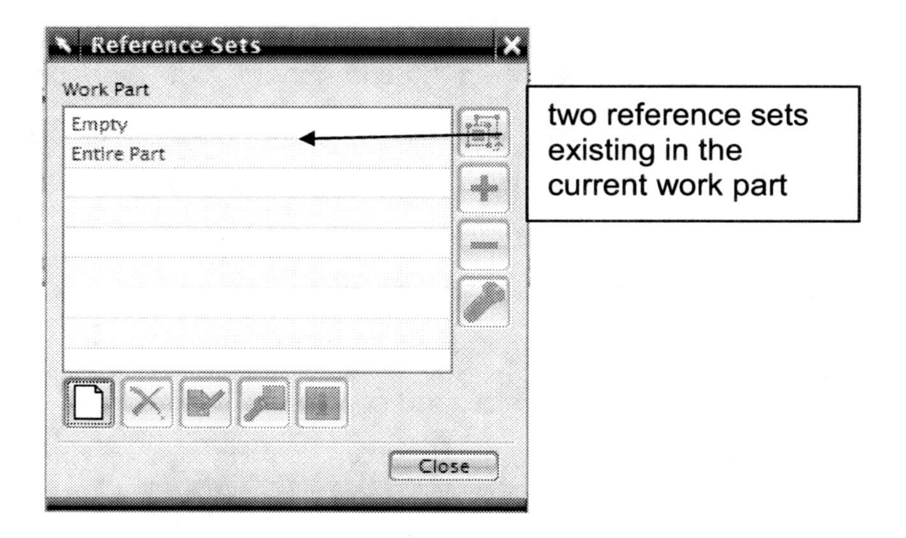

The above dialog shows in the listing window that the current work part already has two reference sets: Empty and Entire Part. Every part file contains these two reference sets by default:

- **Empty** (no geometry, component part is not visible in the assembly)
- **Entire Part** (this is the default condition, contains all objects, datums, sketches, drafting objects, etc)

The dialog has options for Create, Delete, Rename, Edit Attributes, and Information of reference sets in the current work part. The dialog also contains options to Add/Remove objects to/from the active reference set. And the reference set may contain Attribute information also.

A reference set may contain the following data:

- Name, Orientation, and Origin
- Geometry, Datum Planes and Axes, Coordinate Systems and Component Objects
- Attributes on non-geometric data

Creating Reference Sets

There are three situations where the reference set can be created:

- In the component part file
- In the assembly when the component is the work part
- In the assembly at the time the component part is created

The reference set is always created in the work part.

Which one to use is your choice depending on the design circumstances.

Naming the reference set must follow these rules:

- It must be 30 characters or less; characters after 30 will be truncated
- It must have no imbedded spaces
- The name is not case sensitive. All names are converted to upper case.

Activity 12-4. Creating Reference Sets

In this activity we will create a reference set called **Body** to represent just a solid body and filter out non-solid objects. We will use the completed **Wheel Support Assembly** to illustrate how to create reference sets in some of the components and then how to replace the existing **Entire Part** reference set with the **BODY** reference set in the assembly.

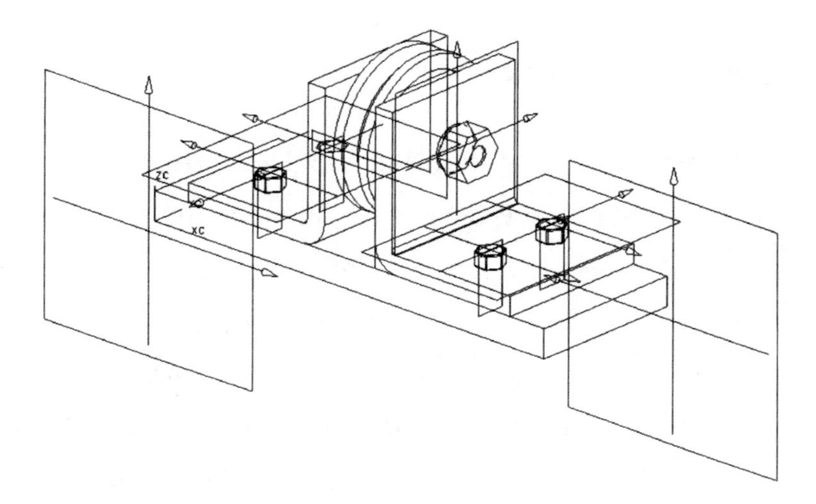

Step 1. Open **Support_Assembly.prt** from the Support directory and save it as **xxx_activity12-4.prt**. Start the **Modeling** application and start **Assemblies** application if necessary.

Step 2. **Open** the **Assembly Navigator Tool (ANT)** window by double-clicking the **ANT** icon in the resource tool bar. Drag the window to a location where the geometry is visible.

Step 3. Create a reference set in the support_angle.

3.1. In the **ANT**, select the component **support_angle** with **MB3** and choose **Make Displayed Part** from the pop-up window.

Note that the assembly disappears from the graphic area and the angle becomes the displayed part. In fact, notice it also becomes the work part.

3.2. Choose **Format → Reference Sets...**

3.3. Choose the **Create** icon in the **Reference Set** dialog.

3.4. Type **body** as the reference set name and choose **OK.**

3.5. Select the solid body of the angle in the graphics area, and choose the **OK** check icon to complete the selection of objects in this reference set. (Insure the selection by reading the status line, which says "Solid Body" as you select the body.) The new reference set appears in the dialog.

3.6. Choose **Close** from the **Reference Set** dialog window since we are not creating another reference set in this component.

Verify the reference set was created.

3.7. Choose **Information → Assemblies → Reference Set**

3.8. The reference set name **BODY** is displayed in the window. Choose OK and verify that the Body reference set contain only one body member. Dismiss the information window if it was selected.

3.9. Choose **Save** from the main menu to save the reference set in the part.

3.10. In the **ANT**, select the component support_angle with **MB3** to pop-up the window and select **Display Parent** and choose **xxx_activity12-4**. (You may or may not change the Work Part at this time—not necessary.)

Note the displayed assembly does not reflect the new reference set immediately. We must replace the reference set in the support angle.

Step 4. **Replace** the Reference Set of the support_angle in the assembly.

4.1. In the **ANT**, choose support_angle with **MB3** in which you created the reference set.

4.2. In the pop-up window, choose **Replace Reference Set** and then choose the **BODY** reference set listed. The reference set, **Entire Part**, of support _angle is replaced with the new **BODY** reference set.

4.3. The assembly display is immediately updated to reflect the change. The datum planes and axes disappear and a sketch disappears.

There are two **support_angle** components in this assembly. Repeat this step for the other **support_angle** component listed in the **ANT**.

Step 5. Create a new Reference Set in the hex_screw.

5.1. In the **ANT**, select **hex_screw** with **MB3** and choose **Make Displayed Part** from the pop-up window.

5.2. Choose **Format → Reference Sets…**

5.3. Choose **Create** icon in the Reference Set dialog.

5.4. Type **body** as the reference set name and choose **OK**

5.5. Select the solid body of the screw in the graphics area, and choose **OK**.

5.6. Choose **Close** from the **Reference Set** dialog.

5.7. **Save hex_screw**.

Step 6. **Replace Reference Set** to **BODY** in the assembly.

Step 7. **Save** the assembly.

Step 8. **Create** body reference sets in all other components and Replace their Reference Sets with **BODY**.

Step 9. **Save** the assembly.

Other Benefits of Using Reference Sets

There are other uses for reference sets other than the one demonstrated here. These will be discussed in this section.

1. Representing Components with Simpler Geometry. Create a reference set called SIMPLE and place in it a simplified solid, or some wireframe representing the outline, or the centerline of tubing.
2. Components whose shape in the design state is different in the installed state. An example of this is a spring, which is designed in a nominal condition, then when installed it is either extended or compressed.
3. Right Hand and Left Hand Parts. Reference Sets may be used to control the use of left- and right-handed parts in an assembly. The solid is created in one configuration in a component part file, and then use the Mirror Body command to create its mirror image. Therefore, left- and right-handed solids are created. Two reference sets are created: one named LH-BODY, with the appropriate solid added to it, and the other reference set named RH-BODY with its appropriate solid added to it. (The Mirror Body command is **Insert → Associative Copy → Mirror Body**. See Chapter 11 for detail.)

12.7 Mating Conditions

One of the important assembly modeling issues is how to establish parametric relationship between components. UG NX5 or later version provides two different user interfaces to do this: Mating Conditions and Assembly Constraints. They look similar and have similar functionalities and command names, but only one between the two can be active at one time. You can pick your desired one by choosing **Preferences → Assemblies** from the top menu and selecting it at the bottom of the Assembly Preferences dialog as shown below

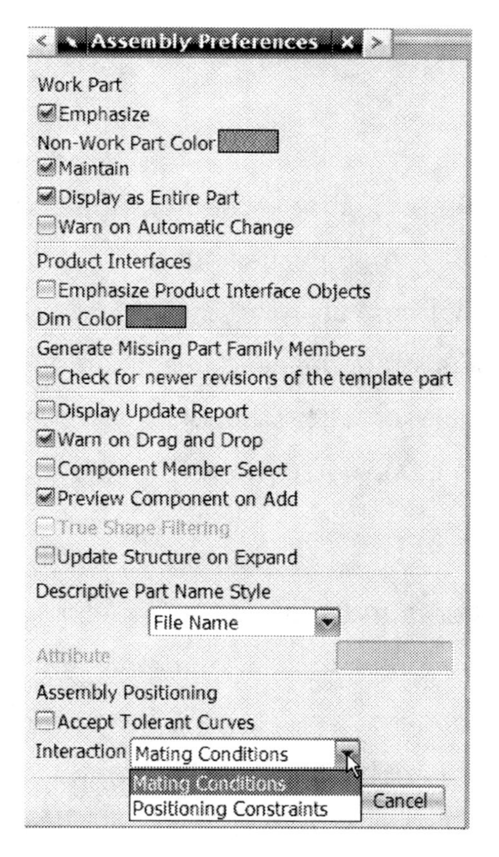

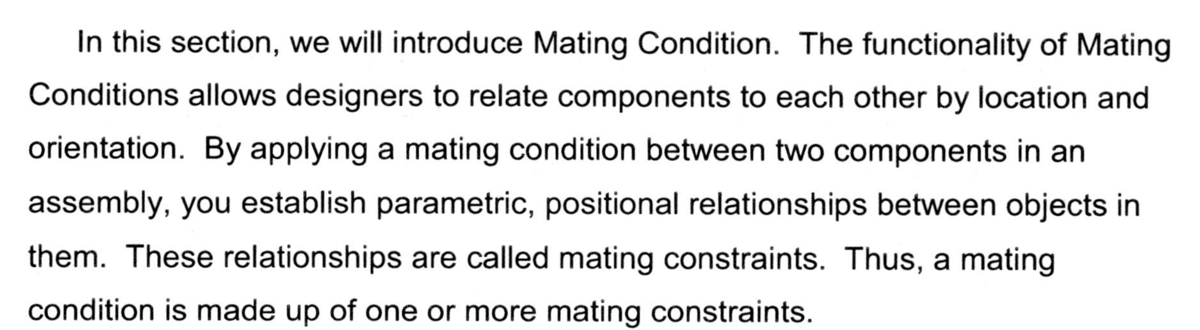

In this section, we will introduce Mating Condition. The functionality of Mating Conditions allows designers to relate components to each other by location and orientation. By applying a mating condition between two components in an assembly, you establish parametric, positional relationships between objects in them. These relationships are called mating constraints. Thus, a mating condition is made up of one or more mating constraints.

The method that we have used so far to locate components within an assembly is the Reposition option. This method does generate a good-looking assembly, but has limitations.

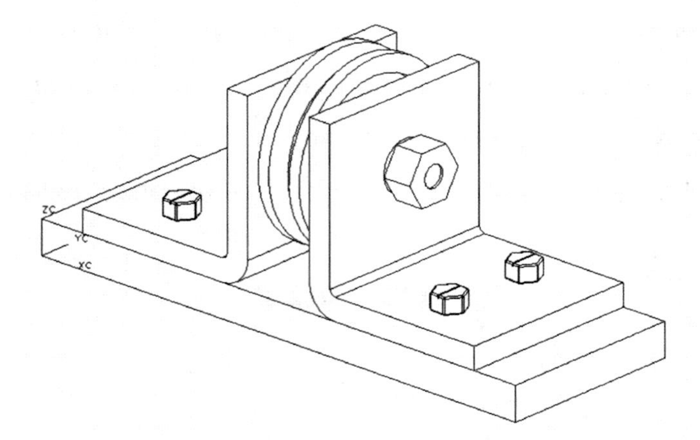

For example, in the Wheel Support Assembly, the screws that hold the support angle to the base are merely positioned there. If the holes in the base were to be relocated, then the holes in the angles and the screws would no longer go through the holes in the base—a misalignment. By creating mating constraints between the holes in both parts and the screw, then any change in location of the holes in the base would move the other holes and screws accordingly.

Mating Conditions differ from the Interpart Modeling which we did earlier using the WAVE Geometry Linker to link geometry in one component to geometry in another component. The purpose for the Geometry Linker is to define the size of geometry and to link that size in a component to the size in another component so that if the size changes in the first component, the size will change in the other component as well. It is true that with the Geometry Linker some mating has taken place, but it has limitations in defining relative position between components. Applying mating conditions allows more flexibility and handling of complex assemblies in this regard.

In general then,

1. We can say that the Geometry Linker is used to link geometry between components to define the size of geometry, and,
2. Mating Conditions are used to define the positioning of components relative to each other.

General Concepts of Mating Constraints

Mating constraints are applied to components when the positional relationship between the components is important and needs to be maintained even though changes are made to the components. So when a component in an assembly is moved to another location, then other components around it will move along accordingly in order to maintain the positional relationship imposed by mating constraints.

When selecting objects to mate, the cue line directs you to select FROM and TO objects. The FROM object is part of the component that is going to move to a new position. The TO object is part of the component that is remaining in its present location.

There are eight mating constraints that are available:

-  Mate—planar objects selected to mate will be coplanar and face opposite each other
-  Align—planar objects selected to align will be coplanar but planes will be in the same direction; centerlines of cylindrical objects will be in line with each other
-  Angle—the objects chosen will have an angle constant between them.
-  Parallel—objects selected will be parallel to each other
- 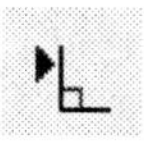 Perpendicular—objects selected will be perpendicular to each other

- Center—objects will be centered between other objects, i.e., locating a cylinder along a slot and centering the cylinder in the slot
- Distance—establishing a +/- distance between two objects
- Tangent—establish a tangent relationship between two objects, one of which is curved: a circle, a sphere, a cylinder or a free form surface

When an object is selected in these constraints, the system is looking at the normal of the object. A face normal in a solid body points away from the solid. When the Mate constraint is applied to two highlighted faces in the below figure, the faces would mate coplanar with the normals opposite each other. The faces do not necessarily have physical contact.

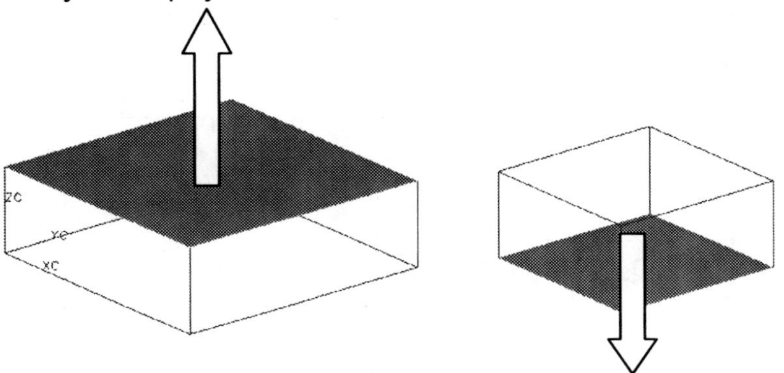

When the same two faces are mated with the Align constraint, the planar face normals are parallel and point in the same direction as shown in the figure below.

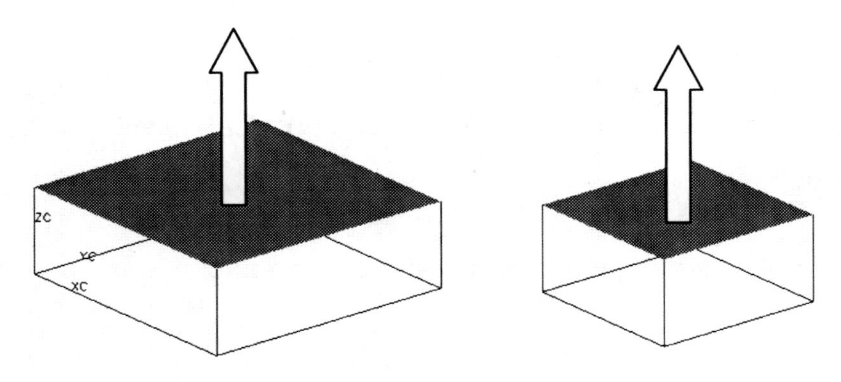

When the Align constraint is used for two cylindrical shapes, two cylindrical faces are aligned such that their centerline vectors point in the same direction as shown below. Alternatively, the Center constraint with 1 to 1 will do the same.

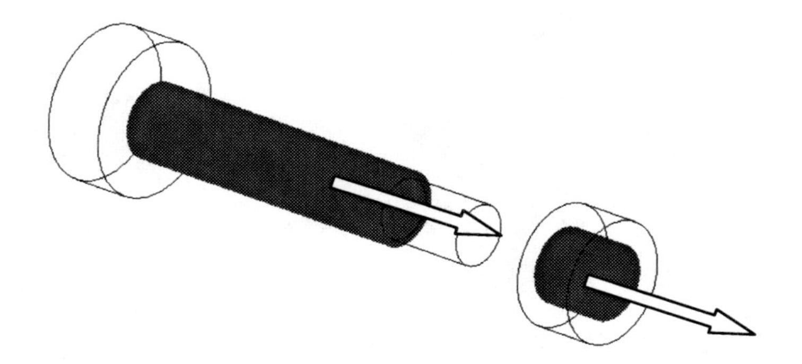

Mating Condition Dialog

Mating constraints may be assigned using one of two available methods.

- Assignment on components already existing in an assembly. In this case, choose **Assemblies → Components → Mate Component**.
- Assignment on components as they are added to an assembly. In this case, choose **Assemblies → Components → Add Component**, and then select "Mate" for the Positioning option.

Both methods can be used to an assembly that has some of the parts already assembled but not mated, and a few parts that need to be added to the assembly and mated at the same time. For both methods, the following Mating Condition dialog appears.

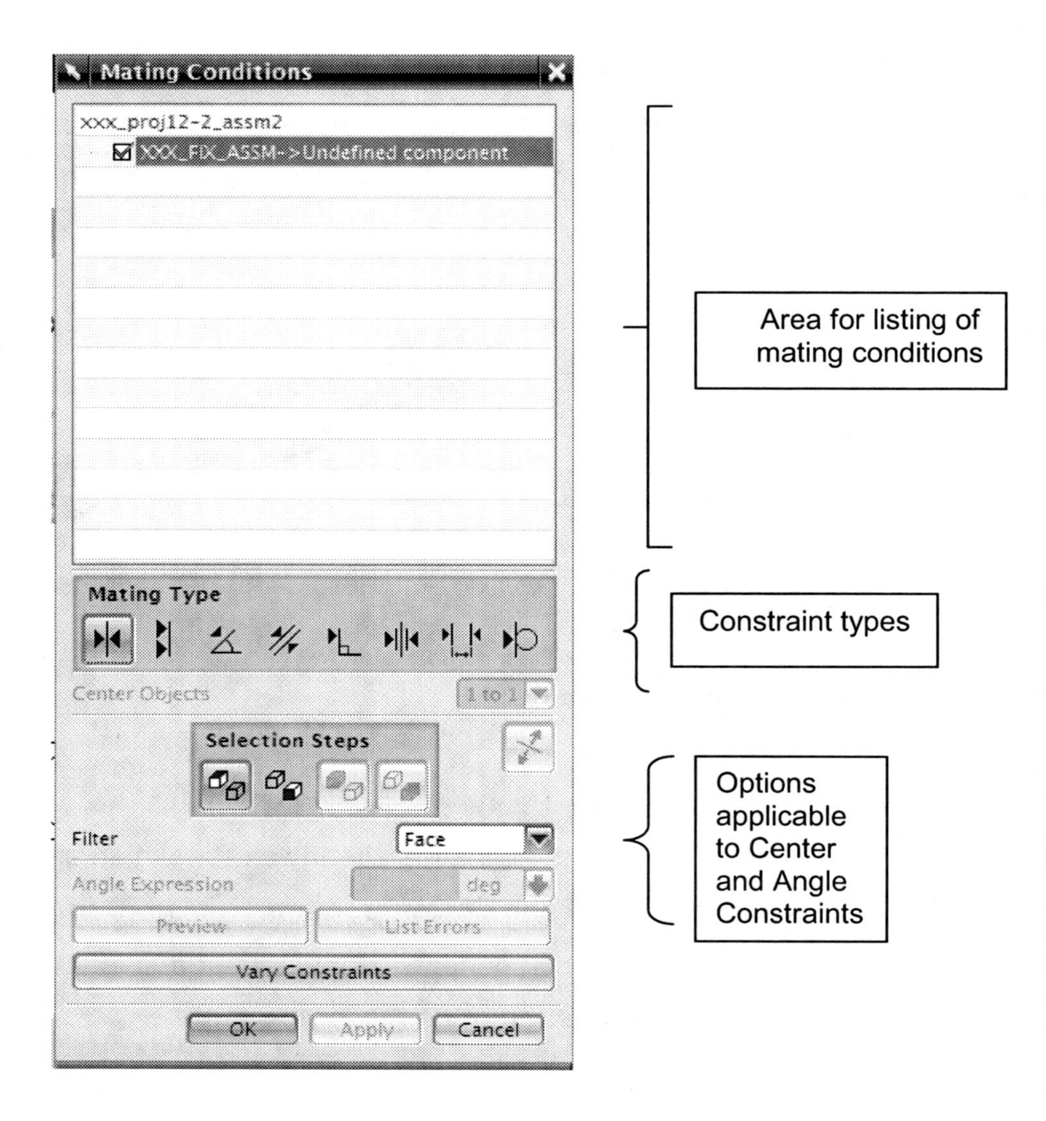

The procedure to assign mating constraints is the same for all of the constraint types:

1. From the dialog, choose the type of constraint you want to apply.
2. From the dialog, choose the object selection filter type (optional).
3. UG prompts in the Cue line to select in the graphic area, an object in the component to be the FROM component (the component that is "moving").
4. UG prompts in the Cue line to select in the graphic area, an object in the component to the TO component (the component that is "stationary").
5. Choose Preview to see the component move in response to your constraint.

6. If satisfied, then choose Apply to actually apply the constraint.
7. If more constraints are to be added, then repeat steps 1-6.
8. When mating constraints are complete, choose OK to dismiss the dialog.

You will have a chance to practice mating conditions and mating constraints n the following activity.

Activity 12-5. Create Mating Conditions

In this activity you will add a new component and mate that component as we add it to the assembly. You will use the same fixture assembly that you have already added two components into in the previous activity.

Step 1. Open part file **xxx_project12-2_assm.prt** from your directory and save it as **xxx_activity12-5.prt**.

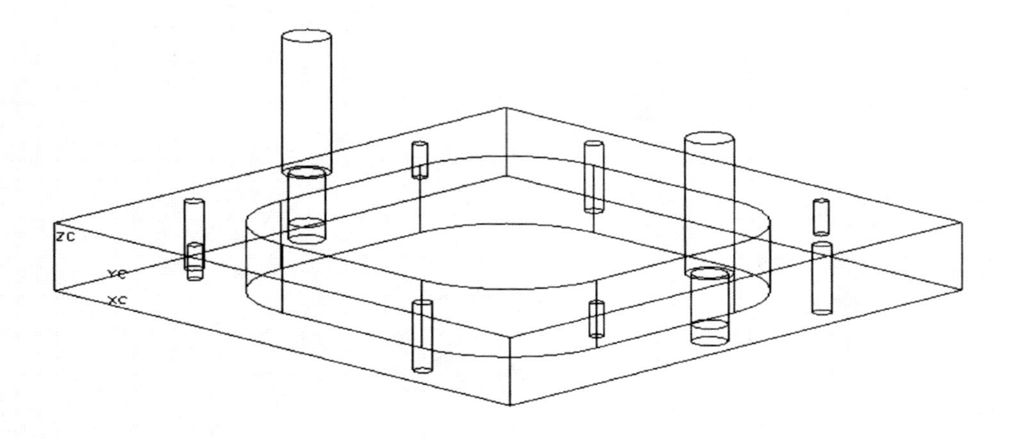

To complete this assembly we need to add the locator blocks.

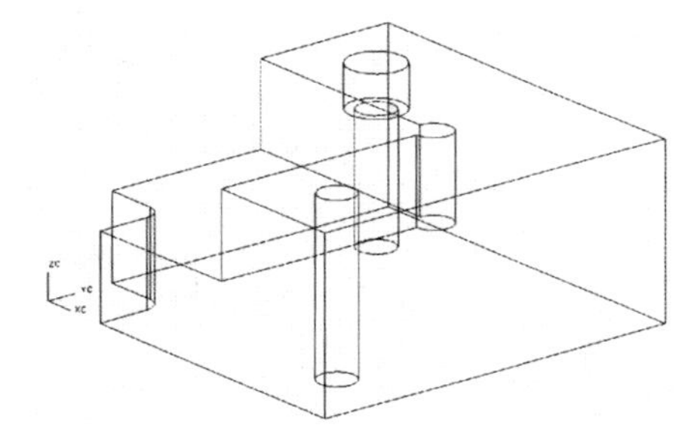

This component is not yet in the assembly, but resides as a piece part in the Fixture directory. As mentioned earlier, you can add components to the assembly and constrain them with appropriate mating conditions at the same time. Note that the assembly should be the displayed part and the work part when the component is added to the assembly and when mating conditions are applied.

You will use the following mating constraints:

- **Mate** using the top face of the baseplate and bottom face of the locator
- **Align** using the holes in each part
- **Align** using the outside planar faces of the baseplate and the locator

The assembly should look like the figure below when you complete these steps.

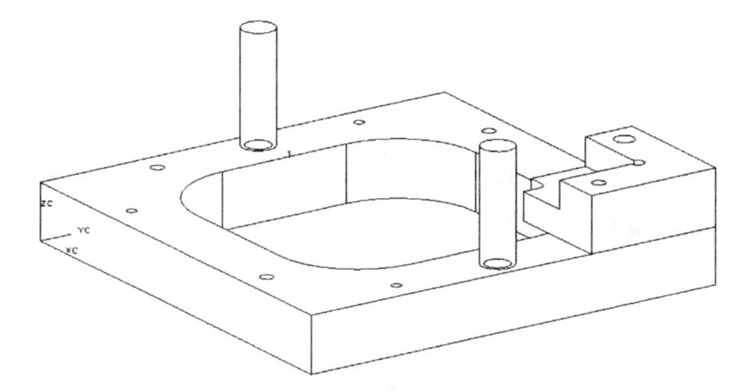

Step 2. Choose **Assemblies → Components → Add Component**

Step 3. The **Add Component** dialog opens. If locator.prt is displayed, highlight it. If locator.prt not present then go down the dialog to Open and browse for the part. When loaded in the Loaded Parts window display and highlights the locator part to be added to the assembly.

The **Add Component** dialog now specifies the following options with pre-specified option values:

- Component name: The component's name within the assembly will be `LOCATOR'. The default assumes that the component's name will be the same as the part file name.
- Reference Set: **Entire Part**.
- Multiple add: **None** is the default and since you are not adding multiple locators, this is appropriate.
- Positioning: This is where you will specify that you want to **Mate** the component as it comes in.
- Layer options: Original layers.

Step 4. From the **Positioning** pulldown, choose **Mate** then choose **OK** to accept the options in the dialog. The **Mating Conditions** dialog appears, ready for you to specify mating constraints. First, you want to constrain the locator to the top face of the baseplate.

Step 5. Choose the **Mate** icon as the first mating constraint.
Step 6. Select the **Filter** pulldown and choose the Face option if necessary.

Although it is not necessary to specify a filter, it is a good habit since this makes it easy to select your intended type of geometry.

Notice that the **From** icon is activated on the Selection Steps icon sequence and the Cue Line is prompting to select a FROM component.

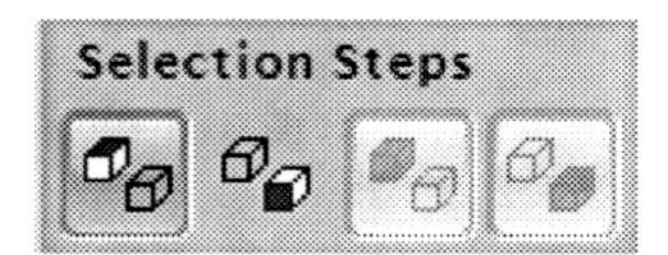

Step 7. The **FROM** object is the locator so select the bottom face of the locator. Notice after the **FROM** selection, the system automatically is ready for the **TO** selection: the Selection Steps dialogue has progressed to the **TO** icon and is ready for the **TO** component object to be selected.

Select the top face of the baseplate.

You will notice several indicators that come up now in the graphic window:

- Constraint indicator arrows, indicating the remaining degrees of freedom,
- The Status line says "Three degrees of freedom remaining".
- The planar normal vectors are pointing in opposite directions, indicating the Mating constraint.
- The designated constraint is listed in the window in the top part of the Mating Conditions dialog.

Step 8. Choose the **Preview** button to have a look at the response to the constraint you added. The change is not permanent until you choose **OK** or **Apply.**

To correct an improperly specified constraint, choose **Unpreview** from the Mating selection dialog, and then re-specify the constraint.

Step 9. If the constraint is correct, select **Apply** to accept the constraint so that more constraints may be added. (The locator part may not be in the correct position, but is on the correct face. Additional constraints will locate it properly.) Now we will use the Align constraint to insure the locator is oriented in the correct way.

Step 10. Choose the **Center** icon as the second mating constraint. The default Center Objects option is 1 to 1.

The Cue Line is prompting to select the **FROM** object.

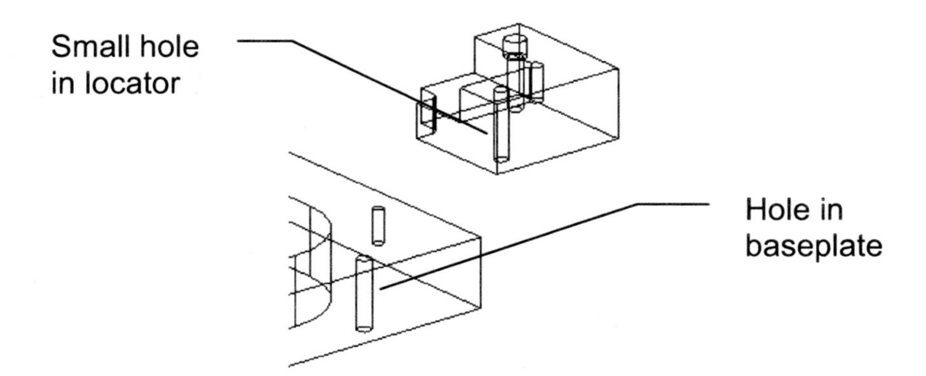

Step 11. Choose the cylindrical face of the hole in the locator as the **FROM** position.

The Cue Line is prompting to select the **TO** object.

Step 12. Choose the cylindrical face of the hole in the baseplate.

Step 13. Visually note that the correct objects were chosen and if so, choose the **Preview** button to watch the component jump to the mating condition.

Step 14. If it looks fine, choose **Apply** to make the constraint permanent.

The Status line still says there is one degree of freedom remaining —and the arrows indicate it is rotation. Another constraint is required to prevent the locator from rotating out of position. There are choices about which object to constrain. In this case you might **Center** the other bolt holes together, or you might **Align** the parallel outside planar faces of the locator and the baseplate.

Step 15. Use the **Align** icon and choose the two parallel outside planar faces of the locator and the baseplate for From and To objects, respectively. After you apply this Align constraint, the Status line will say, "Mating condition is fully constrained" since the degree of freedom remaining is zero.

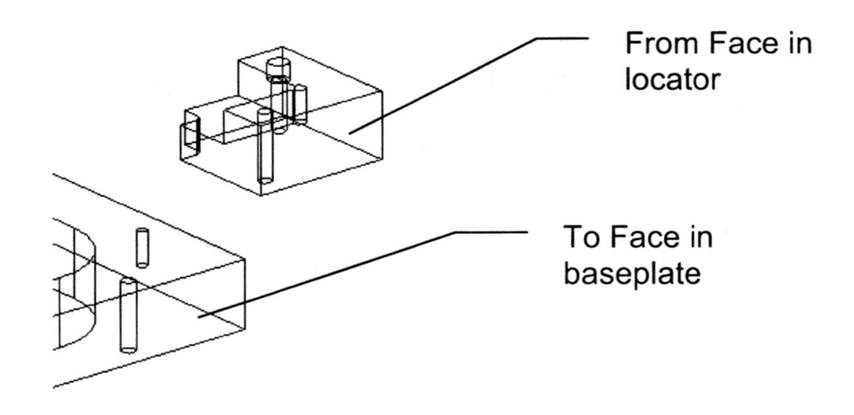

Step 16. **Save** your assembly file and continue to Project 12-3.

Project 12-3. Adding Components and Mating Conditions

This is continuation of the previous activity. Add the locator to the other three corners of the baseplate and apply mating conditions as you add the locator to each corner. The fixture assembly should look like below when completed. If necessary, use Reference Sets to filter out unwanted objects in the assembly.

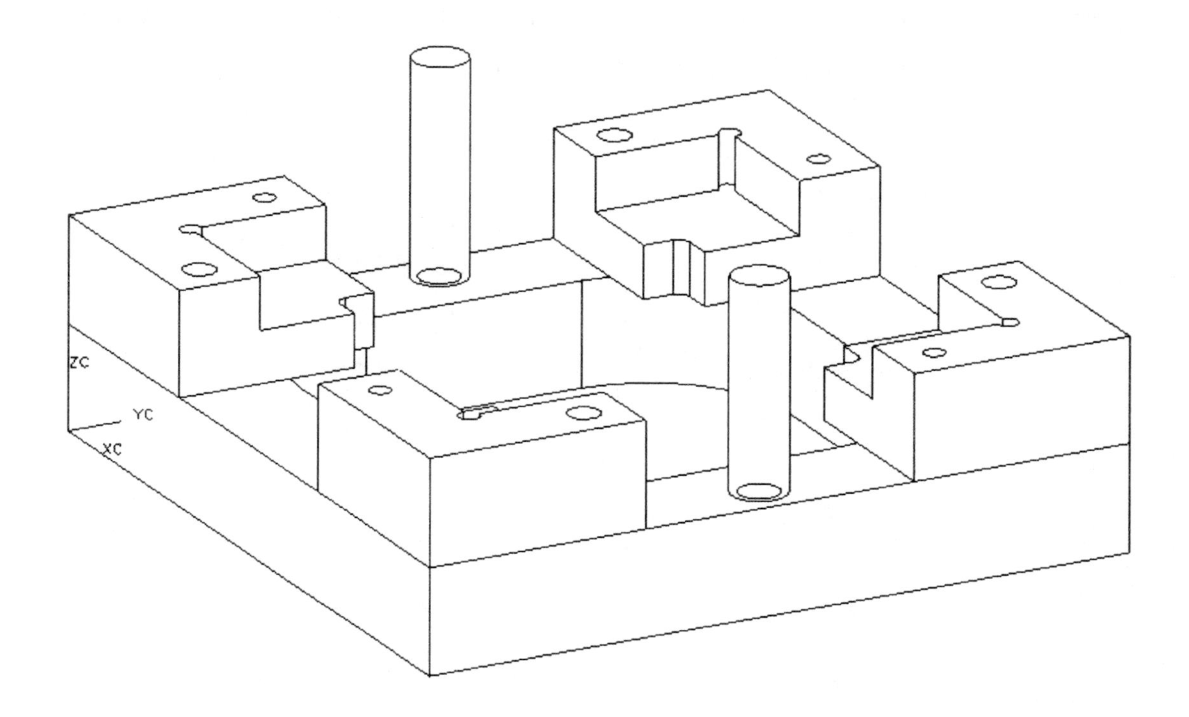

Save the completed part as **xxx_project12-3_assm.prt**. This completes the section on Mating Conditions. Important issues when applying Mating Conditions is to decide which constraints are most appropriate for your design intent. You need watch your selections closely, monitoring the Cue Line to make sure where you are in the process, and then making sure that you obtain the results you want. The mating constraint is not permanent until you choose Apply or OK.

Project 12-4. Caster Assembly

The project shown below allows you to practice mating condition operations covered in this chapter. The caster assembly consists of five components as shown below. All five part files exist under directory Chapter12-partfiles/caster. So this will be a bottom-up assembly modeling exercise. If necessary, use Reference Sets to filter out unwanted objects in the assembly.

Start by creating your own assembly part file (units: inches), name it **xxx_project12-4_assm**, and then add the parts as component objects. When mating the wheel and the axle, center them between the forks.

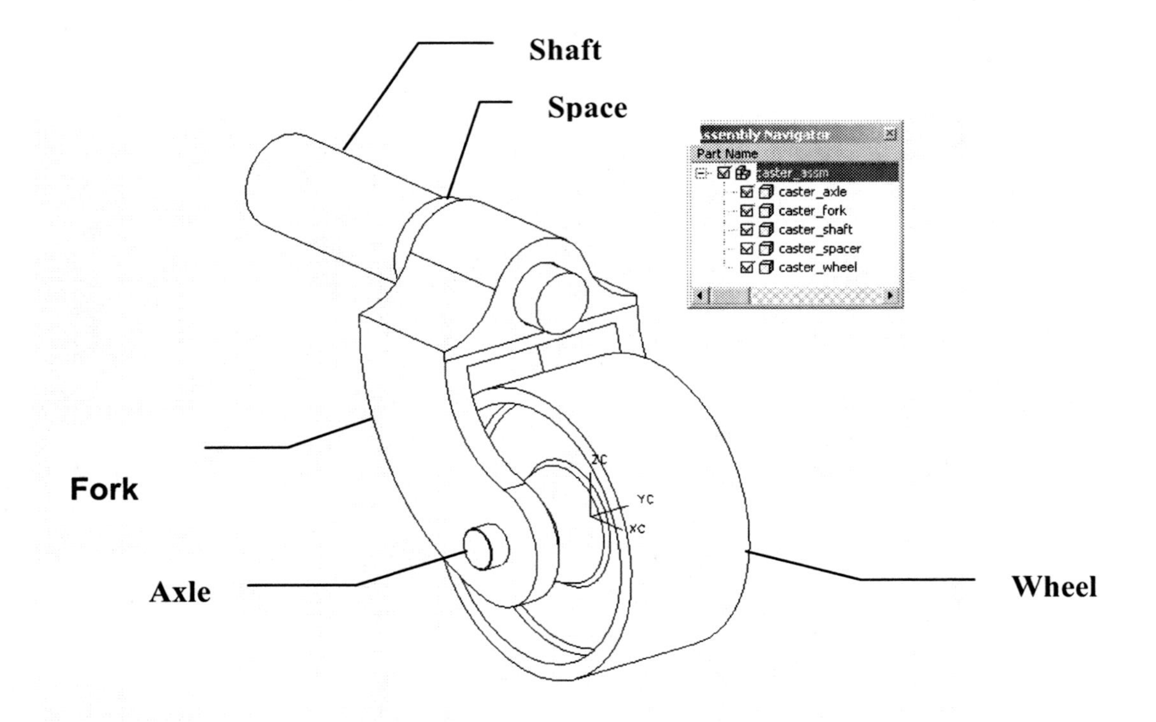

There are more projects in Appendixes A and B to practice various issues on assembly modeling discussed in this chapter.

Exercise Problems

12.1. List and discuss some of Unigraphics Assembly capabilities.

12.2. Define an assembly.

12.3. Define a Unigraphics assembly.

12.4. What is the name of the object that resides in the assembly part file that establishes the associative link to the piece part file?

12.5. How does Unigraphics identify an assembly part file?

12.6. Discuss differences between component object and component part file.

12.7. What parameters of the part file does the component object define?

12.8. Define Bottom-Up modeling approach.

12.9. Define Top-Down modeling approach.

12.10. Discuss the practical application of the Bottom-Up and Top-Down approaches.

12.11. What functionality is used to link the position of a component to another component?

12.12. What functionality is used to link geometry of a component to geometry in another component?

12.13. What purpose do Reference Sets perform in Assemblies?

12.14. Explain the difference in using Reference Set versus Layer Control to accomplish the purpose for Reference Sets.

12.15. What is the functionality that must be enabled by the system manager before the WAVE Geometry Linker may be used?

12.16. Which sets of commands are used to create an assembly Bottom-Up?

12.17. Which sets of commands are used to create an assembly Top-Down?

12.18. Describe briefly how to apply Mating Conditions.

12.19. Contrast the differences in using Mating Conditions vs. Reposition to locate components relative to each other in an assembly.

12.20. What unique characteristics does a sub-assembly have?

Chapter 13. Master Model

In this chapter, we will introduce the concept and purposes of the Master Model method, which can assist the design process from initial design through manufacturing of a part in a company. This concept centers on the individual model part created by a designer. To understand the benefits of the Master Model method in the design process, it is helpful to review the traditional approach to the design process. We will compare the traditional method to the Master Model method by following a part from initial design through the production of a part.

Section 13.1 will review the traditional and Master Model methods in the design process. Section 13.2 will discuss the benefits of using the Master Model. Section 13.3 will describe how the Master Model is accomplished. Section 13.4 will put the Master Model into practice by doing two activities.

13.1 The Master Model Concept

We described the product design process in Section 1.1 of Chapter. In the section, we identified six steps of the design process that are needed to bring an idea through to a product design ready for manufacturing.

At the initial phase of the design process, the designer is given a need for a new product and a set of criteria and parameters that the product must satisfy to meet that need. The designer creates a design part, hands off to analysis groups, back to the designer to refine the design even more, also handing off to a draftsman to complete a drawing of the part, and to other groups that may need to see the design part as well (manufacturing, sales, marketing, customers, material planners). As stated in Chapter 1, it is an iterative process involving a cross-functional team of the company.

In traditional engineering and manufacturing companies the design process described above is a serial process where each functional discipline does its portion in turn after other disciplines have completed their portion. On the other hand, when the Master Model method in the design process is adopted, the cross-functional team can begin doing their portion of the design process at the same time (concurrently) as the other members of the team are doing theirs. Team members can actually see the digital model while the designer is still creating and designing, and see updates as the design progresses, and all may start doing their portion of the process much sooner than in the serial type process. Another statement in Chapter 1 was that computers don't change the "nature of design process", but computers do improve efficiency and productivity. The Master Model method enhances those factors even more enabling products to come to market much faster than the traditional method.

The Master Model is enabled by the assembly modeling functionality in which each cross-functional team member creates an assembly file where the designer's part is added *as a component object in an assembly. Then each team performs its own task in its own assembly part file but not in the original design part file.* Recall from Section 12.2 of Chapter 12 that the component object in this assembly file does not contain all the design information but only the pointer linking to the designer's file. Then a drafting team adds drafting information to its own assembly file that contains the component object of the designer's part; likewise, any CAE analysis, NC programming, documentation, and others, is done on separate assembly files instead of the original designer's part file. More importantly, these assembly files can be created before the design is completed and the work of other disciplines may begin before completion of the design. This process is no longer serial but is concurrent. Design changes later added to the design's part file will be immediately reflected upon those assembly files as the component object points to the designer's part file.

This is the fundamental concept behind the Master Model: all downstream disciplines that utilize the original designer's model must use a copy of the model as a component in an assembly to perform their disciplines; and all work is done concurrently and in cooperation with the original designer.

Graphically, the Master Model concept is visualized as below.

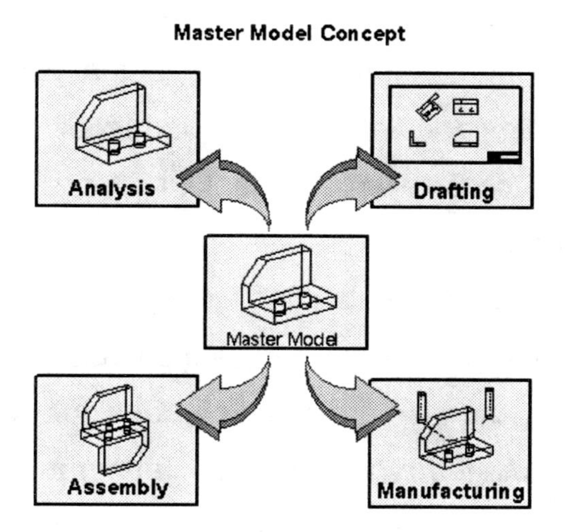

The original model created by the designer becomes the "Master Model" from which all copies are made for other disciplines. In assembly modeling terminology, the Master Model is the component part file and the component part file is added as a component object in other part files, to become an assembly.

13.2 Benefits of the Master Model

Why use Master Model methods? In addition to the concurrent engineering benefits discussed above, there are several benefits that all contribute to bringing a better-designed part to market faster.

- The original model retains the integrity of the designer's intentions and cannot be corrupted by downstream operations or other designers, draftsmen, or other disciplines by not letting others change the original model. The original designer maintains full control of the design without sacrificing the ability of others to aid in the process to bring a part to production.
- Many disciplines may begin work and planning on a part before the part is completely designed and released for final production. Other disciplines may copy and see the part as it progresses through the design stage and

begin planning what operations need to be done to perform the tasks of their discipline.

- For example, manufacturing operations require a considerable amount of planning and setup work before a part may actually be put on a machine and cut. The planning and setup work include tooling, fixture design, NC programming and etc. By allowing a look at the preliminary outline of a part, and subsequent looks as the design develops, manufacturing planners can begin the initial planning necessary to set up to cut the part.
- CAE analysis engineers may begin deciding what analysis may be needed on this part and also provide a part designer with useful feedback at the early design stage for possible design upgrade.
- When a particular part is used in conjunction with other parts to build an assembly of parts to be used in a product, then the Master Model allows many designers to concurrently create models of the various parts in an assembly and to share those parts in an assembly maintained by yet another designer. All designers will have access to use other designers' Master Models for their design to ensure that parts that mate together will indeed fit together. Thus, accuracy and quality of the finished parts is enhanced, and time to market is reduced.
- In assembly modeling, the copy of the Master Model is not a duplicate of the original part but in reality a virtual copy (the pointer in essence). That is, the Master Model is not duplicated onto the disk of other users, but remains on the disk of the Master Model designer thus taking up no additional disk space. Each other designer has a virtual "copy" to use as background for mating and designing their parts. Thus, total usage of storage disk space in the Master Modeling is reduced over the traditional method, which may use several duplicates, causing redundancies and potential inconsistencies.

The Master Model method, with the usage of assembly modeling, facilitates the industry goal of concurrent engineering, helping the various disciplines working together concurrently. Concurrent engineering greatly reduces the time to market from initial concept through manufacturing with minimal effort to update design changes.

13.3 How is the Master Model Accomplished?

As mentioned above, the Master Model is accomplished using the tool in Unigraphics called Assembly Modeling and components to create copies of the original design. It is a powerful tool that allows virtual copies to be made of the parts and still retain a link to the original model. The link allows the original to be changed as the design progresses and the copies to be updated so other disciplines always have the most up-to-date model from which to work.

The process to accomplish the Master Model is as follows:

- Each discipline creates an assembly part file using their naming convention, and
- Adds the Master Model part to its assembly Bottom Up as a component object as discussed in Chapter 12 Assembly Modeling. The usual command sequence **Assemblies → Components → Add Component** is used to add the designer's part file.

That's all there is to it. The component in the assembly part file may be drafted, or analyzed, or NC programmed by other disciplines. Each discipline has the ability to save the objects created in conjunction with their discipline; but they do not have the ability to make changes to the original model and save those changes because they don't have write privilege to the Master Model.

In the previous diagram, each of the part files shown around the Master Model part file is a separate assembly part file. When the Master Model part is revised, its component objects in those assembly part files can be updated with minimal or no associativity loss. For instance, when the Master Model has been drafted and annotated, the update to the Master Model will not only update the views on the drawing of the Master Model, but will also update all the associated dimensions and other annotations.

13.4 Master Model in Practice

This section illustrates the Master Model in practice using an example. Drafting is chosen as another discipline for the illustration. In the following example, a Master Model part was already created by a designer, and the drafting department created from the Master Model an assembly file to store drafting information such as drawings for the Master Model part. The designer will make a design change in the Master Model part and we will see how the change in the Master Model will be reflected in the drawing file in the drafting department.

Activity 13-1. Exploring the Benefits of Master Model

Step 1. **Open** part file **shoulder_screw_dwg.prt** from the directory Chapter 12/partfiles/Support.

Step 2. Choose **Start → Drafting**.

Step 3. Inspect the shoulder screw for its dimensional values, particularly the length of the threaded portion of the shaft. **(3.250 – 2.500 = .750)**

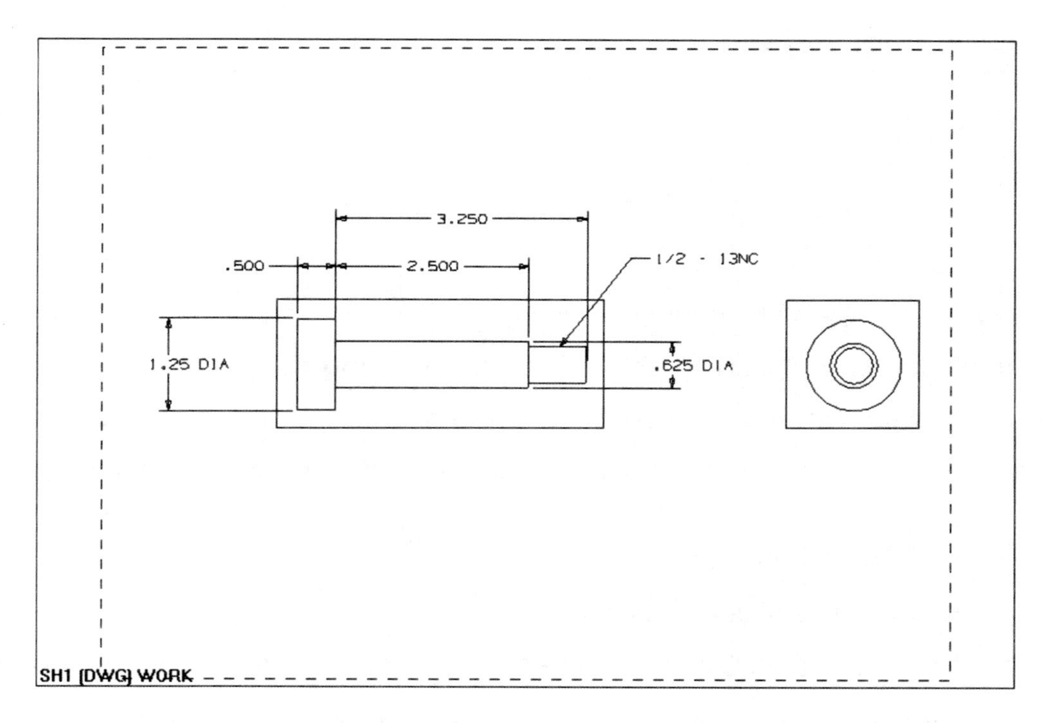

This looks like a normal detail of a component part, but examining it closely as in the following step will reveal that it is a component object in the assembly file showing drafting information.

Step 4. Choose Assemblies → Reports → List Components

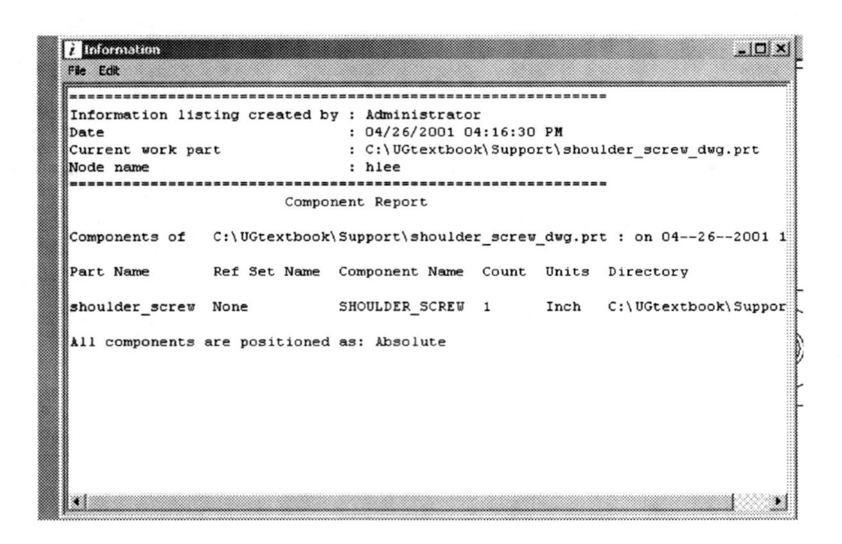

```
Information listing created by : Administrator
Date                           : 04/26/2001 04:16:30 PM
Current work part              : C:\UGtextbook\Support\shoulder_screw_dwg.prt
Node name                      : hlee

                    Component Report

Components of   C:\UGtextbook\Support\shoulder_screw_dwg.prt : on 04--26--2001 1

Part Name       Ref Set Name  Component Name  Count  Units  Directory

shoulder_screw  None          SHOULDER_SCREW  1      Inch   C:\UGtextbook\Suppor

All components are positioned as: Absolute
```

The Information window appears showing the assembly file structure for the file **shoulder_screw_dwg.prt**. It shows that the file **shoulder_screw_dwg.prt** contains one component object named **shoulder_screw.prt**. It also shows the pointer information where the Master Model **shoulder_screw.prt** is located.

Dismiss the Information window.

Step 5. Retrieve the Master Model file.

5.1. Choose **File → Open** and select **shoulder_screw.prt** from the list of the files in the same directory. The **Open Part** window appears. Choose the button "Change Displayed Part".

Step 6. Enter the **Modeling** application by **Start → Modeling**.

Step 7. Edit the length of the threaded shaft on the screw.

7.1. Choose **Edit→ Feature → Edit Parameters**, select **BOSS(2)** from the list, and choose **OK**. Select **Feature Dialog** and edit the height of the boss to 1.0.

7.2. Choose **OK** three times and the screw updates.

Step 8. Change the displayed part to the drawing.

8.1. From the main pulldown menu **Window**, select **shoulder_screw_dwg.prt** from the list.

Step 9. Choose the **Drafting** application by **Start → Drafting**.

The drawing of the shoulder screw comes up and notice that the bottom left corner next to the drawing name reads (OUT-OF-DATE). This is a notification to the draftsman that something has changed and the drawing needs to be updated.

Step 10. Update the drawing.

10.1. Select the update view icon in the **Drawing Layout** toolbar. The **Update Views** dialog appears. Select **All** and choose **OK**.

Step 11. Watch the views update and notice the dimensions update to reflect the longer shaft.

Step 12. Retrieve the support assembly and see the update to the shoulder screw in the assembly. The shoulder screw was changed at the component part file level, so the assembly should be updated as well.

12.1. Start the **Modeling** application. **File → Open Support_Assembly.prt** from the same directory. In case you have a difficulty in loading components of this assembly, change the file loading option by choosing **File → Options** (Load Options and then choosing an option either "From Directory" or "As Saved" for the Load Method. Usually one of these options works, depending on where the component parts of the assembly are located.

12.2. Verify the shoulder_screw has been updated.

Step 13. Close all parts.

Activity 13-2. Creating a Master Model Assembly for Drafting

This activity will demonstrate the Master Model method by creating an assembly file to be used by the drafting department in Chapter 14.

Step 1. Create a **New** part file **xxx_locator_dwg** (units: inches). Start the **Modeling** application. Start the **Assemblies** application if necessary.

Step 2. Using **Assemblies** add **locator.prt** to this part file to create an assembly. This part is under directory Chapter 12/partfiles/fixture.

- **2.1.** Choose **Assemblies → Components**.
- **2.2.** Choose **Add Component**.
- **2.3.** Choose the **Choose Part File** button to browse directories.
- **2.4.** Choose **locator.prt** from directory Chapter 12/partfiles/Fixture.
- **2.5.** On the **Add Existing Part** dialog set the **Layer Options** to **Original** and **Positioning** to **Absolute** (if not already).
- **2.6.** Choose **OK**.
- **2.7.** On the **Point Constructor** dialog make sure the **XC**, **YC**, and **ZC** are all 0, 0, 0. If not, choose the **Reset** button.
- **2.8.** Choose **OK** to place the locator component at 0, 0, 0.
- **2.9.** After the locator part is on the screen, **Cancel** the dialog.

Step 3. Save and close the assembly with **File → Close → Save and Close**.

The Master Model part, locator.prt, was virtually copied into the assembly file, xxx_locator_dwg.prt, as a component object to it. In the next chapter, Chapter 14 Introduction to Drafting, we will continue to use this assembly file and add the drawing information into the assembly file.

Exercise Problems

13.1. List and discuss the benefits of using the Master Model in the design process.

13.2. List and discuss the traditional method of the design process.

13.3. What other disciplines can benefit from using the Master Model? Discuss the potential benefits.

13.4. Which Unigraphics capability is used to implement the Master Model?

13.5. Which part file in the assembly hierarchy is the Master Model?

Chapter 14. Introduction to Drafting

In this chapter, we will introduce the Unigraphics Drafting application. The goal will be to give the designer/draftsman enough knowledge of drafting tools to create a basic drawing of their design including placing of views, dimensioning, and other annotation.

This chapter consists of seven sections. Section 14.1 gives an overview of the Drafting application. Section 14.2 introduces creating drawings and adding views. Section 14.3 shows you how to add dimensions to the drawing. Section 14.4 discusses customizations to the drawing to meet company standards and preferences. Section 14.5 shows you how to add notes and labels to the drawing. Section 14.6 addresses how to make modifications to the drawing for more customization. Section 14.7 adds the additional functionality of Utility Symbols to make the drawing look more professional.

14.1 Introduction to the Drafting Application

The Drafting application is based on creating views from a solid model. The views will be treated and handled as if they are 2D, but they are closely associated to the original solid model. Some useful features of the Unigraphics Drafting application are:

- Once you identify what view type you want, and then views may be added just by indicating the location with the cursor.
- As orthographic views are added, they will automatically be aligned with the parent view.
- Every view is associated with the solid. Thus, when the solid is changed, the drawing will be updated, including views and dimensions.
- Drafting annotations (dimensions, labels, and symbols with leaders) are place directly on the drawing.
- Drafting annotations (see above) are associated with the solid geometry that was selected and will update automatically if the geometry is changed.

- Fully associative view boundaries are automatically calculated and sized when the drawing is updated.
- The Preferences pull-down dialog allows for extensive customization of annotation and views.

In this chapter we will learn how to create a drawing, add orthographic views, dimensions, other annotations, and edit and customize all of the aforementioned.

14.2 Creating Drawings and Adding Views

We will begin by creating a drawing with views to get a quick start to this topic.

Activity 14-1. Create a Drawing

We will use the master model part file we created in Chapter 13.

Step 1. **Open** part file **xxx_locator_dwg.prt**.

Step 2. Verify the locator part is a component in the part file just opened.

2.1. Choose **Assemblies → Reports → List Components**

Step 3. Choose **Start → Drafting**

Step 4. Modify the Drawing size

Since the default drawing size is an E size we need to make the size smaller to fit our part onto the drawing better.

4.1. The **Sheet** dialog comes up.

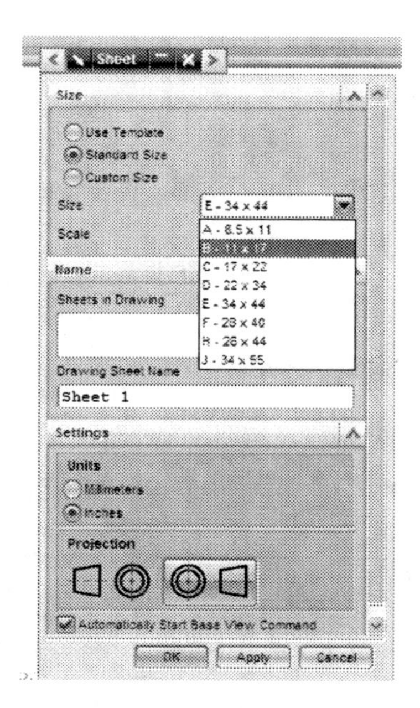

4.2 Choose Pull-down arrow in the middle of the dialog where it shows the drawing size: **E – 34 X 44**

4.3 Choose another drawing size--the **B** size. The **B** size should be large enough for our part.

4.4 Note that the drawing scale may be changed as well, but we'll leave as is. Note the drawing name defaults to SHEET 1. We could edit the name but we'll leave the name as is also.

Choose **OK**, the dialog dismisses and the graphics area shows the updated drawing size.

5.1. The **Base View** dialog bar comes up shown in the upper left corner (or wherever the default is set up) of the graphics window.

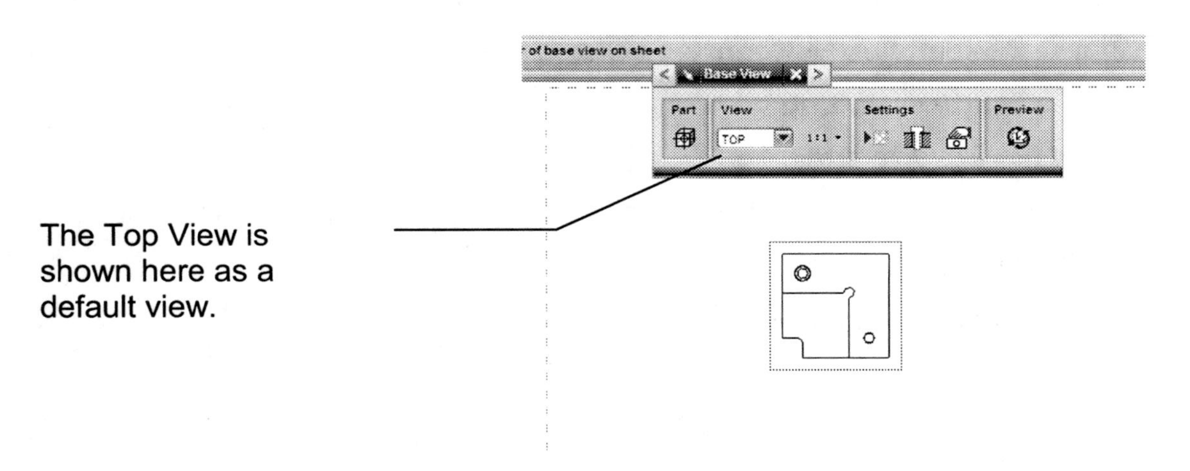

The first view placed on the drawing must be the Base View type. Notice that it is the default selection and does not have to be chosen. The Top view is the default view to be placed. The cursor shows a rectangular border and the Top view of the part and Unigraphics is ready for you to place the view on the drawing.

5.2. Select the location on the drawing sheet to place the view.

> If you have selected the Top view and you intend to place Front and Right views, then select a location in the upper, left corner of the drawing sheet, leaving room for dimensions and other anticipated annotation.

The selection remains active to place more views on the drawing, however, the dialog changes to an orthographic projection of the base view just added.

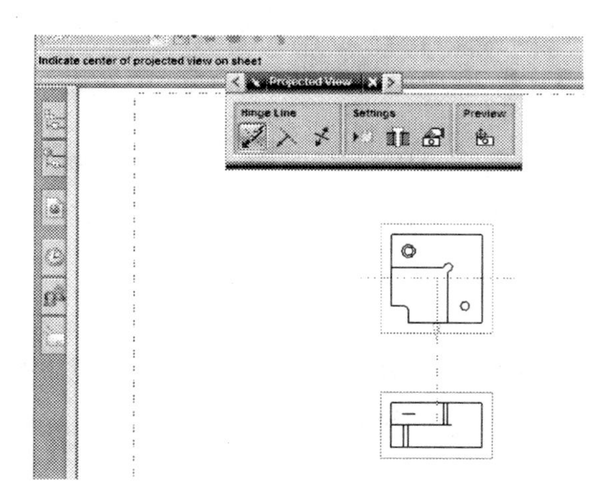

The figure illustrates the placing of the Top view and the cursor ready to place an orthographic view.

5.3. Another rectangular outline appears on the cursor and you may slide the rectangle up and down in a vertical line from the **Top** view, or horizontally from the **Top** view, or a projection at any angle--in other words, an orthographic projection.

5.4. Select a position below the **Top** view, again leaving room for dimensions and other annotations.

The **Front** view is added and another view may be added using the existing parent view, the **Top** view.

However, we want the **Right** view projected off the **Front** view so we must change the Parent view.

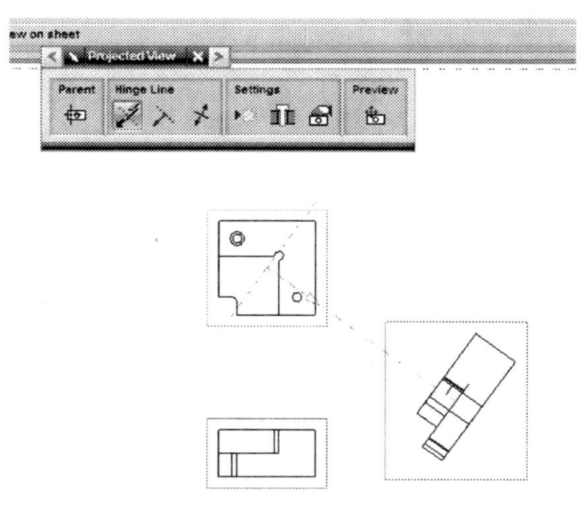

5.5. The ESC key or middle mouse button is used to cancel the placement of views. Select a button and the two views are left as placed.

5.6. Choose the **Projected View** icon . The graphic comes up with another projected view on the cursor, projected from the existing parent view.

5.7. On the **Projected View** dialog bar, choose the **Parent** view icon , the Cue line prompts to Select parent view, and select the Front view as the new parent view.

5.8. A Right view appears on the screen and use the cursor to drag it the proper distance from the Front view.

5.9. It is helpful now to add a **TFR-TRI** view to the upper right corner to help identify the part for other people to aid in visualizing the part.

5.10. Therefore, repeat step 5.1 again, choose the **Base View** icon again, choose **TFR-TRI** from the drop-down window and place the **TFR-TRI** view in the upper right corner of the drawing.

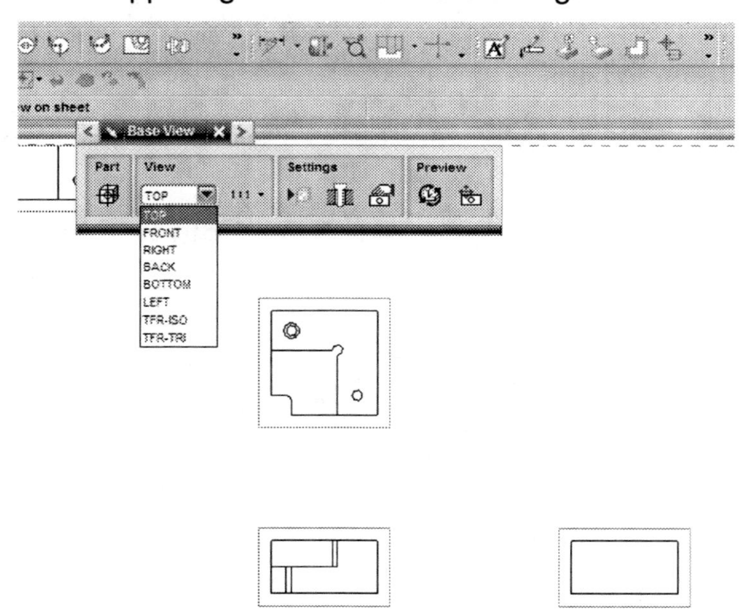

5.11. Choose **ESC** to close the **Projected View** dialog.

The drawing should look something like this when complete.

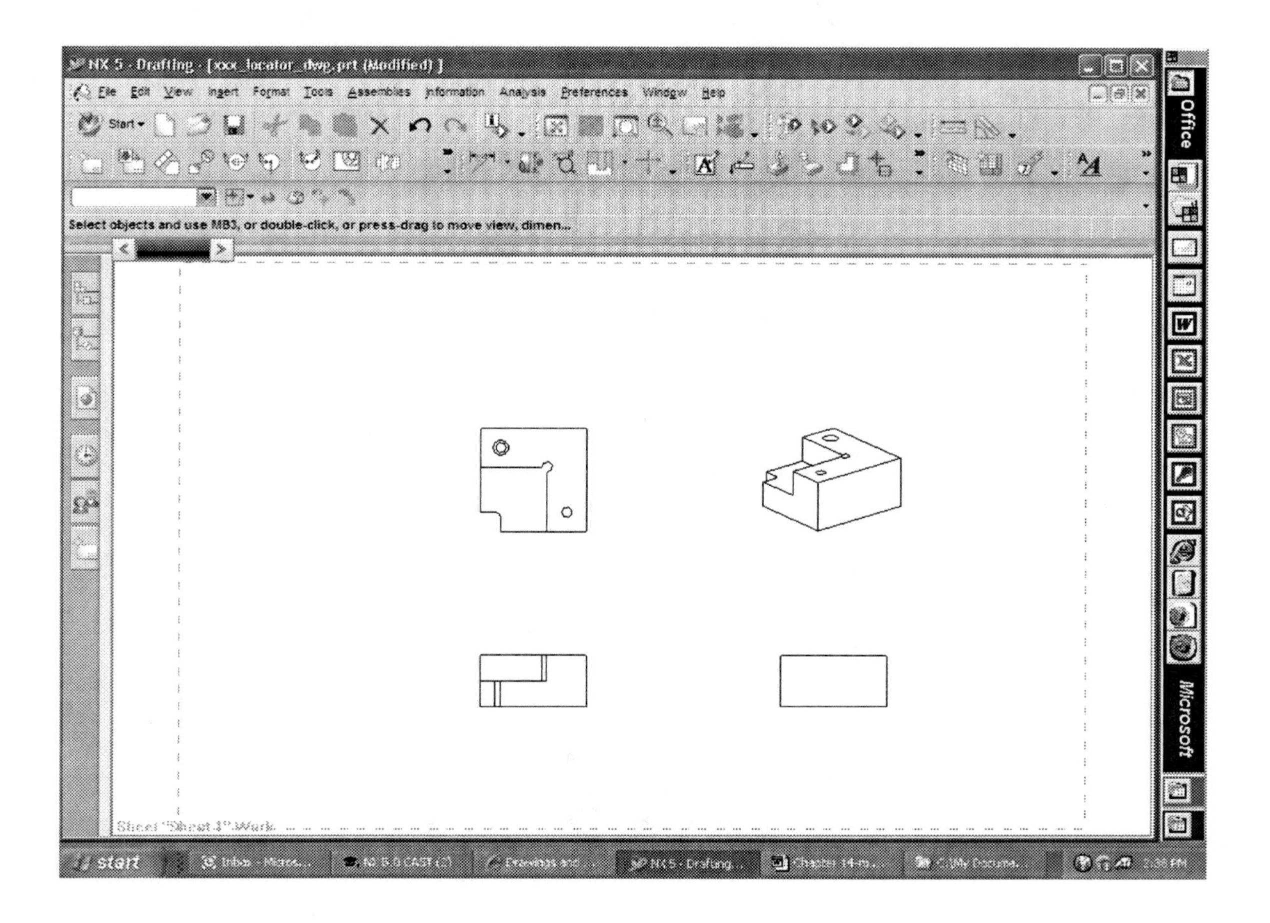

As you can see, views can be placed on a drawing quickly. With knowledge of your part size, views can be placed on the drawing accurately as well.

14.3 Creating Dimensions

The next step we will put on some dimensions. The **Dimension** menu is found in the **Insert** pull-down menu. Notice the **Insert** dialog changed when we entered the Drafting application.

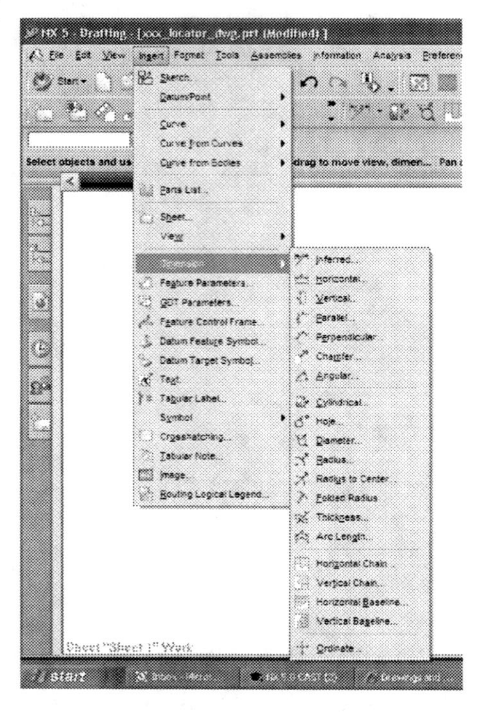

All the dimension types are listed on this cascade menu and may be selected to do your dimensioning. However, it may be cumbersome to go through the **Insert → Dimension → (dimension type)** each time you want to change dimensions. Instead, it is suggested that the toolbars be utilized to quickly choose the dimension types you wish to use.

(Note the default position of this toolbar is horizontal above the graphic screen. If you do not see the same set of icons as above, you may need to customize the Dimension tool bar with **Tools → Customize...**)

From the Dimension toolbar, each dimension type may be selected to bring up a dialog for that dimension type. Each dialog is defaulted to allow quick creation of dimension and placement. This can facilitate quick referencing by other disciplines to start planning for their involvement in the design process. We have shown only the more common dimension types; however, you may want to **Customize** your toolbar with the dimension types you use most often.

Activity 14-2. Dimensioning the Part

The Dimension toolbar is up and ready to select the dimension types on it.

Step 1. Before we start dimensioning we want to turn off the rectangular border around each view. Choose **Preferences → Drafting → View** tab and toggle off **Display Borders**. While you are there toggle off **Delay View Update** if it is checked. Choose **OK**.

Step 2. Choose the **Inferred** dimension type .

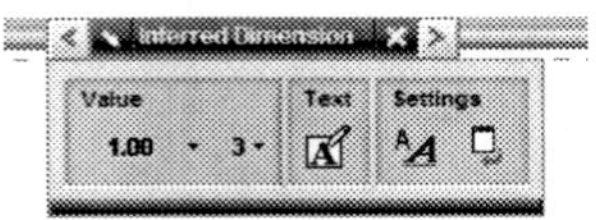

This dimension bar that pops up contains the formatting of a dimension. It is defaulted such that dimensioning may begin without worrying about the setups for now.

The **Inferred** dimension type allows most kinds of dimensions to be placed on the drawing and the type dependent only on the objects you select, where you select them, and sometimes the order. So, it can replace Horizontal, Vertical, Parallel, Perpendicular, Radius, and so forth, depending on how you select. The other dimension types are still available and may be used in difficult situations for the system to interpret which type you want to use.

For now we'll demonstrate the **Inferred** type.

The **Cue Line** is prompting to select the first object or a dimension to edit. Let's dimension the overall length of the part in the **Top** view. We will note that points or the edge may be selected.

2.1. In the **Top** view, select a vertical line on the left side of the part and then, similarly, select a vertical line on the right side of the part.

2.2. Notice the dimension is placed on the cursor and you may drag it up and down on the dimension lines to select a place to locate it.

2.3. Select a location and place the dimension. Notice that even though you could drag the dimension up and down, the number stayed centered between the dimension lines. That is a default already set up.

2.4. The system is ready to create another dimension. All you have to do is select the geometry you want to dimension.

2.5. Infer is still active. This time select a horizontal edge. Note a dimension appears. Drag the dimension to locate it and place it. This got the same result—one pick. If it works for you just pick one edge.

2.6. Select additional geometry and place horizontal dimensions on the views to dimension the part correctly.

Note that if you select a wrong object at any time before placing the dimension, use ESC to cancel the selection. Use MB2 or the roller wheel to cancel the dimension. You should use the TFR-TRI view to verify selections.

Step 3. Create vertical dimensions.

This dimension type works exactly the same way as the horizontal type. Select geometry as appropriate to place vertical dimensions on the views. When finished your drawing may look something like this.

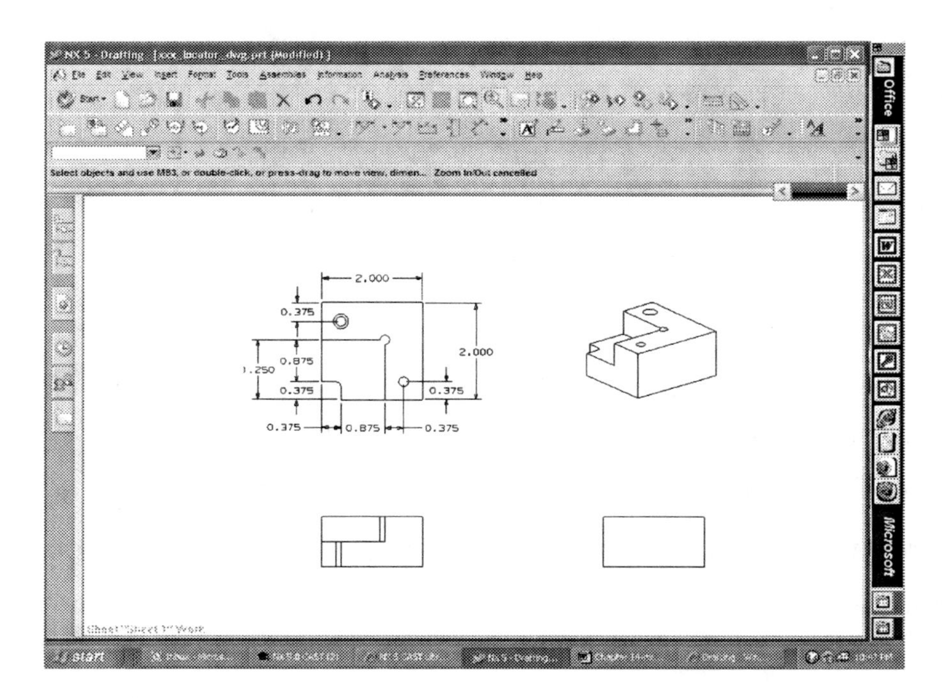

Next, we'll add some diameter and radius dimensions.

Step 4. Continue to use the Inferred dimension type to select holes .

4.1. Choose the hole in the lower right corner of the part. The origin of the dimension is on your cursor and you can locate it around the hole.

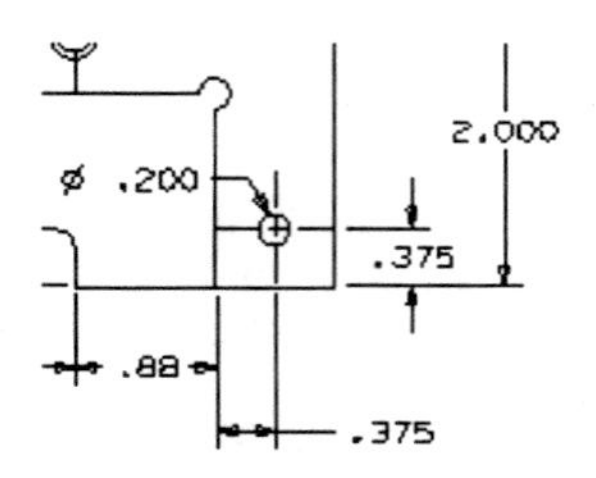

If you want to change the dimension location, select the dimension with MB1 and continue to press MB1 until you move it to the desired location.

Placement Options

There are placement style options on the highlighted icon below in the dimension bar called the Inferred Dimension dialog bar.

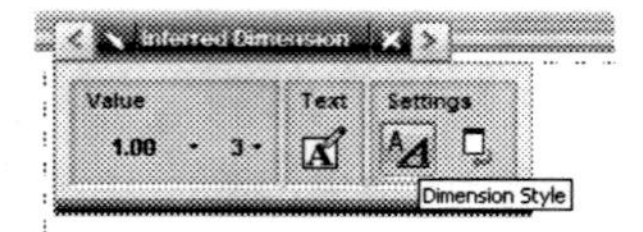

When the Dimension Style icon under Settings is selected the Dimension Style dialog comes up. The top portion is shown.

The option of the dimension with the lightning bolt through it represents Automatic placement for system control of the placement of the origin for dimensions. **Automatic** placement is the default. It means that Unigraphics will infer the placement of the origin only allowing you limited freedom to put the annotation where you want. It will infer the placement of the dimension arrows. This works pretty well for the distance type dimensions: horizontal, vertical, parallel, perpendicular, and cylindrical. But for the diameter, radius, and angular dimensions it may not be sufficient, so we change the option to **Manual** placement as shown below. There are two Manual options.

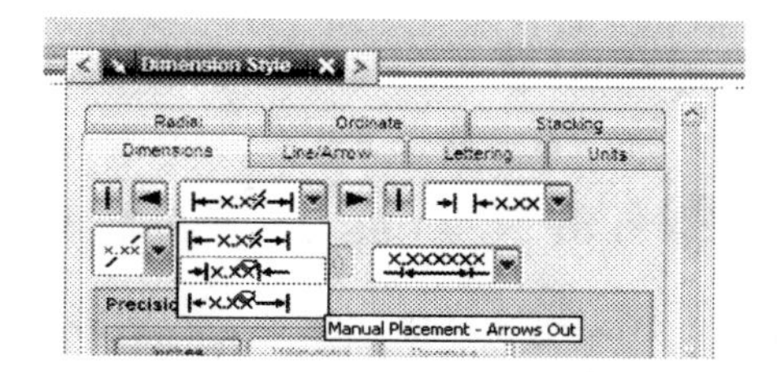

Manual placement with arrows on the outside of the extension lines, or
Manual placement with arrows on the inside of the extension lines.

When manual placement is chosen the location of dimensions and annotation allows complete freedom to place the text anywhere; however, there is a snap point available at the center point between the extension lines to center the text.

We are ready to place that **Hole** dimension again.

Step 5. We are ready to place the hole dimension again.

5.1. First, delete the previous dimension with the undesirable location by choosing **Edit → Delete** and then choosing the dimension and **OK**.

5.2. Go to the **Automatic** button and choose **Manual Placement – Arrows Out**.

5.3. Select the hole in the lower left corner again. Notice now you can drag the origin for the dimension anywhere you want on the drawing.

5.4. The resulting view looks similar to the below one. If you see a diameter symbol "Ø" in place of "DIA", it is due to a different Drafting preference setting, which will be discussed in Section 14.4.

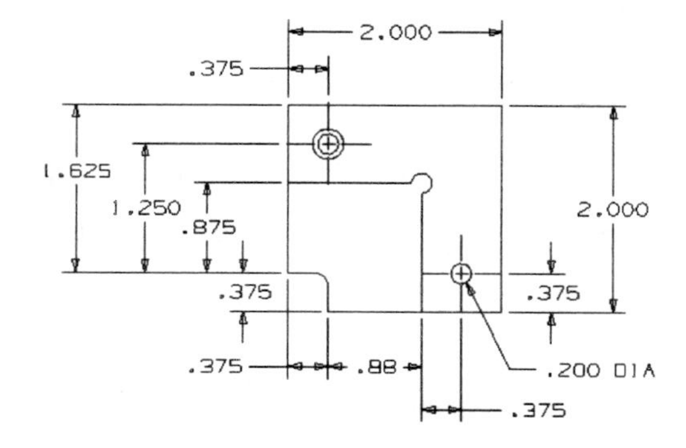

Radius Dimensions

The radius dimensions come in three types:

Radius creates the leader from the dimension toward the center of the radius, but stops before it reaches the center point. Depending on how the Automatic or Manual setting is done the arrowhead may be on the outside or the inside of the arc

Radius to Center creates the leader from the dimension, passes through the arc and ends at the center of the arc with the arrow either on the inside or outside of the arc depending on the Automatic or Manual selection status.

Folded Radius creates an artificial leader from the arc to an artificial point with a fold to indicate the leader is not actual length. This is useful for a large radius that would be out of the view.

Which one among the three to use is up to you and your company standards.

Step 6. Add a radius dimension. If we use the Inferred dimension as we have, a diameter dimension will be placed looking like below.

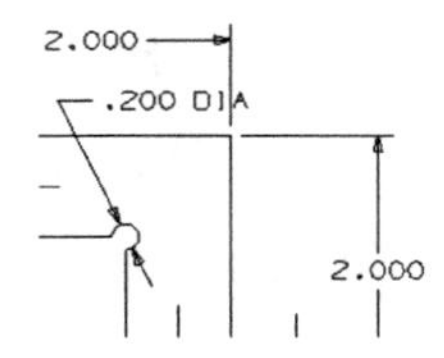

We don't think this is desirable so we will force a radius dimension to be placed.

6.1. Choose the **Radius** icon .

6.2. Select the arc in the corner of the cutout and place the origin of the dimension in a location of your choice. We are still Manual – Arrows Out.

6.3. Choose the **Radius to Center** icon .

6.4. Select the small arc in lower left corner of the part and place the origin of the dimension in a location of your choice.

The resulting view looks similar to the one below.

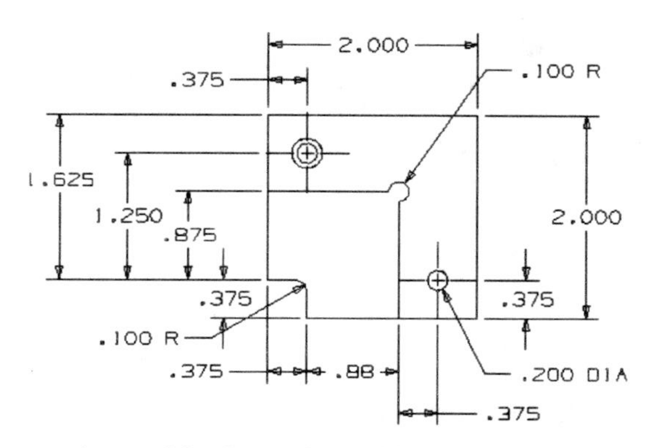

Step 7. Add the Counterbored hole to look like the dimension below.

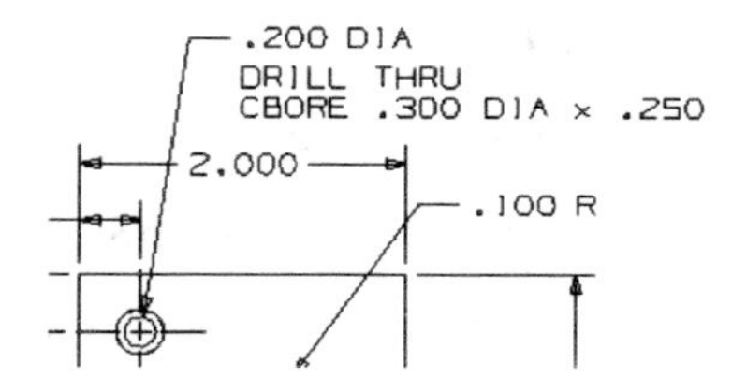

This callout may not be to everyone's company standards but it will illustrate one way to do it.

7.1. Choose the **Hole** dimension type .

7.2. Select the inside diameter of the counterbored hole. Zoom in if needed.

7.3. Notice the dimension is on your cursor ready to be placed on the drawing. But don't place it yet.

7.4. Choose the **Annotation** icon in the Dimension dialog bar to bring up the **Annotation Editor** dialog to append the dimension with annotation.

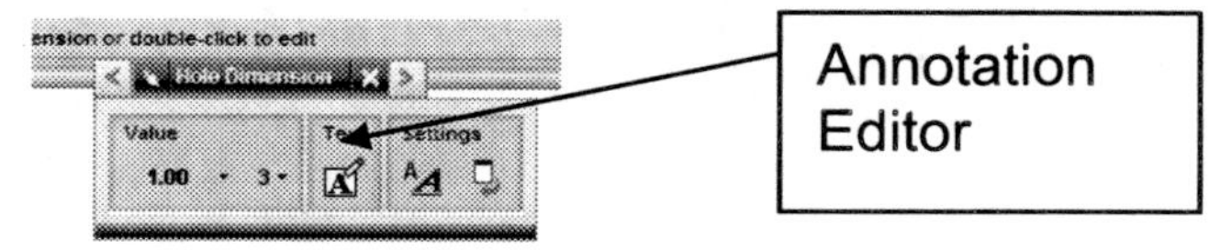

This icon opens the **Annotation Editor** dialog and you may enter pretty much anything you want to add to the standard dimension type.

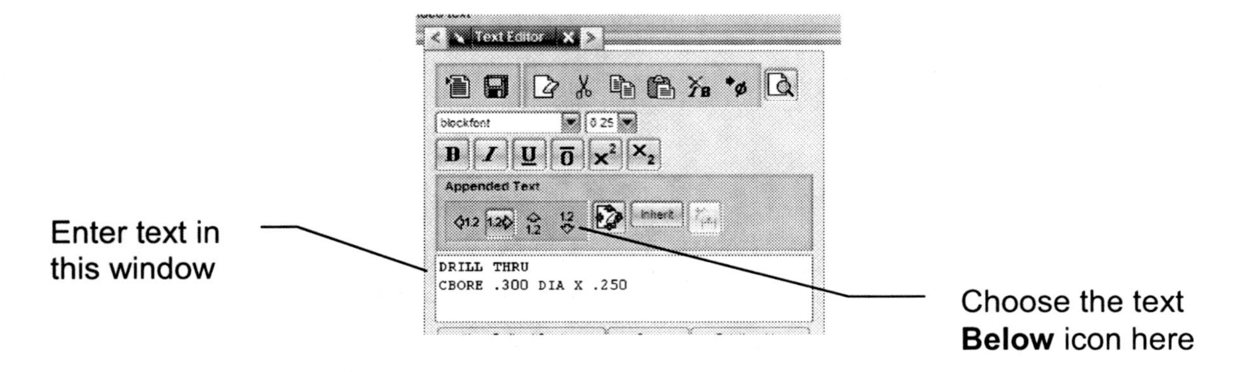

Referring to the figure above of the counterbored dimension, we want to add 2 lines of appended text **Below** the dimension.

7.5. Select the **Below** icon shown above, and then type in the text as shown in the figure above (in all capitals—it is case sensitive).

7.6. Choose **OK** to accept the text and to dismiss the dialog.

Notice the completed dimension is now on your cursor and you can drag it around the screen to place. Also notice the left justification of the text and the right or left side placement of the leader is automatic.

7.7. Use the mouse to place the dimension in a convenient location.

Step 8. **Save** your drawing file.

The specific named dimension types like Horizontal, Vertical, Perpendicular, Parallel, and so forth, while not illustrated here function just the same as the Inferred type except they create that specific type.

They require a first object selection, second object selection and placement of the dimension using the same tool bars as the Inferred. So you may use these when Inferred does not work due to complexity or odd shapes.

14.4 Customizing Drawing

Let's review quickly what we have accomplished so far.

- We created a drawing, changed its drawing size from E to B.
- We placed Top, Front, and Right views on the drawing, orthographically projected from each other.
- We placed a Trimetric view on the drawing to help visualize the part.
- We dimensioned the part and used the Inferred, Hole, and two Radius dimension types.
- We introduced the Annotation Editor where we entered appended text to handle a special case.

We did this to illustrate how quickly and easily a drawing may be created, populated with views, and may be annotated in Unigraphics.

We also did this to illustrate how much of the drawing creation process is automated and defaulted. Most of what we did has defaults built in to allow the placement of views and dimensions quickly and easily. However, the defaults may not fit your company standards, and the defaults don't fit special cases.

This section of the chapter will describe how to change some of the defaults and show you where to find the settings to customize the drawing creation process to your company standards.

Most of the common Unigraphics defaults may be set by the system manager to be permanent changes that become your company defaults and do not have to be changed every time you enter the **Drafting Application**. Then, you only change the settings to fit your special situations as they arise.

There are three methods to change the default settings on the drawing:

- The **Preferences** on the main menu for changes applicable to new placements
- The various dialogs that are used to create dimensions, views, etc.
- The drafting objects may be edited after they are placed on the drawing.

So let's look at these methods and discuss how to change the default settings. The first one is the **Preferences** pull-down off the main menu bar.

Preferences

There are several ways to change the defaults and customize the look of the way drafting is done:

- Change settings in the **Preferences** pull-down main menu, and
- Change settings in the dialogs

We'll look at the **Preferences** pull-down menu first. The **Preferences** menu looks like this:

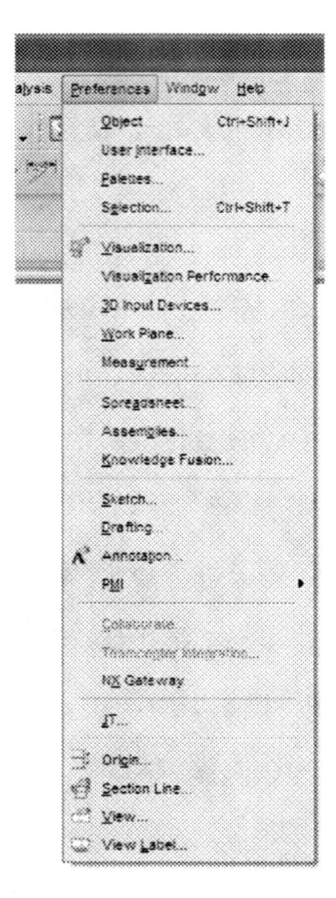

In this section we will look at **Annotation** and **View Display**, and we will look at the changes that may be made in the dialogs.

We will discuss a few general concepts first. As we mentioned earlier, Unigraphics comes with defaults built in for all the settings. You are able to override the **Preference** "factory" defaults with your own or company standards in the following ways:

- Customer Defaults file - contains settings for drafting that take effect whenever you create a new part or choose the Defaults button on a Drafting preferences dialog. The customer defaults files are inside Unigraphics at **File → Utilities → Customer Defaults**. These settings are initially set by Unigraphics, but are allowed to be changed by customers to become the default settings as you create parts.

- Part file Drafting preference settings - You can set preferences within the part file from various Drafting dialogs that affect all of the drafting objects. These settings take precedence over the Customer Defaults file for that part.
- Object specific within a part file - You can selectively change preferences on individual drafting objects from Drafting dialogs. These changes then take precedence over both the **Customer Defaults** file and any part file Drafting settings.

Note here the various levels of setting **Preferences** and the order they get overridden. The **Customer Defaults** file is valuable to set the way you want drafting defaults to be most of the time.

However, as you can see if you make changes in a part file to a drawing, to dimensions, or views, these changes override the settings in the default file. And furthermore, these changes are valid for the entire login session that you are in Unigraphics.

Merely opening a new part file and/or a new drawing do not reset the values back to default settings. The only effective way to reset values is to choose the **Load Defaults** or **Load All Defaults** buttons on the dialogs for which you changed settings.

Let's start with the **Annotation** settings from the main menu, choose **Annotation** from the **Preferences** menu.

Note that these settings are global to the entire drawing and part file and will be applied the next time the setting is used. They override the Customer Default settings so are applied for the entire Unigraphics logon session and subsequent logon sessions.

The dialog for **Dimensions** tab looks like:

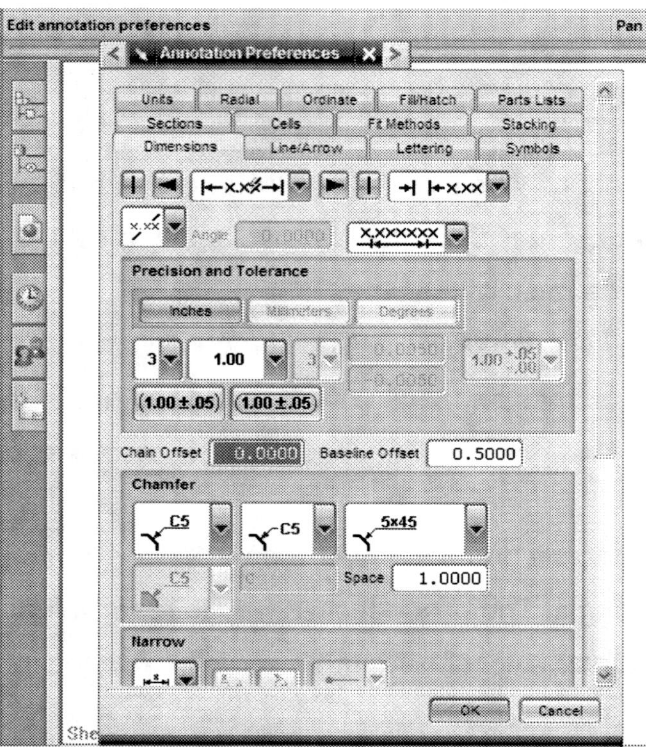

The top two lines of tabs are different preference categories to choose from. Some of the options are explained below.

- **Dimensions**--Lets you set dimension preferences for arrow and line formatting, type of placement, tolerance and precision formatting, dimension text angle, and size relationships for extension line components.
- **Line/Arrow**--Lets you set preferences that apply to leaders, arrows, and extension lines for dimensions and other annotations.

- **Lettering**-- Lets you set preferences that apply to lettering for dimensions, appended text, tolerances, and general text (notes, id symbols, etc.), including lettering size, font, and color.
- **Symbols**-- Lets you set preferences that apply to ID, User Defined, Centerline, Intersection, Target, and GD&T symbols.
- **Units**-- Lets you set preferences for type of units, for the display of linear and angular dimension values as well as dual dimension format.
- **Radial**-- Lets you set preferences for the display of diameter and radius dimension symbols and format.
- **Fill/Hatch**-- Lets you set preferences for cross hatching and area fill.

The **Annotation Preferences** dialog will change with each annotation type selected.

We'll choose the **Dimension** tab and look at a few options to see what can change. The **Dimension** tab dialog is the one shown above and repeated here for convenience:

This line sets whether or not extension lines and arrows are included in a dimension. And it sets the dimension placement options in the Dimension dialog.

Determines the alignment of a dimension value with the leader.

Sets how many decimal places are displayed.

Activates tolerances and the format of the tolerance.

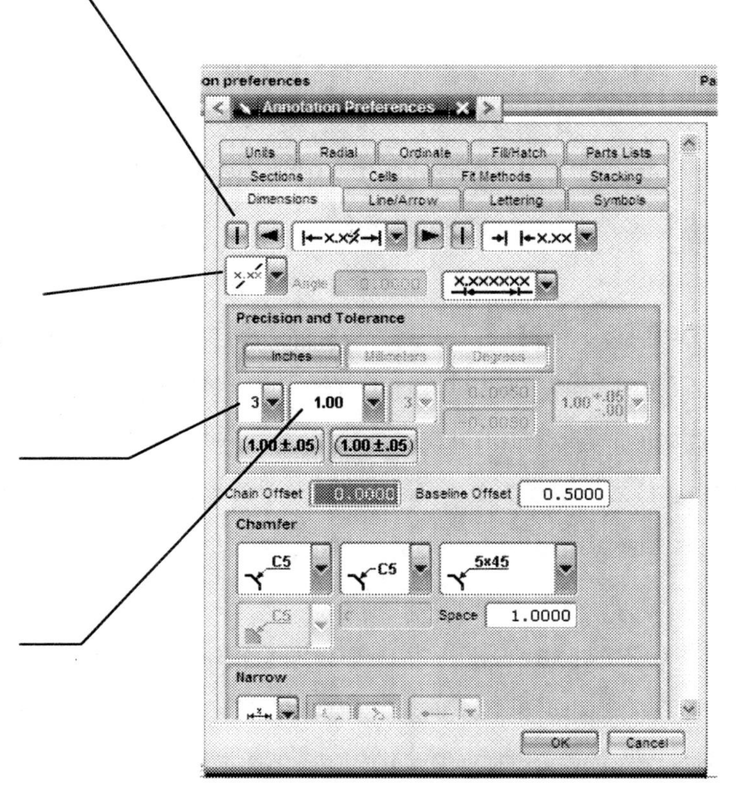

The **Units** button not only allows the choosing of the units to use, but also the formatting of the display of the units.

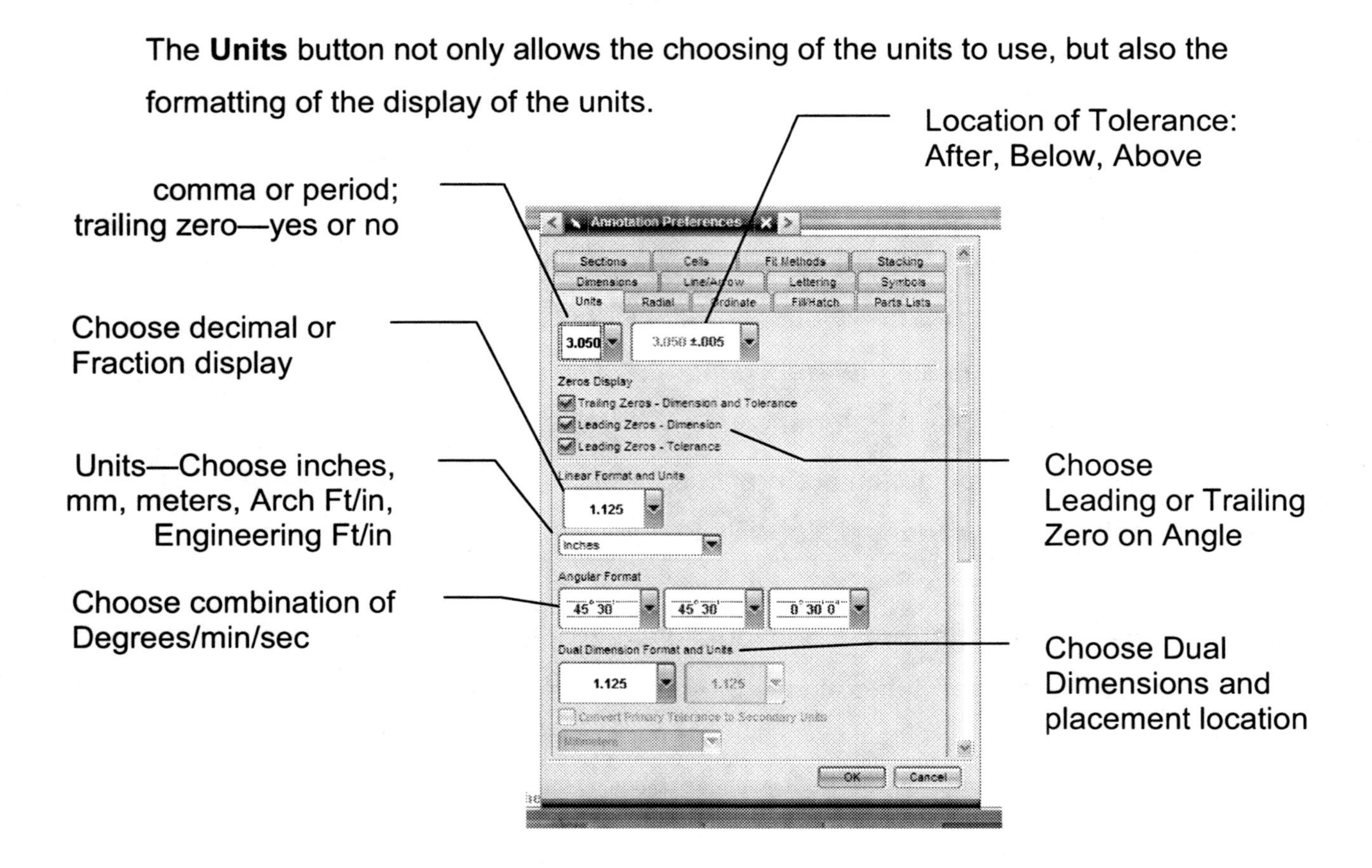

Both of these dialogs are typical of the changes that may be made and the look of all of them. There are too many to cover them all.

The procedure is to make the change(s) you want to make in the dialog, and then choose either **Apply** or **OK** to have the change take effect. The change will be in effect the next time you use the option.

Remember, these settings are global and stay in effect until you change them. When finished with this drawing you may want to set the changes back to the original default values and formats. This may be accomplished by choosing the **Load Defaults** button to change the tab that is active. **Load All Defaults** changes the defaults to Customer Defaults for all of the tabs on this dialog window.

On the other hand, if you want to change the settings of specific dimensions instead of the global setting changes, first choose the specific dimensions in the graphics window and follow the procedure.

Let's look at the **View** dialog off the **Preferences** main menu.

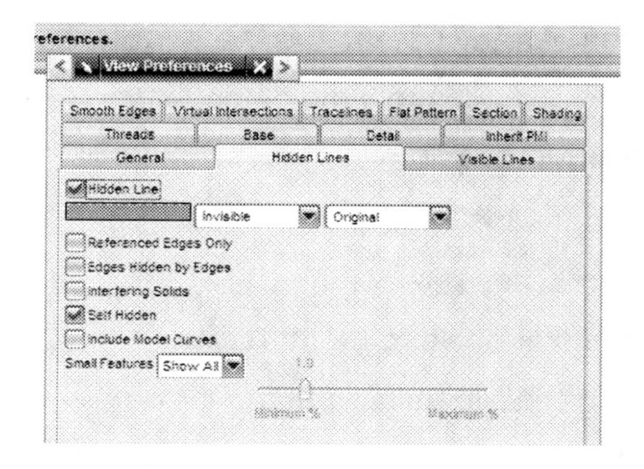

If you desire to make a part look solid in a view by hiding hidden lines, you would change three settings in the above dialog:

- Choose **Hidden Lines** tab,
- Ensure the second pull-down to be **Invisible**,
- Choose **Hidden Line** toggle to put a check mark in the box.
- Choose the **Apply** button for the settings to take affect.

Then all new views added to a drawing would make the part look solid with hidden lines not visible.

In other words, the four tabs in the dialog titled: **Hidden Lines**, **Visible Lines**, **Smooth Edges**, and **Virtual Intersections**, are combined with the window and the two pull-down menus to define how those kinds of curves are displayed in a view.

Selecting the **Hidden Lines** tab changes the options the three windows and the settings are made for **Hidden Lines**. When the **Visible Lines** button is chosen and the settings under the three windows are changed accordingly and you choose the settings for **Visible Lines**. The other buttons work in the same way.

Also insure the **Edges Hidden by Edges** button is toggled with a check mark in it. This button performs the job of erasing hidden edges that are behind other edges (overlapping edges) so there are not two edges on top of each other. This action contributes to the model looking 2D on the drawing.

Activity 14-3. Applying the Hidden Lines View Display

Using the drawing that is already up and the View Preferences for Hidden Lines created a few pages ago, we will apply the settings to a new view.

Step 1. Choose **Preferences → View…**

Step 2. Create the Hidden Line solid look.

- **2.1.** Choose **Hidden Lines** tab,
- **2.2.** Choose second pull-down to be **Invisible**
- **2.3.** Choose **Hidden Line** toggle to put a check mark in the box
- **2.4.** Choose the **Apply** button for the settings to take affect.

Step 3. Add an Isometric view to the drawing.

- **3.1.** Choose the **Base View** icon.
- **3.2.** Select the **TFR-ISO** view from the pull-down menu.
- **3.3.** Select an open area of the drawing to place the view and pick it.

The isometric view is placed and notice it has its hidden edges invisible so it looks solid.

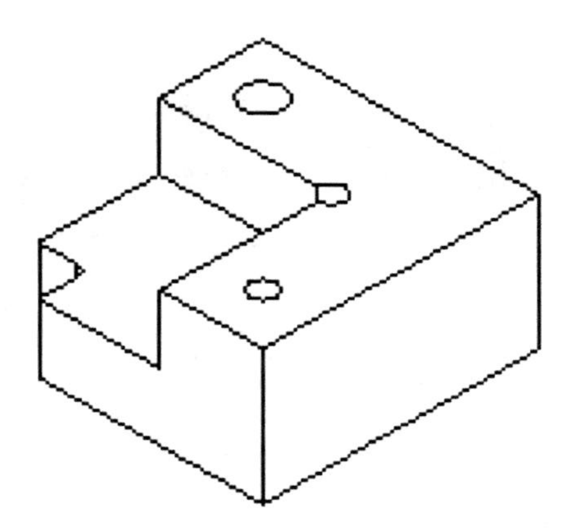

Activity 14-4. Applying the Annotation Changes to Dimension

Step 1. Choose **Preferences → Annotation...**

Step 2. Choose the **Dimensions** tab, if not already chosen.

Step 3. Choose the **Automatic Placement** pull-down and from the pull-down menu that appears choose the **Manual Placement - Arrows In** option.

Step 4. Choose **OK** to accept and dismiss the dialog.

Step 5. Choose the **Horizontal** dimension and note the differences in how the dimension acts on the cursor.

- **6.1.** Select a feature of your choice to dimension. When we placed dimensions previously the placement option was on **Automatic Placement**.
- **6.2.** Select the geometry and note the dimension may be moved about freely to place the dimension.

Modifying the Dialogs

The second method we'll discuss is modifying the dialogs to customize the drafting settings. We discussed and used several of these in Activity 14-2 when we changed the **Auto Placement** setting and the **Annotation Editor** to add appended text. All of these are on the individual dimension types dialogs.

When a dialog is changed like we did above changing the Preferences menu, it is a permanent change until it is changed again. The change overrides any settings in **Customer Defaults** that may be set and it overrides any **Preferences** menu settings that have been made, and in fact some will actually change the **Preference** settings to agree. Also remember, changes to the dialog may be made before you make the selection on the graphics window to place the drafting object.

If you make a mistake in selecting geometry, you must reselect the drafting function you want to do before you reselect correct geometry.

Activity 14-5. Customizing the Drawing

This activity will start from the beginning on a new part and go through the process of creating a drawing and dimensioning to put this information together.

The part we will use is called a shaft support. We will use the Master Model concept to create the drawing and annotate it.

Step 1. Create a new part file.

- **4.1.** In the **File New** window make sure that Modeling tab is selected and the units are inches.
- **4.2.** name the part file **xxx_shaft_support_dwg.prt**.
- **4.3.** Select the Blank template.
- **4.4.** Choose **OK**.

Step 2. Choose **Start → Modeling** and **Assemblies**.

Step 3. Insure the trimetric view is up and change however you like to do that.

Step 4. Create the assembly with the solid body part file.

- **4.1.** Choose **Assemblies → Components → Add Component.**

4.2. Select from the list of parts the **shaft_support.prt** or browse for it by selecting the **Open** icon. This part is located in directory Chapter14-partfiles.

4.3. Choose the pull-down to change the **Reference Set** and select the Reference Set **BODY**.

4.4. Set the coordinates to 0, 0, 0.

4.5. Change the **Layer** setting to **Original** and choose **OK** to accept the dialog.

4.6. Choose **OK**..

4.7. Note that we chose **Original** layer, the solid body is on Layer 10, so change the **Work Layer** to 10 and **Fit** the part in the window.

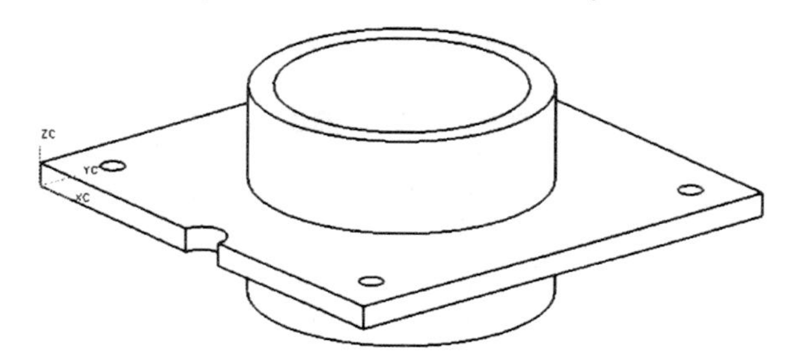

Step 5. Inspect the part and **Save**.

Step 6. Choose **Start → Drafting**.

Step 7. Create the sheet, modify the drawing size and add views.

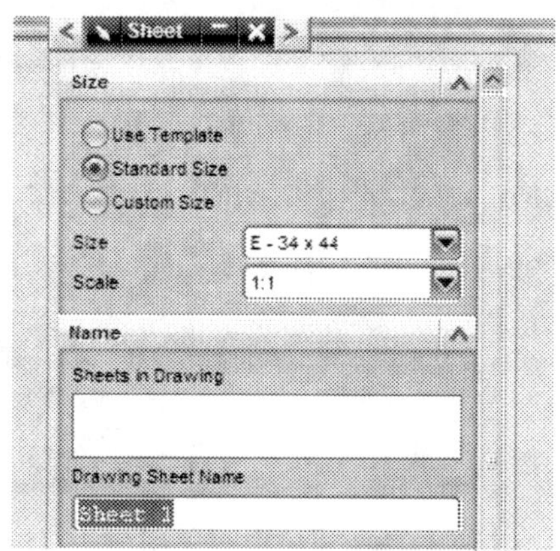

7.1. Please note the 3 buttons to create the sheet: **Use Template**, **Standard Template**, and **Custom Size**. Templates have been discussed in Chapter 1 and the **Drafting** Application has its own set. We will use the

Standard Size option for this class because we want to teach the basics without preset conditions.

7.2. From the **Sheet** dialog, choose the **E – 34 X 44** pull-down menu and choose the **C** size drawing. Note the **Start Base View** box is checked.

7.3. Choose **OK** to accept and dismiss the dialog.

7.4. The base view is placed on the drawing and on the cursor.

7.5. If the preview object is shaded you may want to use the right button on the mouse to select **Style** and change the display to wireframe.

Step 8. Check the **View Preferences** to insure we get dashed hidden lines.

8.1. Choose **Preferences → View**.

8.2. Insure the **Hidden Lines** tab is depressed, insure the selections are Original (gray color), Dash line symbol, Original.

8.3. Insure the **Hidden Line** toggle is checked and the **Edges Hidden by Edges** toggle is checked. Otherwise, make proper changes and choose **OK**.

Step 9. Add the views.

9.1. Choose **Base View** icon .

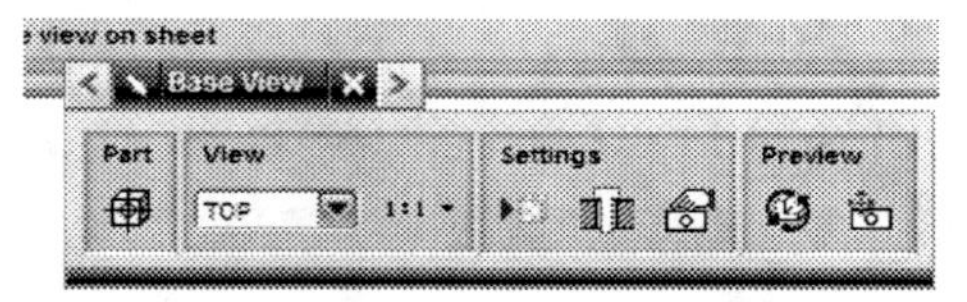

9.2. The **Base View** bar appears and is defaulted to place the **Top** view.

9.3. The **Top** view outline is on the cursor so place the view in the upper left quadrant of the drawing. When the Top view is placed the bar changes.

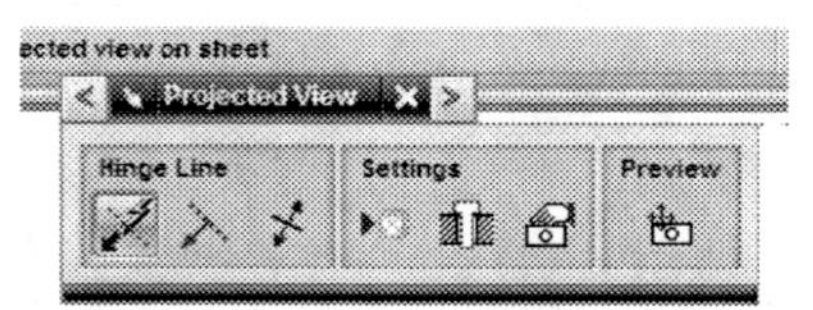

9.4. Another projected view is available to place immediately, so place the **Front** view under the Top view.

9.5. To place the **Right** view we need to change the Parent view, so select **Parent** icon on the bar, and select the **Front** view.

9.6. A view pops up on the cursor, so place the **Right** view to the right of the Front view.

Step 10. Add a solid trimetric view.

Change the **View Preferences** to get **Hidden Lines** invisible.

10.1. Choose **Preferences → View**.

10.2. Choose **Hidden Lines** tab.

10.3. Select **Invisible** from the middle button in the pull-down selections.

10.4. Choose **OK** to accept and dismiss the dialog.

Now back to the **Base View** to add the trimetric view.

10.5. Choose the **Base View** icon and choose the **TFR-TRI** from the list of views in the pull-down.

10.6. Place the trimetric in a convenient location in the upper right quadrant of the drawing.

10.7. Select MB2 to exit from the view placement mode.

Step 11. You may want to turn off the view boundaries.

11.1. Choose **Preferences → Drafting**.

11.2. Toggle the **Display Borders** button off and choose **OK**.

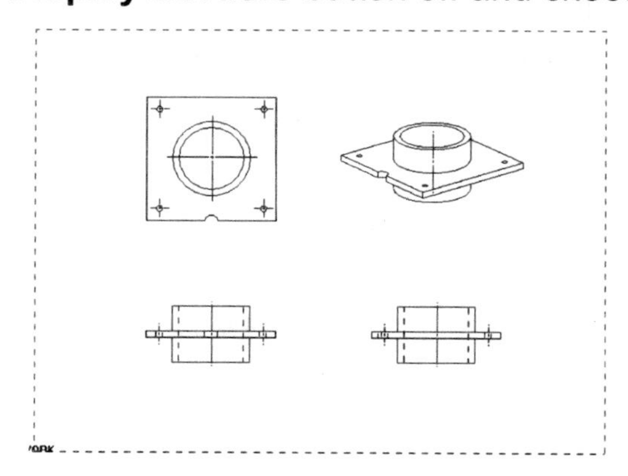

Step 12. Start dimensioning the part.

12.1. Choose the **Inferred** icon.

12.2. Let's start by dimensioning the hole pattern.

12.3. Select the arc centers of the two holes at the top of the part to dimension the distance between the holes. (Or choose centerlines if they are available.) You probably will have difficulty selecting the centerlines of the holes so use the Quickpick pop-up window or you may Zoom in on the area to aid in selecting the Linear Centerline.

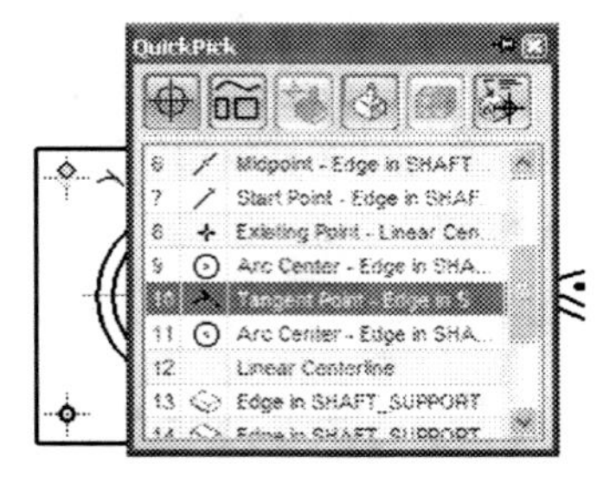

With the two holes selected, before we place the dimension, note that the dimension is not formatted the way we like: single integer dimension, no gap in the leader lines to the centerlines, and dimension over the leader. We fix these before placing the dimension.

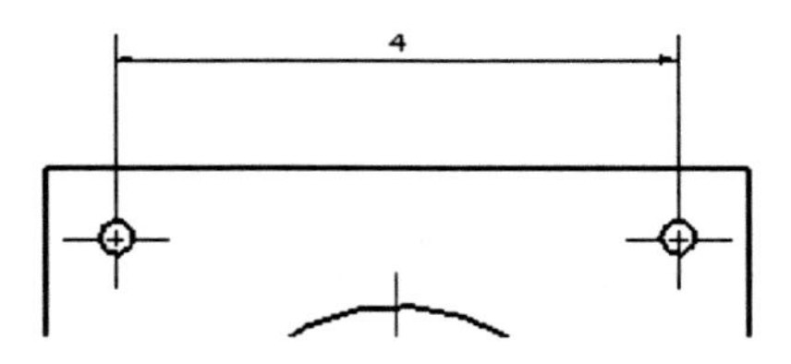

12.4. Changing the preferences. While holding the cursor over the dimension punch **MB3**. On the pop-up, select the **Style** option.

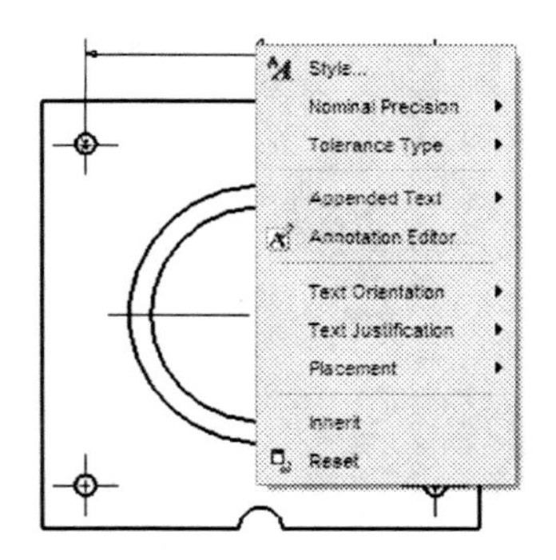

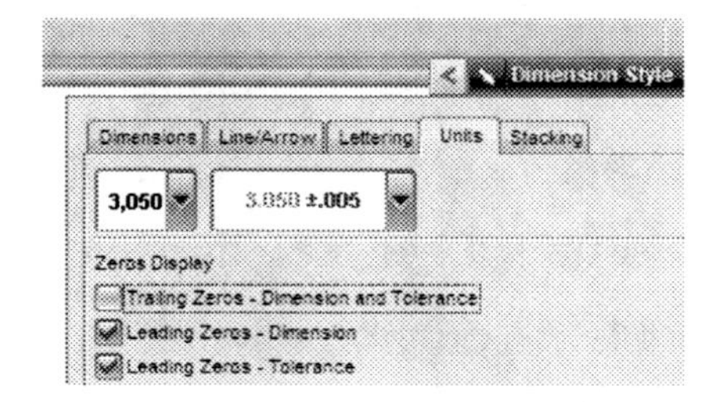

12.4.1. Select the **Units** tab and change the comma delimiter to our decimal point delimiter.

12.4.2. Also on the same tab check the option to turn on the **Trailing Zeroes** option.

12.4.3. Now select the **Line/Arrow** tab. To obtain the gap between dimension leader and hole centerline we change dimensions **H** and **J** to .0625.

12.4.4. Select the **Dimensions** tab and select Text over **Dimension Line** icon with the pull-down and choose the **Text Aligned** option.

12.4.5. Now **OK** the window.

12.4.6. The dimension now looks like we want it. But don't place it yet.

12.5. Let's decide we want to add an Appended Text of the word TYP to account for the two holes at the bottom of the part. From the dimension bar, choose the **Annotation Editor** icon, select the **Below** button, type in **TYP** in the text window of the **Annotation Editor** dialog. (Note you have to first delete our last annotation.)

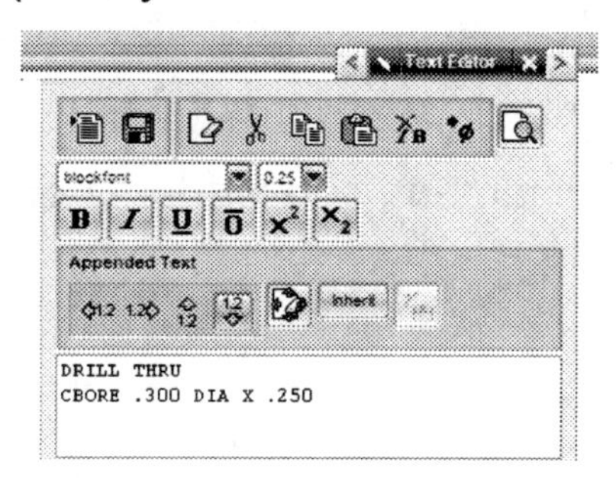

12.6. Choose the **OK** button on the dialog to accept the change.

12.7. Place the dimension on the drawing.

Now we'll add the dimension from the hole to edge of the part.

12.8. Select the left edge of the part and select the left hole.

The appended text TYP is still active and we could place the dimension but let's have the new dimension aligned with the existing dimension and remove TYP.

12.9. The Cue Line prompts us to place the dimension, but first we require a 3 place dimension, so off the **Inferred Dimension** creation bar choose the pull-down menu and the 3.

12.10. We don't want the TYP annotation either so choose the Text Annotation Editor icon or the pop-up menu from MB3, erase the word TYP in the window, choose OK. The annotation disappears from the dimension.

12.11. Then, with the dimension still active on your cursor, select the dimension location that lines up the leader as shown below with the arrow shown.

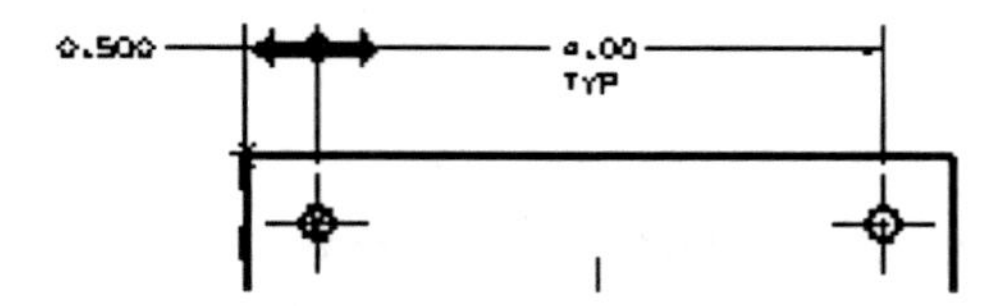

When the lines are aligned the arrows shown above appear indicating alignment, place the dimension with MB1.

12.12. Let's add the dimension for the width of the part. The dialog is still active for **Inferred**, so select the left edge and the right edge of the part and place the dimension.

12.13. Perform similar actions for vertical dimensioning for the vertical direction of the holes. Be aware of the 2- and 3-place dimensions, appended text, alignment, and leader direction.

The Top view now looks like this.

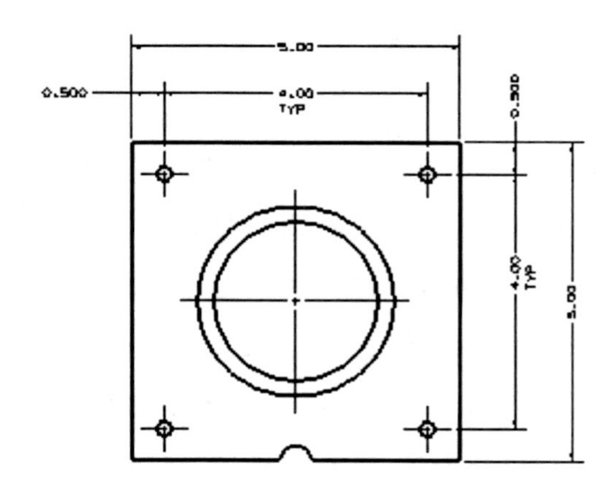

The choice of where to dimension now is pretty much discretionary, but let's choose to finish dimensioning this Top view for ease of illustrating some of the dimensioning points.

Let's add the Diameter dimensions.

12.14. Change the Precision to a 3-place dimension.

12.15. Choose **Preferences → Annotation → Radial**

12.16. The dialog looks like this. We want to change the diameter symbol to DIA and we want the symbol to appear after the number, i.e., .250 DIA. Make the changes as shown in the figure. Choose **OK**.

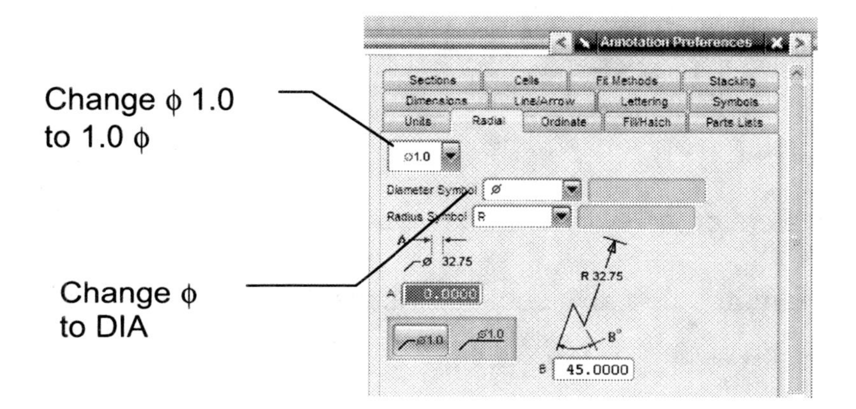

12.17. Inferred is still the active dimension type, however, Inferred is sometimes difficult to select and format on a hole, so select the Hole icon. Then select a .250-dia hole—let's choose the lower, left hole.

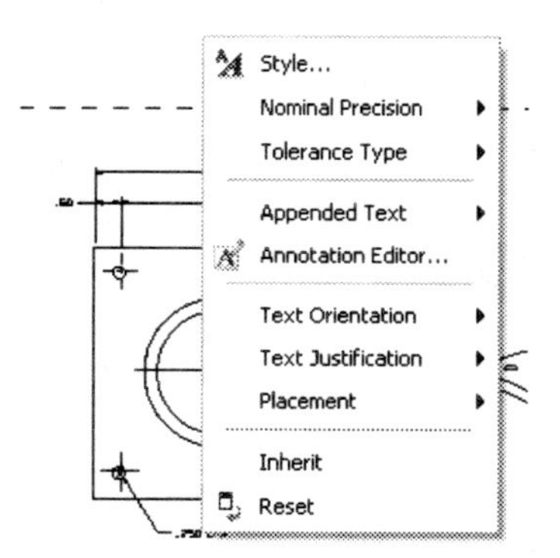

With the dimension on the cursor, another option to change parameters is to select **MB3** to get the above pop-up menu.

12.18. From the pop-up, choose **Style → Dimension** tab, then choose **Manual Placement, Arrows In**.

12.19. Also on the pop-up menu, choose the **Annotation Editor** and add annotation of TYP 4 PLACES Below the dimension.

12.20. Place the dimension in a convenient location to the left of the part.

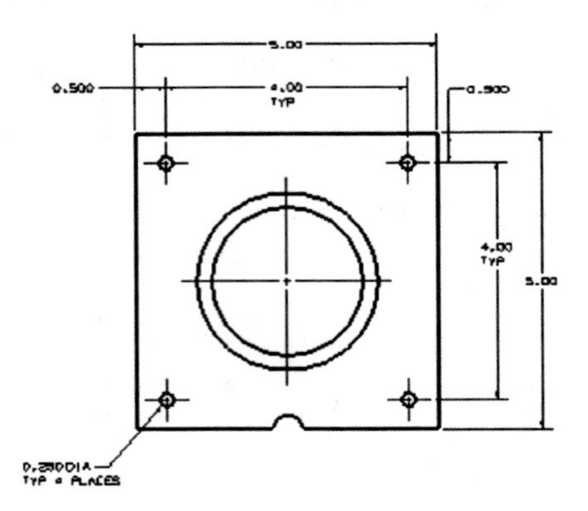

Now let's do the two diameters of the cylinders.

12.21. Choose the **Diameter** icon because we want the full dimension leader through the circles.

12.22. Select the outside circle.

12.23. The settings from the Hole dimension are still good so we don't need to change those.

12.24. The Annotation appendix is still active though, but not needed here, so with the dimension on the cursor, punch MB3 and get the pop-up. The pop-up looks like below, slide down to Appended Text, slide out and down to Clear All. This removes the Appended Text.

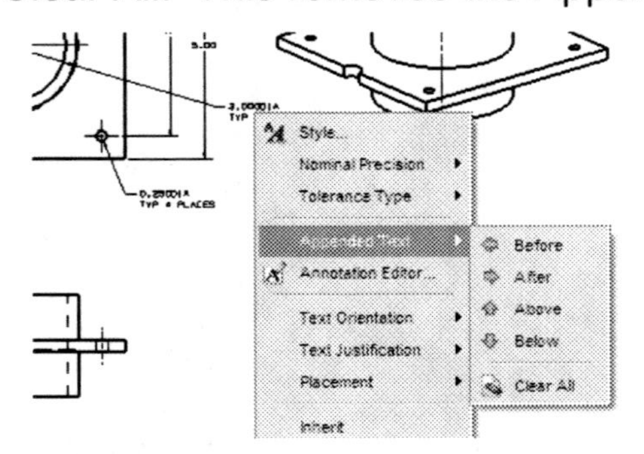

12.25. Now ready so select a spot and place the dimension.

12.26. Select the inside diameter circle and repeat the above steps.

The Top view is looking something like this now. This completes our demonstration of dimensioning on the **Top** view. You may choose to add more as it is not complete.

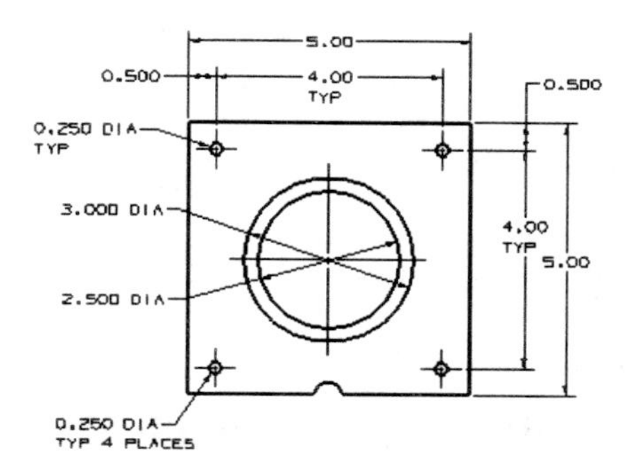

NOTE: For the view below, the lettering height was increased for easier viewing. To edit the lettering select **Edit → Style**. The Class Selection comes up, choose **Select All**, then the **Lettering** tab and type in a new height. We used 0.19.

The **Front** view may be dimensioned in much the same way, only much less needs to be dimensioned. The designer needs to decide what dimensions are important and use dimensions that reflect the designer's needs.

An option a designer/draftsman has in dimensioning diameters is to use the **Cylindrical** dimension type instead of the Diameter type. We will use the **Cylindrical** dimension type just for illustrative purposes here since we already used the **Diameter** dimension.

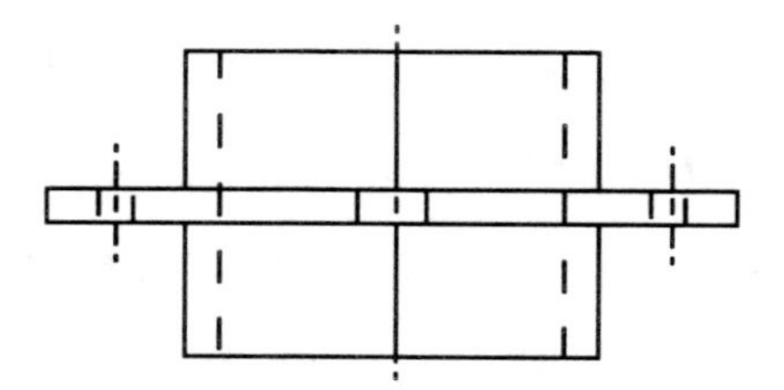

12.27. Choose the **Cylindrical** dimension type . Note that the **Inferred** type may be used, however, sometimes it is difficult to select the correct objects to get the dimension you want.

12.28. Select anywhere on the silhouette of a cylinder and then the other side of the cylinder, then place the dimension.

12.29. Select both silhouettes for the other cylinder and place the dimension.

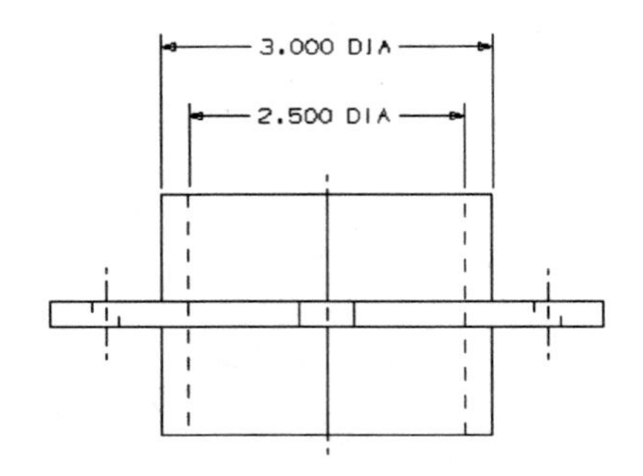

And the result looks something like this.

When you have an option, we'll leave up to you and company practices which dimensioning type you use: **Diameter** or **Cylindrical**.

We'll also leave up to your discretion the rest of the dimensioning of this part. We believe we've given enough tools to complete normal dimensioning. Obviously there are a lot more options, but we believe you've been given enough tools that you will be able to locate and understand how to use the others.

A summary of preferences now for convenience.

As stated, changes to preferences from the **Preference** main menu are permanent for all logon sessions until changed or defaulted to customer defaults.

Preference settings and values changed to the dimension, annotation, etc., dialogs override other settings but only for this logon session and take affect the next time the command is active. If a dialog is already active, no change is made until closed and opened again.

Changes to settings and values using the right-click pop-up menu apply only while the open dialog is active and they apply to that dialog. If dialog is closed then settings revert to the global settings. Notice that the dimension dialog will remain open as long as you continuously select dimension types without closing the dialog. You may select the Inferred type, then directly the Hole type, then directly the Dimension type, etc, without closing the dialog. In this case, the settings made on the right-click pop-up menu will stay in affect until the dialog is closed. Notice in the previous activity from the beginning until the end we did not close the dimension dialog, but selected each dimension type directly. The name on the dialog changed with each dimension type.

Remember, if you want your Annotation preference settings and values to be global you must do them with no dimension dialogs active. If Annotation Preferences are made while a dialog is active the change will not apply to the dimensions that are created while that dialog stays open--only will apply to the next one after dialog is closed and opened again. However, you may go to the Preferences main menu anytime to make other changes with other commands and not interrupt your active action. You will be returned to your action where you left off.

14.5 Notes and Labels

Notes and labels are used to place information on the drawing such as GD&T symbols, the list of Drawing Notes, filling out the Title Block, and explanatory notes about the finishes and manufacturing details that must be passed on to NC Programming or analysis engineers.

For purposes of clarity, we will define **Notes** as stand-alone information without a leader and **Labels** are notes with a leader.

There is not a command titled Notes and Labels; instead they are added with the dialog which is found under the **Insert** menu.

Insert → Text... or the **Text** icon .

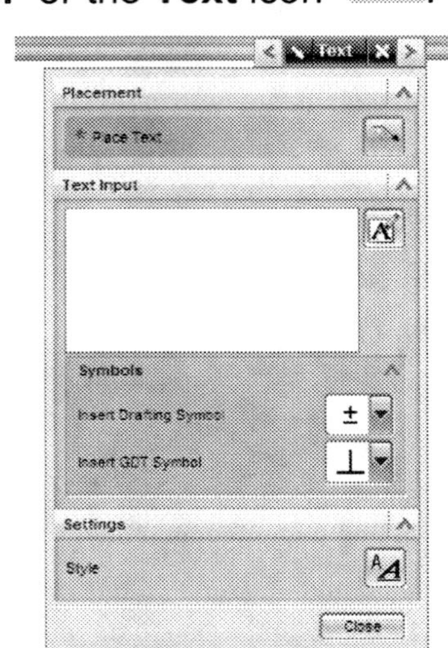

The **Text** dialog bar may now be used to enter text for both notes and labels. It contains the links through the icons to control text size and font, and enter drafting symbols and enter GD & T symbols as well as the other formatting you are accustomed to doing.

The **Text** input window will allow almost unlimited number of characters to be typed in or copied from a file. If all the preferences are set you may just start typing. The typed in text will appear on your cursor and you may place it. If you want to enter symbols or other formatting you select the appropriate icon on the dialog and select them to insert. The **Text Editor** and **Annotation Style** icons are available on the dialog for formatting as well.

Activity 14-6. Adding a Label

The purpose of the activity is to illustrate the functionality of adding Annotation.

We'll use our previous part file **xxx_shaft_support_dwg.prt** and add a label to the Front view. The Front view currently looks something like this.

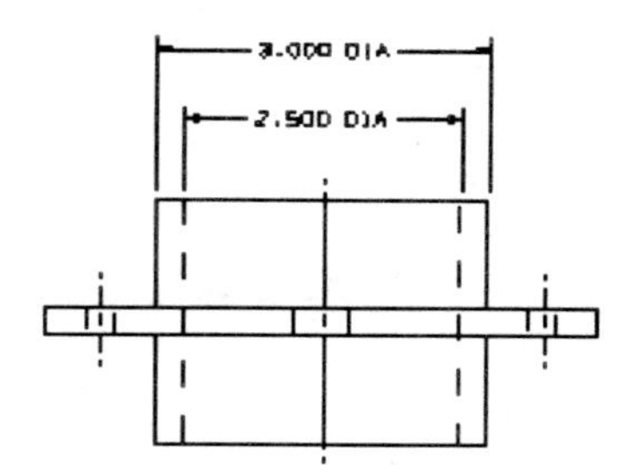

We want to add a fillet note that looks something like this.

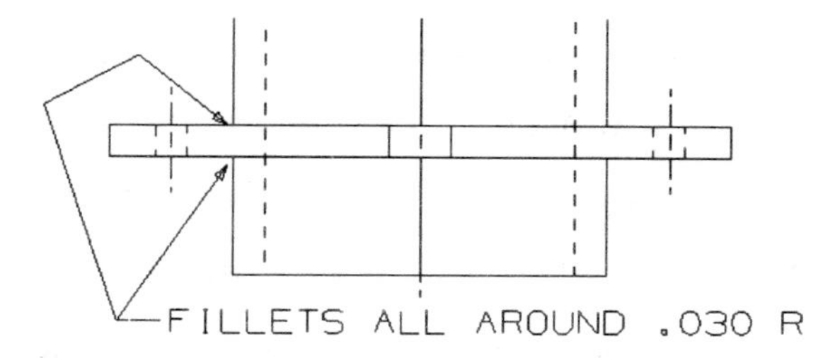

Step 1. Open the **Text** dialog.

1.1. Choose **Insert → Text...** or it's icon .

1.2. Type into the edit window: FILLETS ALL AROUND .030 R (enter capitals—use Caps Lock key).

(Notice the text is on the cursor, note the character size and you may want to increase the character size to .250 or .188 or your standard.)

1.3. To change character size, select **Style** at the bottom of the dialog and type in the new size. Select **OK** and text will change.

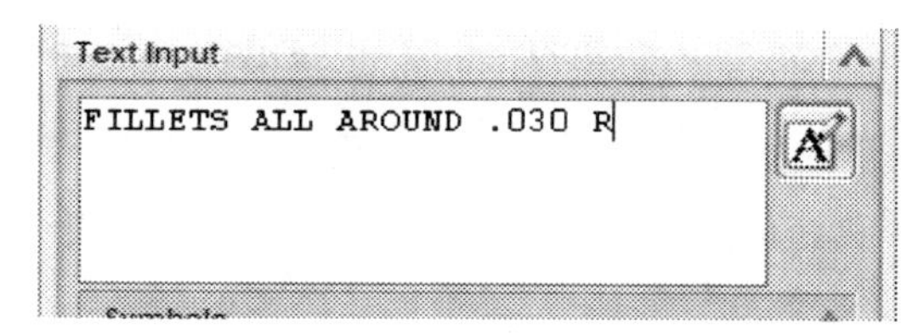

Step 2. Create 2 leaders to point to the areas where the fillet radius is required. Leaders are created by first selecting at the place on an object you want the arrow to be located, then drag the text to where you want it located.

To create the two leaders with intermediate breaks in them will require several steps and using 2 tool bars and the text editing window.

Select the **Leader Tool** icon shown above. The dialog appears.

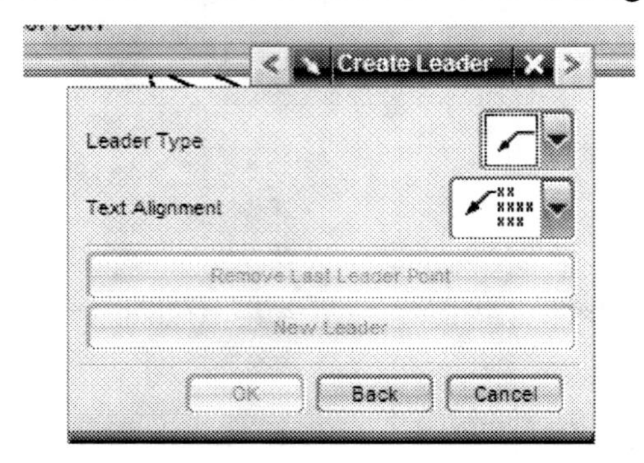

With this dialog active we are allowed to select intermediate points on the leader. We still begin with the location for the arrowhead on the leader.

2.1. Select the Leader Type and Text Alignment you want. Then move the cursor to the intersection as shown to start the leader and select the bottom of the silhouette when it highlights and in the corner as well.

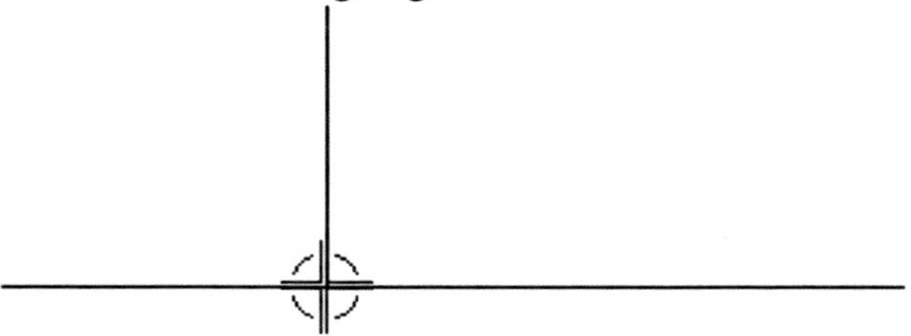

2.2. Move the cursor to a location up and to the left as shown in the figure below and select an intermediate point. An asterisk and the leader will appear at the point. Select a second intermediate point as shown below. You will not see a leader until both ends are selected.

2.3. Move the cursor down below the figure as shown below and select a point there. An asterisk and leader will appear there as well.

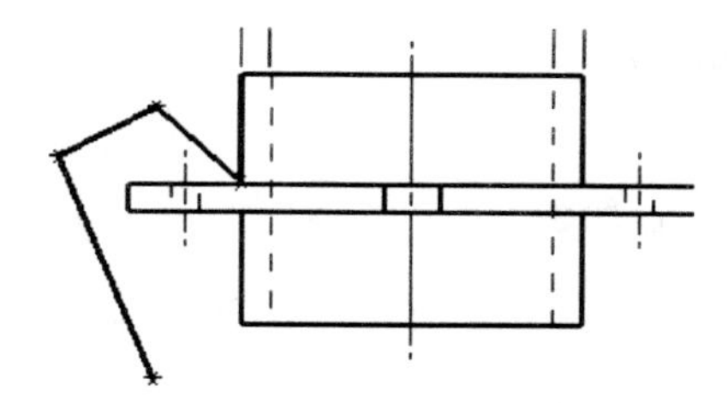

Notice the text itself is not visible while we are in multiple leader mode.

2.4. From the **Create Leader** dialog choose the **New Leader** option.

2.5. The system prompts to select the start of the leader so select the bottom side of the flange and silhouette like we did for the upper corner.

2.6. This leader can go direct to the end of the previous leader so select the end of the leader so they coincide.

2.7. Now we are ready to locate and place the text of the label. Choose **OK** on the **Create Leader** dialog. The text appears. Move the text around with the cursor until the leader off the text is horizontal from the last point and press **MB1** to place text approximately as shown. There is no automatic alignment available.

This completes the activity for creating Annotation.

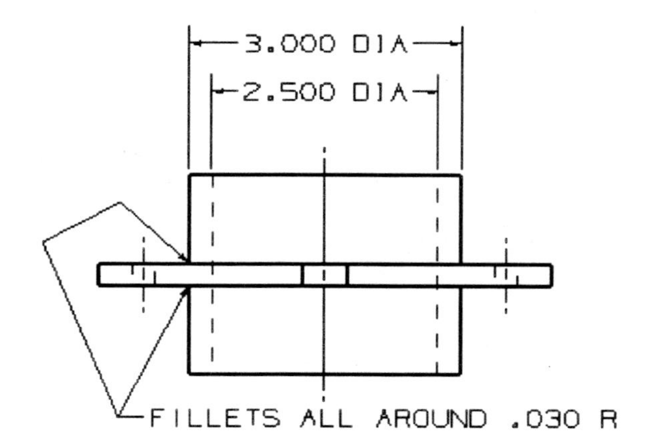

14.6 Modifying the Drawing

In this section we will explore other modifications to the drawing that are common to do. In the first activity we did a change to the drawing by changing its size from the original E size to a C size.

This time we will change the drawing size but we will do it with views already on it, and we will change the name of the drawing.

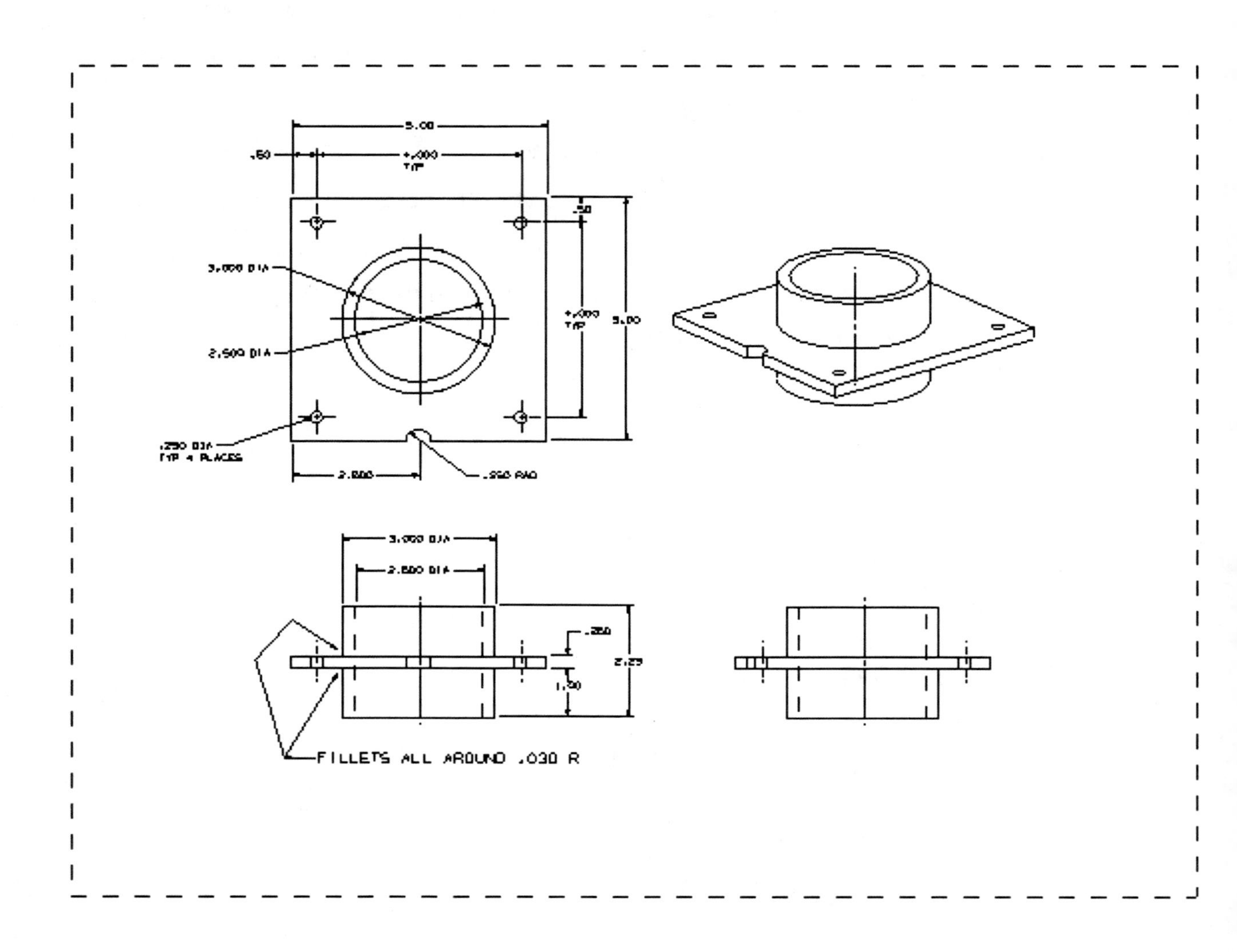

Activity 14-7. Modify a Drawing

We will use our previous drawing to continue this activity.

Step 1. If it's not already open, **Open xxx_shaft_support_dwg.prt**.

The current condition of our drawing is shown on the previous page. Note the nomenclature in the bottom left corner of the drawing: **Sheet1 "Sheet1" Work .**
The first change we will make is to change the name to SHEET1.
You will want to change the drawing name to match your company standards: i.e., 53C12345-1.

Step 2. Choose **Edit → Sheet**.

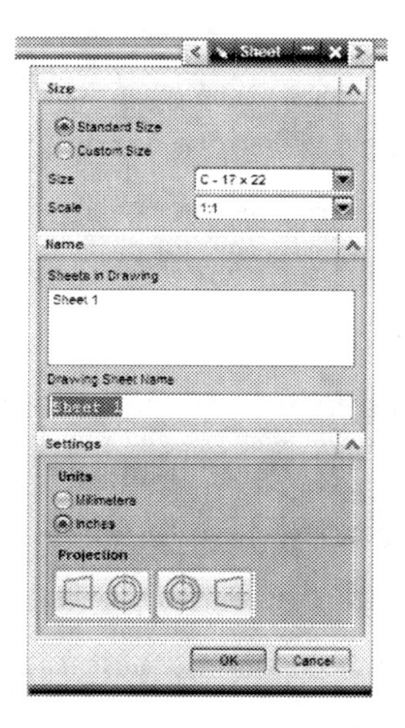

Step 3. Type in the name **shaft support** in the box labeled **Drawing Sheet Name**.

This is not case sensitive—all names will be converted to capitals.

Step 4. Choose **OK** to accept the change. Notice the drawing name change in the lower, left corner.

Now let's change the drawing size and see the effect.

As you see in the figure, a C size drawing is 17 X 22, a B size is 11 X 17, and a D size is 22 X 34, etc.

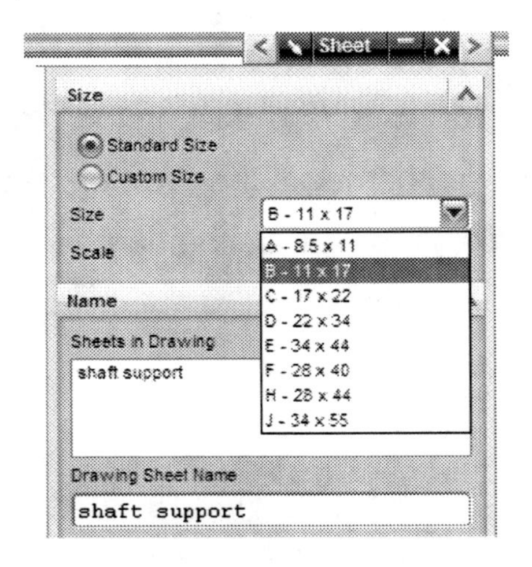

On a Unigraphics drawing, these sizes are measured from the lower left corner of the drawing, the drawing origin. Furthermore, the views placed on the drawing have a drawing **X** coordinate and a drawing **Y** coordinate, so they are a **X,Y** distance from the drawing origin.

When the drawing size is changed the views retain the same **X,Y** coordinate value on the new drawing size. The significance of this is that when the drawing gets smaller, some views a distance from the origin may be off the drawing—out of the boundaries of the drawing (or off the "paper"). This condition of making the drawing smaller might cause problems in that it might be difficult to get the views re-located that are off the drawing.

Conversely, when a drawing gets larger there may be a whole lot of extra room on the drawing with the concentration of views in the lower left quadrant of the new drawing. This condition is not bad; just means you have to move the views around some, but it leaves room for new views and other annotation to be added.

Here's what our drawing would look like if we changed the drawing to a B size.

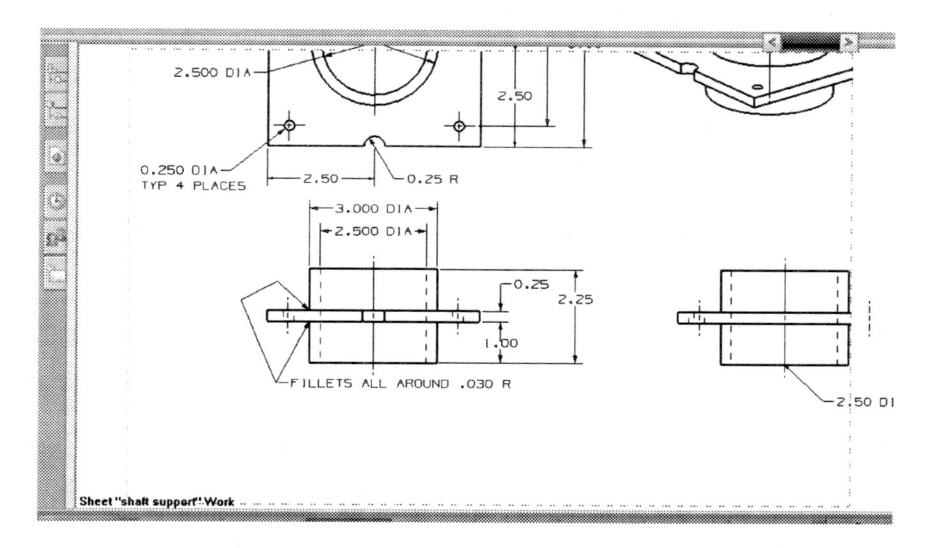

This is not a disaster—at least all the views are visible and they can be moved to be entirely on the drawing. But, obviously, this size is too small for our part.

Let's change the drawing size.

Step 5. Choose the **C – 17 X 22** window pull-down menu and select the **D** size drawing from the menu.

Step 6. Choose **OK** and watch the drawing change.

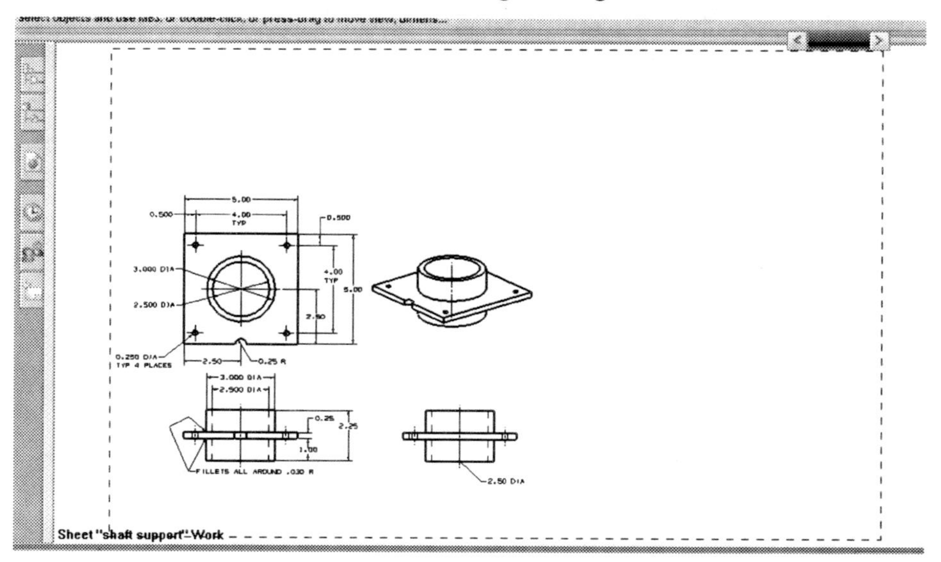

This is perfectly acceptable and probably what we wanted in the first place--to add additional views, perhaps section views, to add drawing notes, and a title block. Of course, some moving of views would be needed as well.

Activity 14-8. Removing a View from the Drawing

Step 1. First let's change the drawing back to the **C** size, so repeat Step 5 and 6 above choosing the **C** size.

Step 2. Choose **Edit → Delete**

The dialog comes up.

A view is deleted the same way any other object is deleted, although being careful where you select.

Step 3. Choose any portion of the border around the view from the graphics area. All of the view border should highlight. If the wrong view is selected, push the **Shift** button and select the border again to unselect any views selected.

Hint: When moving the cursor to select the border the cursor will pass over the border and highlight. If moving too fast, it will de-highlight and geometry and dimensions and such will highlight. Move the cursor off those and outside the border until they are not highlighted anymore, then move slowly from the outside to select the border. Make sure no geometry is selected.

Step 4. When the view is selected the OK button becomes active. Choose **OK**. The view is deleted. Note, in general, that when a view is deleted all drafting objects and view modifications associated to that view are deleted as well.

Step 5. **Save** the part file.

14.7 Creating Utility Symbols

Utility symbols are found under the **Insert** pull-down on the main menu. It looks like this.

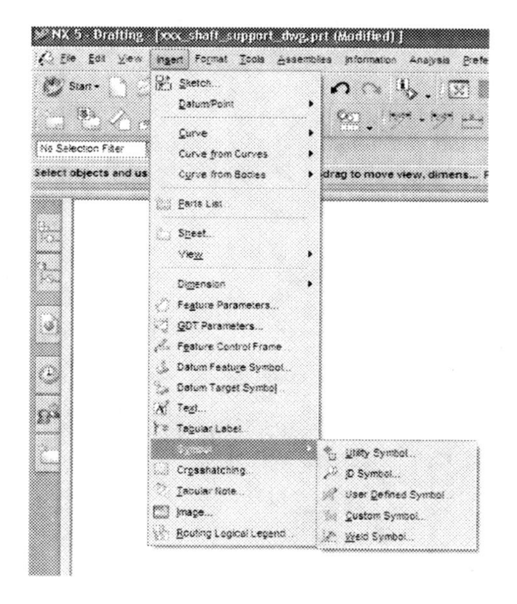

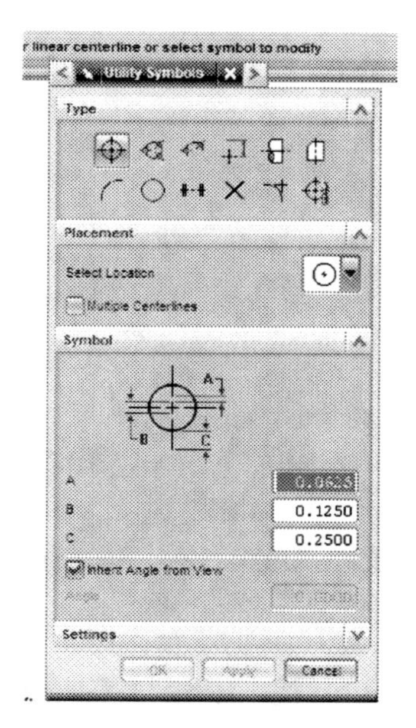

This format of the Utility Symbols dialog looks familiar in the layout with the selection of utility symbols on the top section, point selection choice, then the local display parameters section that defines the symbol.

For instance, the parameters that define the circle centerline that is highlighted are the dimensions of this type of centerline. The distance labeled C is the distance from the circle selected to the end of the centerline.

And why utility symbols are useful is that not only do they look good and are standard symbols on drawings, but also as you would expect, they are associative to the geometry chosen. That means they will update if the geometry changes. If the circle selected to place a Linear Centerline on the circle changes, then the Centerline changes to maintain the parameters set for it. Also utility symbols may be dimensioned to and the dimensions are associative.

Existing utility symbols may be edited on this dialog by selecting the symbol, and then editing the defining parameters. You accept the edits with **Apply**.

Activity 14- 9. Creating a Linear Centerline

We'll continue with our same drawing and concentrate on the cylinders in the Top view.

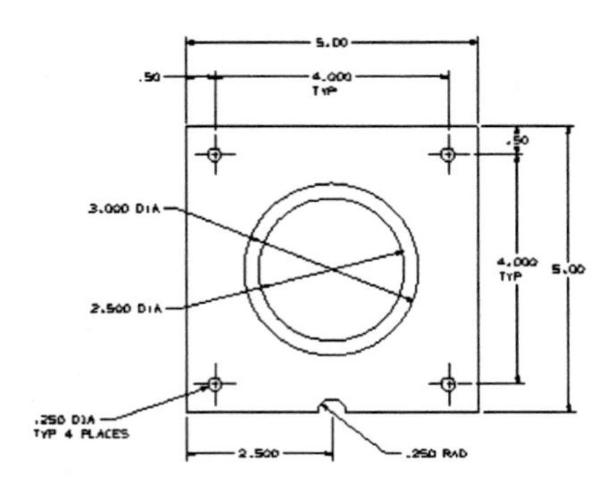

Step 1. Insure that the **xxx_shaft_support_dwg.prt** is the current drawing.

Step 2. Choose **Insert → Symbol → Utility Symbols...** or the icon 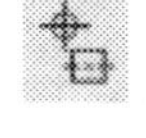 .

Step 3. Choose the **Linear Centerline** icon on the dialog 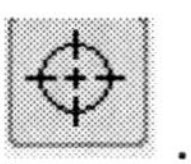 .

Step 4. Choose the outer diameter edge of the cylinder.

Step 5. Choose **Apply**.

The centerline goes on the **Top** view with the centerlines extending past the outer circular edge.

Step 6. Now **Cancel** the dialog.

Next we show how to dimension using the centerlines.

I want to place a vertical and horizontal dimension now but the other dimensions are too close so I need to move some dimensions first to make room. The 5.00 and the 4.000 dimensions need to be moved out and staggered a bit.

Step 7. Place your cursor over one of the **5.00** dimension and it is highlighted. The cursor appears with an arrow pointing out at a 45-degree angle. This means that the dimension may be relocated, so select it with **MB1**, hold **MB1** down and drag the cursor to move up a bit and select a new location out a bit. When the dimension is located where you want it release **MB1** and the dimension is moved. Do a similar relocation for the other **5.00** dimension.

Step 8. Now select each of the **4.000** dimensions and move them out also.

Step 9. Choose the **Inferred** dimension type.

Step 10. To create a horizontal dimension, select a top left edge of the flange at its end point and select the vertical leg of the centerline and place the dimension.

Step 11. Choose the **Vertical** dimension.

Step 12. To create the vertical dimension, select the centerline and a vertical edge of the flange at its end point and place the dimension somewhere convenient.

Step 13. **Save** the part file.

The Top view looks something like this now.

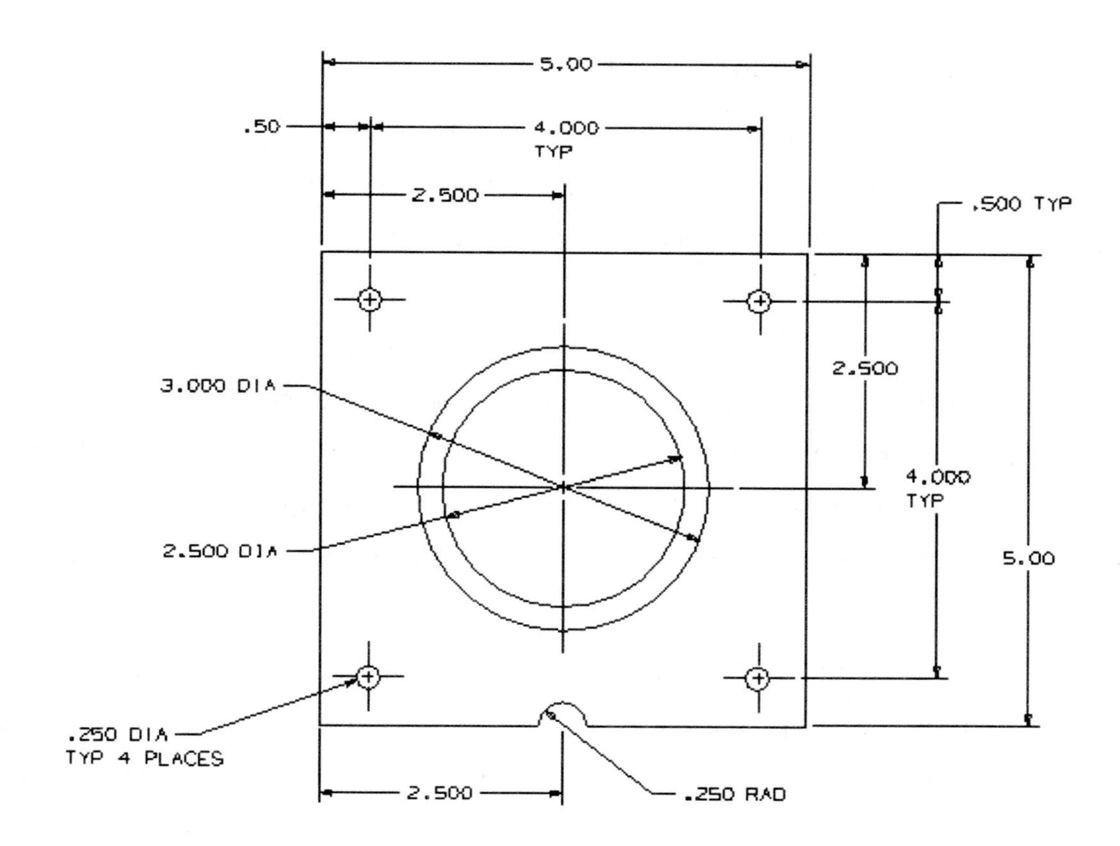

Activity 14-10. Creating a Cylindrical Centerline

We'll create the centerline through the cylinders in the Front view as shown.

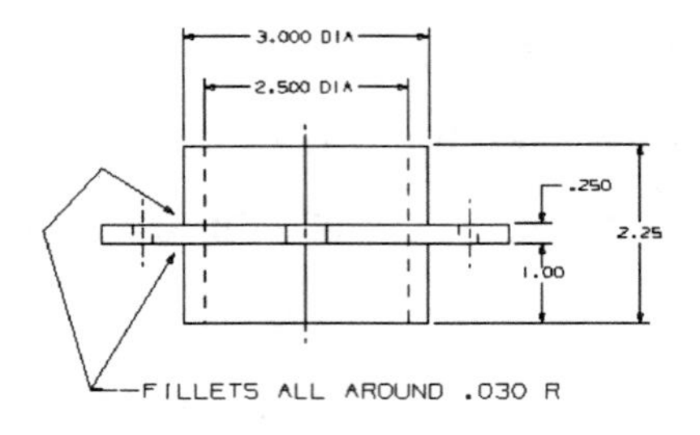

Step 1. Choose the **Utility Symbol** icon .

Step 2. Choose the **Cylindrical Centerline** icon  .

The **Cue Line** is prompting to select a point to define the position for the centerline. You want to select the center point of the circle.

Step 3. Select the horizontal edge at the top of the Front view. It is a circular edge and there is a control point there at the center of the arc. A star will appear at the point selected—insure it is at the center of the circle.

If you have difficulty selecting the circular edge and getting the center point, then change the **Placement / Select Location** option to select the **Arc Center** option.

Step 4. Select the other horizontal edge at the bottom of the part and insure a center point is selected. The centerline is created immediately but inspect it anyway.

This method has the defaults built in to define how far past the edge of the part the centerline extends as defined on the **Utility Symbol** dialog. The defaults may be adjusted to fit your needs by just entering new number into **B** on the dialog and accepting them by choosing **Apply**.

You can also edit the parameters on an existing symbol by selecting the symbol and entering new numbers, and then accepting with **Apply**.

The **Cylindrical Centerline** symbol has another method to place the centerline. It is to change the **Placement / Select Location** option to selecting a **Cylindrical Face** rather than points. When a cylindrical face is chosen the centerline is automatically on the center axis of the cylinder, so all the system needs to know is how far do you want to extend the centerline.

Step 5. Select the **Select Location** pull-down menu and select **Cylindrical Face** icon.

Step 6. Select the inside cylindrical face (it's the only one that goes all the way through), the Cue Line prompts for the end points of the centerline. Pick points outside each end of the cylinder and then you will see the centerline created.

Which option you choose depends on the design intent. The first option associates the centerline to the center points of the circular edges of one of the cylinders and the second option associates the centerline to one cylinder face.

Note that this part is not the best candidate for this second option because two cylinders are present. By selecting only one cylinder and placing the centerline so it looks correct limits the centerline flexibility because it is associated with only one cylinder.

Step 7. **Save** the part file.

Project 14-1. Drafting Baseplate Assembly Components

In this project you will complete dimensioning the Baseplate Assembly component parts. We have already dimensioned the locator part partially. The assignment is to complete dimensioning the locator and to dimension the **locator_pin** and the **fix_baseplate**. These parts are found under directory Chapter14-partfiles. You may create a drawing for the assembly as well. Use the Master Model concept to create a drafting file and use the tools you have learned in this chapter, but not limited to this chapter.

Create two new part files (units: inches) for the drawings of **fix_baseplate** and **locator_pin** as shown below. The naming convention should be **xxx_project14-1_dwg1.prt** and **xxx_project14-1_dwg2.prt**, respectively, where **xxx** is your initials.

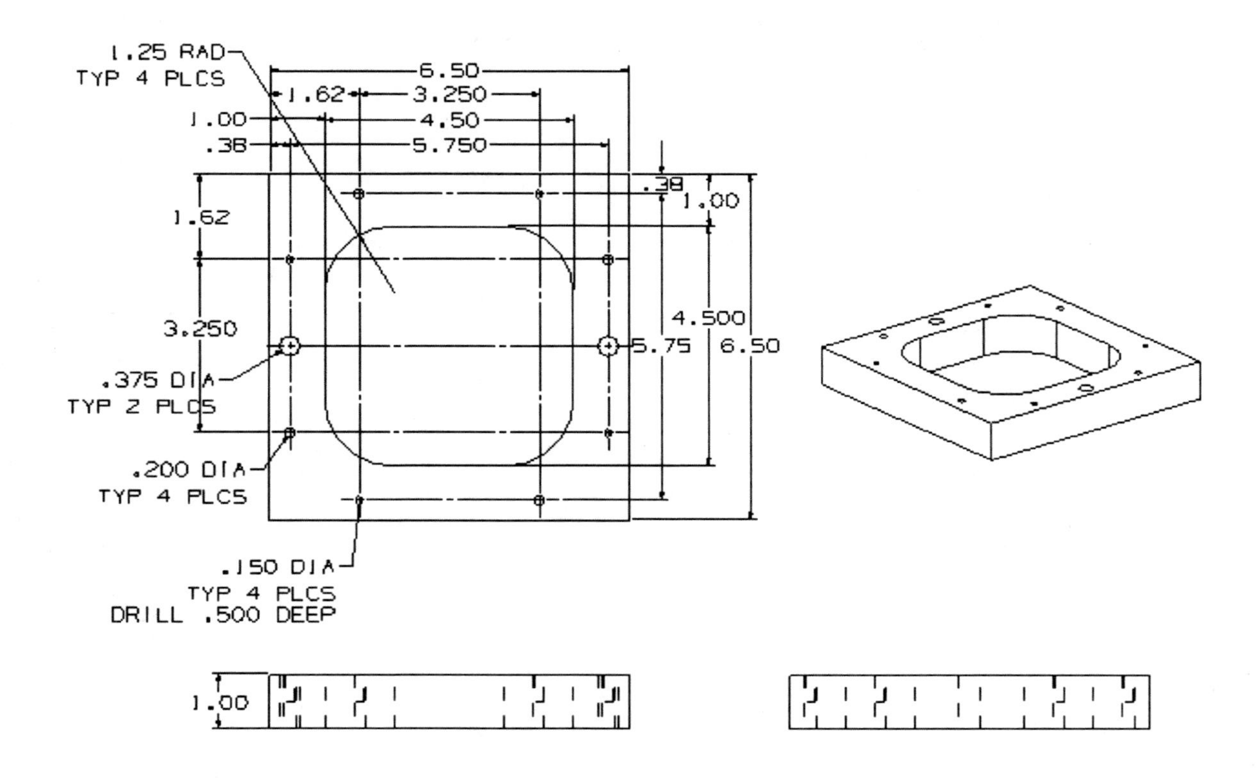

fix_baseplate drawing

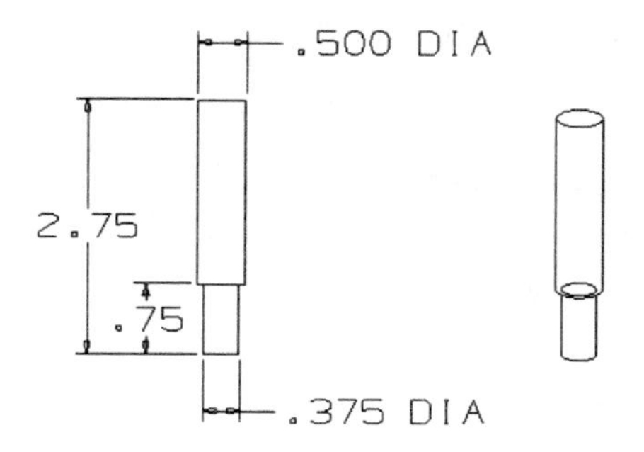

locator pin drawing

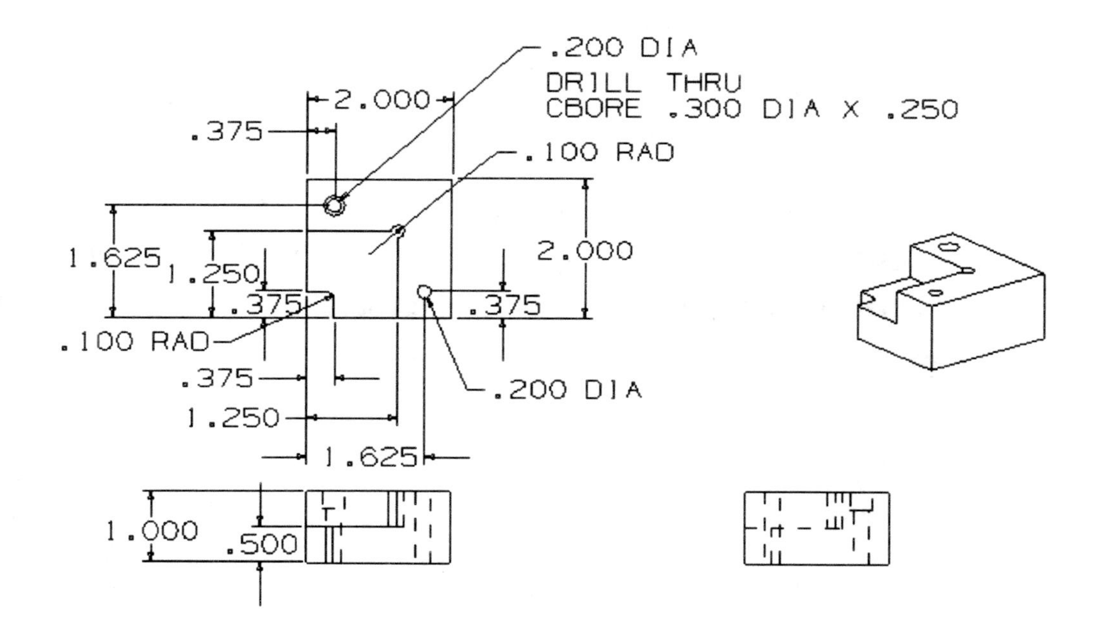

locator drawing

Project 14-2. Drafting Rotator Arm

This project is similar to the previous drafting project with use of the Master Model and dimensions and annotations. The part used for this project is **rotator_arm.prt** under directory Chapter14-partfiles. Alternatively you can use your own part, **xxx_activity10-5.prt**, if you completed **Activity 10-5**.

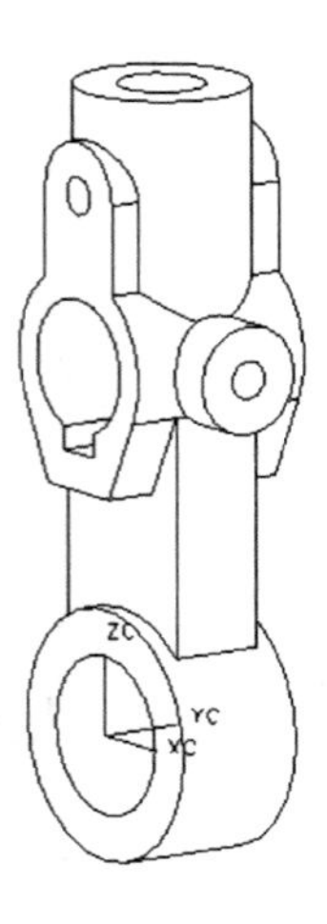

Sample draft drawings are given in the following page. Choose the appropriate size drawing. Note that the part is in millimeters. Save your file as **xxx_project14-2.prt**.

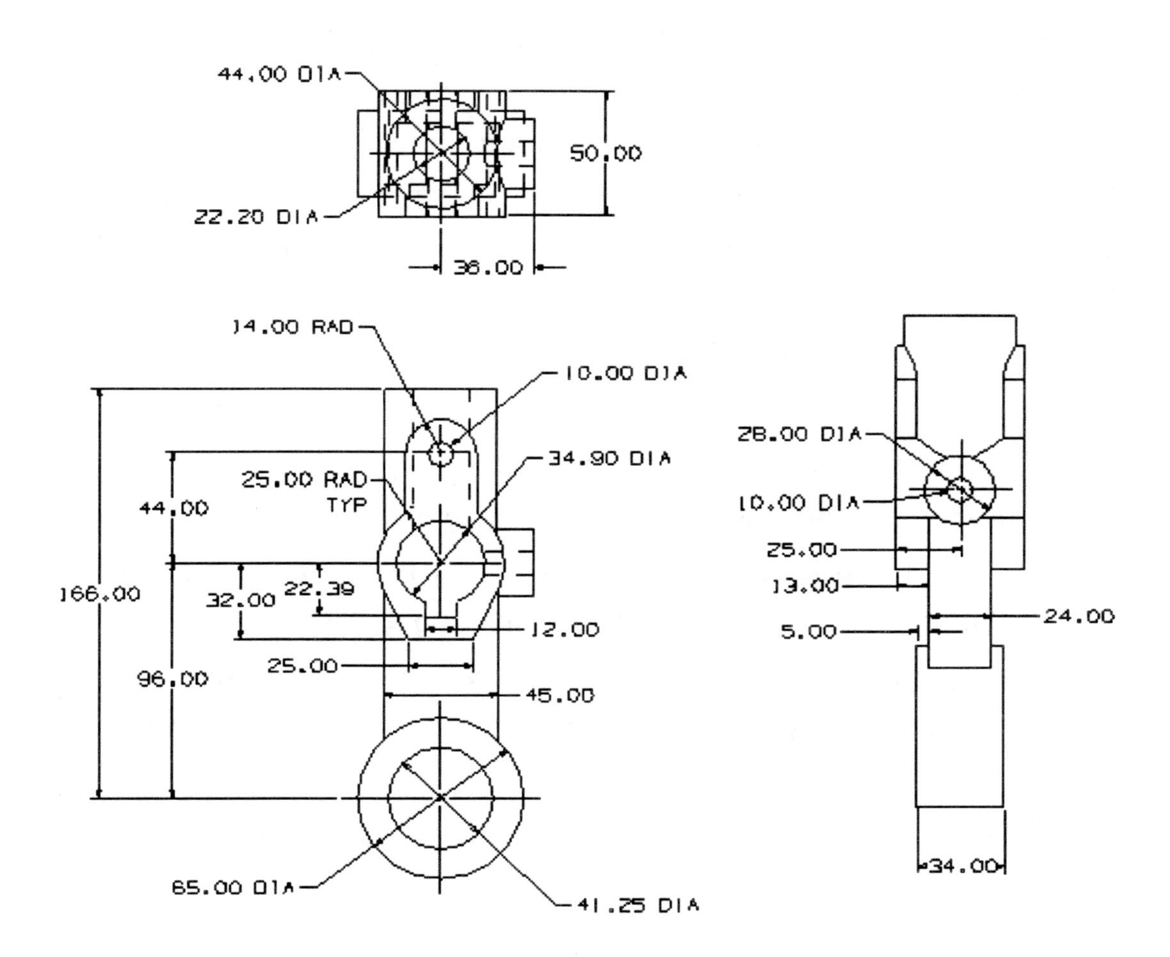

rotator arm drawing

Exercise Problems

14.1. Name the benefits of using Master Model in Drafting.

14.2. Name the benefits of using a solid model for Drafting.

14.3. What name is given to the default drawing in Unigraphics?

14.4. How do you change the drawing sheet name? This name appears at the lower left corner of the graphics window in Drafting application.

14.5. How is the drawing size changed?

14.6. What does the dashed rectangle represent around the drafting graphics window?

14.7. What planning needs to take place before placing views on the drawing?

14.8. Does the above planning need to be exhaustive and detailed?

14.9. What cautions are necessary when a drawing is made smaller?

14.10. Where is the location of 0,0,0 in the drawing coordinate systems?

14.11. What is the significance of where 0,0,0 is located? Why do we care?

14.12. What is the general procedure to add views to a drawing?

14.13. What is a good practice to verify before placing views to insure views are displayed the way you want them to, particularly if not familiar with a part file?

14.14. Generally, what's the difference in changing the customized settings in the Preferences dialog versus changing them in the dialog you are working with?

14.15. What is the difference in a Note and a Label?

14.16. What does Edges Hidden by Edges do?

14.17. What is the general procedure to modify an annotation already on the drawing?

14.18. What action do you take to cancel a selection when placing an annotation when you realize you've made a mistake in what you've picked?

APPENDIX A. Guided Student Assembly Project
Building the Geneva Cam Assembly Model

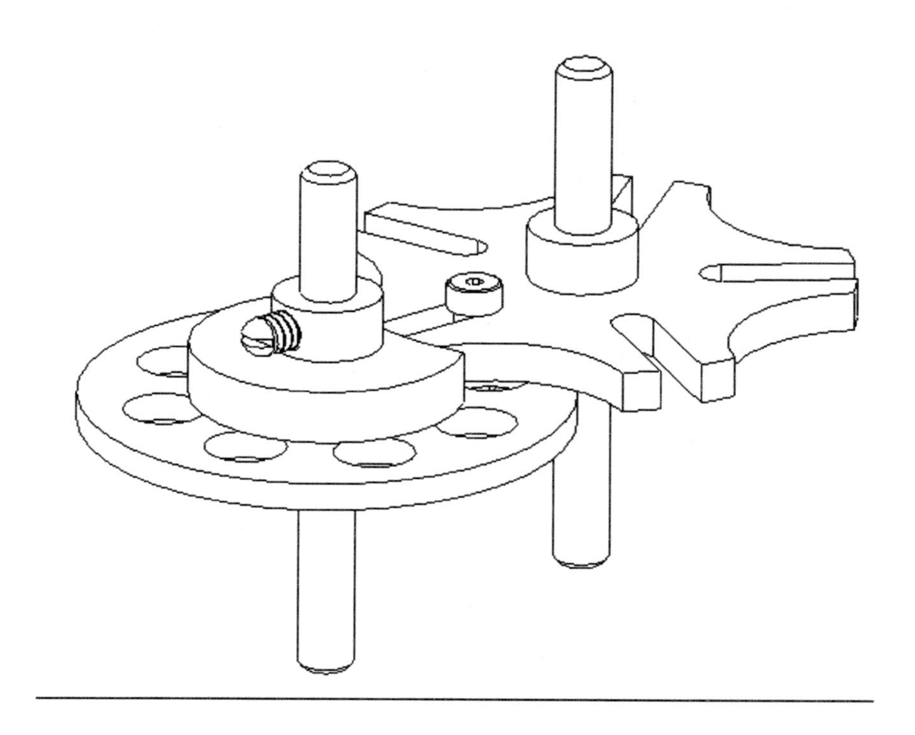

The Geneva Cam assembly consists of 9 unique component parts and total 11 components. Their names, drawings, and dimensions are given in the following pages.

The purpose of this appendix is to give students more opportunities to practice what they have learned this book. In this appendix, students will be guided to model individual parts of the Geneva Cam assembly from a scratch and to put them into the final assembly. Section 1 addresses individual part modeling. Section 2 addresses assembly modeling. Three approaches to assembly modeling are presented. They are repositioning only, using mating conditions, and using interpart modeling.

Geneva Cam Assembly[1]

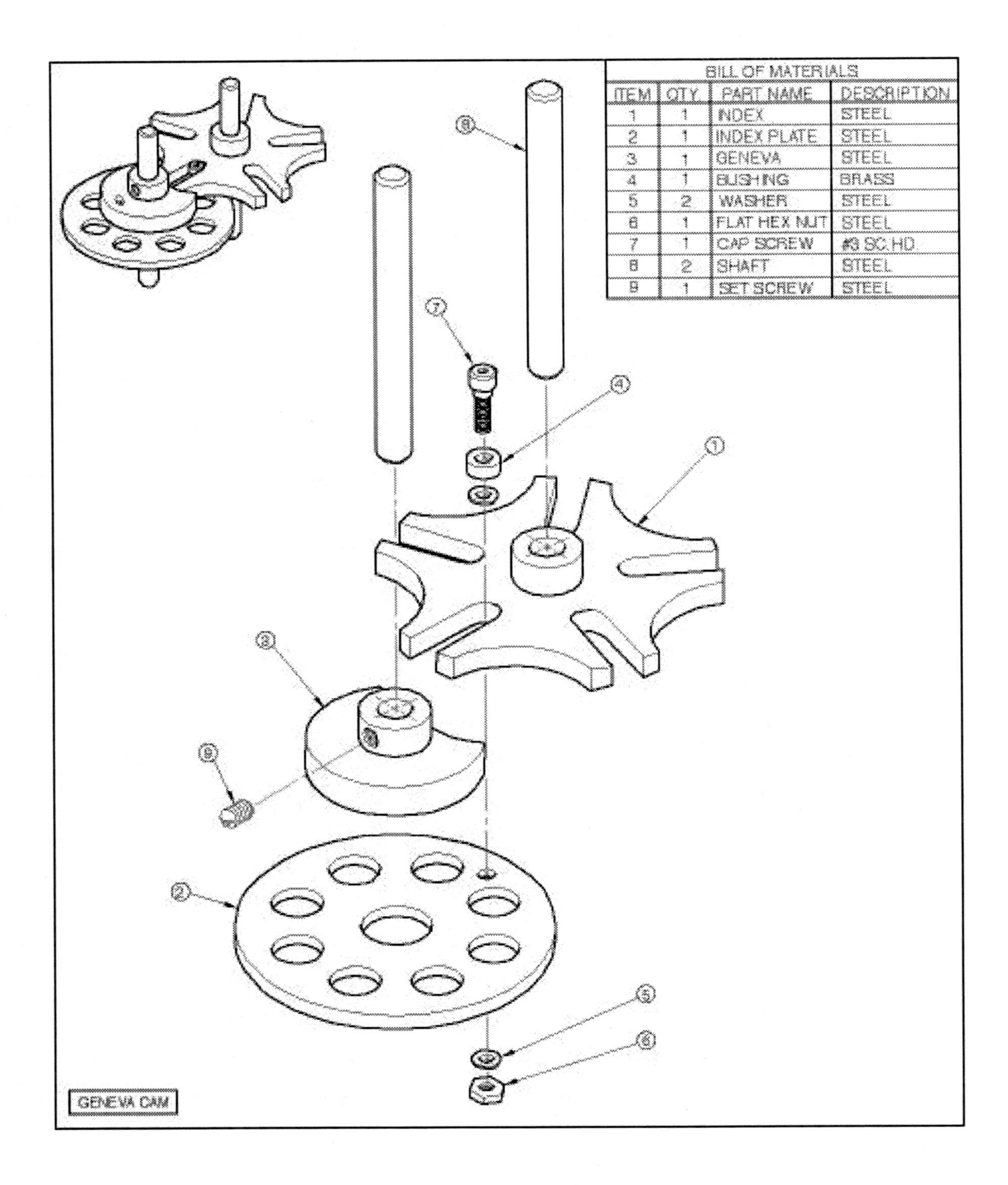

BILL OF MATERIALS			
ITEM	QTY	PART NAME	DESCRIPTION
1	1	INDEX	STEEL
2	1	INDEX PLATE	STEEL
3	1	GENEVA	STEEL
4	1	BUSHING	BRASS
5	2	WASHER	STEEL
6	1	FLAT HEX NUT	STEEL
7	1	CAP SCREW	#3 SC HD
8	2	SHAFT	STEEL
9	1	SET SCREW	STEEL

[1] [1] The drawings for this assembly are taken from Technical Graphics Communication by Bertoline and et al. (1997) (reprinted with permission from the publisher McGraw-Hill)

Key Components of Geneva Cam Assembly

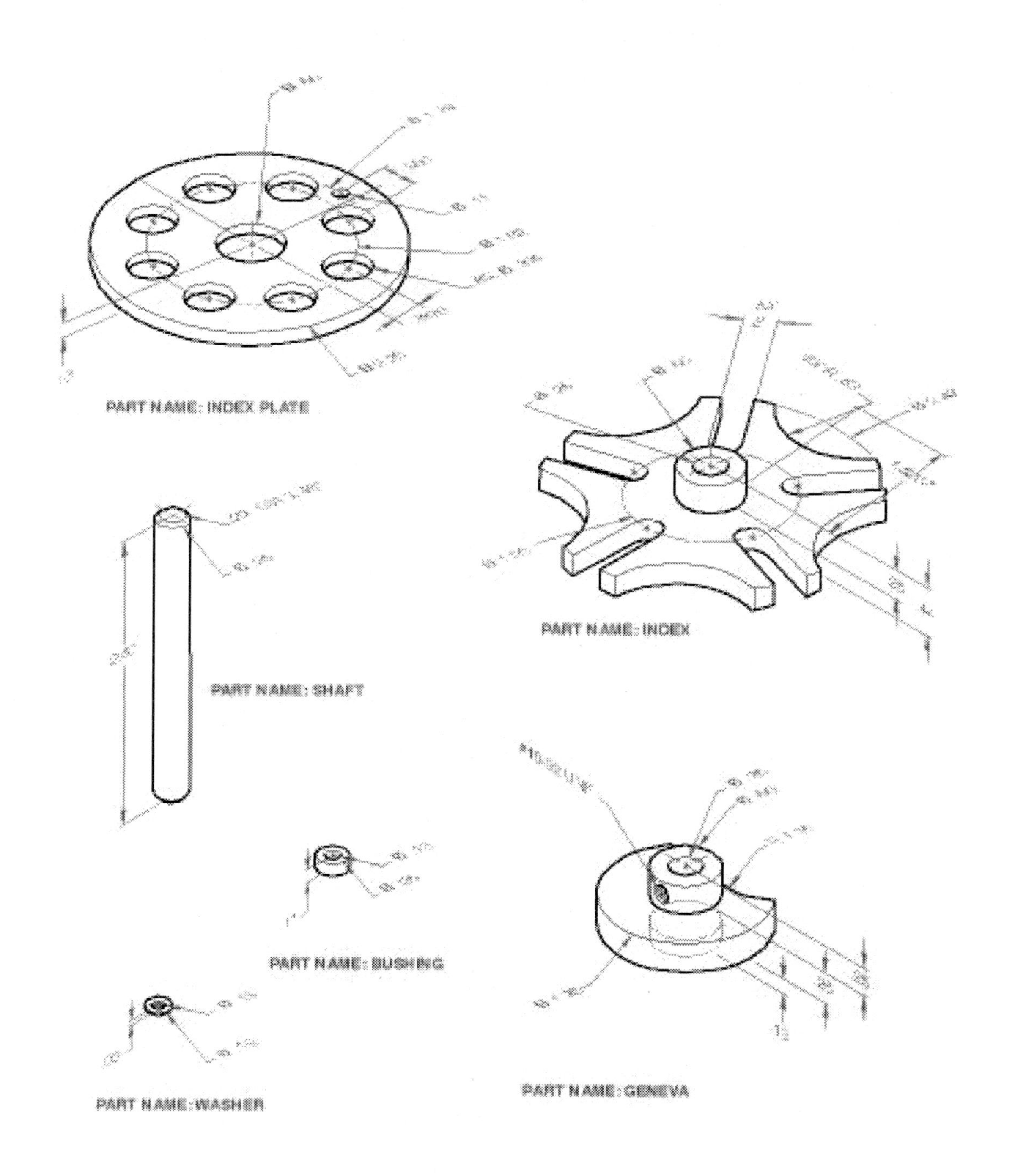

Section 1. Modeling Individual Parts

In this section, you will create individual component pars of the Geneva assembly. Create a new subdirectory (or folder), xxx_AppendixA_Proj1, where xxx is your three-letter name initials. Save these part files under this directory. It is helpful to change the components to different colors to distinguish them easier in the assembly, so we will assign colors as we create the parts as well. Since white is the system color we will avoid our objects being white.

Section 1-1

<u>Creating the Index</u>

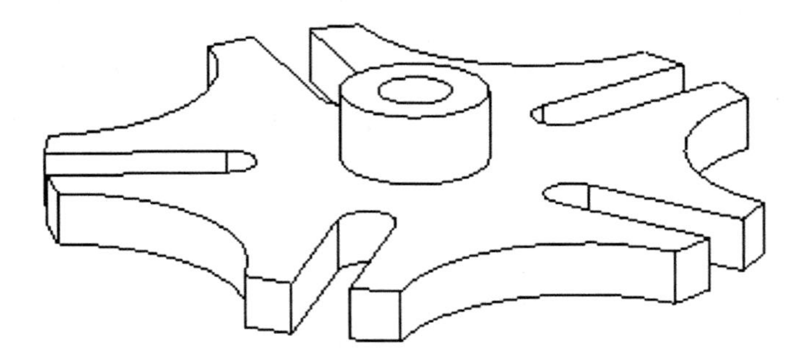

Step 1 – A new part file

Create a new file (using inches) named xxx_Geneva_index.prt and start the modeling application, where xxx is your three-letter name initials. If it is not already, change the solid color to green. The creation color is found at Preferences → Object… and setting Type to Solid Body and Color to Green.

Step 2 – Cylinder

Create a cylinder with diameter 2.47 and height .44-.25= .19 at WCS 0,0,0.

Step 3 – Boss

Create a .25 high, .5 diameter boss on the top face of the cylinder. Use point onto point for placement and choose the arc center of the circle edge of the cylinder.

Step 4 – Thru Hole

Create a .25 diameter through hole on the top face of the boss created in the previous step. Use point onto point placement and choose the arc center of the circle edge of the boss. The placement face should be the top face of the boss, and the thru face is the bottom face of the entire solid.

Step 5 – Datum Plane

Change the work layer to 61 and create a datum plane that goes through the axis of the boss cylindrical face. This datum plane will be used to position a slot that will be created in the next step.

Step 6 – Rectangular Slot

Change the work layer to 1 and create a rectangular slot (do not use a thru slot option). Placement face is the top face of the cylinder created in Step 1. Select the datum plane as the horizontal reference. Use the following parameter for the slot: width= .2, depth= .5 (any number larger than .19 which is the index depth, length=3 (any number larger than 2.47 which is a diameter of the index). Position it using two methods. First, use line onto line by aligning the datum plane to the center line of the slot. Second, use horizontal positioning with distance -1.25/2 between the arc center of the boss and the arc center of the slot half circle. If it is not positioned properly, you may need to change the horizontal distance to the positive number, 1.25/2.

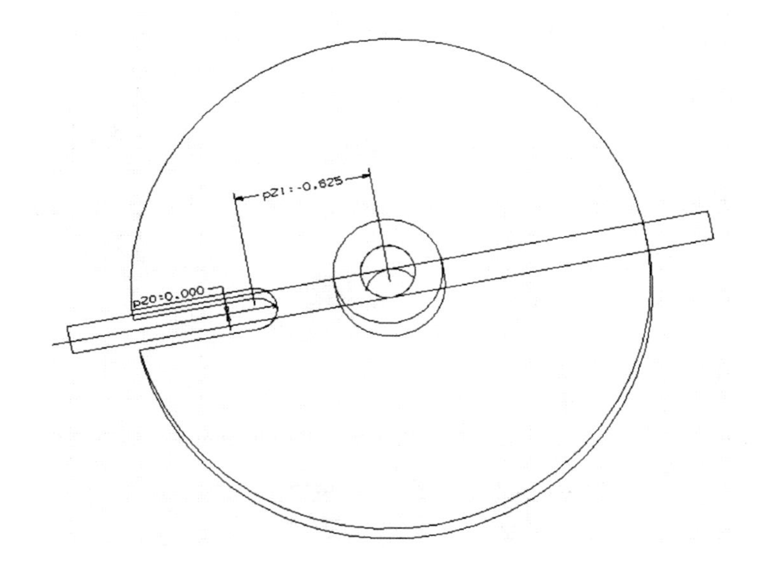

Step 7 – Thru Hole for Circle Indentation

Create a thru hole with diameter 2 x .63 for the side circle indentation for the index. The placement face is the top face of the cylinder.

Position it using two methods. First, use point onto line by aligning the datum plane to the hole center. Second, use horizontal position with distance 1.529 between the arc center of the boss and the hole center. For horizontal positioning, you need select the datum plane as the horizontal reference. If it is not positioned properly, you may need to change the horizontal distance to the negative number, -1.529.

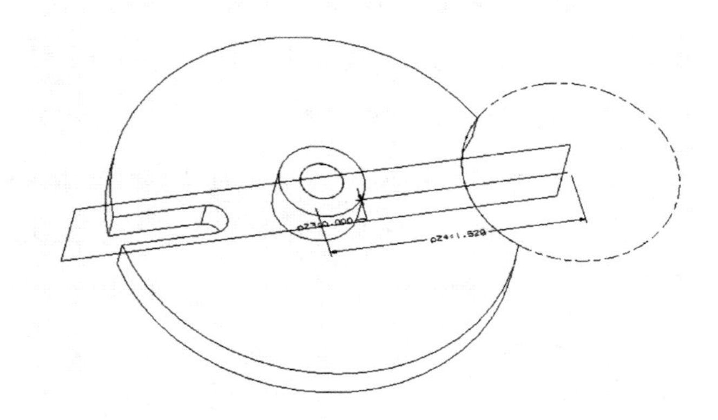

Step 8 – Datum Axis

Change the work layer to 62. Create a datum axis that goes through the axis of the boss. This reference feature will be used to create an array for the next step.

Step 9 – Circular Array

In order to make the complete part, it is necessary to create circular array instances for the rectangular slot and the indented thru hole.
Click on the instance array icon. You will create a circular array of 5 instances, with the angle between being 360/5. You will instance two features, which are the rectangular slot and the thru hole for the side indentation. The rotation axis for the instance will be the datum axis.

You have now completed the index.

Section 1-2

Creating the Index Plate

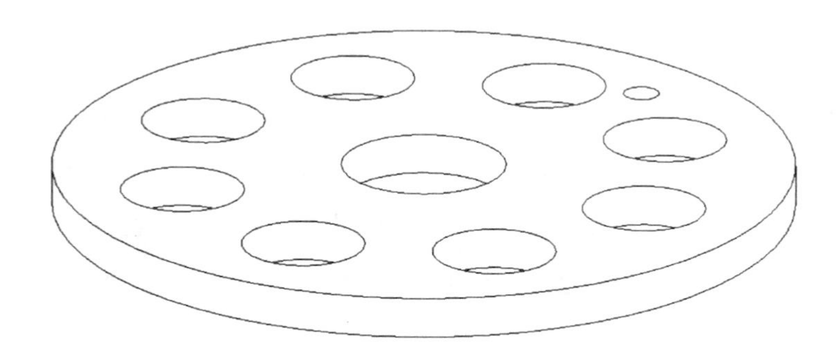

Step 1—New part file

Create a new file named xxx_Geneva_index_plate.prt. Use inches. Change the creation color to cyan.

Step 2 – Cylinder

To begin the index plate, create a 2.25 diameter, .130 high primitive cylinder.

Step 3 – Datum Planes

The next step is to change the work layer to 61 and create two datum planes. First create a datum plane that goes "through face axis" by choosing the outer cylindrical face of the cylinder. Next create a datum plane that is rotated 90° from the first plane.

Step 4 – Center Hole

Create a .5 diameter thru hole in the center of the cylinder you just created.

Step 5 – Outer Hole

The next step is to create one of the holes around the outer part of the plate. To do this first create one thru .375 diameter hole and position it with parallel distance of 1.52/2 between the arc center of the center hole and the center of the current hole and then perpendicular distance from one datum plane of .287 as shown below.

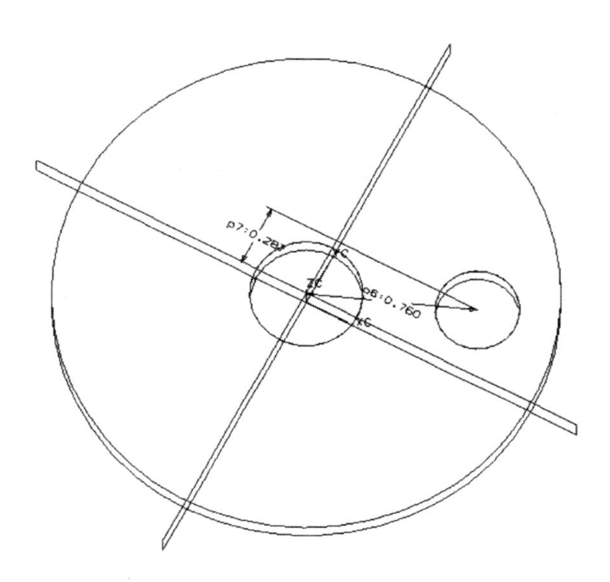

Step 6 – Circular Array

Change the work layer to 62 and create a datum axis that goes through the axis of the face of the cylinder created in Step 2.

Create a circular instance, 8 holes, angle of 360/8. The rotation axis should be the datum axis.

Step 7 – Small Hole

The final step in creating the plate is to make a simple thru hole with diameter .11. Position it with parallel distance of 1.79/2 between the arc center of the center hole and the center of the current hole and then a horizontal distance of .35 from the arc center of one near outer hole (.375) as shown below. Note that when you pick horizontal distance for positioning, you will be prompted to specify **horizontal reference**. Use one datum plane that is parallel to the dimension distance being positioned.

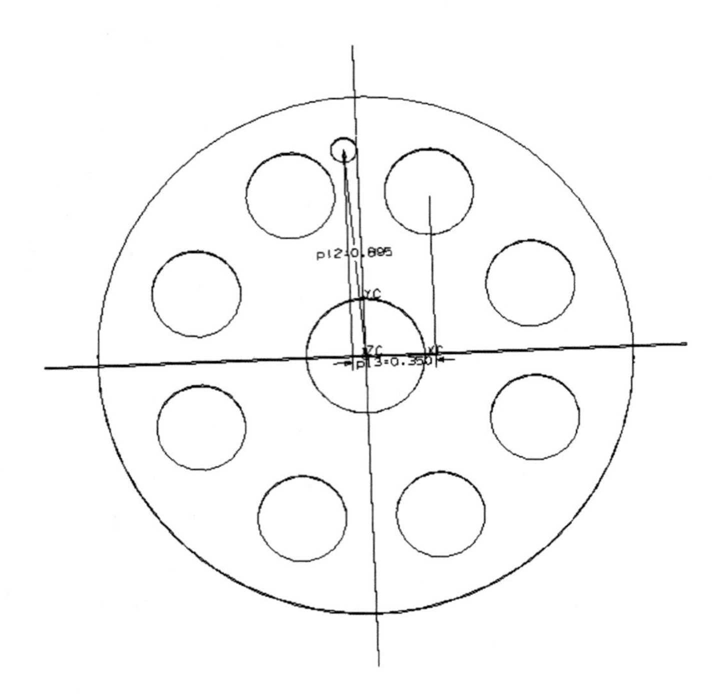

You have now completed the index plate.

Section 1-3

<u>Creating the Geneva</u>

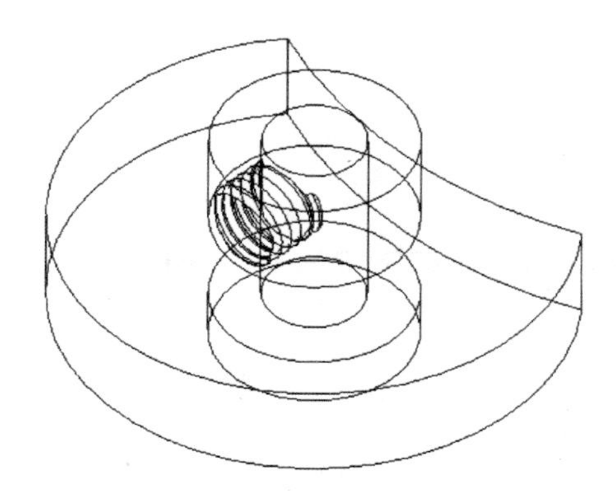

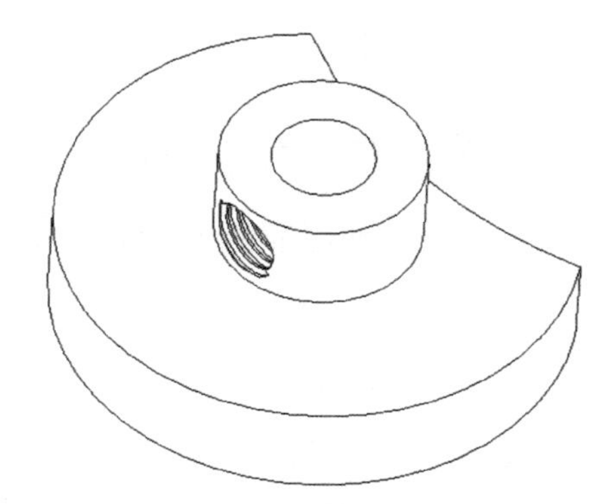

Step 1—New part file

Create a new file (using inches) named xxx_Geneva_Geneva.prt. Change the creation color to red.

Step 2 –Cylinder

The part will begin with a primitive cylinder 1.25 in diameter and .25 high.

Step 3 – Boss on the top

Next, create a .5 diameter, .25 high boss in the center of the top face of the cylinder.

Step 4 – Boss on the bottom

Next, create a .5 diameter, .13 high boss in the center of the bottom face of the cylinder.

Step 5 – Hole

Now create a .25 diameter thru hole that is centered on the top face of the top boss created in Step 3 and goes through to the bottom face of the bottom boss created in Step 4.

Step 6 – Datum Planes

You will need to create three datum planes to use for reference on the next few steps.

First change the work layer to 61 and create the first datum plane DATUMP-1 that goes through the face axis of the cylinder.

Then create the second datum plane DATUMP-2 that goes also through the axis but is rotated 90° from the first plane.

Now create the third plane DATUMP-3 that is offset 1.529 inches from DATUM-1.

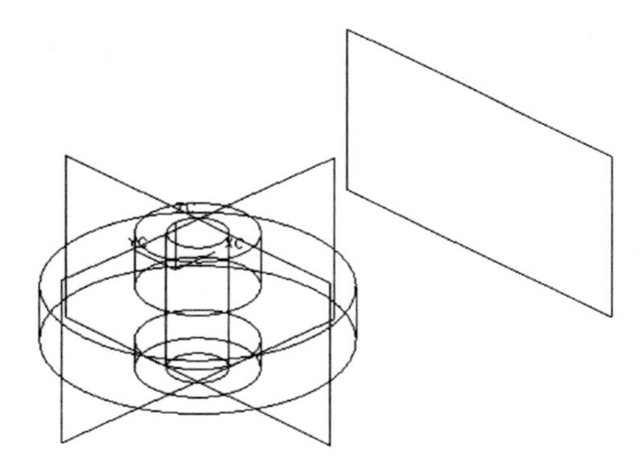

Step 7 – Indentation

You will now create the circular section indentation in the side of the Geneva. To do this, create a 2.5 inch diameter, 1 inch deep hole. Place the hole on the top face of the cylinder (not the boss) and align the center of the hole to the intersection of DATUMP-2 and DATUMP-3, using point onto line positioning.

Step 8 – Side Hole

You will create a hole in the side of the upper boss that the set_screw will go into to prevent the Geneva from spinning on the shaft when assembled.

Create a .19 diameter, 1-inch deep hole and use DATUMP-1 as the placement face. Make certain that the hole goes out in the direction opposite the indentation created in the previous step. Position the hole in the middle height of the top boss.

Step 9 – Thread

The final step in creating the Geneva is to create an internal thread on the side hole that the set_screw will fit into.

Choose Insert → Design Feature →Thread... Choose thread type Detailed. Choose the cylindrical face of the side hole to put the thread on. Select DATUMP-1 as the start face. Choose OK to accept the default parameter values of the thread.

You have now completed the Geneva.

Section 1-4

Creating the Bushing

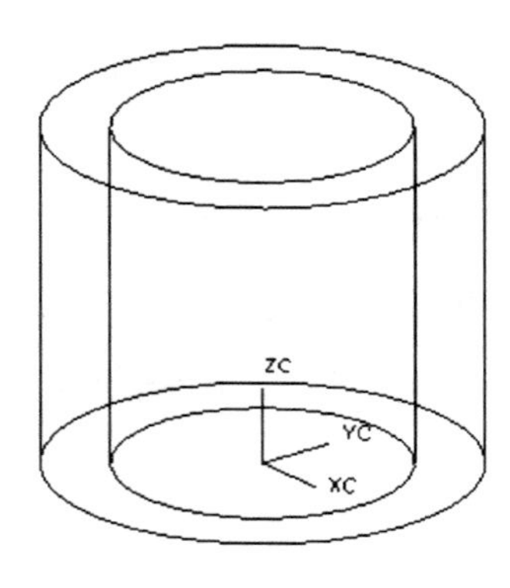

Step 1—New part file

Create a new part file (using inches) named xxx_Geneva_bushing.prt. Change the creation color to aquamarine.

Step 2 – Cylinder

Create a primitive cylinder .19 in diameter and .16 inches high.

Step 3 – Hole

To finish the bushing, create a thru hole .13 inches in diameter on the top face of the cylinder as the placement face. Pick the bottom face of the cylinder as the thru face. Use point-onto-point for the positioning method and choose the arc center of the circular edge of the cylinder.

You have now completed the bushing.

Section 1-5

<u>Creating the Washer</u>

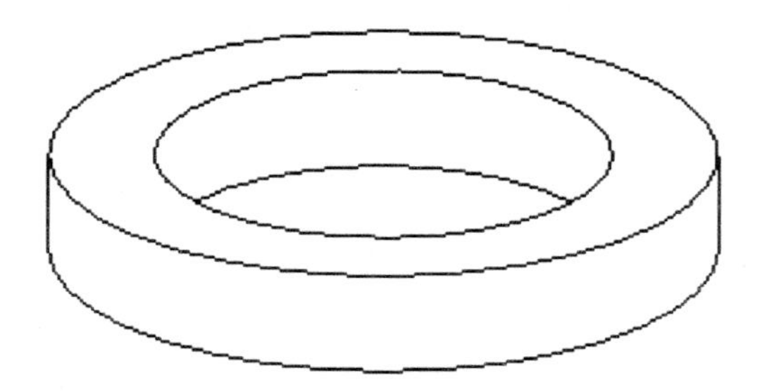

Step 1—New part file

Create a new part file (using inches) named xxx_Geneva_washer.prt. Change the creation color to yellow.

Step 2 – Cylinder

Create a primitive cylinder .19 in diameter and .03 inches high.

Step 3 – Hole

To finish the washer, create a thru hole .13 inches in diameter on the top face of the cylinder as the placement face. Pick the bottom face of the cylinder as the thru face. Use point-onto-point for the positioning method and choose the arc center of the circular edge of the cylinder.

You have now completed the washer.

Section 1-6

Creating the Nut

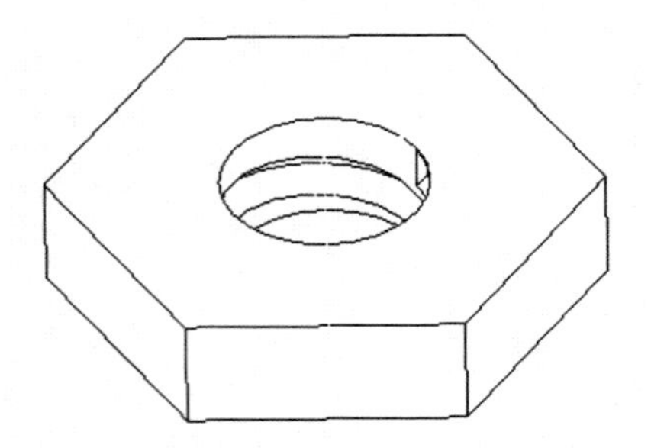

Step 1 – New part file

Create a new part file (using inches) named xxx_Geneva_nut.prt. Change the creation color to green.

Step 2 – Hexagon curves

Change the work layer to 41 (the default layer standard for curve is 41 to 60). Choose Insert → Curve→Polygon…. Enter the number of sides 6 and the inscribed radius .1. When the Point Constructor dialogue appears, choose the Reset button and then OK. The Hexagon curve is created at WCS 0,0,0.

Step 3 – Extrusion

Change the work layer to 1. Extrude the hexagon curves that you just created a distance of .05 inches to create the solid. To pick the curves, you can pick each of them or simply use Chain Curves.

Step 4 – Hole

First you need to create two datum planes to use for reference for positioning the hole to be created.

Change the work layer to 61 and create the first datum plane DATUMP-1 that is a center plane between two parallel side faces of the extruded body.

Create the second datum plane DATUMP-2 that is a center plane between another two parallel side faces of the extruded body.

Create a .875-inch diameter thru hole that is in the center of the hexagon extrusion. Position it applying point onto line twice using datum planes DATUMP-1 and DATUMP-2.

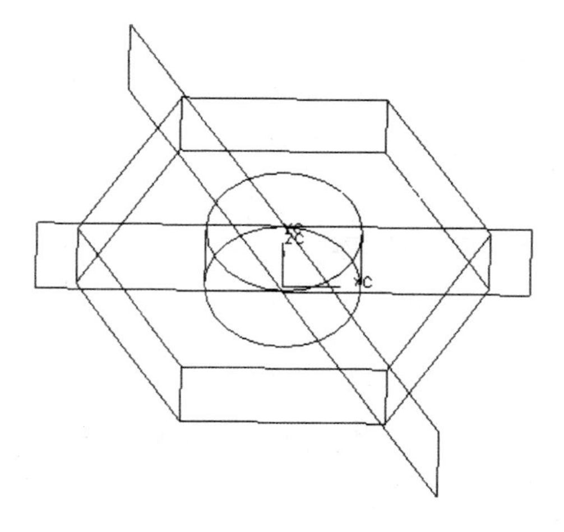

Step 5 – Thread

Create an internal detailed thread. Select the inside face of the hole that was created in the last step. Accept the default parameter values.

You have now completed the nut.

Section 1-7

Creating the Cap Screw

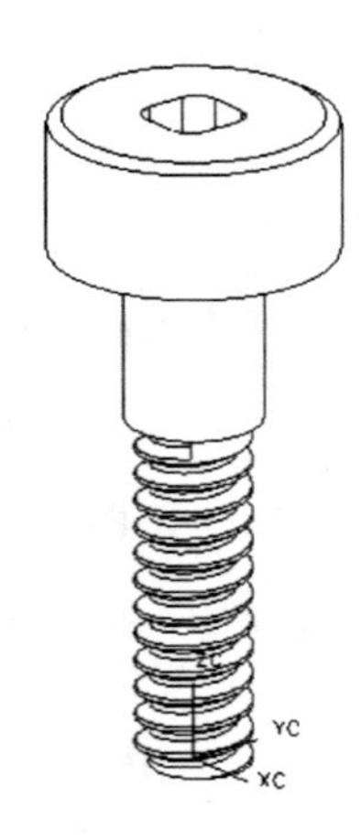

Step 1—New part file

Create a new part file (using inches) named xxx_Geneva_cap_screw.prt. Change the creation color to magenta.

Step 2 – Cylinder

Create a .10 diameter, .32 high cylinder.

Step 3 – Boss

On top of the cylinder you just created, add a boss that is .12 diameter and .16 high. Position it point-onto-point with the arc center of the circular edge of the cylinder.

Step 4 – 2nd Boss

On the top of the boss you just created, make another boss that is .25 diameter, .125 high. Place it point-onto-point with the arc center of the circular edge of the previous boss.

Step 5 – Chamfer

Next, insert a .0125 by 30° chamfer on the top edge of the second boss.

Step 6 – Thread

Create an external thread by choosing Insert → Detail Feature →Thread... Choose thread type Detailed. Choose the cylindrical face of the smallest cylinder to put the thread on. Choose OK to accept the default parameter values of the thread.

Step 7 – Rectangular Pocket

You will create a rectangular pocket at the center of the top face of the screw for the fastening tool to grip on.

First you need to create two datum planes to use for reference for positioning the pocket to be created.

Change the work layer to 61 and create the first datum plane DATUMP-1 that goes through the face axis of the cylinder.

Create the second datum plane DATUMP-2 that goes also through the axis but is rotated 90° from the first plane.

Create a rectangular pocket with the following parameters (length=width=.075, depth=.1, corner radius=floor radius=.02). Use any of the datum planes for Horizontal reference. Position it using line onto line by selecting DATUMP-1 and a centerline of the pocket. Repeat line onto line positioning by selecting DATUMP-2 and the other centerline.

You have now completed the cap screw.

Section 1-8

<u>Creating the Shaft</u>

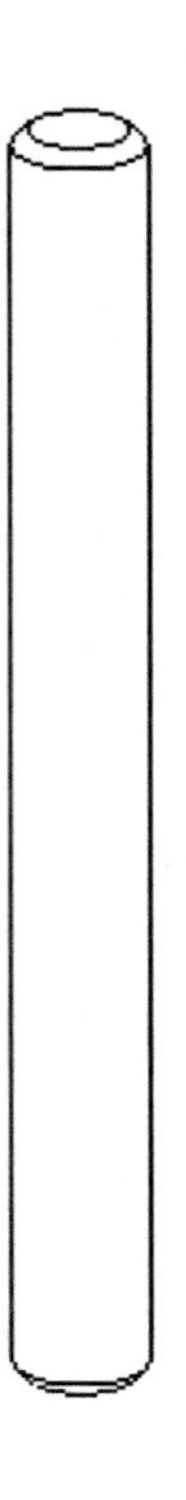

Step 1-- New part file

Create a new part files (using inches) named xxx_Geneva_shaft.prt. Change the creation color to orange.

Step 2 – Cylinder

Create a cylinder that is .25 inches in diameter and 2.5 inches high.

Step 3 – Chamfer

The only other step in creating the shaft is to chamfer the end. To do this select the Chamfer icon, and choose a single offset of .031 inches and pick the two circular edges on both ends of the part.

You have now completed the shaft.

Section 1-9

Creating the Set screw

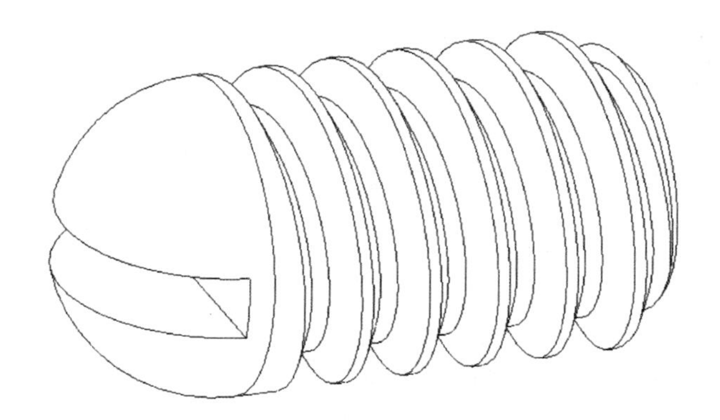

Step 1 – New part file

Create a new file (using inches) named xxx_Geneva_set_screw.prt. Change the creation color to brown.

Step 2 – Cylinder

Create a .19 diameter, .35 high cylinder in the +YC direction.

Step 3 – Chamfer

Create a .02 single offset chamfer on the right end of the cylinder

Step 4 – Datum Planes

Change the work layer to 61 and create 2 datum planes, DATUMP_1 that is in the same plane as the unchamfered end of the cylinder and DATUMP_2 that goes through the face axis of the cylinder. The resulting figure looks below.

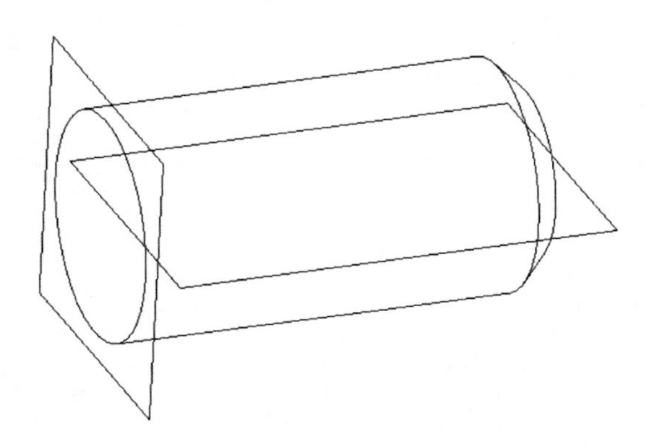

Step 5 – Blend

Blend the unchamfered edge of the cylinder with a radius of .095 inches to round the top of the screw. Accept the default values in the Edge Blend dialog

Step 6 – Rectangular Pocket

Now you need to create a rectangular pocket for a tool to fit into. Its parameters are X length=1, Y length =.04, and Z length =.08. Select DATUMP_1 as the planar placement face. Position the pocket such that the long dimension is line-onto-line with DATUMP_2. You will see the status line saying "Feature position is underspecified" but it will create a desired pocket since X length is sufficiently large.

Step 7 – Thread

Create an external thread on the cylindrical face. Select the thread type Detailed. Select the cylindrical face. If prompted to select a start face, select the top face of the cylinder with a chamfered end. Accept the default parameter values.

You have now completed the set screw. This completes section one.

Section 2. Assembly Modeling

Section 2-1

<u>Assembling Parts With Repositioning only</u>

In this assembly modeling approach, components of the individual parts created in the previous section are assembled such that components do not have any association to each other. Components are just placed in space individually using WCS. Thus, when you move one component in space by choosing Assemblies → Components → Reposition Component..., the rest of components still remain in the same place.

Step 1 – Create a new part file

Create a new file (using inches) named xxx_Geneva_assm_repo.prt and save it under directory xxx_AppendixA_Proj1.

Step 2 – Add the index plate

The procedure to add a part will be the same for each of the components except positioning of the part in space. Choose Application → Assemblies. To add a part to the assembly, choose Assemblies → Components → Add Component.

Click Choose Part File and locate the xxx_Geneva_index_plate.prt file, if the file is not already open. Otherwise, select it on the list of opened parts. Take the default reference set. For the multiple add option, choose no and for the positioning option, choose reposition. The layer option should be original.

Click OK for the 0,0,0 insertion point. The Reposition component window will pop up. If necessary, rotate the index_plate a little about its center point to make positioning of the index easier.

Step 3 – Add the index

Now add xxx_Geneva_index to the assembly. The distance between the center of the half-circle that is inside the slot in the index part and the center of the bolt hole in the plate should be 0. Also, the z coordinate of the bottom of the index should be the same as the z coordinate of the top of the plate.

To do this, first insert the plate at 0,4,0 in WCS. This will make it easier to see what is going on. Rotate the index if necessary so that a slot on the index should point toward the center of the index plate. Use the Reposition Component dialog to rotate the index if necessary. Now we need to move it in the y and z direction to align the hole. Choose the Point to Point icon. From the Point Construction dialog, choose the Arc Center icon. The first selection point is the From point and should be the bottom edge of the half-circle at the end of the slot that is aligned with the y-axis. The second point is the To point and should be the top edge of the small hole in the index plate. Click OK and the plate and the index should be properly aligned.

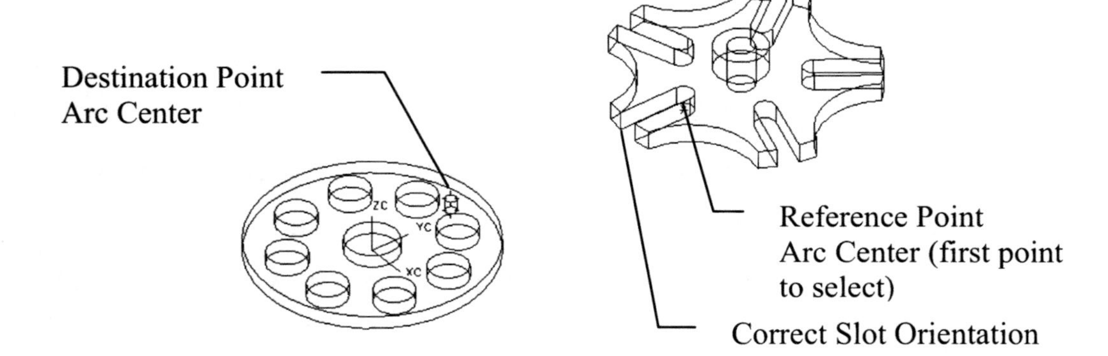

Step 4 – Add the Geneva

Click Add part and place the xxx_Geneva_geneva at the point -2,0,0. The part may need to be rotated 90° about it's center to align the indented round correctly, and then translated such that the bottom face of the body of the Geneva is aligned with the top edge of the hole in the index plate to be finished. Therefore, translate Point to Point of the appropriate arc centers. After the Geneva has been moved be sure and choose OK to confirm the translate operation. The completed position of the three parts should look like below.

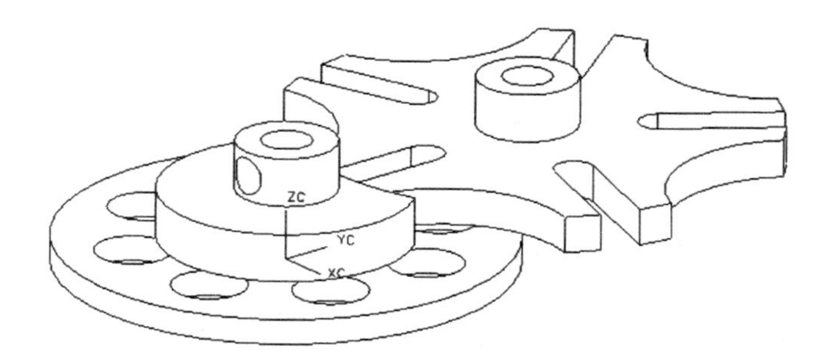

Step 5 – Add the two shafts

When you add xxx_Geneva_shaft, choose multiple add (or just do it twice) and the insertion point for each will be the arc center of the Geneva and of the index. The positioning in the z direction is unimportant just so they look as shown in the drawing. Be sure to choose OK to confirm the move.

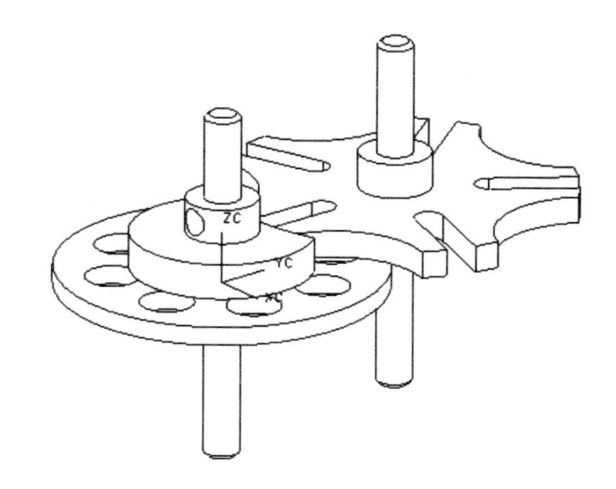

Step 6 – Add the two washers

Add and position the component xxx_Geneva_washer (two units) to the assembly part. Both washers are placed on the index plate: one on top of it and one on the bottom of it. Add the first washer on the top face of the index plate and align the arc center (From point) of the bottom face of the washer with the arc center (To point) of the top edge of the smallest hole in the index plate. This washer will fit in the slot. Remember to confirm the move by choosing OK.

Add the second washer on the bottom face of the index plate and align the arc center of the top face of the washer with the arc center of the bottom edge of the smallest hole in the index plate. And again, choose OK to confirm the move.

Step 7 – Add the bushing

Add and position the component xxx_Geneva_bushing to the assembly part. Add it on the top face of the top washer and align the arc center (From point) of the bottom face of the bushing with the arc center (To point) of the top face of the washer. It too should fit into the slot, rest on top of the washer, and be flush with the top of the index. And don't forget to choose OK to confirm.

Step 8 – Add the cap screw

Add the component xxx_Geneva_cap_screw to the assembly part. The insertion point of the part can be the arc center of the slot in the index. Reposition the screw in the z direction by choosing translate and then selecting the arc center of the bottom face of the screw head and the arc center of the top edge of the bushing. Don't forget the OK.

Step 9 – Add the nut

Adding xxx_Geneva_nut is similar to adding the washer and cap screw. This time the To point for the nut center will be the arc center of the bottom face of the second washer that you added in Step 6. Don't forget the OK. The figure below illustrates the placement of the 2 washers, bushing, nut and cap screw, with the other parts suppressed for clarity. The threads have been suppressed for clarity.

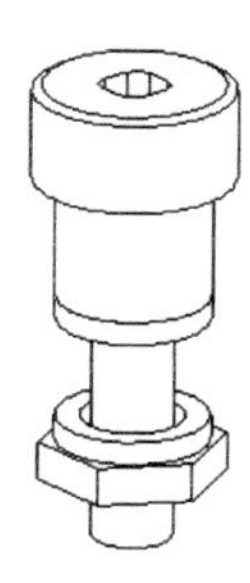

Step 10 – Add the set_screw

The set_screw may be tricky because of the threads and orientation. Add the part file and place it initially at 0, -2, 0 to make placement easier. Rotate the part by 90° if necessary so that is oriented properly with the hole in the Geneva. Now you can translate the part so that the arc center of the threaded end of the screw is coincident with the circumference of the shaft through the Geneva. Suppressing the threads on the set_screw and the hole in the Geneva may make placement easier. It is suggested the WCS origin is relocated to be at the centerline of the set_screw hole and at the circumference of the shaft as shown in the figure below. For this, you need to use the dimensions of proper features to determine the exact location of the WCS origin. For the Reposition translation, the From point is the center of the chamfer end on the set_screw and the To point is the WCS 0,0,0.

Note in the figure below the threads have been Suppressed for clarity and it is suggested that you suppress them for ease of selecting the edges.
Don't forget to Unsuppress when finished.

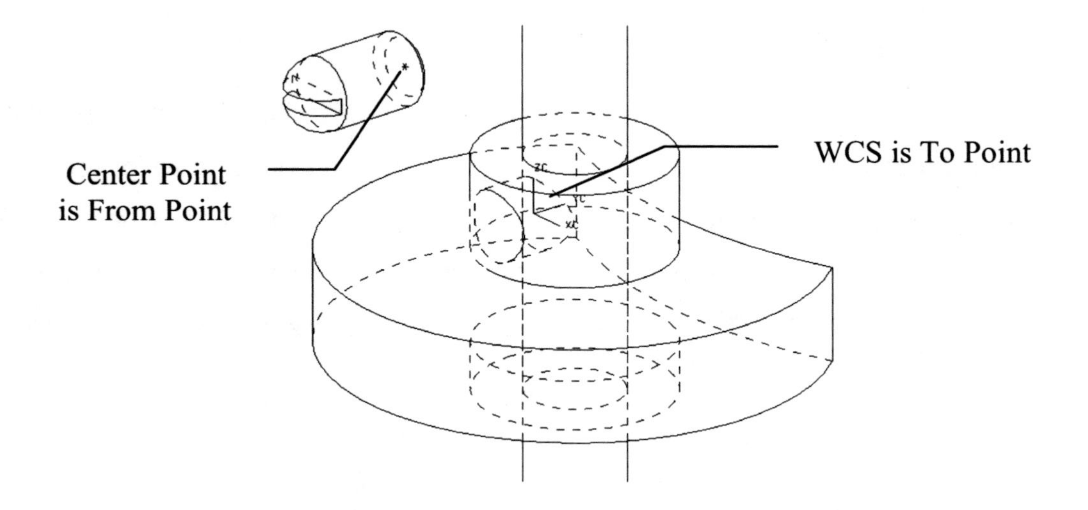

The completed assembly should look like the assembly shown at the beginning of this section.

This completes section 2-1.

Section 2-2

Assembling Parts with Mating Conditions

In this assembly modeling approach, we have a new design intent: components of the Geneva assembly are assembled such that they are spatially associated. This means that when one component moves to another location in space, other components move accordingly to maintain spatial relationships to the moved component. For example, the axis of a nut is "aligned" with the axis of a washer. When a nut moves in space, a washer will automatically move in order to maintain the alignment relationship to the nut.

The procedure of adding a part with mating conditions is the same as in Section 2-1 except that you have to select Mate instead of Reposition for the positioning option in the Adding Existing Part dialogue.

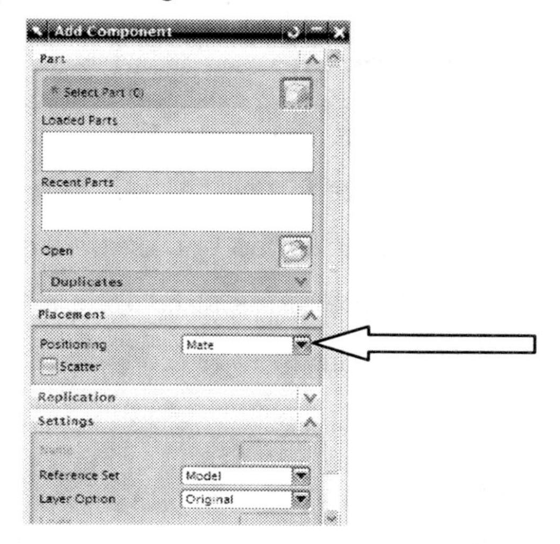

The Mating Conditions dialog displays when you select Mate as the positioning option. This dialogue shows the two parts that you are currently mating, and the mating constraints that have been specified thus far between the two parts. You can delete constraints from the list if you apply them incorrectly, as well as being able to add more constraints to other sets of parts. The process of mating parts together involves a step-by-step selection of the different type of mating constraints to be satisfied until the part is placed as needed. Particular types of mating constraints that you will mainly use in this Geneva assembly include Mate (planar objects selected to mate are coplanar and face opposite each other) and Align (centerlines of cylindrical objects will be in line with each other) or Center 1 to 1.

Step 1 – Create a new part file

Create a new file (using inches) named xxx_Geneva_assm_mate.prt and save it under directory xxx_AppendixA_Proj1.

Step 2 – Add the index plate

This step will be identical to Step 1 of the previous section. It adds the first part into the assembly and orients it properly.

Step 3 – Add the index

Add the part just as before, but make sure to click on Mate for the Positioning option. You may need to rotate the piece 180° to make it look right. Align the centerline of the cylindrical face at the end of the slot in the index to the centerline of the cylindrical face of the small hole in the index plate. Next mate the bottom face of the index with the top face of the index plate.

Step 4 – Add the Geneva

Add the Geneva to the assembly part. Align the axis of the Geneva with the axis of the index plate and mate the bottom face of the Geneva with the top face of the index plate. Rotate the part 90° to orient it correctly.

Step 5 – Add the two shafts

Add the shaft to the assembly part. The only mating constraint necessary is to align the axis of the shaft with the axis of the hole that it is being inserted into. Reposition both shafts such that they are about centered vertically in the hole.

Step 6 – Add the two washers

Add two units of washers to the assembly part. Add the first washer with the following mating constraints: mate the bottom face of the washer with the top face of the index plate and align the axis of the washer with the axis of the smallest hole in the index plate.

Add the second washer with the following mating constraints: mate the top face of the washer with the bottom face of the index plate and align the axis of the washer with the axis of the smallest hole in the index plate.

Step 7 – Add the bushing

Add the bushing to the assembly part. Mate the bottom face of the bushing with the top face of the washer and align the axis of the bushing with the axis of the small hole.

Step 8 – Add the cap screw

Add the cap screw part and choose Mate at the positioning. Mate the bottom surface of the screw head cylinder to the top surface of the bushing and align the axis of the screw to the axis of the bushing.

Step 9 – Add the nut

Add the nut and choose Mate at the positioning. Mate the top surface of the nut with the bottom surface of the second washer and align the axis of the nut with the axis of either the washer or the small hole in the plate.

Step 10 – Add the set_screw

Align the axis of the screw with the axis of the hole in the Geneva. You can use either Align or Center 1 to 1 constraint for this. If your set screw is upside down, click the alternate solution button. Position the screw relative to the hole such that the end of the screw is just contacting to the shaft in the center of the Geneva (you can use the Tangent constraint for this: the set screw flat bottom face is tangent to the shaft face).

Step 11 – Verify the design intent

Reposition (Assembly → Components → Reposition Components) the index plate with Delta X=5, Delta Y=5, and Delta Z=3 and observe how other components react. Do other components move along?

This completes Section 2-2.

Section 2-3.

Assembling Parts with Interpart Modeling

In this assembly modeling, we have a different design intent that is more challenging than the previous ones. In addition to spatial relationships among the components as accomplished in Section 2-2 Mating Conditions, the new design intent is that components are dimensionally associated. This means that parts that fit together directly with one another should be associative to make sure that they will still fit together after a design change. For example, a change of a hole diameter will cause to change the diameter of the shaft that goes into the hole so that there is fit between two parts.

There are two methods available to accomplish this type of design intent: WAVE Geometry Linker and Interpart Expressions. These will be presented to allow you to experiment on both methods. The first method in Section 2-3 Part A was covered in the Chapter 12 on Assemblies. The second method in Part B has not been covered in the main text but included here for demo and is optional. Interpart Expressions can be complicated because the chain effect of design change of the involved parts is less obvious and should not be used lightly.

A. Using Linked Geometry

A method of making an assembly interpart modeled is to use linked geometry. This means that existing geometry in an assembly is used in the subsequent creating of additional components. This can be easier and less confusing to follow than the interpart expressions method, but can only be used in a limited number of cases. This method is used when you want to create geometry from the geometry of an existing component. This will be illustrated in the example following.

Create a new directory (or Folder) xxx_AppendixA_Proj2. Create duplicates of all individual part files created in Section 1 and save them under this new directory. Repeat this for xxx_Geneva_assm_mate.prt and rename it as xxx_Geneva_assm_geolink.prt. You are going to make changes in these duplicated files, leaving the original files intact.

Step 1 – Create geometry to link from

Before you can use geometry to link, it must already exist in the assembly file.

Change the file open option by choosing File→Options →Load Options and choose the Directory option in the dialogue. Open xxx_Geneva_assm_geolink.prt.

Step 2 – Link the geometry

The first part to link will be the shaft component. First, remove one shaft component from the assembly.

Now create a new component in the assembly and save it as xxx_Geneva_shaft_geolinked_1.prt. It will initially be empty. Now make the new component part the work part. Open the wave geometry linker dialogue by choosing Assembly → Wave Geometry Linker. Click on the circular edge of one of the holes that the shaft fits into and click OK. This geometry is now listed as a linked feature in xxx_Geneva_shaft_geolinked_1.prt. .

Step 3—Create the component part

Now create a shaft by extruding the linked feature (i.e., geometrically linked circle edge) using start distance –1.25 and end distance 1.25. Next chamfer the ends as before (single offset of .031 inches).

Repeat Steps 2 and 3 to create the other shaft.

Step 4—Create other parts similarly.

Create the washer in the same manner by linking to the geometry (the small hole circle edge of the index plate) and extruding with an offset.

Step 5 – Verify the design intent.

Change the diameter of the small hole of the index plate and observe if the washer hole, which is geometrically linked to the hole, also changes to still maintain fit.

B. Using Interpart Expressions

Interpart Expressions (IPE) are used when the design intent indicates that it is beneficial to relate one component to another component in an assembly. This is accomplished through creating a link between a feature expression in one component and a feature expression in another component. The linking allows the value of that expression in one component to control the value of the expression in the other component. In the case of the Geneva cam we've been working with an example would be linking the diameter of the hole in the Index part to the diameter of the Shaft that passes through it.
Note that the IPE capability is defaulted to be turned off in the ug_english.def and ug_metric.def files and it must be turned on before IPE may be used.

General Procedure to Create Interpart Expressions

1. From Design Intent identify the components and features you wish to link by expressions.
2. Make each of the affected components the work part and rename the p# for the expression you wish to use to a meaningful name
3. Make one of the components the work part, select the affected expression and create the link to appropriate expression created in Step 2.
4. That completes the linking by expression of the components. Repeat Steps 1 through 3 for each pair of components you wish to create interpart expressions.

Step 1 – Prepare Part Files

To get started we'll take the process step by step and also look at the big picture of the process. Create a new directory (or folder) xxx_AppendixA_Proj3. Create duplicates of all individual part files created in Section 1 and save them under this new directory. Copy the xxx_Geneva_assm_mate.prt and rename it as xxx_Geneva_assm_intexp.prt. Do not copy the xxx_Geneva_assm_repo.prt. You are going to make changes in these duplicated files, leaving the original files intact.

Change the file open option by choosing File → Options → Load Options and choose the From Directory option in the dialogue. Open xxx_Geneva_assm_intexp.prt. The entire Geneva cam should load into the assembly.

In general, to proceed with IPE the first step is to identify those interpart dimensional relationships between component parts that can share the same values and thus can link to the same expression names fulfilling the design intent for the fit of the parts. Below are some of such examples in our Geneva cam.

Part -and-	**Part**	**Relationship**
Index Plate	Geneva	Center hole = diameter of bottom boss
Geneva	Index	Outer radius = notch radius
Index Plate	Cap Screw	Small hole radius = thread portion radius
Index Plate	Washer	Small hole radius = inner radius
Index Plate	Nut	Small hole radius = inner diameter
Index	Bushing	Slot Width = Diameter of Bushing
Geneva	Shaft	Center Hole Diameter = Outer Diameter
Geneva	Index	Center Hole Diameter = Center Hole Diameter
Geneva	Set_screw	Side Hole Diameter = Screw Diameter

Next identify the key dimensions in the assembly that can drive interpart relationships among component parts, such as listed above. Some of the examples are listed as follows:

geneva_diameter=.63*2

index_height=.44-.25

index_center_hole = 0.25

index_plate_center_hole_dia=.50

index_plate_small_hole_dia=.11

index_plate_thickness=.13

notch_dia=2.47

set_screw_dia=.19

shaft_dia=.25

slot_width=.2

We will relate the Shaft diameter to the Index hole diameter.

Step 2 – Rename the Expressions in Each Component

At the current status of the assembly and components the names above do not exist. Instead, the components are defined by their p# numbers. The next step then is to identify the feature expressions in each component that you want to link and change the p# to a meaningful name. The names above are realistic names that may be used in their respective components.

Note that names are not an absolute requirement for creating an IPE; it is just a convenience to remember which expressions do what. Interpart Expressions may be created using p#.

The procedure is to make the component the Work Part, open the Expression dialog, identify the parameter you wish to use in the IPE and rename it from p# to a meaningful name. Repeat for affected components.

In our case, the Shaft diameter and the Index hole diameter follow these steps.

Make the Shaft the work part.

Choose the Tools → Expression, select the p# that is the diameter of the shaft, choose Rename and enter a name for the diameter, such as shaft_dia. Hit OK and then Apply.

Make the Index the work part. In the same way, change the expression for the center hole to a new name, such as center_hole. Hit OK and Apply.

Make the work part the assembly again.

Step 3 – Create the Interpart Expression to Components

The setup is complete now and ready to create the actual linked expressions.

Make the Shaft the work part.

Open the Expression dialog.

Select the diameter expression, shaft_dia, or whatever you chose to call it.

The expression will appear in the text window (where the arrow is above) and you may edit it.

Click the cursor in the window and delete out the number value, but not the name. Choose the Create link button, select the Index from the list, hit OK and select the diameter expression we created earlier. Choose OK.

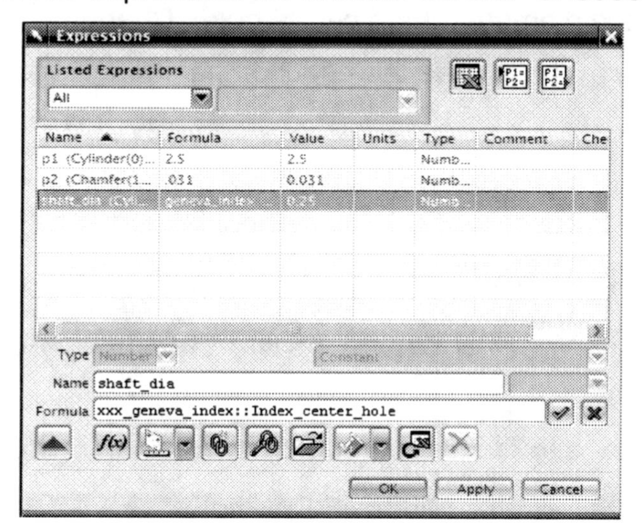

The line in the Formula window looks as shown. Hit the Enter key and the line will be accepted and become the new expression for the hole. Choose the Apply to confirm the whole procedure.

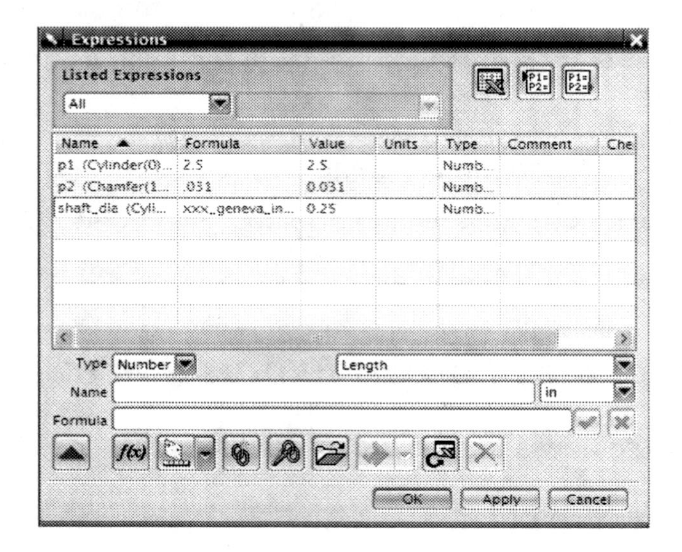

Thus, the Interpart Expression has been created.

Step 4 – Verify the design intent

Make the assembly file the work part to look at the part in entirety again.

We related the parts such that the hole in the index is the driver now of the relationship. We can see if it all works properly by making the Index the work part and changing the diameter of the center hole. Change it to .2 or to .3, hit the Enter key and choose Apply. See the effect of the change on the shaft and index.

Follow-Up

Did you notice anything? There are two shafts. Both shafts changed. Now the other shaft doesn't fit the hole through the Geneva component anymore.

Another optional task is to create an interpart expression in the same manner for the Geneva component so it is related to the shaft.

This completes Section 2-3.

Appendix B. Additional Design and Assembly Projects

This appendix includes additional design projects for five assemblies[1]. Each of the assemblies consists of several components. In order to complete each assembly modeling, you need to first create a solid model for each component using various combinations of features that are covered in the first 11 chapters of this book. Then you need to assemble them using the assembly modeling and master model concepts that are addressed in Chapters 12 and 13. Chapter 12 presents three approaches to assembly modeling. They are repositioning, mating, and interpart modeling. We suggest you to practice all three approaches and compare them with respect to different levels of associative and parametric relations among different components. As a result, different approaches cause different effects of design changes to models. You will find some non-critical dimensions, particularly on threads, missing in the drawings. Make default standard parameter values of threads or make your best guess so that the resulting assemblies look as drawn.

Project B-1. Double Bearing Assembly
Project B-2. Double Block Assembly
Project B-3. Wheel Support Assembly
Project B-4. Shock Assembly
Project B-5. Butterfly Valve Assembly

---

[1] The drawings for these assemblies are taken from Technical Graphics Communication by Bertoline and et al. (1997) (reprinted with permission from the publisher McGraw-Hill)

Project B-1. Double Bearing Assembly

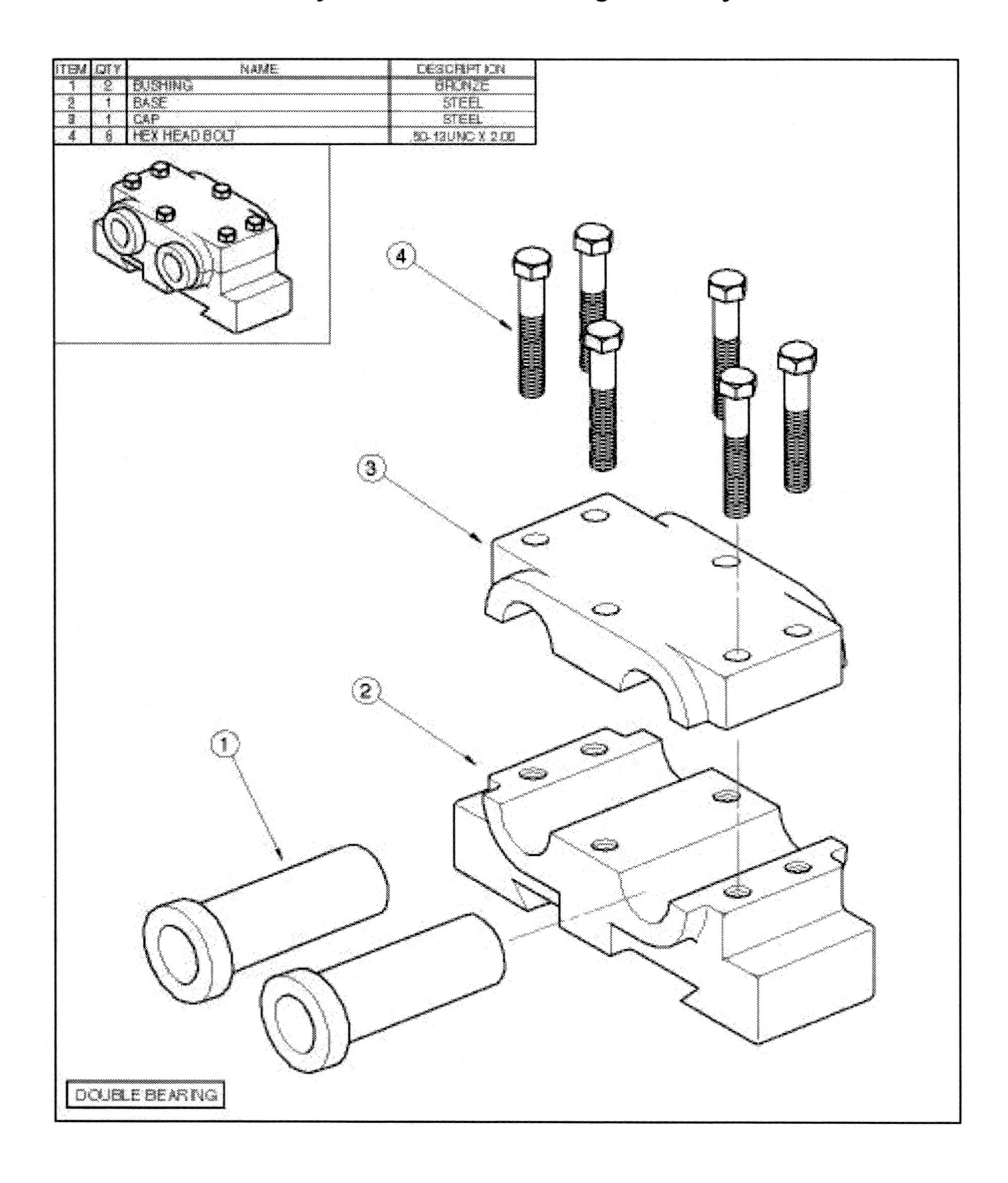

ITEM	QTY	NAME	DESCRIPTION
1	2	BUSHING	BRONZE
2	1	BASE	STEEL
3	1	CAP	STEEL
4	6	HEX HEAD BOLT	.50-13UNC X 2.00

Project B-1. Continued

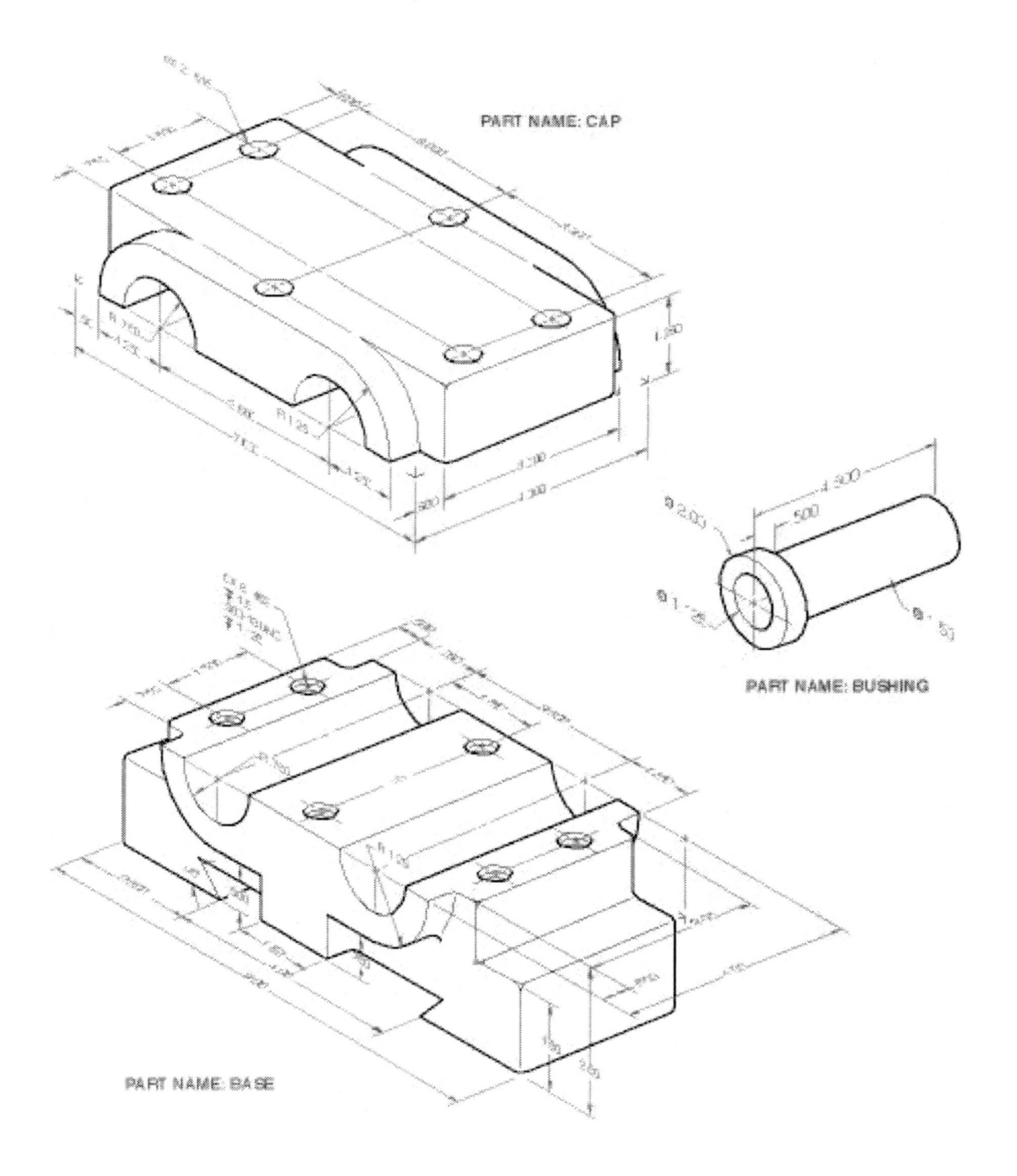

Project B-2. Double Block Assembly

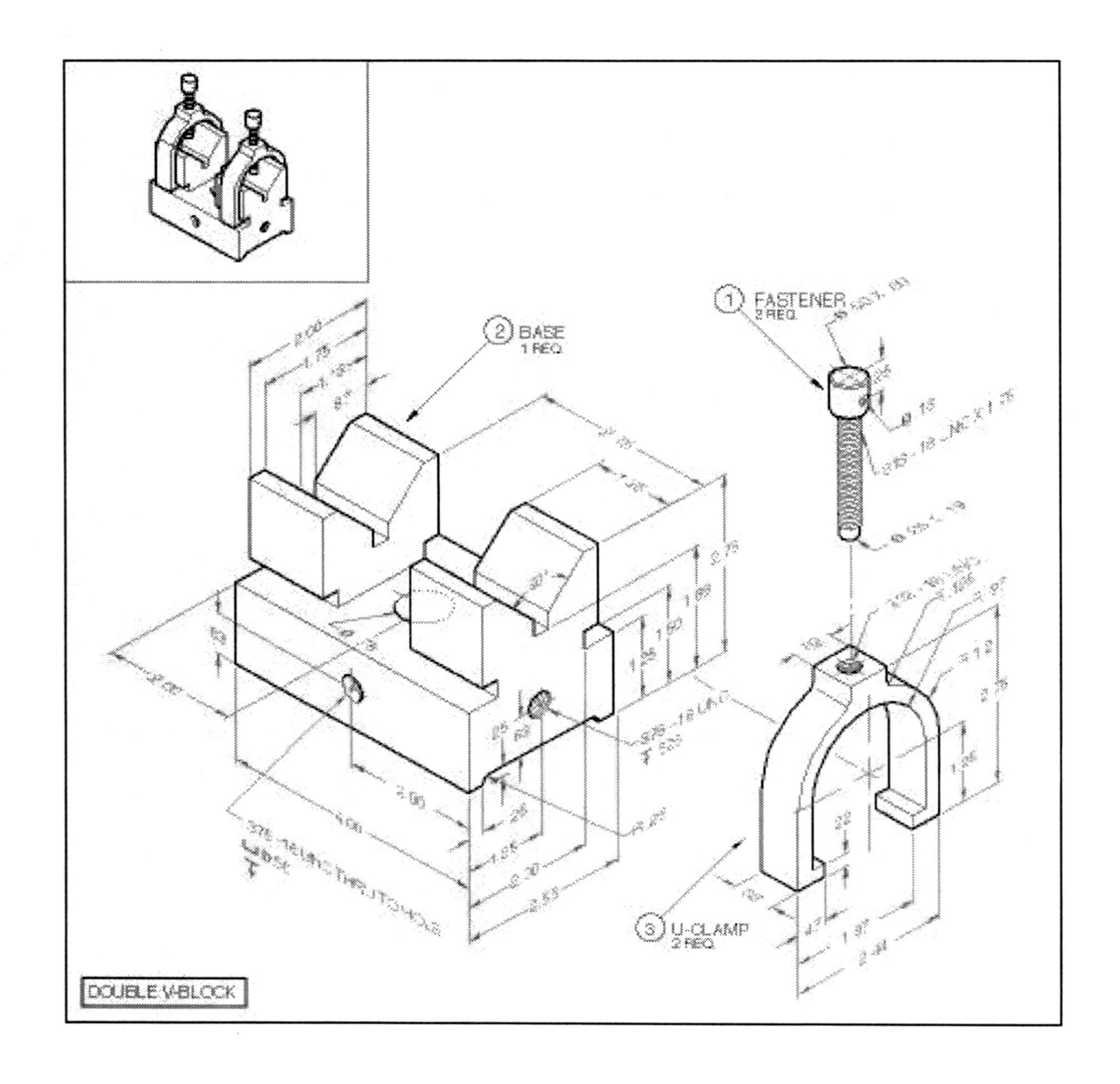

Project B-3. Wheel Support Assembly

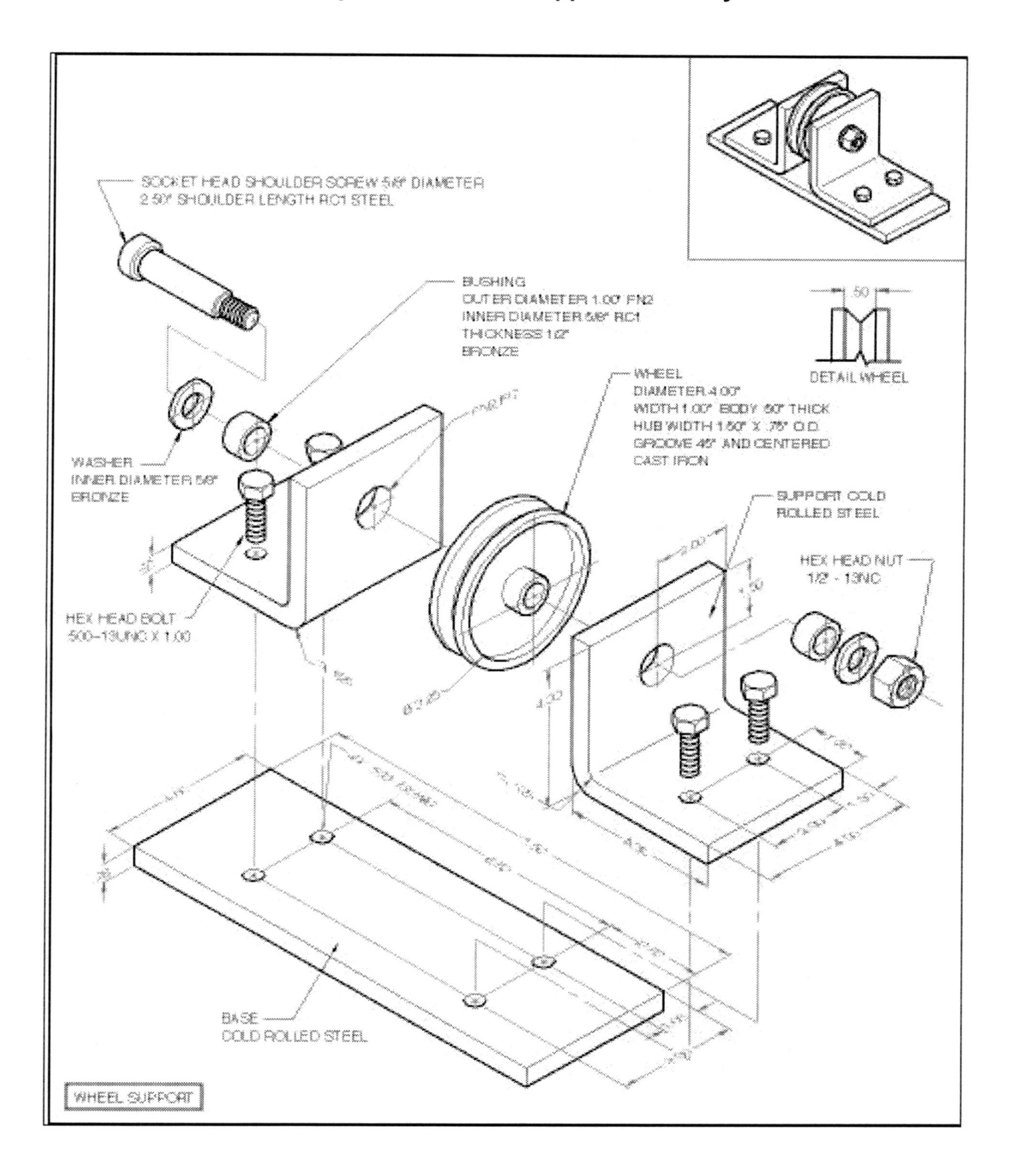

Project B-4. Shock Assembly

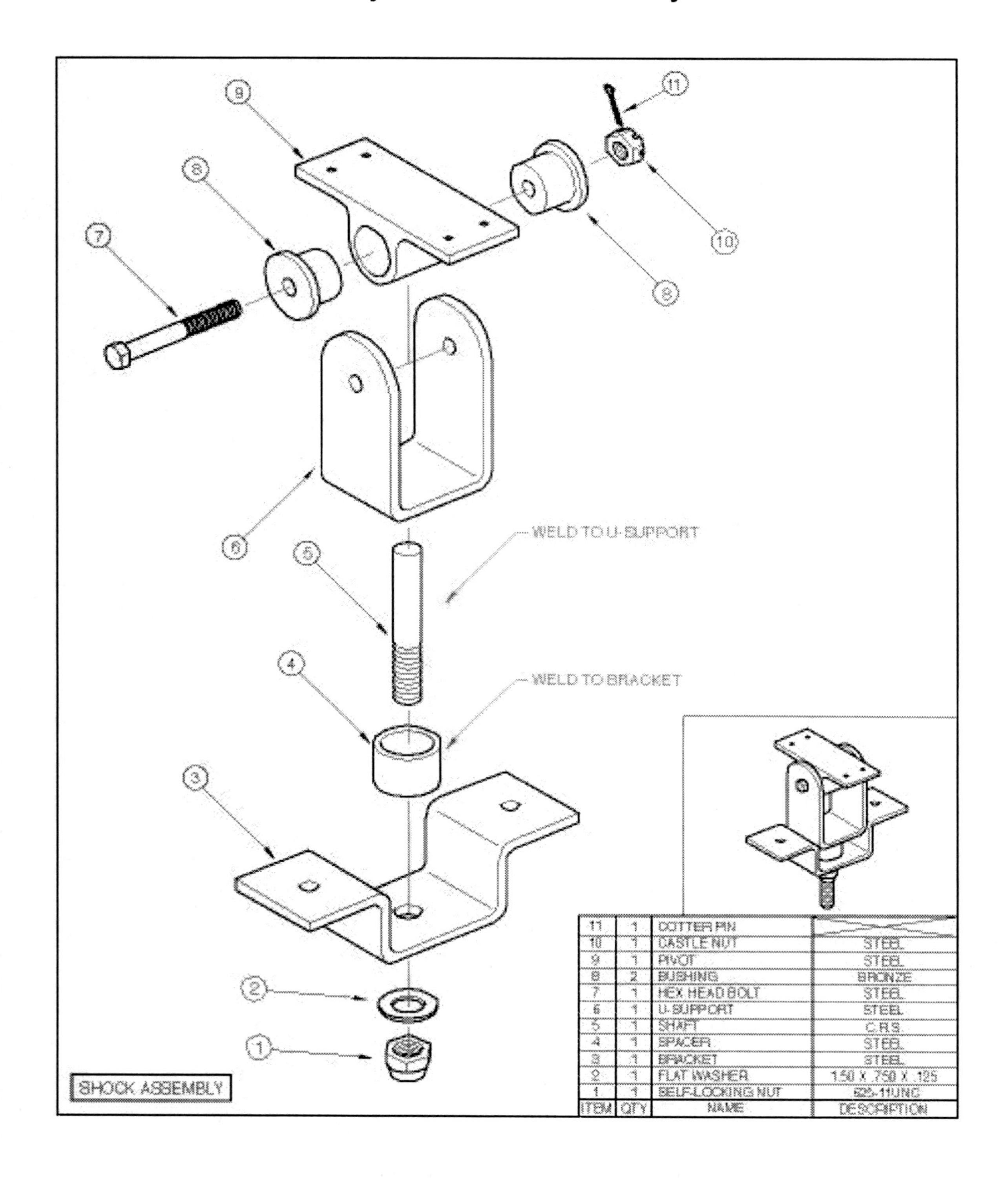

ITEM	QTY	NAME	DESCRIPTION
11	1	COTTER PIN	
10	1	CASTLE NUT	STEEL
9	1	PIVOT	STEEL
8	2	BUSHING	BRONZE
7	1	HEX HEAD BOLT	STEEL
6	1	U-SUPPORT	STEEL
5	1	SHAFT	C.R.S.
4	1	SPACER	STEEL
3	1	BRACKET	STEEL
2	1	FLAT WASHER	1.50 X .750 X .125
1	1	SELF-LOCKING NUT	625-11UNC

Project B-4. Continued

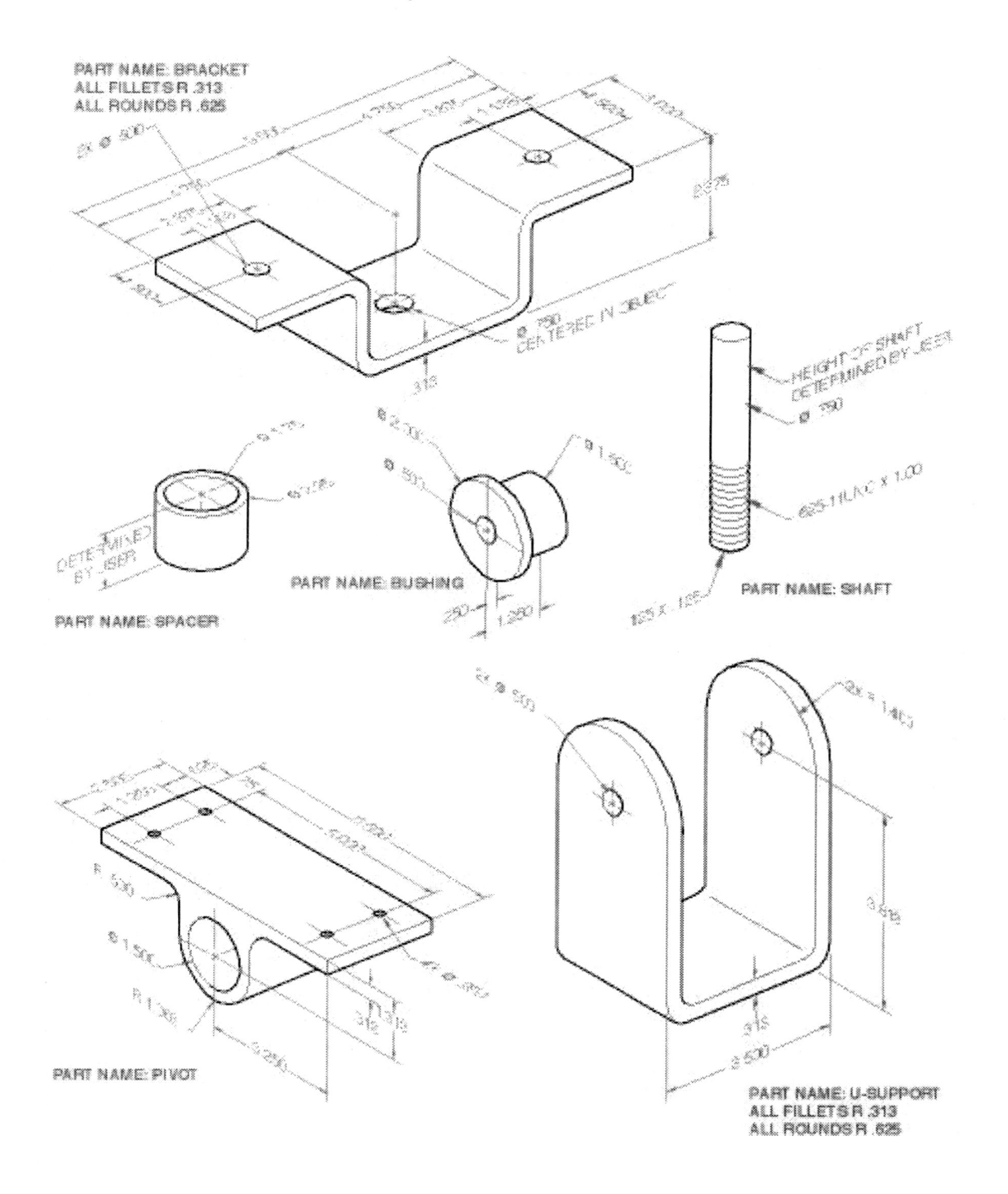

Project B-5. Butterfly Valve Assembly

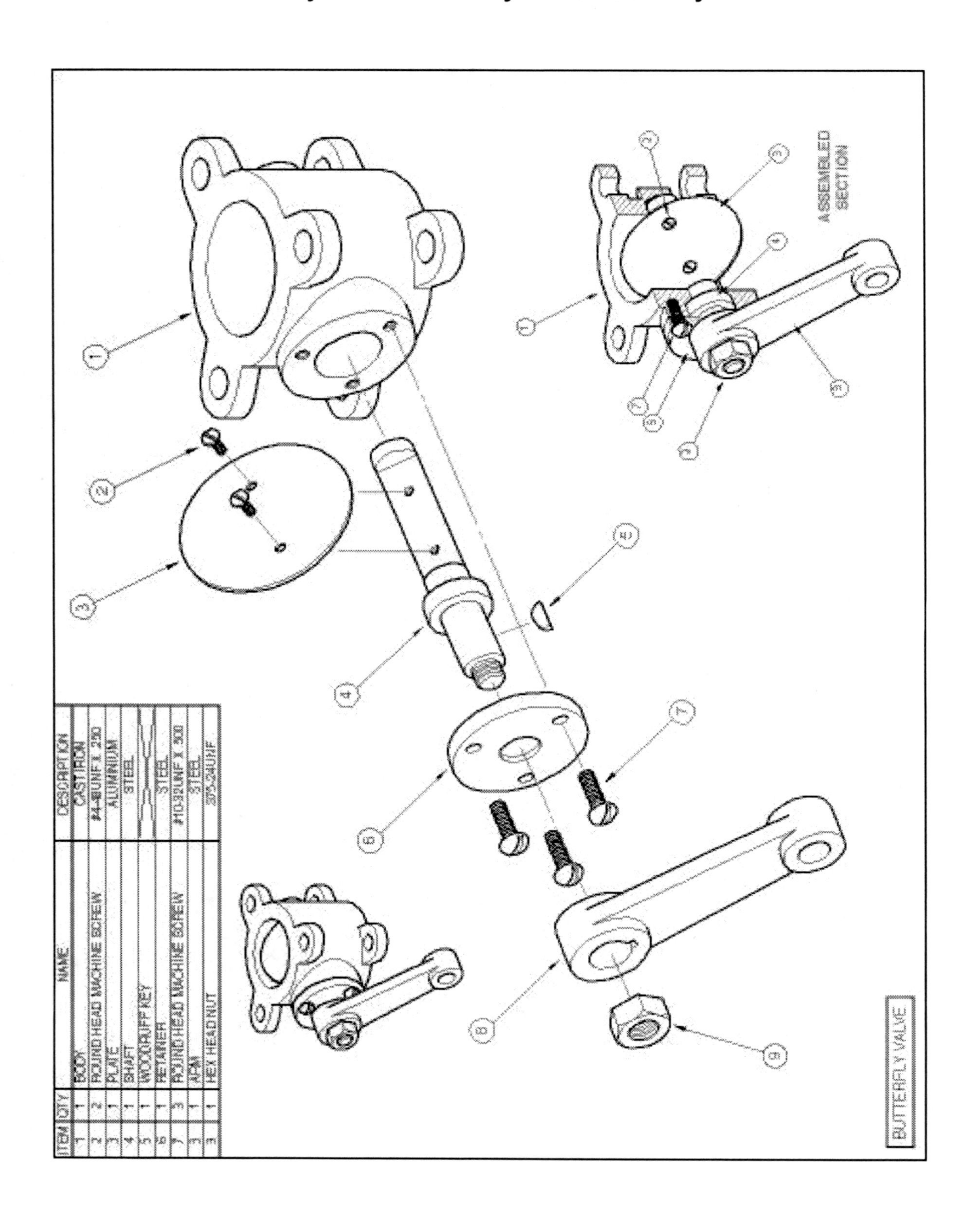

Project B-5. Continued

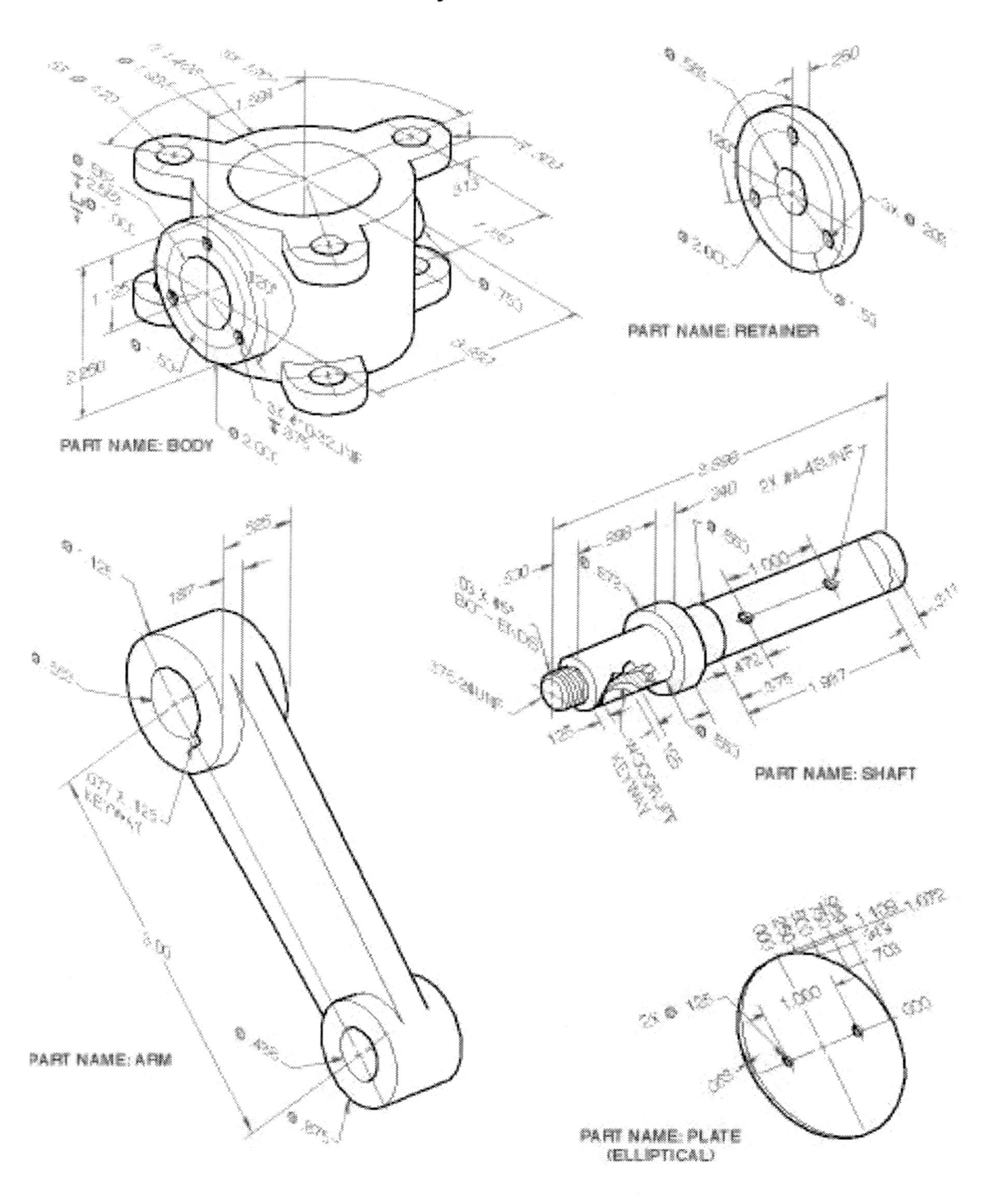

Appendix C. Glossary

ABS	Absolute coordinate system.
Absolute Coordinate System	Coordinate system in which all geometry is located from a fixed or absolute zero point.
Active View	One of up to 50 views per layout in which you can directly work.
Anchor point	The intersection point of a general conic's two end tangent vectors.
Angle	In Unigraphics, an angle measured on the X-Y plane of a coordinate system is positive if the direction that it is swept is counterclockwise as viewed from the positive Z axis side of the X-Y plane. An angle swept in the opposite direction is said to be negative.
Apparent intersection	The intersection between a curve and the projection of another curve that lies in a different plane. The curves must be nonparallel and the projection occurs along the ZC axis.
Approximate Rho	Approximate Rho basically involves treating an arc or spline as if it were a conic, and then computing the rho based on that premise. This can be useful in a number of cases. For example: You may have a spline that resembles a conic, but has an inflection and a lot of data. Being able to get the rho value gives you a head start in replacing the spline with a simple conic.

	You may have a group of cross section curves and want a section surface. If you can get the rho value for the section curves you can build a law function to create a section surface with varying rho. Use the Information function to find the Approximate Rho for an arc or spline.
Arc	An incomplete circle; sometimes used interchangeably with the term "circle".
ASCII	American Standard Code for Information Interchange. It is a set of 8-bit binary numbers representing the alphabet, punctuation, numerals, and other special symbols used in text representation and communications protocol.
Aspect Ratio	The ratio of length to height which represents the change in size of a symbol from its original.
Assembly	A collection of piece parts and sub-assemblies representing a product. In Unigraphics, an assembly is a part file that contains components.
Assembly Part	A Unigraphics part file that is a user-defined, structured combination of sub-assemblies, components and/or objects.
Associativity	The ability to tie together (link) separate pieces of information to aid in automating the design, drafting, and manufacture of parts in Unigraphics.
Attributes	Pieces of information that can be associated with Unigraphics geometry and parts such as assigning a name to an object.
Auxiliary View	In Unigraphics, a view that shows the true size and shape of a

	part's face.
Bezier spline	A single segment B-spline.
Block Font	A Unigraphics character font that is the default font used for creating text in drafting objects and dimensions.
Boundary	A set of geometric objects that describes the containment of a part from a vantage point.
Bottom-Up Modeling	Modeling technique where component parts are designed and edited in isolation of their usage within some higher level assembly. All assemblies using the component are automatically updated when opened to reflect the geometric edits made at the piece part level. See Top-down Modeling.
Bridging Curves	A method of creating a b-curve by blending or bridging any two given curves at two specified points on the curves.
B-spline	Abbreviation of nonuniform rational B-spline. Curve created from construction points. See NURBS.
Body	Class of objects containing sheets and solids (see Solid Body and Sheet Body).
CAD/CAM	Computer Aided Design/Computer Aided Manufacturing.
Canned Layout	One of the five layouts available to the user. These include: L1 - Single View, L2 & L3 - Two Views, L4 - Four Views, and L6 - Six Views.
Canned View	One of the canned views available to the user. These include (but are not limited to) the following: TOP, FRONT, RIGHT, LEFT, BOTTOM, BACK, TFR-ISO (top-front-right isometric),

	and TFR-TRI (top-front-right trimetric).
Category, Layer	A name assigned to a layer, or subset of layers. A category, if descriptive of the type of data found on the layers to which it is assigned, will assist the user in identifying and managing data in a part file.
CGM	Computer Graphics Metafile. An ANSI standard format for picture files which can be easily moved between different operating systems and can be read by many viewing programs and plot despoolers.
Chaining	A method of selecting a sequence of curves which are joined end-to-end.
Child	A feature that depends on another for its existence. For example, a hollow cannot exist without a solid, such as a block, that the hollow can be formed in. Also called a "dependent".
Circle	A complete and closed arc, sometimes used interchangeably with the term "arc".
Class Selection	A list of options that allows the user to select objects by various methods.
CL File	Cutter Location File.
CL Point	Cutter Location Point.
Column	A single file array of points used to create a surface. Usually lined up along the V direction of the surface. See Rows and Columns for additional details.

Component	A collection of objects, similar to a group, in an assembly part. A component may be a sub-assembly consisting of other, lower level components.
Component Part	A separate Unigraphics part file that the system associates with a component object in the assembly part.
Cone Direction	Defines the cone direction using the Vector Constructor (common tool) or Constructor.
Cone Origin	Defines the base origin using the Point Constructor (common tool) or Constructor.
Conic	A conic or conic section is a curve that can be formed by intersecting a cone and a plane (parabola, hyperbola, ellipse).
Constraints	Refers to the methods you can use to refine and limit your sketch. The methods of constraining a sketch are geometric and dimensional.
Construction points	Points used to create a spline. Construction points may be used as poles (control vertices), defining points, or data points. See Poles, Defining Points, and Data Points.
Contiguous	End-to-end, as in contiguous curves.
Continuity Checks	Continuity describes the behavior of curves and surfaces at their segment boundaries. The two types of continuity usually dealt with in Unigraphics are mathematical continuity, denoted Cn, where n is some integer, and geometric continuity, denoted Gn. Within Unigraphics these can be loosely defined by the following: Gn indicates the true degree of continuity between two

geometric objects. For example, G0 means the two objects are connected; G1 means they are smoothly connected up to one differentiation, or are tangency continuous. G2 means they are smoothly connected by up to two differentiations, or are curvature continuous; etc. Gn continuities are representation (parameterization) independent.

Cn indicates the degree of continuity between two segments of a b-curve or a b-surface in the NURB representation. Generically, C0 means the two segments are G0 connected. C1 means they are G1 connected; etc. But, C0 does not mean the two segments are just G0 connected -- they could actually be G1 or G2 connected, and so on.

The key point is that Gn is for real physical continuity, while Cn is one mathematical representation of it, which may not be faithful. Since NURB is an industry standard for freeform geometry, Unigraphics uses it. But we always try to have Cn represent the same degree of continuity as Gn, to avoid cases where a curve is G1, but has C0 junction, etc.

Quoting from the ICAD Surface Designer Reference manual: "C0 continuity implies that a common point exists between two adjacent segments (i.e., the segments are touching). C1 implies that there is a common point and the first derivatives of the polynomials (i.e., the tangent vectors) are the same. C2 implies that the first and second derivatives are the same. Geometric continuity is less strict than mathematical continuity. G0 and C0 are equivalent, that is, the segments are positionally continuous. G1 implies that the tangent vectors are equal in direction, but not magnitude. G2 implies the curvature is the same, but the second derivatives are not."

Control Point	A position on existing geometry. Control points are: existing points, endpoints of conics, endpoints and midpoints of open arcs, centers of circles, midpoints and endpoints of lines, and endpoints of splines. The control point for a closed circle is its center, while the control points for an open arc are its end and midpoints. A spline has a control point at each knot point.
Control Polygon	A polygon associated with a spline, which can be used to control the shape of the spline.
Control Vertex	A vertex of the control polygon of a spline.
Coordinate System	A system of axes used in specifying positions (CSYS).
Coordinate System Tool	A group of options used for defining coordinate systems in a variety of different ways. This subfunction is also used to change the work coordinate system and to rotate the part.
Counterclockwise	The right-hand rule determines the counter- clockwise direction. If the thumb is aligned with the ZC axis and pointing in the positive direction, counterclockwise is defined as the direction the fingers would move from the positive XC axis to the positive YC axis.
Current Layout	The layout currently displayed on the screen.
Cross Splines	This is specified when creating a b-surface using the Cubic Fit method. Internal cross splines are not displayed by the system and are not selectable Unigraphics objects; rather, the system uses cross splines internally to help in the construction of b-surfaces.

Cubic Fit Surface	A b-surface creation technique in which a sequence of curves running in a roughly parallel direction is selected, and the system then fits" them with a b-cubic b-surface.
Current Layout	The layout currently displayed on the screen. Layout data is kept in an intermediate storage area until it is saved.
Curve	A geometric object; this may refer to a line, an arc, a conic, spline or b-curve.
Curve Extension	Unconstructed portion of an open arc. The part of a circle that has been cut away.
Data Base	A comprehensive collection of information referring to the objects that make up a part.
Data Points	Spline construction points for the Least Squares Fit method. A spline created with this method are fit to the points within a certain tolerance and do not necessarily pass through them.
Datum	A datum is a fixed point that anchors a sketch point to a specified location.
Default	Assumed values when they are not specifically defined.
Defining Face	A face of a solid used to create a Midsurface.
Defining Points	Spline construction points. Splines created using defining points are forced to pass through the points. These points are guaranteed to be on the spline.
Degree	A mathematical concept referring to the degree of the polynomial in the equation defining the surface or spline.

Degree-of-Freedom Arrows	Arrow-like indicators that show areas that require more information to fully constrain a sketch.
Dependent	Same as Child.
Derivative Vector	The first derivative or tangent vector to a curve at a given point.
Design in Context	The ability to directly edit component geometry as it is displayed in an assembly. Geometry from other components can be selected to aid in modeling. Also referred to as edit in place.
Detail View	In Unigraphics, a view that illustrates an enlarged section of another view (i.e., a blow-up).
Die Engineering	A separate Unigraphics module that augments the capabilities of modeling by providing a suite of tools targeted for the design of sheet metal stamping dies.
Dimensional constraint	This is a scalar value or expression which limits the measure of some geometric object such as the length of a line, the radius of an arc, or the distance between two points.
Directory	A hierarchical file organization structure which contains a list of filenames together with information for locating those files.
Display File	A file containing display data for retrieval of a part.
Displayed Part	The part currently displayed in the graphics window.
Drawing	A collection of an unlimited number of views. Stored data includes the reference point for each view.

Edit in Place	See Design in Context.
Emphasize Work Part	A color coding option that helps distinguish geometry in the work part form geometry in other parts within the same assembly.
End Point	An end point of a curve or an existing point.
Extension Surface	A tangential, normal, or angled surface created from an existing base surface using the Extension Surfaces creation method.
Expression	An arithmetic or conditional statement that has a value. Expressions are used to control dimensions and the relationships between dimensions of a model.
Face	A region on the outside of a body enclosed by edges.
FACEPAIR_DEF	The name of the Face Pair feature that is created by two opposing defining faces.
FACEPAIR_SEL	The name of the Face Pair feature that is created with a user-selected surface.
Face Pair feature	The fundamental building block of a Midsurface feature. It contains two lists of opposing faces and the resulting Midsurface.
Family member	A read-only part file created from, and associated with, a template part and family table.
Family table	A table created from a template part, in the UG spreadsheet function, that describes the various attributes of the template part that you can change when you create a family member

Fast Font	A Unigraphics character font designed to provide faster Regenerate operations.
Feature	An all-encompassing term which refers to all solids, bodies, and primitives.
File	A group or unit of logically related data which is labeled or named and associated with a specified space. In Unigraphics, parts, patterns, schematic symbols, CL and UG/Open GRIP source, GRIP intermediate, GRIP execution and Font Object data are all stored as files.
Font Box	A rectangle or box" composed of dashed line objects. The font box defines the size, width and spacing of characters belonging to a particular font.
Fonts, Character	A set of characters designed at a certain size, width and spacing.
Font, Line	Various styles of lines and curves, such as solid, dashed, etc.
Font Object Library	A Unigraphics library containing font object files. Each file includes the necessary information for displaying a particular character font. The font object library can only be accessed through the Font Management option.
Fonts, Line	Various styles of lines and curves, such as solid, dashed, etc.
Font Table	An ordered list of font names representing the character fonts available for the current part.
Free Form Feature	A body of zero thickness. (see Body and Sheet Body)

Gateway Reset	A technique used when recording macros that assures they will always run, regardless of your location within Unigraphics when you launch them. To use Gateway Reset, simply make the first event in the recording of a macro, Application®Gateway. This cancels the current application, and places the macro at a known starting point in the user interface.
Generator Curve	A contiguous set of curves, either open or closed, that can be swept or revolved to create a body.
Geometric Constraint	A relationship between one or more geometric objects that forces a limitation. For example, two lines that are perpendicular or parallel specifies a geometric constraint.
Global Layer Mask	The global layer mask for the entire part. The layer mask defines which layers will be visible and selectable. A view can use either the global or the individual layer mask.
Grid	A rectangular array of implied points used to accurately align locations which are entered by using the screen position" option.
Group (v.)	Grouping is a procedure for conjoining selected objects so that they can be treated as a single object.
Group (n.)	A collection of selected objects which are treated as a single object.
Guide Curve	A set of contiguous curves that define a path for a sweep operation.
Half Angle	The half vertex angle defines the angle formed by the axis of

	the cone and its side.
Hardcopy	In general, a printed copy of computer output - e.g., drawings or listings. More specifically, the output of the hard-copy unit often attached to a Unigraphics system.
Individual Layer Mask	See View Layer Mask.
Inflection	A point on a spline where the curve changes from concave to convex, or vice versa.
Information Window	The window used in listing operations, such as Information.
Interactive Step	An individual menu in a sequence of menus used in performing a Unigraphics function.
Implied intersection	Intersection formed by extending two line segments that do not touch to the position that they cross. The line segments must be nonparallel and coplanar.
Isometric View	Isometric view orientation - one where equal distances along the coordinate axes are also equal to the view plane. One of the axes is vertical.
Isoparametric Trim/Divide	An option which allows you to trim a b-surface in either the U or V isoparametric direction at a specified parameter.
Join Curve	A method of creating a b-curve in which curves (lines, arcs, conics or splines) may be selected for conversion into a b-curve.
Knot points	Points along a B-spline, representing the endpoints of each

	spline segment.
Layer	A layer is a partition of a part. Layers are analogous to the transparent material used by conventional designers. For example, the user may create all geometry on one layer, all text and dimensions on a second, and tool paths on a third.
Layout	A collection of viewports or window areas, in which views are displayed. The standard layouts in Unigraphics include one, two, four or six viewports.
Least Squares Method	This is a technique for fitting a spline to a series of construction points. The sum of the squares of all of the distances between the spline and the construction points is minimized. This method helps reduce the number of points needed to define the spline and helps ensure a certain degree of smoothness for the spline.
List Box	The usually scrolled window found in dialogs from which you select items. Referred to as a changeable window when it serves multi-purposes.
Macro	A series of interactive steps that are organized as a group and called as a unit.
Menu	A list of options from which the user makes a selection.
MIDSRF	The name of the Midsurface feature.
Midsurface	A sheet body which is created halfway between two defining faces. The points and normals of the parent faces are averaged at corresponding parameters. The properties of a midsurface are exactly the same as any other sheet body; the

	only difference being the method of creation.
Midsurface feature	A feature that gives the user the ability to create and manipulate a list of Midsurfaces and its defining Faces Pairs.
Modal	A parameter or a status is said to be modal if it retains the value assigned to it when it was last used, rather than automatically reverting back to some default value after each usage.
Model Space	The coordinate system of a newly created part. This is also referred to as the absolute coordinate system. Any other coordinate system may be thought of as a rotation and/or translation of the absolute coordinate system.
Multi-Patch Sheet	A b-surface sheet that consists of multiple segments.
Native Mode File System	The Native Mode File Management system is a part of the Common File Interface (CFI) that allows Unigraphics to interface with the native file system on your computer. The Native File Mode allows you to use standard operating system commands and directory structures to manipulate UGII files.
Name (of an expression)	The name of an expression is the single variable on the left hand side of the expression. All expression names must be unique in a part file. Each expression can have only one name. See Expression.
Neutral Point	The starting point, or zero position, of the Navigation Cursor, as specified when the Navigate option is invoked and MB1 is pressed and held down.

NURBS	Acronym standing for Non Uniform Rational B-spline, often referred to as B-spline in the documentation.
Objects	All geometric entities within the Unigraphics environment.
Offset Surface	A Unigraphics surface type created by projecting (offsetting) points along all the normals of a selected surface at a specified distance.
Ordered Point Constructor	A list of options (methods) used to select a large number of point objects using either any of the three available chaining methods or the standard Point Constructor method which allows for specification of required points in turn.
Origin	The point X = 0, Y = 0, Z = 0 for any particular coordinate system.
Orthographic Projection	Projects views from a parent view (also called `base view'). The projected views are automatically aligned and scaled to match the parent view. By choosing the projection angle you want, selecting an existing model and indicating the parent view, you can construct a drawing containing various orthogonal views by simply choosing the view to be projected and the direction of the projection.
Parallel Projection	In a parallel projection, the lines are all parallel to a specified vector. The object is projected onto the view plane by passing parallel lines through each point at the object and finding their intersections with the view plane.
Parametric design	Concept used to define and control the relationships between the features of a model. Concept where the features of the model are defined by parameters.

Parent	A feature on which one or more feature(s) depend for existence. For example, if a block has been hollowed, the block is the parent of the hollow feature.
Part or Model	A collection of Unigraphics objects which together may represent some object or structure.
Part Family	A template part and its family table and family member parts.
Part Modality	Variables that are not automatically changed by the system throughout the life of a particular part (e.g. character size and line width). These variables are filed with the part and will be in effect each time the part is retrieved from the library until explicitly modified by the user and filed with the part.
Partially Loaded Part	A component part which, for performance reasons, has not been fully loaded. Only those portions of the component part necessary to render the higher level assembly are initially loaded (the reference set).
Parts List	A rectangular array of text containing a list of an assembly's contents together with a brief description of them.
Patch	Sections of a free form sheet body. Sheets can consist of a single patch or multiple patches.
Pattern	A collection of objects from a master part file that can be placed (merged) into your current part file as many times as necessary. A pattern is a single object, both from your point of view (e.g. selection, transformations, etc.) and from the system's point of view.
Peak	A point on a spline where the local radius of curvature is

	maximum.
Perspective Projection	In a perspective projection, all lines emanate from a common point; called the center of projection. An object is projected onto the view plane by passing those lines through each point at the object and finding their intersections with the plane.
Point Set	A distribution of points on a curve between two bounding points on that curve.
Point/Slope Continuity	A method for comparing checkpoints on curves or surfaces to corresponding points on curves or surfaces.
Point Constructor	A list of options (methods) by which positions can be specified in Unigraphics. Also called Point Common Tool and Point Constructor.
Plotter	A device which typically uses a pen to draw a permanent copy of the displayed image.
Poles	Spline construction points. Splines created using poles gravitate towards the points but do not actually touch them (except at the endpoints). Poles offer better control of the curvature of the spline than do defining points. Also called control vertices.
Polynomial Cubic B-Surface	A b-surface consisting of an array of associative cube patches which can be exported to most other CAD/CAM/CAE applications.
Position continuous	Referring to a spline that has no breaks in it.
Project	A method to project certain types of geometry onto surfaces

Points/Curves	and planes. Also allows you to move or copy the object to be projected.
Read-only Part	A part for which the user does not have write access privilege.
Real Time Dynamics	Produces smooth pan, zoom, and rotation of a part, though placing great demand on the CPU.
Reference Set	A named section or partition of a Unigraphics part created in the Assemblies application.
Reference View	A view on a drawing in which you cannot directly work.
Refresh	A function which causes the system to refresh the display list on the viewing screen. This removes temporary display items and fills in holes left by Blank or Delete.
RGB	Short for red-green-blue, representing a mixing model or method of describing colors, used with light-based media (such as monitors). RGB uses the additive primaries method, mixing percentages of red, green, and blue to get the desired color. Adding no color produces black, and adding 100 percent of all three colors results in white. Interactive options exist for RGB values in the Unigraphics Color Palette and High Quality Image Output Setup options.
Right-Hand Rule, Conventional	The right-hand rule is used to determine the orientation of a coordinate system. If the origin of the coordinate system is in the palm of the right fist, with the back of the hand lying on a table, the outward extension of the index finger corresponds to the positive Y axis, the upward extension of the middle finger corresponds to the positive Z axis, and the outward extension of the thumb corresponds to the positive X axis.

Right-Hand Rule for Rotation	The right-hand rule for rotation is used to associate vectors with directions of rotation. When the thumb is extended and aligned with a given vector, the curled fingers determine the associated direction of rotation. Conversely, when the curled fingers are held so as to indicate a given direction of rotation, the extended thumb determines the associated vector.
Rho	Rho is the projective discriminant, a scalar value that controls the "fullness" of each conic section. See Approximate Rho.
Row	A single-file array of points used to create a free form sheet body. Defines the U direction of the sheet. See Rows and Columns for additional details.
Screen Cursor	A marker on the screen which the user moves around using some position indicator device. Used for indicating positions, selecting objects, etc. Takes the form of a full-screen cross.
Selection Ball	The screen cursor contains a selection ball when necessary. A portion of an object must fall within this area to be selectable. The size of the selection ball is set in Preferences. The default radius is .125 inches and is centered at the cursor.
ShapeSCAN	A Unigraphics application for reverse engineering that allows you to rapidly model physical parts based on digitized point clouds and for which the corresponding mathematical models do not exist.
Sheet body	An object consisting of one or more faces not enclosing a volume. A body with zero thickness (see Body).
Side face	A list of opposing faces found in a Face Pair.

Single-Patch Sheet	A b-surface sheet that consists of only a single segment.
Sketch	A collection of geometric objects that closely approximates the outline of a particular design. You refine your sketch with dimensional and geometric constraints until you achieve a precise representation of your design. The sketch can then be extruded or revolved to obtain a 3D object or feature.
Sketch Coordinate System (SCS)	The SCS is a coordinate system that corresponds to the plane of the sketch. When a sketch is created the WCS is changed to the SCS of the new sketch.
Slope	The slope in the X-Y plane is the ratio of the Y change in relation to the X change between two points on the line.
Smart body	All bodies created using curves including solid and sheet. A smart body will automatically update when you modify any of its creation parameters (that is, the defining curves, Rho, or the Row and Column degrees).
Solid body	An enclosed volume. A type of body (see Body).
Spline	A smooth free-form curve.
String	A contiguous series of lines and/or arcs connected at their end points.
Studio for Design	A Unigraphics module that provides tools specifically tailored for Industrial Design applications. This includes basic tools for the initial concept stages, such as the creation and visualization of virtual designs, and progresses ultimately through the production of primary and secondary surfaces.

Sub-assembly	A part that both contains components and is itself used as a component in higher-level assemblies.
Surface	The underlying geometry used to define a face on a sheet body. A surface is always a sheet but a sheet is not necessarily a surface (see Sheet Body).
Sweep of Arc	The number of degrees covered by an arc.
System	The Unigraphics system.
Tangent continuous	Indicating a smooth spline. A tangent continuous spline has no "kinks" or "folds" in it.
Template Part	A Unigraphics part file constructed in such a way as to allow a family of parts to be built based on it.
Temporary Part	An empty part that is optionally created for any component parts that cannot be found during the loading of an assembly.
Tolerance	The allowable deviation from a standard, especially the range of variation permitted in maintaining a specified dimension in machining a piece; the total permissible variation in a size or location dimension; the difference between the permitted minimum and maximum sizes of a part.
Top-down Modeling	Modeling technique where component parts can be created and edited while working at the assembly level. Geometric changes made at the assembly level are automatically reflected in the individual component part when saved. See Bottom-Up Modeling.
Trim	To shorten or extend a curve.

Trimetric View	A viewing orientation which provides you with an excellent view of the principal axes. In Unigraphics II, the trimetric view has the Z-axis vertical. The measure along the X-axis is 7/8 of the measure along Z, and the measure along the Y-axis is 3/4 of the measure along Z.
UG/Open GRIP	Graphics Interactive Programming - A high-level language that provides the user with an optional means of operating the system. Almost any operation which can be performed interactively in Unigraphics can also be performed by executing the commands of a UG/Open GRIP program. It performs mathematical computations, includes branching and looping capabilities, and allows the use of separately compiled sub-programs.
Undulations	Reversals of curvature in a spline.
Units	The unit of measure in which you may work when constructing in Unigraphics. Upon log on, you may define the unit of measure as inches or millimeters.
Upgraded Component	A component that was originally created pre-V10, but has been opened in V10 or later and been upgraded to remove duplicate geometry.
User Exits	Allows you to automatically run UG/Open GRIP programs at certain points in Unigraphics.
User Table	A list of all valid Unigraphics user numbers and passwords. When a user logs on, the user number and password will be compared to those in the table.

User Tool	Custom built dialogs from which you can launch macros and UG/Open API and GRIP programs.
Vertex	A defining point in certain geometric constructions. The vertices of a spline are the intersections of adjacent line segments forming the control polygon of the spline; the vertices of a polygon are the intersection points of adjacent sides of the polygon.
View	A particular display of the model. View parameters include view orientation matrix; center; scale; X,Y and Z clipping bounds; perspective vector; drawing reference point and scale.
View Dependent Edit	A mode in which the user can edit a part in the current work view only.
View Dependent Geometry	Geometry created within a particular view. It will only be displayed in that view.
View Dependent Modifications	Modifications to the display of geometry in a particular view. These include erase from view and modify color, font and width.
View Layer Mask	A layer mask specific to a view. The mask will determine which layers are visible for that view. Use the individual (global) mask to control layer visibility for specific (entire) views.
View of the cursor	Perspective that allows you to select positions in a three dimensional environment. Indicating a screen position with respect to the *view of the cursor* specifies a location in a plane parallel to the viewing screen. This is different than the screen position in the Point Constructor which always specifies a position in the XC-YC plane.

WCS	Work Coordinate System.
WCS, Work Plane	The work coordinate system (WCS) is the coordinate system singled out by the user for use in construction, verification, etc. The coordinates of the WCS are called work coordinates and are denoted by XC, YC, ZC. The XC-YC plane is called the work plane.
Where-Used Report	Contains a list of all part files in which a specified component part exists.
Work Layer	The layer on which geometry is being constructed. You may create objects on only one layer at a time.
Work Part	The part in which you create and edit geometry. The work part can be your displayed part or any component part that is contained in your displayed assembly part. When displaying a piece part, the work part is always the same as the displayed part.
Work View	The view in which work is being performed. When the creation mode is view dependent, any construction and view dependent editing will occur only in the current work view.
Z Clipping Planes	Front and back clipping planes parallel to the X-Y plane of the view. Geometry is clipped to these planes. Only geometry which lies between the two planes is displayed.
XC axis	X axis of the work coordinate system.
YC axis	Y axis of the work coordinate system.
ZC axis	Z axis of the work coordinate system.